RIVINGTON'S BUILDING CONSTRUCTION

Major Percy Smith

With an introduction by Lawrance Hurst

VOLUME 3

DONHEAD

First published in 1875 by Rivingtons
This edition revised and reprinted in 1904 by Longmans, London

This reprinted edition © Donhead Publishing Ltd 2004

Simultaneously published in the United Kingdom and
Massachusetts, USA by Donhead

Donhead Publishing Ltd
Lower Coombe
Donhead St Mary
Shaftesbury
Dorset SP7 9LY
Tel. 01747 828422
www.donhead.com

New introduction to this 2004 edition © Lawrance Hurst

ISBN 1 873394 66 7

A CIP catalogue for this book is available from the British Library

Printed in Great Britain by J. H. Haynes & Co. Ltd. Sparkford, Somerset

Donhead Publishing would like to acknowledge the help of Mike Chimes
at the Institute of Civil Engineers, London, in loaning an original copy of
the work for this facsimile reprint.

British Library Cataloguing in Publication Data

Rivington's building construction
1. Building
690
ISBN 1873394667

Library of Congress Cataloguing in Publication Data
A catalog record for this book has been requested.

NOTES

ON

BUILDING CONSTRUCTION

ARRANGED TO MEET THE REQUIREMENTS
OF THE SYLLABUS OF THE BOARD OF EDUCATION,
SOUTH KENSINGTON

PART III.

MATERIALS

Sixth Edition, Revised (1904)

LONGMANS, GREEN, AND CO.

39 PATERNOSTER ROW, LONDON

NEW YORK AND BOMBAY

1904

NOTES ON BUILDING CONSTRUCTION

Arranged to meet the requirements of the Syllabus of the Board of Education, South Kensington.

In Four Parts. Medium 8vo. Sold separately.

Part I.   *With* 695 *Illustrations.*   1 os. 6d. *net.*

Part II.   *With* 496 *Illustrations.*   1 os. 6d. *net.*

Part III. **Materials.** *With* 188 *Illustrations.*   18s. *net.*

Part IV.   **Calculations for Building Structures.** *With* 597 *Illus-trations.* 13s. *net.*

BY THE SAME AUTHOR.

BUILDING CONSTRUCTION. (Forming a volume of LONGMANS' ADVANCED SCIENCE MANUALS.) With 385 Illustrations and an Appendix of Examination Questions. Crown 8vo, 4s. 6d.

LONDON, NEW YORK, & BOMBAY: LONGMANS, GREEN, AND CO.

PREFACE TO PART III.

THESE Notes are intended to furnish a Student with information amply sufficient to enable him to pass the Honours Examination of the Board of Education, so far as a knowledge of Building Materials is concerned.

They have, however, been extended somewhat beyond what is actually necessary for this purpose by the addition of Tables and information of a practical nature, which it is hoped may be useful to young Engineers. Architects, and others engaged in the design and erection of structures of different kinds.

It was considered that a work upon materials, written merely to meet the requirements of students in the earlier stages of the Science Examinations, would be unsatisfactory.

Such a work would contain very elementary information on the subject. It would be so condensed that it would not give a fair idea of the great differences which exist in the characteristics and qualities of even ordinary building materials; and being thus narrowly restricted, it would tend to encourage the pernicious practice of cramming.

In order to keep the bulk of the work within reasonable bounds, it has been necessary strictly to limit the scope of the Notes.

It will be well, therefore, to state exactly what they are meant to contain and what is purposely excluded.

They deal with the nature, characteristics, qualities, and defects of the materials used in Building and Engineering works ; and they describe the methods of examining and testing such materials.

The information given is restricted to that required by an Engineer, Architect, or Builder, in order to select and understand the materials with which he has to deal.

The principal varieties of building materials used in Great Britain and Ireland are described or noticed, but no reference is made to materials used only abroad—in India or the Colonies.

Descriptions of the manufacture of materials, or of the methods by which they are procured, have been excluded, except in so far as some such knowledge is necessary for an intelligent appreciation of the characteristics of the material.

The actual cost of materials has also, as a rule, been excluded. This varies from time to time, and must be ascertained from the annual Price Books.

The methods of measuring and valuing materials must also be studied in works devoted to those subjects.

It was originally intended to include in Part III. the information regarding stresses in parts of Structures required for Stages 2 and 3 and Honours.

The bulk of the volume, however, renders it necessary to reserve these subjects for another Part, which contains, as far as possible, all the information relating thereto that is required for the Examinations of the Board of Education in Building Construction.

NOTES ON BUILDING CONSTRUCTION.

Note to Part III

THE following List contains the names of the books which have been referred to and consulted in the preparation of these Notes.

Information derived from them has been acknowledged as far as possible upon the pages where it is given.

The writer is indebted also to many friends and to others for valuable particulars regarding special points.

On all sides,—from scientific and professional men, from quarry owners, manufacturers, and merchants,—the information asked for has been most willingly given.

The writer is glad to have this opportunity of expressing his thanks for the valuable assistance he has thus received, and for the very kind manner in which it has always been afforded to him.

Abney's Chemistry of Building Materials.
Anderson's Strength of Materials.
Ansted's Practical Geology.
Barlow's Strength of Materials.
Bauerman's Metallurgy of Iron.
Beare, Professor H., on Building Stones, Proc. Inst. C.E.
Bernays' Lectures on Chatham Dockyard Works.
Bloxam's Metals.
      „   Chemistry.
Box on Heat.
Britton on Dry Rot.
Brown's Forester.
Burnell on Limes and Cements.
Burns's Guide to Bricklaying, Plastering, etc.
Clark on Roads and Streets.
    „   Manual of Rules and Tables.
Cooke's Aide Mémoire.
Couche's Railways.
Cresy's Encyclopædia.
Dana's Mineralogy.
Davidson on House Painting.
Davies' Slate and Slate Quarrying.

De la Beche's Report on the Geology of Cornwall, Devon, and Somerset.
Dent's Chemistry of Building Materials.
Dobson and Mallet on Brick and Tile Making.
Downing's Construction.
Ede's Management of Steel.
Experiments on Steel by a Committee of Civil Engineers.
Fairbairn's Application of Iron to Building Purposes.
 ,, Iron Manufacture.
 ,, Useful Information for Engineers.
Galton's Hospitals.
Gillmore on Limes, Cements, and Mortars.
 ,, on the Compressive Resistance of Freestone, etc.
Gordon's Lead Poisoning of Water and its Prevention.
Greenwood on Steel and Iron.
Gwilt's Encyclopædia of Architecture by Papworth.
Hartwig's Sea and its Living Wonders.
Haupt's Military Bridges.
Hill's Lectures on Machinery used by Engineers.
Holtzappfel's Mechanical Manipulation.
Hull's Building and Ornamental Stones.
Humber on Water Supply.
Hunt's Guide to the Museum of Practical Geology.
 ,, Handbook to the Exhibition, 1862.
 ,, Mineral Statistics.
Hurst's Architectural Surveyor's Handbook.
 ,, Tredgold's Carpentry.
Hutton's Practical Engineer's Handbook.
Kirkaldy's Experiments on Iron and Steel.
Knapp's Technology.
Knight's Dictionary of Mechanics.
Laslett's Timber and Timber Trees.
Latham on Wrought Iron Bridges.
Latham's Sanitary Engineering.
Lipowitz on Manufacture of Portland Cement.
Lyell's Geology.
Matheson's Works in Iron.
Miller's Organic Chemistry.
Molesworth's Pocketbook of Engineering Formulæ.
Mushet on Iron and Steel.
Newlands' Carpenter's and Joiner's Assistant.
Page's Economic Geology.
Parkes' Hygiene.
Percy's Metallurgy.
Pole on Iron.
Rankine's Applied Mechanics.
 ,, Civil Engineering.
 ,, Useful Rules and Tables.
Reid and Lipowitz, Practical Treatise on Manufacture of Portland Cement.
 ,, on Concrete.
 ,, on Portland Cement, its Manufacture and Uses.
Redgrave, Calcareous Cements.

Report on the Exhibition of 1871.
Report on the Exhibition of 1876.
Report of the Royal Commission on the Selection of Stone for building the New Houses of Parliament.
Report of Commissioners appointed to inquire into the application of Iron to Railway Structures.
Report of Tests on the Strength of Structural Material made at Watertown Arsenal.
Richardson's Timber Importer's Guide.
Roorkee Treatise, Civil Engineering.
     „   „   Applied Mechanics.
Seddon's Builder's Work.
Sheffield Standard List.
Smith's Lithology.
Spon's Illustrated Price Book.
    ,,   Workshop Receipts.
Stevenson on Harbours.
Stoney on Stresses.
Traill on Mild Steel.
Tredgold's Carpentry.
Unwin's Iron Bridges and Roofs.
    „   Elements of Machine Design.
    ,,   Testing of Materials of Construction.
Ure's Dictionary of Arts, Manufactures, and Mines.
Vicat on Cements. Translated by Capt. J. T. Smith, F R S
Whichcord's Observations on Kentish Ragstone.
Wilkinson's Practical Geology of Ireland.
Woodward's Recent and Fossil Shells.
Wray's Application of Theory to Construction.
Wray on Stone.
Proceedings of the Chemical Society.
    Do.   Institution of Civil Engineers.
    Do.   American Society of Civil Engineers.
    Do.   Institute of Engineers.
    Do.   Institution of Mechanical Engineers.
    Do.   Iron and Steel Institute.
    Do.   Institute of Naval Architects.
    Do.   Society of Arts.
    Do.   Philosophical Society of Glasgow.
Professional Papers of the Corps of Royal Engineers.
The Professional Journals relating to Engineering, Architecture, Building, etc.
Circulars and Catalogues of several Manufacturers and Merchants.
List of Quarries (under the Quarries Act) and of Mines, prepared by H.M. Inspectors of Mines.

Note for Students.

The Syllabus of the Board of Education makes reference to the following subjects in connection with the examinations in materials:—

Stage 1.

The nature and properties of sand, lime, and cement; the composition of mortar or concrete . . . ; the properties of bricks, stones, tiles, and slates; the various kinds of timber in ordinary use; the constituents of cast iron, wrought iron, and steel, and the essential or characteristic differences of their properties.

Stage 2.

The best-known building stones, their quarrying, bedding, cutting, and dressing; characteristics of timber, its conversion and seasoning; the nature, qualities, and weights of various kinds of slates.

Stage 3.

The methods of testing cement, timber, iron, and steel.

Terra-cotta and artificial stone, their manufacture and uses.

Masonry. Character of various stones used in building, and localities where found; how to test for quality and bed; fitness of various stones for different atmospheres; weight generally, and approximate strength.

Timber: its seasoning, diseases, cause of decay, and means of preserving it.

Cast iron, wrought iron, and steel: properties, uses, strength, weight, and preservation; preservation of iron, timber, etc.; various kinds of glass and glazing.

Candidates for the Honours Examination should study the whole volume, with the exception of the tables, lists of brands, recipes, and other similar matters, most of which are in very small print, and are intended chiefly for use in practice.

Note to the Sixth Edition.

In this Edition the Chapters generally have been revised, and in some cases extended.

The Chapter on Iron has been revised, with special reference to the use of Mild Steel.

Attention has been directed to modern methods of Portland Cement manufacture, and of Concrete mixing, while reference is made to the Standardisation of Rolled Steel Sections.

CONTENTS OF PART III.

CHAPTER I.

STONE.

b

CONTENTS.

CHAPTER II

BRICKS, TILES, TERRA COTTA, ETC.

CONTENTS.

CONTENTS.

CONTENTS.

CONTENTS.

CONTENTS.

CHAPTER IV.

METALS.

CONTENTS.

CONTENTS.

CONTENTS.

CHAPTER V.

TIMBER.

General.—GROWTH OF TREES—Annual Rings—Medullary Rays—Sapwood—Heartwood—Felling—Squaring—CHARACTERISTICS OF GOOD TIMBER—DEFECTS IN TIMBER — Heartshakes—Starshakes — Cupshakes—Rind Galls—Upsets—Foxiness—Doatiness—Twisted Fibres —CLASSIFICATION OF TIMBER—Pine Wood—Leaf Wood—Soft Wood —Hard Wood—*Classification of Fir Timber*—Pine—Fir or Spruce.

MARKET FORMS OF TIMBER.—Log—Balk—*Fir*—Hand Masts—Spars— Inch Masts—Balk Timber—Planks—Deals—Whole Deals—Cut Deals—Battens—Ends—Scaffold and Ladder Poles—Rickers. *Oak* —Rough Timber—Sided Timber—Thick Stuff—Planks—*Waney Timber*—*Compass Timber*.

Description, Appearance, Characteristics, and Market Forms of Different Kinds of Timber. Pine Wood or Soft Wood. —NORTHERN PINE—Appearance—Varieties in use—*Balk*—Dantzic

CONTENTS.

CONTENTS.

CONTENTS.

CONTENTS.

CONTENTS.

CHAPTER IX.

MISCELLANEOUS.

Glue.—Manufacture—Characteristics—Preparation—Uses—Strength—*Glues to resist Moisture—Marine Glue.*

Size.—Manufacture—*Double Size—Patent Size—Kilvin Dry—Clear Cole—Parchment Size—Gold Size*—Oil Gold Size—Burnish Gold Size—Japanners' Gold Size.

Knotting.— *Ordinary*— First Size — Second —*Patent Knotting—Hot Lime.*

Paste.—Recipes for four varieties.

Gold Leaf.—Market Forms—*Pale Leaf Gold—Dutch Gold—Gold Paint.*

Putty.—*Painters' and Glaziers'*—Hard—Very Hard—Soft—*Plasterers' Putty—Thermo-plastic Putty.*

Rust Cement.—Manufacture—*Quick-setting—Slow-setting.*

Laths.—*Plasterers'*—Thickness—Market Forms—*Metal, Slate or Tiling Laths.*

Vulcanised Indiarubber.

Tar.—*Coal Tar*—Naphtha—Creosote—Pitch—*Wood Tar*—Stockholm—Archangel—American—Pitch—*Mineral Tar.*

Creosote.—Hygeian Rock.

Felt.—Asphalted—Sarking—Inodorous Bitumen—*Fibrous Asphalte—Hair Felt—Cement for Felt—Tarring Felt.*

Asbestos.—Raw—Concrete Coating—Roofing—Sheathing—Felt.

Willesden Fabrics.—Paper—Canvas—Wire Wove Roofing—Emery—Silicate Cotton.

Nails.—*Fine—Bastard—Strong—Tenpenny—Fourpenny,* etc.

Cast — Malleable — Hand-wrought — Cut — Patent Machine-wrought.

Rose Nails.—*Rose Sharp Points*—Fine—Canada—Strong. *Rose Flat Points*—Fine—Strong. *Rose Clench.*

Clasp Nails.— *Wrought*— Fine— Strong— *Cut.* Brads —*Flooring—Cabinet—Glaziers' Sprigs.*

Clout Nails—Fine—Strong—*Countersunk.*

Wire Nails—Pointes de Paris.

CONTENTS.

APPENDIX.

Physical Properties of Materials, and Loads and Stresses to which they are subjected.

NOTES ON BUILDING CONSTRUCTION
Arranged to meet the requirements of the Syllabus of the Board of Education, South Kensington.

PART II.

EXTRACT FROM CONTENTS.

RIVINGTON'S SERIES OF

NOTES ON BUILDING CONSTRUCTION

Arranged to meet the requirements of the Syllabus of the Board of Education,
South Kensington.

Part IV. —CALCULATIONS FOR BUILDING STRUCTURES.

EXTRACT FROM CONTENTS.

CHAPTER I.

STONE.

General Remarks.—In the following Notes no attempt will be made to describe the appearance and characteristics of all the different kinds of stone used in this country.

Such a task would be almost endless, and it would also be unprofitable. No description upon paper would give a practical idea of the appearance of the different varieties, and moreover the aspect and qualities of stone from the same quarry vary as different beds are reached.

It is therefore proposed to describe the characteristics which are common to most building stones, and to point out the qualities that are necessary to ensure a good material for building or engineering work.

A knowledge of these will form a guide in selecting stone for such purposes from any quarry, new or old, whether in this country or abroad.

This having been done, a few of the best known British building stones will be described, in order that the student may have some idea of their peculiarities and uses.

Tables will be added, giving the names of the principal quarries in the country, which will serve to impress upon the student the numerous varieties of stone which exist, and the localities in which they occur.

It is hoped that these Tables will be of use to the practical man, but, in order that they may be so, it will be advisable to describe exactly how they were prepared.

They include all the quarries reported upon by the Royal Commissioners who selected the stone for the Houses of Parliament, except a few which have since ceased work.

This list was extended by adding to it the names of the principal quarries given in the official report on Mineral Statistics, by Mr. Hunt.

Next are added a few important quarries mentioned in Hull's *Building Stones*, De la Beche's Report, Wray on *Stone*, Gwilt's *Encyclopædia of Architecture*, and some known to the author personally.

The list thus formed was completed as far as possible by comparison with the specimens in the Museum of Practical Geology and with those in another good collection of building stones.

The list was then sent to a great many different parts of the country, to be checked and supplemented by professional men having local knowledge, and also to a London stone merchant of great experience.[1]

With regard to any important stones of which the author had no personal knowledge, special information was obtained from experienced men on the spot.

The Tables are arranged to show the geological formations from which the different varieties of stone are obtained.[2]

These Notes do not, however, enter at all upon the subject from a geological point of view; the relative position of the different geological strata must be ascertained from works specially devoted to that subject.

Any reference to the quarrying, working, or cost of stone has also been avoided.

CHARACTERISTICS OF BUILDING STONE.

In selecting a stone for a building or engineering work, inquiry and investigation should be made to ascertain whether it possesses certain important characteristics mentioned below :—

Durability, or the power of resisting atmospheric and other external influences, is the first essential in a stone for almost any purpose.

The durability of a stone will depend upon its chemical composition, its physical structure, and the position in which it is placed ; and the same stone will greatly vary in its durability according to the nature and extent of the atmospheric influences to which it is subjected.

To make sure that a stone will " weather,"—that is, will wear well under exposure to the weather—many points have to be inquired into.

Chemical Composition.—The chemical composition of the stone should be such that it will resist the action of the atmosphere, and of the deleterious substances which, especially in large cities, the atmosphere often contains.

[1] For this edition the lists of Stone and Slate Quarries have been again revised by the aid of the List of Quarries (under the Quarries Act) and of Mines, prepared by H. M. Inspectors of Mines, together with other sources of information.

[2] The quickest way of finding a stone in the Tables is to look it out in the Index at the end of this volume.

These destroying substances are taken up by the moisture in the air, or by the rain, and are thus conveyed into the pores of the stone.

The sulphur acids, carbonic acid, hydrochloric acid, and traces of nitric acid, in the smoky air of towns,[1] and the carbonic acid which exists even in the pure atmosphere of the country,[2] ultimately decompose any stone of which either carbonate of lime or carbonate of magnesia forms a considerable part.

The oxygen even in ordinary air will act upon a stone containing much iron, and the fumes from bleaching works and factories of different kinds very soon destroy stones whose constituents are liable to be decomposed by the particular acids which the fumes respectively contain.

In addition to the direct chemical action of the sulphuric and sulphurous acids upon the constituents of stones, sulphates are sometimes formed by them which crystallise in the pores of the stone, expanding and throwing off fragments from the surface.

The durability of a stone depends, therefore, to a great extent upon the relation between its chemical constituents and those of the atmosphere surrounding. A stone which will weather well in the pure air of the country may be rapidly destroyed in the smoke of a large town.

Nature and Extent of Atmospheric Influence.—The same stone will weather very differently according to the nature and extent of the atmospheric influences to which it is subjected.

From what has been said above, it is evident that most stones will stand a pure atmosphere better than one which is charged with smoke, or with acids calculated to attack the constituents of the stone.

It is also evident that the stone will be less attacked in dry weather than during rain; the destructive acids cannot penetrate so deeply, and the frost has no influence whatever when the stone is dry.

The number of days on which there is rain in any district has therefore a great influence on the durability of stone used in that district.

---

[1] Dr. Angus Smith calculated that 15,000 tons of carbonic acid were daily evolved in Manchester. The air contained from ·04 to ·08 per cent of carbonic acid; the rain from 1·4 to 5·6 grains of sulphuric acid per gallon, and as much as 1¼ grain of hydrochloric acid.

[2] Dr. Angus Smith found ·03 per cent of carbonic acid in the pure air of the mountains of Scotland.

Wind has a considerable effect upon the durability of stone.

A gentle breeze dries out the moisture, and thus favours the lasting qualities of stone.

A high wind, however, is itself a source of destruction; it blows sharp particles against the face of the stone, and thus grinds it away. Moreover, it forces the rain into the pores of the stone, and may thus cause a considerable depth to be subject to the effects of acids and frost.

" Variation of temperature, apart from the action of frost, is also a cause of decay, the expansion and contraction due to it causing the opening of undetected natural joints, but its effect must be comparatively slight as a destructive agent."[1]

The Position of a Stone in a Building may very much influence its durability.

The stone in that side of any building which faces the prevailing rain is, of course, more liable to decay than it is in the other sides.

Any faces of stone that are sheltered altogether from the sun and breeze, so that the moisture does not quickly dry out, are very liable to decay.

This may be noticed especially in buildings of an inferior stone situated in a bad atmosphere. In these it will be seen that the soffits of arches and lintels, the shady sides of window jambs, and parts of carvings which the sun never gets at, are always the first portions of the building to decay.

Any stone exposed to very different degrees of heat on its different faces is liable to crack from unequal expansion and contraction.[1]

The Physical Structure of a stone is of the greatest importance, for upon it depends greatly its power of resisting the action of the atmosphere.

White chalk and marble are of the same chemical composition —both nearly pure carbonate of lime—yet the latter, especially when polished, will resist an ordinary atmosphere for a long time, while the former is rapidly disintegrated and destroyed.

Hence stones which are crystalline in structure are found to weather better than those that are non-crystalline.

No stone intended for the exterior of a building should have a porous surface, otherwise the rain conducts the acids from the atmosphere into the pores of the stone, which soon becomes decomposed.

[1] Wray *On Stone.*

Again, in winter the wet penetrates the pores, freezes, expands, and disintegrates the surface, leaving a fresh surface to be similarly acted upon, until the whole stone is gradually destroyed.

If the chemical composition and remaining qualities of two stones are the same, then the stone which has the closer and finer grain of the two is likely to be more durable than the other.

It is important that a stone should be homogeneous in its structure. If the grains and the cement uniting them are both of lasting material, the stone will be very durable. If the grains be easily decomposed and the cementing material remains, the stone will become spongy and porous, and then liable to destruction by frost. If the cementing material is destroyed, the grains will fall to pieces.

It is important that the stone should contain no soft patches or inequalities; unequal weathering leaves projections which catch the rain, etc., and hasten decay.

Facility for Working.—The readiness with which stone can be converted by the mason into the various shapes in which it is required for different kinds of work is of importance from an economical point of view.

The characteristics of a stone in this respect will depend in some cases upon its hardness, but will also be influenced by the soundness of its texture; by its freedom from flaws, shakes, vents, etc.; and also by its natural cleavage and other peculiarities.

A soft stone of even grain and without distinct beds would naturally be selected for carved work, while a hard stone in thin layers, easily separated, would be well adapted for building good and economical rubble masonry. (See Part I.)

Hardness.—The hardness of stone is often of importance, especially if it is to be subjected to a considerable amount of wear and friction, as in pavements. It is, moreover, important when the stone is to be used for quoins, dressings, and other positions where it is required to preserve a sharp angle or "*arris.*" Hardness combined with toughness is also essential in good road metalling, which should not, however, be liable to splinter or to grind readily into dust.

It does not follow because a stone is hard that it will weather well; many hard stones are more liable to atmospheric influence than those of a softer texture, whose chemical composition is of a more durable nature.

Stone used for work exposed to the action of water should be

hard ; running or dripping water soon wears away the surface
The blocks of stone in marine works are subject to serious injury
not only from the impact of the waves themselves, but from the
sand and stones thrown against them by the force of the sea.

Strength.—The strength of stone should be ascertained if it is
to be subjected to any excessive or unusual stresses.

Stones in ordinary building or engineering works are generally,
under, compression, occasionally subject to cross strain, but never
to direct tension.

It is generally laid down that the compression to which a stone
should be subjected in a structure should not exceed $\frac{1}{10}$ to $\frac{1}{20}$ of
the crushing weight as found by experiment.

Practically, however, the compression that comes upon a stone
in any ordinary building is never sufficient to cause any danger
of crushing.

The greatest stress that comes upon any part of the masonry
in St. Paul's Cathedral is hardly 18 tons per square foot. In St.
Peter's, Rome, it is about 15 tons per square foot.

By a reference to page 83 it will be seen that these stresses
would be safely borne even by the softer descriptions of stone.[1]

The weakest limestones that exist will bear a compression of
60 tons per foot, while the resistance of ordinary building stones
ranges from 100 to 500 tons per square foot, and in the case of
granites and traps rises as high as 700 to 1200 tons per square foot.

It is possible, however, in some forms of arches, in retaining
walls, and in other structures, that a considerable pressure may be
concentrated upon certain points, which are liable to be crushed.

Weight.—The weight of a stone for building has occasionally
to be considered.

In marine engineering works it is often advisable to use heavy
stones to resist the force of the sea.

A light stone would be best adapted for arches, while heavy
stones would add to the stability of retaining walls.

Appearance.—The appearance of stone is often a matter of
importance, especially in the face work of conspicuous buildings.

In order that the appearance may be preserved, a good weathering stone should of course be selected, free from flaws, clayholes, etc.

All varieties containing much iron should be rejected, or they
will be liable to disfigurement from unsightly rust stains caused

---

[1] The student will bear in mind the reduction in the compressive strength of piers
of masonry or brickwork as compared with the single blocks of which they are con-
structed. See pp. 80, 116.

by the oxidation of the iron under the influence of the atmosphere.

Stones of blotched or mottled colour should be regarded with suspicion. There is probably a want of uniformity in their chemical composition, which may lead to unequal weathering (see p. 5).

Position in Quarry.—In order to obtain the best stone that a

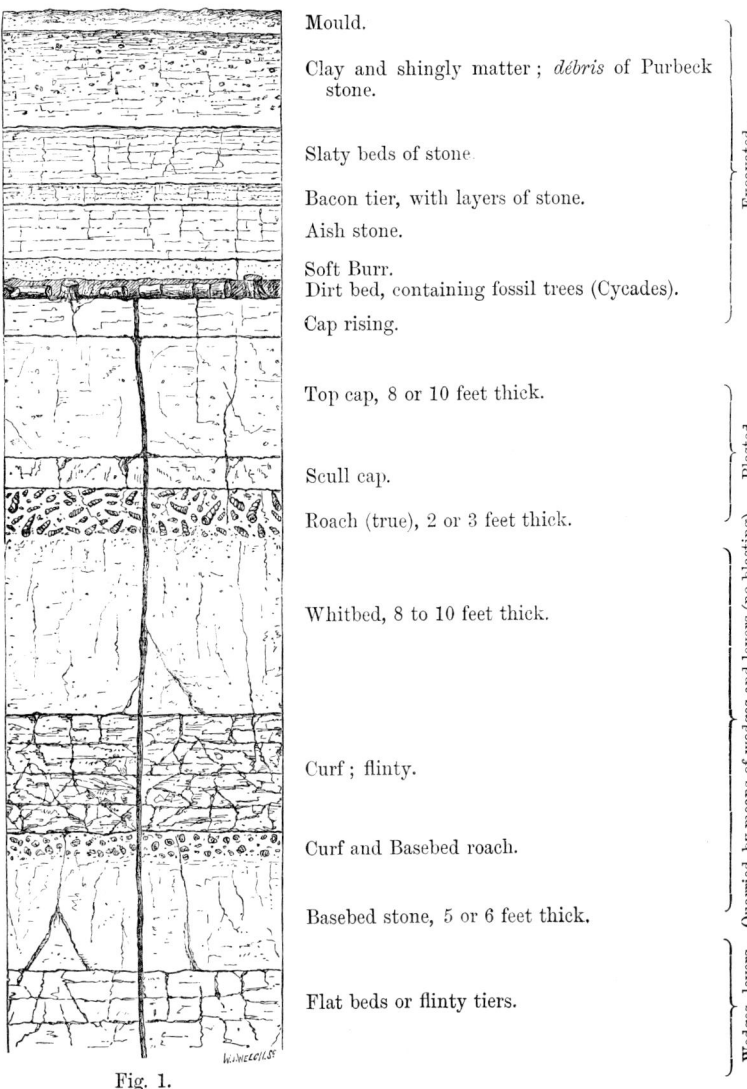

Mould.

Clay and shingly matter ; *débris* of Purbeck stone.

Slaty beds of stone.

Bacon tier, with layers of stone.

Aish stone.

Soft Burr.
Dirt bed, containing fossil trees (Cycades).

Cap rising.

} Excavated.

Top cap, 8 or 10 feet thick.

Scull cap.

Roach (true), 2 or 3 feet thick.

} Blasted.

Whitbed, 8 to 10 feet thick.

Curf ; flinty.

Curf and Basebed roach.

Basebed stone, 5 or 6 feet thick.

Flat beds or flinty tiers.

} Quarried by means of wedges and levers (no blasting).

} Wedges, levers.

Fig. 1.

quarry can furnish, it is often important that it should be taken from a particular stratum.

It frequently occurs that in the same quarry some beds are good, some inferior, and others almost utterly worthless for building purposes, though they may all be very similar in appearance.

To take Portland stone as an example. In the Portland quarries there are four distinct layers of building stone.

Fig. 1 is a section showing approximately how the strata in a Portland quarry generally occur.

Working downwards, the first bed of useful stone that is reached is the *True* or *Whitbed Roach*—a conglomerate of fossils which withstands the weather capitally. Attached to the *Roach*, and immediately below it, is a thick layer of *Whitbed*—a fine even-grained stone, one of the best and most durable building stones in the country ; then, passing a layer of rubbish, the *Bastard-Roach*, *Kerf*, or *Curf* is reached, and attached to it is a substantial layer of *Basebed*.

The *Bastard-Roach* or *Basebed-Roach* and the *Basebed* are stones very similar in appearance to the *True Roach* and *Whitbed ;* but they do not weather well, and are therefore not fitted for out-door work.

Though these strata are so different in characteristics, the good stone can hardly be distinguished from the other even by the most practised eye.

Similar peculiarities exist in other quarries.

It is therefore most important to specify that stone from any particular quarry should be from the best beds, and then to have it selected for the work *in the quarry* by some experienced and trustworthy man.

The want of this precaution led to the use of inferior stone (though from very carefully chosen and good quarries) in the Houses of Parliament.

Seasoning.—Nearly all stone is the better for being seasoned by exposure to the air before it is set.

This seasoning gets rid of the moisture, sometimes called "quarry sap," which is to be found in all stone when freshly quarried.

Stone should, if possible, be worked at once after being quarried, for it is then easier to cut, but unless this moisture is allowed to dry out before the stone is set, it is acted upon by frost, and thus the stone, especially if it be one of the softer varieties, is cracked, or, sometimes, disintegrated.

The drying process should take place gradually. If heat is applied too quickly, a crust is formed on the surface, while the interior remains damp, and subject to the attacks of frost.

Some stones (see p. 59) which are comparatively soft when quarried, acquire a hard surface upon exposure to the air.

Natural Bed.— All stones in walls, but especially those that are of a laminated structure, should be placed " on their natural bed,"—that is. either in the same position in which they were *originally deposited* in the quarry, or turned upside down, so that the layers are parallel to their original position, but inverted. If they are placed with the layers parallel to the face of the wall, the effect of the wet and frost will be to scale off the face layer by layer, and the stone will be rapidly destroyed.

In arches, such stones should be placed with the natural bed as nearly as possible at right angles to the thrust upon the stone, —that is, with the " grain" or laminæ parallel to the centre lines of the arch stones, and perpendicular to the face of the arch.

In cornices with undercut mouldings the natural bed is placed vertically and at right angles to the face, for if placed horizontally, layers of the overhanging portion would be liable to drop off. There are, in elaborate work, other exceptions to the general rule.

It must be remembered that the beds are sometimes tilted by upheaval subsequent to their deposition, and that it is the original position in which the stone was deposited that must be ascertained.

The natural bed is easily seen in some descriptions of stone by the position of imbedded shells, which were of course originally deposited horizontally. In others it can only be traced by thin streaks of vegetable matter, or by traces of laminæ, which generally show out more distictly if the stone is wetted.

In other cases, again, the stone shows no signs of stratification. and the natural bed cannot be detected by the eye.

A good mason can, however, generally tell the natural bed of the stone by the " feel" of the grain in working the surface.

A stone placed upon its proper natural bed is able to bear a much greater compression than if the laminæ are at right angles to the bed joints.

Sir William Fairbairn found by experiment that stones placed with their strata vertical bore only $\frac{6}{7}$ the crushing stress which was undergone by similar stones on their natural bed.[1]

Agents which destroy Stones.—The two principal classes of agents which destroy stone have already been described.

They are—Chemical agents, consisting of acids, etc., in the atmosphere; and Mechanical agents, such as wind, dust, rain, frost, running water, force of the sea, etc.

There are other enemies to the durability of stones, which may just be glanced at, viz.—

> Lichens.
> Worms or Molluscs.

LICHENS.—In the country lichens and other vegetable substances collect and grow upon the faces of stones.

These are in many cases a protection from the weather, and tend to increase the durability of the stone. The fine rootlets spread themselves over the surface and into the interstices, covering the face from the action of wind and weather.

In the case of limestones, however, the lichens sometimes do more harm than good, for they give out carbonic acid, which is dissolved in rain water, and then attacks the carbonate of lime in the stone.

MOLLUSCS.—The *Pholas dactylus* is a boring mollusc found in sea water, which attacks limestone, hard and soft argillaceous shales, clay, and sandstones. It also attacks wood, but granite has been found to resist it successfully.

These animals make a number of vertical holes close together, so that they weaken and eventually destroy the stone.

By some it is supposed that they secrete a corrosive juice,[2] which dissolves the stone ; others consider that the boring is mechanically done by the tough front of the shell covering the Pholas.[3] These animals are generally small, but sometimes attain a length of five inches—the softer the rock the bigger they become. The shale beds, on which was founded the quay wall at Kirkcaldy, were so perforated by Pholades that they crushed under the superincumbent pressure, and a settlement resulted.[4]

The most notable instance of injury done by *Pholades* is at Plymouth breakwater, where, in consequence of their attacks, the limestone blocks had to be replaced by granite.[4]

The *Saxicava* is another small mollusc, found in the crevices of rocks and corals, or burrowing in limestone, the holes being sometimes six inches deep. It has been known to bore the cement stone (clay-ironstone) at Harwich, the Kentish Rag at Folkestone, and the Portland stone used at Plymouth breakwater.

---

[1] Rankine, *Civil Engineering.*
[2] Hartwig's *The Sea and its Living Wonders.*
[3] Woodward's *Recent and Fossil Shells.*
[4] Stevenson *On Harbours.*

EXAMINATION OF STONE.

Speaking generally, in comparing stones of the same class, the least porous, most dense, and strongest, will be the most durable in atmospheres which have no special tendency to attack the constituents of the stone.

Fracture.—A recent fracture, when examined through a powerful magnifying glass, should be bright, clean, and sharp, with the grains well cemented together. A dull, earthy appearance betokens a stone likely to decay.

Tests.—In examining a stone it may be subjected to various tests, some of which afford a certain amount of information as to its characteristics.

Resistance to Crushing.—The strength of the stone as regards resisting compression may be ascertained by crushing specimens of suitable form (see pp. 81-84).

This is not a very important test, for the reasons given at page 6, but some authorities consider that it affords an idea of the powers of the stone with regard to resisting frost.

Absorption.—A more important guide to the relative qualities of different stones is obtained by immersing them for twenty-four hours, and noting the weight of water they absorb. The best stones, as a rule, absorb the smallest amount of water.

The Table at p. 85 shows the amount of water absorbed in twenty-four hours by several of the most important English stones, some known to be durable, and others the reverse. This will afford a useful guide in judging of the quality of any new stone after ascertaining its powers of absorption.

Brard's Test.—Small pieces of the stone are immersed in a concentrated boiling solution of sulphate of soda (Glauber's salts), and then hung up for a few days in the air.

The salt crystallises in the pores of the stone, sometimes forcing off bits from the corners and arrises, and occasionally detaching larger fragments.

The stone is weighed before and after submitting it to the test. The difference of weight gives the amount detached by disintegration. The greater this is, the worse is the quality of the stone.

The action of the salt was supposed at one time to be similar to that of frost, but Mr. C. H. Smith has pointed out that it is essentially different, inasmuch as water expands in the pores as it freezes, but the salt does not expand as it crystallises.

Acid Test.—Simply soaking a stone for some days in dilute solutions containing 1 per cent of sulphuric acid and of hydrochloric acid, will afford a rough idea as to whether it will stand a town atmosphere.

A drop or two of acid on the surface of the stone will create an intense effervescence if there is a large proportion present of carbonate of lime or carbonate of magnesia.

Mr. C. H. Smith's Test was proposed for magnesian limestone, but is useful for any stone in determining whether it contains much earthy or mineral matter easy of solution.

" Break off a few chippings about the size of a shilling with a chisel and a smart blow from a hammer ; put them into a glass about one-third full of clear water ; let them remain undisturbed at least half an hour. The water and specimens together should then be agitated by giving the glass a circular motion with the hand.   If the stone be highly crystalline, and the particles well cemented together, the water will remain clear and transparent, but if the specimens contain uncrystallised earthy powder, the water will present a turbid or milky appearance in proportion to the quantity of loose matter contained in the stone.   The stone should be damp, almost wet, when the fragments are chipped off."

The best way of carrying out this test is to pulverise the stone and then treat it as above described.   The heavy particles will sink to the bottom and the earthy turbid matter will settle more slowly.

Practical Way of ascertaining Weathering Qualities.—The durability of a stone to be obtained from an old established quarry may generally be ascertained by examining buildings in the neighbourhood of the quarry in which the stone has been used

If the stone has good weathering qualities, the faces of the blocks, even in very old buildings, will exhibit no signs of decay ; but, on the contrary, the marks of the tools with which they were worked should be distinctly visible.

Exposed cliffs or portions of old quarries, or detached stones from the quarry, which may be lying close at hand, should also be examined, to see how the stone has weathered.

In both cases care should be taken to ascertain from what stratum or bed in the quarry, the stones have been obtained.

Quarrying.—This is too large a subject to be entered upon in these Notes.

It will be sufficient to remark that in quarrying stone for building purposes there should be as little blasting as possible, as it shakes the stone, besides causing considerable waste.

Care should be taken to cut the blocks so that they can be placed in the work for which they are intended with their natural beds at right angles to the pressure that will come upon them.

If this is not attended to, the blocks will be built in in a wrong position, or great waste will be incurred by converting them.

Scientific Classification.—The different kinds of stone used for building and engineering works are sometimes divided into three classes :—1. The Siliceous. 2. The Argillaceous. 3. The Calcareous ; according as flint (silica), clay (formerly called " argile "), or carbonate of lime,[1] forms the base or principal constituent.

Practical Classification.—In describing the physical characteristics of stones, for practical purposes it will be better to classify them as follows :—

1. Granites and other igneous rocks.
2. Slates and Schists.
3. Sandstones.
4. Limestones.

[1] *Calcium Carbonate.*

GRANITE AND OTHER IGNEOUS ROCKS.

Granite is, as its name implies, a stone of crystalline granular structure.

True or Common Granite.—There are several varieties of stone practically known as granite, but true granite consists of crystals of quartz and felspar mixed with particles of mica.

Composition.—The quartz is a very hard glassy substance in grey or colourless amorphous lumps, occasionally in crystals.

The felspar should be crystalline and lustrous, not earthy in appearance; its grains are of different shapes and sizes, and their colour may be white, grey, yellowish pink, red, or reddish brown.

The mica is in dark grey, black, brown, flexible, semi-transparent glistening scales, which can easily be flaked off with a knife.

Granite generally contains more felspar than quartz, and more quartz than mica.

The colour of the stone depends upon that of the predominating ingredient, felspar.

" An average granite may be expected to contain from two to three fifth parts of crystals of quartz or crystalline quartz; about the same, more or less, of felspar, also partly crystalline and chiefly in definite crystals; and the remainder (one-tenth part) of mica. But the mica may form two or three tenths, and the quartz three-fifths or more, while the proportion of the felspar, as well as the particular composition of the felspar, both vary extremely."[1]

The durability of the granite depends upon the quantity of the quartz and the nature of the felspar.

If the granite contains a large proportion of quartz, it will be hard to work; but, unless the felspar is of a bad description, it will weather well.

The felspars that occur most commonly in granite are potash felspar (*orthoclase*) and a lime and soda felspar (*oligoclase*).

Sometimes both these varieties are found in the same stone.

Of the two, potash felspar is more liable to decay than the other.[2]

Mica is easily decomposed, and it is therefore a source of weakness.

[1] Ansted's *Practical Geology.*   [2] Wray.

If the mica or felspar contain an excess of lime, iron, or soda, the granite is liable to decay.

" The quantity of iron, either as the oxide or in combination with sulphur, must affect the durability of granite, as well as of all other stone.

" The iron can generally be seen with a good glass, and a very short exposure to the air, especially if assisted in dry weather by artificial watering (better still, if 1 per cent of nitric acid be added to the water), ought to expose this.

" The bright yellow pyrites crystallised in a cubical form appear to do little harm. The white radiated pyrites (marcasite), on the contrary, decompose quickly.

" Where the iron stains are large, uneven, and dark coloured, the stone may fairly be rejected, at any rate for outside work.

" When the discoloration is of a uniform light yellow, it is probable that little injury will be done to the stone in a moderate time, and unless appearance is a matter of great importance, such granite would not be rejected.

" In the red granites, the discoloration from iron does not show so easily, but still sufficiently to guide the engineer if bad enough to cause rejection." [1]

The quality of granite for building purposes depends upon its durability, and upon the size of the grains. The smaller these are, the better can the granite be worked, and the more evenly will it wear.

"In using granite for ornamental purposes, the coarser-grained stones should be placed at a distance from the eye, the finer-grained stones where they can be easily inspected. Without attention to this point, very little better effect is produced than by a stone of uniform colour." [1]

Syenite and Syenitic Granite are generally included by the engineer and builder under the general term granite.

True Syenite consists of crystals of quartz, felspar, and hornblende, the latter constituent taking the place of mica in ordinary granite. It derives its name from the granite of Syene, in Upper Egypt, though it has been shown that the latter is really a syenitic granite of the composition mentioned below.

Syenitic Granite consists of quartz, felspar, mica, and hornblende, the last-named constituent being added to those of ordinary granite.

[1] Wray.

Characteristics.—The syenites and syenitic granites are generally of darker colour than ordinary granite, caused by the grains of hornblende.

" The syenitic granites are on the whole tougher and more compact than the ordinary granites, take on a fine polish, and are exceedingly durable.

" They occur less abundantly in nature; but their rarer use most frequently arises from the darker tints imparted to them by the hornblende." [1]

The following varieties of granite may be briefly noticed, though they are of no great importance in connection with building and engineering works:—

TALCOSE GRANITE contains, in addition to the ingredients of common granite, *talc*, a material which scales off in thin flakes, having a whitish colour and unctuous feel.

Such granites are said not to weather well.

PROTOGENE contains talc instead of mica.

CHLORITIC GRANITE contains chlorite, an olive-green mineral, generally granular, and of a pearly lustre.

SCHORLACEOUS GRANITE contains pieces of *schorl*, " a black, hard, brittle, mineral crystallised in masses or long crystals, sometimes columnar, and radiating from a centre." [2]

GRAPHIC GRANITE is composed of long parallel prisms of quartz and felspar, the ends of which when broken across look like the letters of cuneiform inscriptions.

This granite contains very little mica, and is not much used for building purposes.

PORPHYRITIC GRANITE is the name given to those varieties in which large, distinct, independent crystals of felspar occur at random interspersed through the mass.

These crystals are sometimes called " horse's teeth."

Quarrying and Dressing.—Granite is quarried either by wedging or by blasting. The former process is generally reserved for large blocks, and the latter for smaller pieces and road-metal.

It is better to have the blocks cut to the desired forms in the quarries; first because it is easier to square and dress the stone while it contains the moisture of the ground or " quarry-sap ; " also because the local men, being accustomed to the stone, are able to dress it better and more economically, and part of the work can be done by machinery, generally to be found at the principal quarries. Moreover, the bulk of the stones being reduced by dressing, the cost of carriage is saved, without much

[1] Page's *Practical Geology.*   [2] Wray *On Stone.*

danger of injuring the arrises in transit, as the stone is very hard.

Uses to which Granite is applied.—Granite is used chiefly for heavy engineering works, such as bridges, piers, docks, lighthouses, and breakwaters, where weight and durability are required. It is also used especially for parts of structures exposed to blows or continued wear, such as copings of docks, paving, etc. The harder varieties make capital road metal.

In a granite neighbourhood the stone is used for ordinary buildings; but it is generally too expensive in first cost, transport, and working, and is therefore reserved for ornamental features, such as polished columns, pilasters, heavy plinths, etc.

The granular structure and extreme hardness of granite render it ill adapted for fine carving, and its surface is entirely destroyed by the effects of fire.

Varieties in Common Use.—Granite is found in Aberdeenshire, Kirkcudbrightshire, Argyleshire, and the Islands of Mull and Arran. Also in Cornwall, Devonshire, Leicestershire, Cumberland, and the islands of Guernsey and Jersey. The Irish granites occur chiefly in the counties of Wicklow, Wexford, Donegal, and Down.

The SCOTCH GRANITES are most esteemed for beauty and for durable qualities, especially those from the two great districts of Aberdeen and Peterhead—the stone from the former is generally grey, and that from the latter red.

The other best known varieties of Scotch granites are those from *Rubislaw, Stirling Hill, Dalbeattie, Ross of Mull, Kemnay, Kintore, Cruden,* etc.

The CORNISH AND DEVONSHIRE granites, sometimes called moorstones, have not so high a character. They contain a large proportion of felspar, which in some cases weathers very badly. The potash felspar of these granites, when decomposed, turns into Kaolin or porcelain clay.

The LEICESTERSHIRE GRANITES are, generally speaking, syenites—very hard and tough, difficult to dress, and therefore not much used for building purposes. They are well adapted for paving sets, and make capital road metal.

JERSEY AND GUERNSEY GRANITE is also syenitic. It is a good weathering stone, very hard, durable, used for paving purposes, but rather apt to become slippery.

The IRISH GRANITES are very numerous. Grey varieties are obtained from Wicklow and Dublin. Those of a reddish tint from Galway. A good bluish grey granite comes from Castle Welland, County Down; Counties Donegal and Mayo produce good red granites. Several colours and varieties come from Carlow. Newry supplies a greenish syenite.[1]

FOREIGN GRANITES. Large quantities of imported granite are used in this country both for heavy engineering works and paving, etc., much of which is quarried in Norway. No details of localities can here be given, but amongst other Scandinavian Quarries may be mentioned that of Fredrikshald, Idefiord.

---

[1] Wilkinson's *Practical Geology of Ireland.*

The following Tables give a list of some of the principal GRANITE QUARRIES in Great Britain and Ireland. In these and the following Tables of Sandstones and Limestones, the names of certain quarries have been retained, where their connection with well-known works or buildings appeared to render such a course desirable, although their names may not appear in the latest official *List of Quarries in the United Kingdom of Great Britain and Ireland and the Isle of Man.*

LIST OF SOME OF THE PRINCIPAL GRANITE QUARRIES IN GREAT BRITAIN AND IRELAND.

NAME OF QUARRY OR LOCALITY.	SITUATION.	COUNTY.	COLOUR OF STONE.	WEIGHT PER FOOT CUBE IN LBS.	REMARKS.
English Granites.					
ANGLESEA	Holyhead	Anglesea	White	...	This stone, sometimes called a granite, is really a quartz rock. Used for Holyhead Breakwater. Too hard to work for ordinary buildings.
BARDON HILL	Leicester	Leicestershire	Greenish	...	Road metal, much used in Midlands.
BLACKENSTONE, Dartmoor	Moreton Hampstead	Devonshire	Grey	...	Paving in Exeter and buildings. A porphyritic granite, with large brown crystals.
BOSS	St. Brenard	Cornwall	Bluish grey	...	Used in London Bridge, British Museum, etc.
CARNSEW	Penryn	Do.		...	Very fine grain. Putney Bridge.
CASTLE HILL	Groby	Leicestershire.			
CHARNWOOD	Loughborough	Do.	Dark green, containing very little pink	186·8	Syenite.
CHEESEWRING	Linkinhorne	Cornwall	Grey	183·4	Used at Westminster Bridge; Rochester Bridge; Birkenhead, Southampton and other Docks; Great Bases Lighthouse, Ceylon; Plymouth Forts; Devonport Docks; new Eddystone Lighthouse.
CLEE HILL	Ludlow	Shropshire		...	Basalt. Road metal and sets.
CLIFF HILL	Markfield	Leicestershire.	Dark green, containing some pink	176·0	Syenite.
COLCERROW	Par.	Cornwall		...	Used at Plymouth Breakwater and Lighthouse; Keyham; Portsmouth; Chatham; Pembroke Docks; Exeter Market, etc.
DARBISHIRE'S	Penmaenmawr	Carnarvonshire		...	Diorite. Road metal, sets, etc.
DE LANK	Bodmin	Cornwall	Grey	...	Very large stones, fine and close grained. Used in works at Milford, Portland, Devonport; lighthouse on the Smalls; Blackfriars Bridge; Wax Chandlers' Hall, City.
ENDERBY HILL	Narboro'	Leicestershire			
FREEMATOR	Tavistock	Devonshire.		...	
GROBY	Bardon	Leicestershire	Pink and green	173·4	Syenite. For paving sets and road metal.
GROSMONT	Grosmont	Yorkshire			Whinstone Mines.
GUNNISLAKE	Gunnislake	Cornwall	White and grey	...	Milford and Middlesborough Docks; Devonport Town Hall.
HAYTOR, or High Tor	Bovey Tracey	Devonshire.	Grey	...	London Docks, etc.
HEALE, Dartmoor	Plymouth	Do.	Do.	...	Nelson Column; Royal Exchange; H.M. Dockyards; Crystal Palace.
HECKWOOD	Tavistock	Do.		...	Keyham Dockyard.

HERM . . .	Herm .	Near Guernsey .	Grey . . .	187	Syenite. Contains felspar, quartz, hornblende, and a little mica; fine grained; weathers well; hard. Used chiefly for paving.
LAMORNA and Penzance	Penzance .	Cornwall .	Do. . .	...	Used at Devonport, Keyham, and other dockyards; Portland; Down; Alderney; Fortifications, St. Ives.
LA MOYE .	...	Jersey .	Pink and Grey .	162·7	Syenite. Used chiefly for paving sets and road metal.
LUNDY ISLAND .	Bideford .	Devonshire .	Grey . .		
LUXULYAN .	Liskeard .	Cornwall.			
MAEN . .	Constantine	Do. .	Dark green .	...	
MARKFIELD .	Leicester .	Leicestershire .	Pale blue .	170	A syenite used for paving and road metal.
MERRIVALE BRIDGE	Whitchurch	Devon .			Good stone.
MOUNTSORREL .	Loughborough .	Leicestershire .	Pinkish brown .	164·1	Syenite. Used chiefly for paving and road metal.
NEW MILL . .	Madron .	Cornwall .	...	...	"Stone of excellent quality, and can be obtained of very large size."[1]
PENMAENMAWR .	Penmaenmawr .	Carnarvonshire .			Diorite. Road metal, sets, etc.
PENRYN . .	Penryn .	Cornwall .	Grey . .	182·7	Used at Devonport, Keyham, Portsmouth, Chatham, Deptford Dockyards; Portland, Dover, and Alderney breakwaters; works at Hull, Liverpool, Birkenhead, etc.
PEW TOR .	Tavistock .	Devonshire .	Reddish brown and grey	167	Known as "Tamar" granite. Large blocks in Plymouth Breakwater; Duke of Bedford's buildings, Tavistock.
POLKANUGGO .	Stythians .	Cornwall.	Pink.		
PORT JOHN .	...	Jersey .	...		
ROWLEY REGIS .	Rowley Regis .	Staffordshire .			Basalt. Several quarries.
ROYAL OAK .	Princetown .	Devonshire.			
SHAP FELL, Wasdale Head	Shap, in Penrith	Westmoreland .	Pink or reddish brown. Dark and light.	168·5	A porphyritic granite, containing large flesh-coloured crystals of felspar (orthoclase); takes a fine polish; converted by machinery. Mausoleum, Lowther Castle; columns in Hull Museum; columns, St. Pancras Station; posts, western area, St. Paul's; pedestal, Palmerston's Statue, Westminster. New Dock, Malta.
SHEFFIELD .	Penzance .	Cornwall .	Greenish grey .	...	
ST. AUSTELL .	St. Austell .	Do. .	...	...	Much porcelain clay obtained from the decomposing granite here.
ST. BLAZY .	Par . .	Leicestershire.	Grey . .		
STONEY STANTON	Stoney Stanton				
TRETHWY .	Par . .	Cornwall.	...		Porphyritic. Used for Duke of Wellington Sarcophagus, St. Paul's, etc.
TROWLESWORTHY	...	Devon .	Red . .	...	The only red granite in West of England.
VALE CASTLE .	...	Guernsey .	Grey ; blue .	187	Syenite and diorite.

[1] Guide to Museum of Practical Geology.

TABLE OF GRANITES—*Continued.*

NAME OF QUARRY OR LOCALITY.	SITUATION.	COUNTY.	COLOUR OF STONE.	WEIGHT PER FOOT CUBE IN LBS.	REMARKS.
Scotch Granites.					
ARDSHEAL	Ballachulish	Argyleshire.			
AVOCHIE	Huntly	Aberdeenshire.			
BLACKHILL	Cruden	Do.	Red	166·6	Takes a beautiful polish; turned by machinery into columns, etc.
BONAW	Loch Etive	Argyleshire.			
CAIRNCRY	Aberdeen	Aberdeenshire.			
CLINTERTY	Newhills	Do.	Reddish grey.		
CORRENNIE	Cluny	Do.	Salmon, grey, red		Chiefly quartz and felspar with specks of mica or horneblende; Glasgow Municipal Buildings.
COVE	Nigg	Kincardineshire	Dark grey		Used chiefly for kerbs and sets.
CRAIGNAIR, Dalbeattie	Dalbeattie	Kirkcudbrightshire	Grey	213	Liverpool; Birkenhead; Newport; Swansea Docks; Liverpool Borough Bank; Buildings in Manchester.
CREETOWN	Newton-Stewart	Do.	Do.		Several quarries.
DANCING CAIRN	Buxburn	Aberdeenshire	Usually grey; sometimes red		Buildings in Aberdeen. Used in London for kerbs, paving, etc. Composed of quartz, orthoclase, oligoclase, and mica.
DYCE	Kirkhill	Do.	Grey.		
FURNACE	Furnace	Argyleshire.			
HIGH ROCK	Breadalbane	Perthshire	Dark red	166·0	Fine grain. Used chiefly for monumental work.
HILL O' FARE	Banchory	Kincardineshire	Do.		
INVERARAY	Inveraray	Argyleshire	Grey		Docks, Hull; Newcastle; Leith; Forth Bridge, etc.
KEMNAY	Kemnay	Aberdeenshire			
KIRKMABRECK	Creetown	Kirkcudbrightshire.			
LONGHAVEN	Ellon	Aberdeenshire.			
OLDTOWN	Oldtown	Do.	Do.		
PERSLEY	Aberdeen	Do.	Red	165·9	Pillars, Carlton Club; good colour; coarse grained. Composed of red orthoclase, albite, black mica, quartz. Good to polish.
PETERHEAD	Peterhead	Do.			Liverpool Docks; Westminster Bridge; Northern Lighthouses; Albert Memorial.
ROSS OF MULL	Bunessan	Argyleshire	Pink, red, and grey		Buildings in Aberdeen. Used in London for kerbs, paving, monumental work, etc.; Bell Rock Lighthouse.
RUBISLAW	Aberdeen	Aberdeenshire	Grey		
SCLATTIE	Buxburn	Do.	Do.		

Name	Locality	County	Colour	Sp. gr.	Remarks
STIRLINGHILL	Peterhead	Aberdeenshire	Red	165·9	Duke of York's Column; columns in St. George's Hall, Liverpool, and Fishmongers' Hall. Coarse grained. Good to polish. Much used in London for paving and sets.
TILLYFOURIE	Cluny	Do.	Bluish grey	...	Very like the granite from Kemnay.
TOM'S FOREST	Kintore	Do.	Grey		
TORMORE	Ross of Mull	Argyleshire.	Do.	...	Buildings in Aberdeen. Used in London for kerbs, paving, etc. Composed of quartz, orthoclase, oligoclase, and mica.
TYREBAGGER	Newhills	Aberdeenshire.			
Irish Granites.					
ALTNAVEIGH	Altnaveigh	Armagh.			
BALLYBREW	Karneystown	Wicklow.			
BALLY KNOCKAN and Golden Grove	Blessington	Do.	Greyish white	169·1	Works freely; fit for ornamental work. Much used in Dublin; Kingstown Station.
BALLYNACRAIG	Newry	Down.			
CARNSORE POINT and Killiney Hill	Carnsore Point	Wexford	Grey	...	Large blocks; good quality. Kingstown Harbour; Thames Embankment; buildings in Dublin. Very scarce.
CARRIN HILL	Killeavy	Armagh.			
CASTLEWELLAN	Castlewellan	Down	...	...	Used for base and pedestal of Albert Memorial.
CROREAGH	Croreagh, Newry	Do.			
DALKEY	Dalkey	Dublin	Do.	169·6	A good stone for building, road metal and paving sets; hard to work. St. Werburgh's and St. Paul's Churches.
DONEGAL	Dungloe	Donegal	Reddish	...	Dungloe Church.
FURLOUGH	Galway	Galway	Do.	...	A porphyritic granite containing large crystals of pink orthoclase.
GARVARY WOOD	...	Donegal Wicklow.	Pink and black.		National Bank of Ireland and other public buildings in Dublin.
GLENCREE	Enniskerry	Wicklow.	...	...	
GORAGHWOOD	Goraghwood	Armagh.			
KINGSTOWN	Dublin	Dublin	Grey	171	Kingstown Railway Station. Hard stone to work.
LOUGH	Carlow	Carlow	Cream colour, white, yellow	...	Fine compact texture; curiously spotted. New Church, Carlow.
MOOR	Newry	Down	Grey	168	
NEWRY	Do.	Do.	Do.	170	A hard good stone for general building purposes; very durable; one of the best granites in Ireland. Much used in North of Ireland.
PARNELLS	Arklow	Wicklow.			
ROSTREVOR	Rostrevor	Down.			
SHANTALLA.	Shantalla.	Galway	Brownish red	...	Polished slabs, pillars, carved Celtic crosses.
WARRENPOINT ROAD	Newry	Down.			

Igneous Rocks other than Granite.—There are several rocks which more or less resemble the granite in their characteristics, and are generally associated with it in the classification of building stones. These rocks are, however, seldom used for building or engineering works, except in the immediate neighbourhood of the place where they are found.

The Porphyries " generally occur as dykes and eruptive masses intersecting the older schists and slabs, and are usually much fissured and jointed, and for this reason incapable of being raised in massive monoliths like the granite." [1]

There are two principal varieties found in Great Britain. Each consists of a general mass or base, through which are scattered crystals varying in size from small grains to $\frac{3}{4}$ inch in length. The stone breaks with a smooth surface and conchoidal fracture.

Felstone Porphyry consists of a base, which is an intimate mixture of quartz and orthoclase, known as *Felsite*, with independent crystals of felspar.

Quartziferous Porphyry has a base consisting of a granular crystalline compound of quartz and felspar, with individual crystals of felspar and quartz. [1]

Characteristics.—" Both varieties appear in many tints—red, flesh-coloured, fawn-coloured, black, bluish-black, and bluish-green ; and both varieties may contain, in subordinate quantities, other crystals than those enumerated above. [2]

" Incapable of being raised in large blocks, they are polished only for minor ornaments ; their principal use in Britain being for causeway-stones and road metal, for which their hardness and toughness render them specially suitable.

" Though chiefly used for road material, in some districts they are employed in the building of country mansions, farm sheds, and walls ; and when properly dressed and coursed make a very fair structure (especially the fawn-coloured sorts), and are perfectly indestructible." [2]

Some of the darker varieties are too sombre for building purposes, except when used for ornamental purposes to relieve surfaces of lighter stone.

ELVAN (a term originally peculiar to Cornwall and Devon) is found in dykes or veins traversing the granite or slate ; the dykes varying in width from a few feet to 300 or 400.

It usually differs from granite in the absence of mica and in the fineness of its grain. It sometimes contains schist.

" It is much used as a building stone in Cornwall, and is found to be very durable," [3] also as road material in competition with Guernsey granite.

Stone locally known as Elvan is also met with in County Wexford.

GNEISS is composed of the same constituents as granite, but the mica is more in layers, and the rock has therefore a stratified appearance.

The rock splits along the layers with facility, and breaks out in slabs from a few inches to a foot in thickness.

---

[1] Quartz porphyry is quarried in Carnarvonshire, at Llanbedrog, Trefor, Nevin, and Portnant.   [2] Page.   [3] Wray.

It is used both as a building material in the bodies of walls (with dressings of brick, or more easily dressed stone) and for flagging.[1]

MICA SCHIST, sometimes called *Mica Slate*, is composed chiefly of mica and quartz in thin layers : the mica sometimes appears to constitute the whole mass.

Its colour is grey or silvery grey, and it has a shining surface, owing to the quantity of mica present.

It breaks out in thin even slabs, and the more compact varieties are used for flagging, door and hearth stones, and furnace linings.[1]

HORNBLENDE SCHIST, or *Hornblende Slate*, is usually black, composed principally of hornblende, with a variable quantity of felspar, and sometimes grains of quartz.

It resembles mica schist, but has not so glistening a lustre, and seldom breaks into thin slabs. It is tougher than mica schist, and is an excellent material for flagging.[1]

TRAP ROCKS.—*Greenstone*, also called *Trap* or *Whinstone*, is a mixture of felspar and hornblende.

It has sometimes a granular crystalline structure, and at other times it is very compact without apparent grains.

It is generally of a greenish colour, but varies in tint from light-greenish grey to greenish-black or black.

It is extremely hard and tough, and makes capital road metal—is very often split up by joints, so that it is well suited for paving setts, but not for large blocks. Its colour is too sombre for the walls of houses.

Some of the stratified varieties are dangerous as building stones, being liable to decomposition on exposure to the weather, even where there is no frost.

Varieties in Common Use.—*Penmaenmawr Stone* from N. Wales is largely used throughout the country for paving setts. It is very easily split by cutting a fine line with an axe in the direction required, and then giving the stone a few smart taps with a hammer.[2] Quarries, *Penmaenmawr* and *Darbishire's*.

Bardon Hill Stone from Leicestershire is also much used for road metal in the central counties.

Stone of this description is also found in *Cornwall*, near *Edinburgh*, in Argyleshire, at *Carlin Knowse* and other places in Fifeshire, and also in County Wexford.

Whinstone is found in Wigtownshire, near Selkirk, in Kincardineshire, near Haddington, near Edinburgh, at Falkirk, in Perthshire, Fifeshire, Inverness-shire, Ross-shire, and other places in Scotland.

BASALT resembles greenstone, but is composed of lime felspar, augite,[3] olivine, and titano-ferrite.[4]

[1] Dana and Wray.   [2] Seddon.
[3] Black and greenish-black crystals of anhydrous silicate of magnesia.
[4] Titanic iron.

It occurs in dykes or sheets penetrating or lying between older rocks, or upon the surface, and is sometimes stratified, sometimes columnar.

" It varies in colour from greyish to black. In the lighter coloured felspar predominates ; in the darker iron or a ferruginous augite."[1] It is often of a dark green.

This stone affords a great resistance to crushing, and is eminently adapted for paving curbs, etc.

Rowley Rag is a basalt found in Staffordshire, and used for road material, paving sets, and also for making artificial stone.

This material is found also in the counties of Armagh, Antrim, and Londonderry.

SLATES AND SCHISTS.

CLAY SLATE.—The ordinary slate used for roofing and other purposes is an argillaceous rock, compact and fine grained. It was originally a sedimentary rock, but it will no longer divide along the planes of bedding, but splits readily along planes called " planes of slaty cleavage."

This facility of cleavage is one of the most valuable characteristics that slate possesses, as it enables masses to be split into thin sheets, whose surfaces are so smooth that they lie close together, thus forming a light and impervious roof covering.

These planes of cleavage are caused by intense lateral pressure.

Planes of cleavage are either coincident with the layers of deposit or lie at angles with them. When they are in the same plane, or nearly so, the rock is converted into slabs for paving ; or planed, if it is soft enough, and made into cisterns, etc. The reason that it cannot be made into roofing slates is that the lamina of the bedding and the lines of cleavage run into each other and render the surface rough and uneven.

There is another line of imperfect cleavage, which will yield to the chisel. Along this line the blocks of slate are split up longitudinally. It is along this line that fracture occurs when a slate is accidentally broken. The split along this line is called by quarrymen the " *Plerry.*"

Quarrying.—The rock is worked in " Floors," or tunnels one above another.

Powder is used to detach the blocks, which are plerried into widths suitable for making the best-sized slates ; then split into thicknesses of about three inches ; cut by circular saws into suitable lengths ; split by skilful hands with the aid of thin inch chisels ; and squared, either by machinery or by hand.

---

[1] Dana's *Mineralogy.*

Cambrian slates are not sawn, because natural joints occur at distances about equal to the length of the slates. They are generally squared by hand.

Slate rock becomes more compact and the blocks are generally larger and more valuable the deeper they are from the surface ; but the rule does not always hold good, and there is apparently a limit to it. The blocks are split more easily when fresh from the quarry.

CHARACTERISTICS OF SLATES.

Hardness and Toughness.—A good slate should be both hard and tough.

If it is too soft it will absorb moisture, the nail holes will become enlarged, and the slate will be loose.

If it be brittle it will fly to pieces in the process of squaring and holing, or at any rate will break on the roof if any one walks over it, which is often necessary when the roof is being repaired.

A good slate should give out a sharp metallic ring when struck with the knuckles—it should not splinter under the slater's zax —should be easily " holed " without danger of fracture, and should not be tender or friable at the edges.

Colour.—The colour is not much guide to the quality of a slate. Some people think, however, that the black varieties absorb moisture, and decay.

The colours of slates vary greatly. Those most frequently met with are dark blue, bluish-black, purple, grey, and green.

Red, and even cream-coloured slates have been found.

Some slates are marked with bands or patches of a different colour—*e.g.,* dark purple slates often have large spots of light green upon them. These are generally considered not to injure the durability of the slate, but they lower its quality by spoiling its appearance.

Absorption.—A good slate should not absorb water to any perceptible extent.

The amount of absorption may be ascertained by the test given at page 28.

Grain.—A good slate should have a very fine grain.

The grain of the rock is easily seen, and the slates are cut so that the grain is in the direction of their length, in order that if a slate breaks when on the roof it will not become detached, but will divide into longitudinal pieces, which will still be held by the nails.

Veins are dark marks running through some slates. They are always objectionable, but particularly when they run in the direction of the length of the slate, for it will be very liable to split along the vein.

Pyrites.—Crystals of iron pyrites are often found in slates, especially in those from Scotland, etc.

They are often considered objectionable. It should, however, be borne in mind that there are two varieties of pyrites, of the same chemical composition but of different crystalline form, and very different in their resistance to atmospheric influence.

Ordinary Iron Pyrites, consisting of yellow brassy crystals, generally cubical, weathers well. The crystals have been found perfectly bright and firm in their places in roofs 100 years old, even in the atmosphere of Glasgow.

White Iron Pyrites (or marcasite), on the other hand, is easily decomposed, and slates containing it ought to be rejected. This form of pyrites is generally dull and wanting in lustre, and is therefore not easily seen.

Sizes.—The slates sent to the market are squared in the quarry—sometimes roughly by hand, sometimes by machinery—to certain sizes, which are distinguished by different names,[1] as shown in the following Table.

In buying and selling slates in this country, a "thousand" is generally understood to mean 1200 or 1260.

The Table shows the weight of 1200 slates of each description ; it also shows the number of yards covered per "thousand" of 1200 slates, and the cost per square.

Strength.—The crushing strength of some British and American slates is given in the Table of Building Stones, p. 81. The transverse strength of a specimen of Monson (U.S.) slate tested at Watertown Arsenal was found to give a modulus of rupture of 7671 lbs. per square inch, and a shearing strength of 2192 lbs. per square inch. Compare also the moduli of rupture for Building Stones, p. 84, and for Brick, p. 121

---

[1] These names are used in the building trade, but not much in the quarries.

The following Table of the Different SLATES in use has been taken from the list of the Oakeley Slate Quarries Company. The prices were those current in January 1904, but vary of course according to the state of the market.

NAMES.	Sizes.	FIRST QUALITY. Computed Weight.	MEDIUM QUALITY. Computed Weight.	Sizes.	SECOND QUALITY. Computed Weight.	Sizes.	THIRD QUALITY. Computed Weight.	1 M. of 1200 Allowing 3 inch lap will cover, about
		t. c. q.	t. c. q.		t. c. q.		t. c. q.	
Empresses . . .	26 × 16	4 0 0	4 0 0	..	..	..	..	170 Sq. Yds.
Small Empresses.	26 × 15	3 15 0	3 15 0	..	..	..	..	160 do.
Princesses . . .	24 × 14	3 5 0	3 5 0	24 × 14	4 0 0	..	..	136 do.
Duchesses . . .	24 × 12	2 15 0	2 15 0	24 × 12	3 10 0	..	..	116 do.
Small Duchesses.	22 × 12	2 10 0	2 10 0	22 × 12	3 5 0	..	..	105 do.
Marchionesses .	22 × 11	2 5 0	2 5 0	22 × 11	3 0 0	..	..	97 do.
Countesses . .	20 × 10	1 15 0	1 15 0	20 × 10	2 7 2	20	2 10 0	78 do.
Wide Viscountesses	18 × 10	1 12 2	1 12 2	18 × 10	2 2 2	..	..	69 do.
Viscountesses . .	18 × 9	1 7 2	1 7 2	18 × 9	1 17 2	18	2 0 0	62 do.
Wide Ladies . .	16 × 10	1 7 2	1 7 2	16 × 10	1 17 2	..	..	60 do.
Broad Ladies . .	16 × 9	1 5 0	1 5 0	16 × 9	1 12 2	..	..	54 do.
Long Ladies . .	16½ × 8½	1 5 0	1 5 0	16½ × 8½	1 12 2	16	1 15 0	52 do.
Ladies	16 × 8	1 2 2	1 2 2	16 × 8	1 10 0	..	..	48 do.
Wide Headers . .	14 × 12	1 10 0	1 10 0	14 × 12	1 17 2	..	..	60 do.
Headers	14 × 10	1 5 0	1 5 0	14 × 10	1 15 0	14	1 5 0	50 do.
Small Headers .	13 × 10	1 2 2	1 2 2	13 × 10	1 7 2	..	..	46 do.
Small Ladies . .	14 × 8	1 0 0	1 0 0	14 × 8	1 5 0	13	1 0 0	40 do.
Narrow Ladies .	14 × 7	0 17 2	0 17 2	14 × 7	1 2 2	..	..	36 do.
Doubles	13 × 7	0 15 0	0 15 0	13 × 7	1 0 0	..	..	32 do.
Wide Doubles . .	12 × 8	0 17 2	0 17 2	12 × 8	1 0 0	..	..	32 do.
Small Doubles .	12 × 6	0 14 0	0 14 0	..	..	..	..	25 do.
Singles	10 × 8	0 15 0	0 15 0	..	..	..	..	26 do.

HEXAGON, GOTHIC, DIAMOND, ROUND, OR OTHER FANCY SLATES made to order in First Quality only.

QUEENS—*First Quality.*—24, 26, 28, 30, 32, and 34 inches long and various breadths, assorted as made, without specifying quantities of each, at per ton of 20 cwt. of 112 lbs., 90s. and upwards.

Medium Quality.—As above, 75s. and upwards.
Second Quality.— Do. 60s. do.

The weights given are for Portmadoc slates. Slates from Bangor and Penrhyn are somewhat heavier, and Westmoreland slates heavier still.

Rags vary in size, but average 36×24 inches. Imperials 30×24 inches. Queens, Rags, and Imperials are sold by weight.

THE FOLLOWING PRICES ARE FOR THE BEST (OAKELEY COMPANY'S) PORTMADOC SLATING.

NAMES.	SLATING PER SQ. 2½ IN. LAP Zinc nails.	Copper nails.	LABOUR AND NAILS. Zinc.	Copper.	STRIP AND RELAY.
	£ s. d.	£ s. d.	s. d.	s. d.	s. d.
Ladies	1 13 6	1 15 0	8 3	10 6	12 0
Countesses	1 18 0	2 0 0	6 6	8 0	11 0
Duchesses	1 17 6	2 0 0	6 0	8 6	10 6
Queens or Rags	2 10 0	2 12 6	10 0	12 6	15 0
Imperials	2 10 0	2 12 6	10 0	12 6	15 0
Westmoreland	...	4 5 0	13 6	16 6	18 6
Eureka Green	2 7 6	2 10 0	7 0	9 6	12 6
Permanent Green . . .	1 18 6	2 0 0	7 0	9 6	12 6
Whitland Abbey . . .	2 10 0	2 17 6	7 0	9 6	12 6

If 3 inches lap, extra per square, 1s. 6d.
William's patent slate ridge and hips, with 2 inch circular roll and flat sides, per foot run, 1s. 6d.
Robinson's ditto, 10d. ; do, 1½ inch ls. ; do, 3 inch roll 1s. 4d.
Hips and ridges without roll, 14 inches girt ditto, 8d.
For fixing with white lead and copper screws, 4d.
1 inch, 1¼ inch, and 1½ inch copper nails per lb., 10d.

Quality.—The market qualities of slates are classed in the quarries according to their straightness, smoothness of surface, fair even thickness, and in the Cambrian quarries according to the presence or absence of discoloration.

Slates are generally divided into 1st and 2d qualities; in some cases a " medium quality " is quoted.

All slates for good work should be hard, tough, non-absorbent, of uniform colour, free from patches, from veins, iron, cross-grain, and with smooth and even surfaces.

Thickness.—The thickness increases with the area of the slate, and the rule for the proportionate thickness varies in different quarries but for Welsh slates is somewhat as follows :—[1]

	THICKNESS.	
	1st Quality.	2d Quality.
Duchesses and Marchionesses . . .	$\frac{3}{16}$ inch	$\frac{1}{11}$ inch.
Countesses and large Ladies . . .	$\frac{1}{6}$,,	$\frac{1}{4}$,,
Doubles	$\frac{3}{20}$,,	$\frac{3}{13}$,,

The best qualities of Welsh slates generally split easily into even sheets with smooth surfaces, and holding their thickness close up to their edges, even after being squared.

Irish and Scotch slates are often of very uneven thickness, being thicker in the middle than near the edges, and very much stouter and more substantial than Welsh slates of the same area.

Slates are sometimes split too thin, so that they are not strong enough for roofing purposes.   The Ffestiniog quarries have produced (for exhibition as specimens of perfection of cleavage) slates 5 feet to 10 feet long, 6 inches to 12 inches broad, only $\frac{1}{16}$ inch thick.[2]

Tests.—The following rough tests are generally recommended, but they are not of a practical character, nor can they be relied upon.   Experience is required to judge of a slate by the eye.

1. Weigh the slate carefully when dry, steep it in water for 24 hours, run the water off, and weigh again--any difference of weight will show the amount of absorption.

2. Stand the slate in water up to half its height—if it be of bad quality the water will rise in the upper half, but in a good slate no sign of moisture will be seen above the water-line.[3]

3. Breathe on the slate.   If a clayey odour be *strongly* emitted it may be inferred that the slate will not " weather." [4]

[1] Wray.   [2] Hunt.   [3] Gwilt.   [4] Dempsey.

DIFFERENT FORMS OF SLATE.

Slate Slabs.—Besides the small thin slates used for roofing, large and thick slabs, and even blocks of slate, are quarried out and used for many purposes connected with building and engineering works.

Slate in these forms is particularly useful on account of its strength. "The strength of slate 1 inch thick is considered equal to that of Portland stone 5 inches thick,"[1] and "its resistance to shearing is said to be greater than that of any other stone."[2]

Slate slabs are easily obtained of any length under 6 or even 8 feet, and containing from 10 to 30 superficial feet.

Their thickness ranges from 1 inch to 3 inches.

Larger slabs may be obtained by paying extra. The Exhibition of 1862 contained one sent by the Llangollen Slate Company which measured 20 × 10 feet, and weighed $4\frac{1}{2}$ tons ; also several from the Ffestiniog quarries of the Welsh Slate Company averaging 14 feet by 7 or 8 feet.[3]

They may be procured either self-faced—that is, as they are split from the blocks—rough sawn, quarry planed, or polished.

The edges are sawn square, planed, filed, or rounded.

Such slabs may be fitted with great accuracy, and are used for cisterns, urinals, troughs, mantelpieces, baths, window and door-sills, skirtings, flooring, wine-bins, steps, landings, etc.

Slate Blocks, containing as much as 2 or 3 cubic feet, can easily be obtained.

In Wales and other slate districts they are sometimes used for the walls of buildings, and slate in scantlings is substituted for much of the wood work, *e.g.*, in door and window frames.

Slate is also sent out from the quarries in the form of steps, sills, etc.

The same material is used for making ridge rolls of different patterns for roofs, dowels for heavy masonry, etc. etc.

Enamelled Slate is prepared by painting slate slabs, baking them, colouring to pattern, covering them with a coating of enamel, rebaked and rubbed down several times alternately, and then polished.

It is often made to represent different varieties of marble, and is much in request for chimney-pieces and other purposes for which marble is used, also for sanitary purposes.

Varieties in use.—There are many slate quarries throughout Great Britain and Ireland, on the Continent, also in Canada and the United States.

Some American slates have been imported during late years, but the great bulk of the slates used for building are from home quarries.

WELSH SLATES.—The finest slates found in the United Kingdom come from Wales.

---

[1] Papworth, 657. [2] Wray.

[3] Hunt's *Handbook, Exhibition* 1862.

The slates from the *Silurian* formations of Merionethshire, Montgomery-shire, etc., are generally of a blue or grey colour, and of beautiful cleavage, splitting very thin, and sawn square by machinery.   The best-known quarries are those in the Ffestiniog district, such as the Oakeley, Llechwedd, Maen-offeren, Votty, New Welsh Slate Mines, and others.

The slates of the *Cambrian* formation in Carnarvonshire are of varied colours—blue, purple, green, and dark grey.   They are more siliceous than the Lower Silurian slates, and not so easily cleaved.   They are therefore thicker and heavier, but they are very hard and ring well when struck. Their edges are not sawn, for the reasons given above.   The best-known quarries are those of Penrhyn and Dinorwic.

Many of the quarries produce also slabs of first rate quality.

ENGLISH SLATES are generally thicker and coarser than those from Wales— hard, tough, and very durable.   The best known are the green slates from Westmoreland and Lancashire, and the slabs from Delabole in Cornwall.

SCOTCH SLATES are also thick and coarse, and generally contain a large pro-portion of iron pyrites, which, however, does not interfere with their good weathering qualities.

The best-known quarries are those of Ballachulish, Easdale, and Cullipool. They are generally blue.

IRISH SLATES.—Many of the best qualities resemble the Welsh varieties, others are thicker and coarser.

Among the best-known Irish roofing slates are those from the Killaloe, Victoria, or Ormonde Quarries.   Slabs of a high quality are exported from Valencia in county Kerry, suitable for billiard tables, creamery fittings, etc., together with roofing and school slates.

TABLE OF SLATE QUARRIES.

QUARRY OR LOCALITY.	SITUATION.	COUNTY.	REMARKS.
Welsh Slates.			For general remarks on the characteristics of Welsh slates, see pp. 28-30.
ABERLLEFENNY .	Corris . .	Merionethshire .	Slate mine.
ALEXANDRA .	Bryngwyn .	Carnarvonshire.	
BRAICHGOCH .	Corris . .	Merionethshire .	Do.
BRYN EGLWYS .	Towyn . .	Do.	Do.
CEFN . . .	Cilgerran . .	Pembrokeshire.	
CEFNDU . .	Llanberis . .	Carnarvonshire.	
CILGWYN . .	Nantlle . .	Do.	
DINORWIC . .	Llanberis . .	Do.	
DIPHWYS CASSON .	Blaenau Fes- tiniog . .	Merionethshire .	Do.
DOROTHEA . .	Nantlle . .	Carnarvonshire.	
ERA . . .	Machynlleth .	Montgomeryshire	Do.
GLANRAFON .	Glanrafon .	Carnarvonshire.	
GLYNRHONWY .	Llanberis . .	Do.	
LLANDILO . .	Llandilo . .	Pembrokeshire.	
LLECHWEDD .	Blaenau Fes- tiniog	Merionethshire .	Do.
MAENOFFEREN .	Do. Do.	Do.	Do.
MOELFERNA	Glyndyfrdwy .	Do.	Do.
MOEL TRYFAN .	Bryngwyn .	Carnarvonshire.	
NEW WELSH SLATE . .	Blaenau Fes- tiniog	Merionethshire .	Do.
NEW VRONHEULOG	Nantlle . .	Carnarvonshire.	
OAKELEY . .	Blaenau Fes- tiniog	Merionethshire .	Do.
PENRHYN . .	Bethesda . .	Carnarvonshire.	
PENYRORSEDD .	Nantlle . .	Do.	
RHIWBACH . .	Blaenau Fes- ‑ tiniog	Do.	Do.
RHOSYDD . .	Do. Do.	Merionethshire.	Do.
SOUTH DOROTHEA	Nantlle . .	Carnarvonshire.	
TALYSARN . .	Do. . .	Do.	
VICTORIA . .	Bettws Garmon	Do.	
VOTTY and Bo- WYDD . .	Blaenau Fes- tiniog	Merionethshire .	Do.
WHITLAND ABBEY	Danderwen .	Pembrokeshire .	
WRYSGAN . .	Blaenau Fes- tiniog	Merionethshire .	Do.
WYNNE . .	Chirk . .	Denbighshire .	Do.
English Slates.			For general remarks on the characteristics of English slates, see p. 30.
ADDISTONE . .	Torver . .	Lancashire.	
APPLETHWAITE .	Applethwaite .	Westmoreland.	
BANNERSIDE .	Torver . .	Lancashire.	
BROAD MOSS .	Tilberthwaite .	Do.	

TABLE OF SLATE QUARRIES—*Continued.*

QUARRY OR LOCALITY.	SITUATION.	COUNTY.	REMARKS.
English Slates— *Continued*			
BURLINGTON .	Kirkby . .	Lancashire.	
CAULDRON . .	Kentmere .	Westmoreland.	
DELABOLE, OLD .	St. Teath .	Cornwall.	
EAST CORNWALL .	St. Neot . .	Do.	Slate mine.
ELTERWATER .	Elterwater .	Westmoreland.	
HODGE CLOSE .	Tilberthwaite .	Lancashire.	
HONISTER . .	Honister . .	Cumberland.	Do.
KIRKSTONE .	Ambleside .	Westmoreland.	
KLONDYKE .	Tilberthwaite .	Lancashire.	
LARCOMBE . .	Diptford . .	Devonshire.	
LAUNCESTON SLATE	South Petherwin	Cornwall.	
MOSS RIGG . .	Tilberthwaite .	Lancashire.	
OKEHAMPTON .	Wiveliscombe .	Somersetshire.	
PARK WOOD .	Kingsnympton	Devonshire.	
PARROCK END .	Tilberthwaite .	Lancashire .	Slate mine.
TILBERTHWAITE .	Do.	Do.	Several quarries.
TORVER . .	Torver . .	Do.	Do. do.
TRACEBRIDGE .	Wellington .	Somersetshire.	
TREBOROUGH .	Treborough .	Do.	
TROUTBECK PARK	Windermere .	Westmoreland.	
YEOLMBRIDGE .	Yeolmbridge .	Devonshire.	
Scotch Slates.			For general remarks on the characteristics of Scotch slates, see p. 30.
ABERFOYLE .	Aberfoyle .	Perthshire.	
BALLACHULISH .	Ballachulish .	Argyleshire.	
BALVICAR . .	Seil Island .	Do.	
BELNAHUA . .	Belnahua Island	Do.	
BREADALBANE .	Luing Island .	Do.	
BRECKLET . .	Ballachulish .	Do.	
CRAIGLEA . .	Logiealmond .	Perthshire.	
CULLIPOOL . .	Luing Island .	Argyleshire.	
EASDALE . .	Easdale Island .	Do.	
GLENALBIN .	Kilninver .	Do.	
PORT MARY .	Luing Island .	Do.	
Irish Slates.			For general remarks on the characteristics of Irish slates, see p. 30.
BENDUFF . .	Leap . .	Cork.	
DRIMOLEAGUE .	Drimoleague .	Do.	
DROMASTA . .	Dromasta .	Do.	
FOURCOIL . .	Clonakilty .	Do.	
GARRYBEG (Killaloe)	Nenagh . .	Tipperary.	
MADRANA . .	Skibbereen .	Cork.	
ORMONDE SLATE .	Carrick-on-Suir	Kilkenny.	
VALENCIA . .	Valencia Island	Kerry.	
VICTORIA SLATE .	Carrick-on-Suir	Tipperary.	

Stone Slates.—So called " slates," being merely thin slabs of stone which splits into thin layers along the planes of bedding, are found in various parts of the country, and used for roofing purposes. They are tilestones rather than true slates.

Among others may be mentioned the Collywiston and Stonesfield " slates," found in several quarries of the oolitic limestone formation, near Stamford in Northamptonshire, Stow-on-the-Wold in Gloucestershire, and at Purbeck.

They are good non-conductors of heat, so that they keep a house cool in summer and warm in winter ; but they are very heavy, especially when soaked with wet, and therefore require roofs of heavy scantlings to support them.

SERPENTINE.

Serpentine derives its name from the mottled appearance of its surface, which is supposed to resemble the skin of a serpent.

Composition.—Pure serpentine is a hydrated silicate of magnesia, but it is generally found intermixed with carbonate of lime, with steatite or soapstone (also a silicate of magnesia), or with diallage, a foliated green variety of hornblende and dolomite.

Colour.—The prevailing colour of serpentine is generally a rich green or red, permeated by veins of the white steatite.

Some varieties have a base of olive-green, with bands or blotches of rich brownish-red or bright-red, mixed with lighter tints, or olive-green, with steatite veins of greenish-blue ; some are red, studded with crystals of green diallage ; some clouded, and some striped.

Characteristics.—Serpentine is massive or compact in texture, not brittle, easily worked, and capable of receiving a fine polish. It is so soft that it may be cut with a knife.

It is generally obtained in blocks from 2 to 3 feet long, and it has been found that " the size and solidity of the blocks increase with their depth from the surface." [1]

Uses.—This stone is greatly used in superior buildings for decorative purposes. It is, however, adapted only for indoor work, as it does not weather well, especially in smoky atmospheres, for it is liable to attack by hydrochloric and sulphuric acids. The red varieties are said to weather better than those of a greenish hue, and it is stated that those varieties especially which contain white streaks are not fit for external work.

[1] Hunt.

It is much used for indoor work, such as tables, shafts, pilasters, jambs for chimney-pieces, and ornaments of different kinds.

Varieties in Common Use.—ENGLISH.—*Lizard Serpentine*, from the Lizard promontory in Cornwall, is perhaps the best known and most extensively used in this country.

There are three varieties of this Serpentine to be found in the locality.

1. " The principal mass, like that of some other districts, is of a deep olive-green, but this is variegated by veins or bands and blotches of rich brownish-red or blood-red, mixed with lighter tints." [1]

·" The best places for obtaining the red-striped varieties which we have seen, occur at the Balk near Landewednack, at the Signal Staff Hill near Cadgwith, at Kennack Cove, and on Goonhilly Downs."

2. " A variety, with an olive-green base, striped with greenish-blue steatite veins, is found . . . near Trelowarren." [2]

3. " An especially beautiful variety is found at Maen Midgee, Kerith Sands, in which the deep reddish-brown base is studded with crystals of diallage, which, when cut through and polished, shine beautifully of a metallic green tint in the reddish base." [2]

Anglesea.—Greenish and reddish serpentines are found at Llanfechell and Ceryg-mollion ; and a serpentinous marble at Tregola, near Llanfechell and near Holyhead.

SCOTCH.—Serpentine rocks occur in several localities in Scotland.

That of *Portsoy, in Banffshire,* " is very rich and varied in colour. It passes from soft green to deep red, and is variegated with veins of white steatite."

Serpentine is also found in the *Ochil Hills, Aberdeenshire,* at *Killin* in *Perthshire,* and in the *Shetland Isles,* where it forms the matrix of the chrome iron ore.

IRISH.—*Connemara* (Co. Galway) furnishes a serpentine in large blocks, ·commonly known as *Connemara Marble* or *" Irish Green "* marble. It is of two kinds.

The first is of a deep uniform shade of dark-green, but the other is mottled, and made up of bands and stripes of greens of different shades, interlaced with white streaks.

The principal quarries are near *Ballinahinch, Letterfrack,* and *Clifden.*

Other green serpentines are found at Crohy Head, and Aughadovey in Donegal, and near Lough Gill in county Sligo.[3]

ANCIENT.—*Vert Antique* is a name applied to many varieties of green serpentinous rock used by the ancient Romans. " These ornamental stones, exported from the ruins of buried cities, have been recut and polished, and are now used in the internal decorations of modern buildings." [4]   A detailed description of the different varieties will be found in Professor Hull's *Treatise on the Building and Ornamental Stones of Great Britain and Foreign Countries.*

SANDSTONES.

Composition.—Sandstones consist generally of grains of quartz

---

[1] Hull's *Building and Ornamental Stones.*
[2] *Report on the Geology of Cornwall, Devon, and Somerset,* by Sir H. de la Beche
[3] Hull.   [4] Wray.

—*i.e.* sand—cemented together by silica, carbonate of lime, carbonate of magnesia, alumina, oxide of iron, or by mixtures of these substances.

In addition to the quartz grains there are often other substances, such as flakes of mica, fragments of limestone, argillaceous and carbonaceous matter, interspersed throughout the mass.

As the grains of quartz are imperishable, the weathering qualities of the stone depend upon the nature of the cementing substance, and on its powers of resistance under the atmosphere to which it is exposed.

Sometimes, however, the grains are of carbonate of lime, embedded in a siliceous cement; in this case the grains are the first to give way under the influence of the weather.

Colour.—Sandstones are found in great variety of colour—white, yellow, grey, greenish-grey, light brown, brown, red, and blue of all shades, and even black.[1]

The colour is generally caused by the presence of iron.

Thus carbonate of iron [2] gives a bluish or greyish tint ; anhydrous sesquioxide [3] a red colour ; hydrated sesquioxides [4] gives various tints of brown or yellow, sometimes blue and green. In some cases the blue colour is produced by very finely disseminated iron pyrites, and in some by phosphate of iron.

Classification.—The sandstones used for building are generally classed as follows, either practically according to their physical characteristics, or scientifically according to their geological position or the nature of their constituents.

PRACTICAL CLASSIFICATION.—*Liver Rock* is the term applied, perhaps more in Scotland than in England, to the best and most homogeneous stone which comes out in large blocks, undivided by intersecting vertical and horizontal joints. In Yorkshire it is known as " *Nell.*"

Flagstones are those which have a good natural cleavage, and split therefore easily into the thicknesses appropriate for paving of different kinds. The easy cleavage is generally caused by plates of mica in the beds.

Tilestones are flags from thin-bedded sandstones. They are split into layers—sometimes by standing them on their edges during frost,—and are much used in the North of England and in Scotland as a substitute for slates in covering roofs.

Freestone is a term applied to any stone that will work freely or easily with the mallet and chisel—such, for example, are the softer sandstones, and some of the limestones, including Bath, Caen, Portland, etc.

Grits are coarse-grained, strong, hard sandstones, deriving their name from the millstone grit formation in which they are found. These stones are very valuable for heavy engineering works, as they can be obtained in large blocks.

SCIENTIFIC CLASSIFICATION.—The geological formations from which the different varieties of sandstone are obtained are shown in the Tables, pp. 39-48.

---

[1] Thus Mansfield stone is pale salmon colour ; Red Corsehill, a brick red ; Robin Hood, pale blue ; Pennant, dark blue.

[2] *Ferrous Carbonates.* [3] *Ferric Oxide.* [4] *Ferric Hydrates.*

but any further notice of their classification from a geological point of view would be out of place in these Notes. With regard to their constituents, they may be divided into the following classes :—

Micaceous Sandstones are those which contain a very large proportion of mica, distributed over the planes of bedding.

Calcareous Sandstones contain a large proportion of carbonate of lime.

Felspathic Sandstones contain a large proportion of felspar, generally produced by the disintegration of granite or other felspathic rocks. The weathering qualities of these depend upon the nature of the felspar. (See p. 13.)

Metamorphic Sandstones are those which have been subjected to heat. They are too hard to work for building purposes, but are very suitable for breaking into road metal.

Tests.—*Fracture.*—The recent fracture of a good sandstone, when examined through a powerful magnifying glass, should be bright, clean, and sharp, the grains well cemented together, and tolerably uniform in size. A dull and earthy appearance is the sign of a stone likely to decay.

Brard's and Smith's Tests.—A sandstone may be subjected to Smith's test or to Brard's test, described at page 11.

Weight and Absorption.—Recent experiments "led to the conclusion that any sandstone weighing less than 130 lbs. per cubic foot, absorbing more than 5 per cent of its weight of water in 24 hours, and effervescing anything but feebly with acid, is likely to be a second-class stone, as regards durability, where there is frost or much acid in the air ; and it may be also said that a first-class sandstone should hardly do more than cloud the water with Mr. Smith's test."[1]

Grain.—It is generally considered that the coarse-grained sandstones, such as the millstone grits, are the strongest and most durable. This, however, seems doubtful ; at any rate, some of the finer-grained varieties are quite strong enough for any purpose, and seem to weather better than the others.

" It appears probable that for external purposes the finer-grained sandstones, laid on their natural bed, are better than those of coarser grain." [1]

Thickness of Layers.—In selecting sandstone for undercut work or for carving, care must be taken that the layers are thick ; and it is of course important that stones should rest in most cases (see p. 38, Part I.) on their natural beds.

Uses.—The hardest and best sandstones are used for important ashlar work ; those of the finest and closest grain for carving ; rougher qualities for rubble ; the well-bedded varieties for flags.

" Some of the harder sandstones are used for sets, and also for road metal, but they are inferior to the tougher materials, and roads metalled with them are muddy in wet, and very dusty in dry weather." [1]

[1] **Wray.**

Principal Varieties in common use.—A few of the best known sandstones will now be described, after which a list will be given of some of the principal quarries in Great Britain and Ireland.

Bramley Fall.—The original stone known under this name was a moderately coarse-grained sandstone of the millstone grit formation, from Bramley, near Leeds. It held a very high character for durability and strength.

It was found in large blocks, and was specially suited and used for heavy engineering works.

Thin stones of good quality cannot be produced from the best beds of the quarries without great waste. When therefore such stones are specified, they are sometimes supplied from the upper beds, which are of inferior quality.[1]

Since the introduction of railways the original Bramley Fall quarries have almost ceased to be worked, but a great deal of similar stone is found to the north of Leeds, and is sold under the same name, which has become a generic name for the class of stone wherever it may be quarried.

As a rule the stone sold under this name has considerable strength and durability, but in some cases an excess of grains of potash-felspar makes it weather badly.

"Owing to its cheapness—and also to a want of knowledge that the best stone rises in large masses—many gentlemen specify their stones for templates, pad stones, bases, steps, and landings and copings to be worked out of Bramley Fall only 7 or 8 inches thick. This mistake has caused some quarry men and producers to substitute inferior top rock for good stone, because the inferior top stone frequently rises in thinner lifts."[2]

Bramley Fall stone has been used for the most massive engineering structures in the country. Its weathering qualities may be observed in Kirkstall Abbey, near Leeds, which was built with this stone in the twelfth century.

Yorkshire Sandstones.—There are so many quarries producing stone of very similar quality and characteristics, classed under this head, that it would be useless to describe them in detail.

These stones come chiefly from the coal measures and millstone grit series; a few come from the new red sandstone formation.

In consequence of the large number of quarries in Yorkshire, the stone is commonly known as *Yorkshire stone*, but a great deal of similar stone is found in the adjacent counties.

Of these stones the finer grained are suitable for building purposes, while the grits are more adapted for heavy engineering works.

The sandstones from the millstone grit or coal measures are considered to offer the greatest resistance to injury by fire, for which reason Minera stone was selected for the National Safe Deposit Co.'s buildings.[3]

A few of the quarries are mentioned in the Tables, pp. 39 to 48. There are several round about the principal towns.

The best flags and landings come from near Bradford and Halifax.

Scotgate Ash.—This stone is produced from quarries at Pateley Bridge near Harrogate. Used chiefly in staircase work, steps, landings, pavements, etc.

The quarries produce landings of any size up to 150 feet superficial, steps up to 20 feet in length, sets, paving and building stones.

Some of the stone is white, some of a light green tint, and a bed called the ragstone is specially recommended by the proprietors for heavy engine bases, foundations, etc.

Forest of Dean Stone.—This very useful stone is found in the coal measures near Lydney and Coleford in Gloucestershire.

There are three distinct series or beds of considerable thickness. Of these the upper series consists of a soft, easily worked stone of various degrees of hardness. The second series is harder than the first, and the third harder than the second, and of a finer grit. Both the second and third series can be quarried in blocks of any size.

---

[1] Mr. Trickett in *Building News*, 25th June 1871.

[2] Mr. Trickett in *Builders' Weekly Reporter*, 23d June 1875.

[3] Wray.

The first and second series are of a grey colour, the third is bluer. Some of the stone has a brownish tint.

The stone weathers well if placed upon its natural bed. Some used in the churches of Newland, Staunton, and Mitcheldean, that has been exposed 400 years, still retains the tool-marks as sharp as ever, but this of course was from the best quarries, carefully selected.

There are a great many quarries in the hands of different proprietors. It is unnecessary to give their names.

The stone is admirably adapted for building, or for heavy engineering work such as bridges and docks.

Where used.—It has been used in the construction of Cardiff, Newport, Gloucester, and Swansea docks ; Folly Bridge, Oxford ; Cardiff Castle and National Provincial Bank, Marlborough ; Cardiff new Barracks ; port of Llandaff ; interior of St. John's and Exeter Colleges ; Taylor and Randolph's buildings, Oxford ; Eastun Castle and Witley Court, Doncaster, etc. etc.

Mansfield Stone is one of the best known and most important building stones in the country.

It is a siliceous dolomite (see p. 58), and is found near Mansfield, Nottinghamshire, in the Permian system, between the new red sandstone and the carboniferous series.

There are several beds found in the quarries, which differ considerably from one another both in composition and texture.

There are, however, two principal varieties of the stone sent into the market, the white and the red, both of them good for building purposes.

Of these varieties the red is considered more durable than the white. Both kinds last well in a clear atmosphere. They are all admirably adapted for the finest ashlar work, turned columns, mouldings, carvings, etc.

WHITE MANSFIELD.—There are several beds of this stone. The top bed of all has a coarser grain than the others. The second and third beds supply a very good fine-grained stone, fit for the finest ashlar work ; while the lowest bed is much harder than the others, and is well adapted for stairs, paving, landings, etc.

RED MANSFIELD is more generally of uniform quality and appearance. The stones of the darkest colour are considered the best.

This stone is quarried by wedges, without blasting. It is procurable in blocks weighing as much as 10 tons, and from 4 to 5 feet thick.

It can be sawn at the quarries into blocks and slabs, or turned on a lathe into columns of any moderate diameter.

WHERE USED.—*Red.*—Bilton House, Trafalgar Square, flagging of terrace ; Hyde Park, Albert Memorial, squares of flagging of terrace ; Burlington House, ashlars, columns, and niches ; St. Pancras Hotel and Station ; voussoirs of arch in main entrance, plinth of hotel, corbels, etc.

White.—Town-hall, Mansfield, Clumber Lodge, etc.

Craigleith Stone is perhaps the most durable sandstone in the United Kingdom. It consists of quartz grains united by a siliceous cement, with small plates of mica. It contains 98 per cent of silica and only 1 per cent carbonate of lime. It is found near Edinburgh, and is used extensively in that city, and also exported.

Many other sandstones of nearly equal importance to the above are mentioned in the following Tables :—

TABLE OF SOME OF THE PRINCIPAL SANDSTONES & QUARRIES IN GREAT BRITAIN AND IRELAND.

NAME OF QUARRY OR LOCALITY.	SITUATION	COUNTY.	COLOUR OF STONE.	WEIGHT PER FOOT CUBE IN LBS.	WHERE USED, AND REMARKS.
British Sandstones.					
Sandstones from the Cretaceous formation.					
BARGATE, or Godalming	Godalming	Surrey	Dark brown ferruginous	163·8	Bridges; churches; paving at cavalry camp, Aldershot. Several quarries.
TUNBRIDGE WELLS	Tunbridge Wells	Kent	Variegated brown	118·1	A compact sandstone. Used for churches and other buildings in Tunbridge Wells and neighbourhood.
COLLEY	Reigate	Surrey	Greenish light brown	103·1	From Lower Greensand—called Reigate stone—formerly known as Gatton. A light stone. Absorbs water; does not weather well. Two beds in the quarry; the upper contains flints, is hard and strong; the lower free from flints; fit for ashlar dressings. Very important that it should be laid on natural bed. Hampton Court, Windsor Castle, Croydon Almshouses and Town Hall. Many churches in Surrey.
GODSTONE	Godstone	Do.	...	112·0	From Upper Greensand. Known as "*Firestone.*" A soft calcareous sandstone. Worked in slabs about 10 inches thick; used to form the floors of glass furnaces; blocks of stone, etc.; and for cleaning hearths; therefore sometimes called *hearthstone.*
HASSOCK (from Kent Rag)	Maidstone	Kent	Light yellow	113 to 115	A sandy limestone. See Remarks, page 66. Several quarries.
Mansfield Stones.					
MANSFIELD Red	Mansfield	Nottinghamshire	Reddish brown	148·6	See Remarks, page 38. Several quarries.
Do. White	Do.	Do.	Whitish brown	146·6	See Remarks, page 38. Several quarries.
Oolite and Lias Sandstone.					
AISLABY	Whitby	Yorkshire	Light brown	126·7	Abbey and Docks, Whitby; University Library, Cambridge; Scarborough and Bridlington Piers.
GROSBY	Do.	Do.	Do.		
PARK QUARRY	New Malton	Do.	Whitish brown	...	Used for building purposes, and for burning lime, though called a sandstone.
PARK SPRING	Farnley, Leeds	Do.	Light ferruginous brown	151·1	Beds very irregular. Used for St. George's Hall, Bradford; Commercial Buildings, Leeds, from the old quarry. Much of the stone known as "Park Spring" comes from other quarries.

TABLE OF SANDSTONES—*Continued.*

NAME OF QUARRY OR LOCALITY.	SITUATION.	COUNTY.	COLOUR OF STONE.	WEIGHT PER FOOT CUBE IN LBS.	WHERE USED, AND REMARKS.
British Sandstones—(CONTINUED). *New Red Sandstone.*					
BEGGARS WELL	Alton	Staffordshire	Red	...	St. Giles, Blackburn; Alsager Church, etc.
BELTON	Drayton	Shropshire	White and light red	...	Buildings in Shrewsbury.
BILLINGE	...	Cheshire	Blue and white	...	Fine close-grained stone. Durable.
BRIDGE	Grinshill	Shropshire	Light cream	122·5	
CARLISLE	...	Cumberland	Warm red	...	
CEFN	Ruabon	Denbighshire	Yellow	128	A good stone. Used at Wrexham and Warrington Barracks.
CORNCOCKLE	Lockerbie	Dumfriesshire	Terra-cotta red.		
CORSEHILL	Annan	Do.	Dark red, and some beds bright pink	154	Close grained; weathers well; works easily; fit for ashlar and well adapted for carving and for columns. Hand in Hand Insurance Office and St. James's Hall; James Street Station, Liverpool; all buildings in Annan.
COVE	Kirkpatrick	Do.	White	...	Good quality; yields large blocks, similar to those used at Alton Towers.
CRUMPWOOD	Alton	Staffordshire			
GATELAW BRIDGE	Thornhill	Dumfriesshire.	Cream	135·8	Easily worked, but venty. Aste Hall; Richmond and Caterick Bridges; Skelton Castle; Darlington Town Hall, etc.
GATHERLY MOOR	Richmond	Yorkshire			
GRINSHILL	Shrewsbury	Shropshire	White or light yellow; reddish brown	140·0	Flaggy at top; large blocks below. Churches in Shrewsbury of eleventh century. Public buildings in most of the Shropshire towns.
HELSBY	Helsby	Cheshire.	Lightish brown grey	133·1	Trentham Hall, Drayton Manor; Town Hall, Derby; Meer Hall, Cheshire; Exchange, Sheffield; Warwick Gaol. Several quarries.
HOLLINGTON	Uttoxeter	Staffordshire			
LAZONBY	Lazonby	Cumberland	Warm red	138·0	Mostly used for flags and landings.
LOCHARBRIGGS	Locharbriggs	Dumfriesshire.			
MANLEY	Frodsham	Cheshire	Yellow	127	A strong-grained stone, very durable, rather porous; very deep blocks; suitable for columns. Used at Chester Castle, Town Hall, Market Hall.
NEWBIGGIN	Corby, nr.Carlisle	Cumberland	Red	154	A good stone. Weatheral Railway bridge.
OSMOTHERLEY	Osmotherley	Yorkshire	Dark brown	....	Used locally. Chain Bridge, Stockton.
PARK QUARRY	Toxall	Staffordshire	Light grey	124·6	Ruins at Toxall; St. George's Church, Birmingham; Sandwell Hall, Birmingham.
PENKRIDGE	...	Do.	Red and mottled	...	West Front, Lichfield Cathedral, County Court, Walsall, and sundry churches.

			Colour		Remarks
PENRITH	Penrith	Cumberland	Warm red	··	Used in neighbourhood, Brougham Castle, etc.
RUNCORN	Runcorn	Cheshire	Darkish red	129·3	Good and bad stone in alternate layers; large blocks. New church and new bridge, Manchester; Liverpool Docks.
Do.	Do.	Do.	Light red variegated	134·8	Docks and Corn Exchange, Liverpool; Stourport Bridge; Worcester and Bewdley Bridges; Gloucester Bridge. Grindstones.
STANLEY	Bewdley	Shropshire	Grey and red	141·4	
STOURTON	Birkenhead	Cheshire	White and yellow.		
WARMANBIE	Annan	Dumfriesshire.			
Upper Permian or Lower Trias.					
ASPATRIA	Aspatria	Cumberland	···	··	Several quarries.
ST. BEES HEAD	St. Bees Head	Do.	Red		Viaduct over river Eden.
Coal Measure Sandstones and Grits.					
ABERCARNE and NEWBRIDGE	Newport	Monmouthshire	Dark bluish grey	167·9	Old churches and buildings in vicinity; Cardiff and Newport Docks.
ACKWORTH	Pontefract	Yorkshire	Light brown	160	Weathers well; suitable for copings.
AUCHINLEA	Cleland	Lanarkshire	···	131·7	Carlisle Railway Station; Western Infirmary, Glasgow.
BARNARD CASTLE	Barnard Castle	Durham	Yellow	140·1	Locally known as Shaw Bank stone.
BINNIE	Uphall	Linlithgowshire	Brownish grey	166	University Club, Edinburgh; Bank; principally used now for monumental work.
BOLTON WOOD	Bradford	Yorkshire	Light yellow		Hard; durable; large blocks. Town Hall, Manchester.
BOTHWELL PARK	Fallside	Lanarkshire.			Recommended for sills, copings, strings, etc.
BRACKENHILL	Pontefract	Yorkshire	Brown	··	May be obtained in large blocks. Used for London Docks; bridges N.W.R. Numerous quarries.
BRADFORD	Bradford	Do.	Light brown	142·2	
BRAMLEY FALL	Leeds	Do.	Light ferruginous brown	142·2	See remarks, p. 37. Entrance gates and lodge to Euston Station, London; Millwall Docks.
BRITANNIA	Bacup	Lancashire.	Dark greenish brown and yellow	140	A good stone; much used. St. Nicholas Square; Clayton Memorial Church; High Level Bridge and Police Court, Newcastle; also for grindstones. There are the "main stone" and "bottom" beds.
BRUXTON	Gosforth	Northumberland			
CALIFORNIA	Hunshelf	Yorkshire.	White.		Used for flagstones, landings, etc.; Apothecaries' Hall, Liverpool.
CALVERLEY WOOD	Calverley	Do.	Light brown		
CATLOW	Nelson	Lancashire	···		Locally known as Horsforth stone. Used at Blackwall, Jarrow, and London Docks.
COLLEGE	Ackworth	Yorkshire.			
COXEY WARREN	Leeds	Do.	···		

TABLE OF SANDSTONES—Continued.

NAME OF QUARRY OR LOCALITY.	SITUATION.	COUNTY.	COLOUR OF STONE.	WEIGHT PER FOOT CUBE IN LBS.	WHERE USED AND REMARKS.
British Sandstones—(CONTINUED). *Coal Measure Sandstones and Grits.*					
COWGLEN	Dunfermline	Fifeshire	...	140·2	Leith New Docks.
COXBENCH	Derby	Derbyshire	Brownish yellow	...	Two beds, one finer grain than other. St. Pancras Station; Market Hall, Derby.
CRAIGLEITH	Edinburgh	Edinburgh	Whitish grey	141·8 to 145·9	One of the best sandstones in the country; quartz grains, siliceous cement, plates of mica; 98 per cent silica, only 1 per cent carbonate of lime; very hard and durable; good for ashlar and all building purposes. Used for nearly all the principal buildings in Edinburgh; also for piers of Southwark Bridge, and many buildings in London.
CROMFORD	Cromford	Derbyshire.			
CROSSLAND HILL	Huddersfield	Yorkshire	Light brown	155·2	Weathers well.
CROSS PLATTS	South Owram	Do.			
CULLALOE	Aberdour	Fifeshire.			
CUNLIFFE	Blackburn	Lancashire	Light bluish grey	...	Good for slabs and steps; very durable.
DARFIELD	Barnsley	Yorkshire.			
DARLEY DALE, or Stancliffe	Bakewell	Derbyshire	Light ferruginous brown	148·2	Abbey in Darley, Stancliffe Hall, Birmingham Grammar School, Birmingham and Nottingham Station. Several quarries.
DENWICK	Alnwick	Northumberland	...	...	Alnwick Castle. Largely used in Newcastle, Sunderland, and locally.
DUKE'S QUARRIES	Whatstandwell	Derbyshire	Red, varied with green, brown, and grey	144·5	Penitentiary, Millbank; Victoria Docks, London; Birmingham and Leicester Gaols; bridges on N.W.R.
DUNMORE, etc. etc.	...	Stirlingshire	Whitish grey and light brown	132	A good weathering stone. Much used in Glasgow and other large towns.
ECCLESHILL	Bradford	Yorkshire	Light brown	166	Medium hardness; sawn for steps. Much used in Bradford.
ELLAND EDGE	Halifax	Do.	Light grey brown	153·3	Fine micaceous grit and flagstone; a very good stone. Used for paving and landings; Victoria Station, Manchester.
ELSWICK	Newcastle	Northumberland	Light brown grey	140	Very hard. Used for St. Mary's Tower; Infirmary; Barracks, Newcastle.
FINSDALE	Morley	Yorkshire.			

					Remarks
FISHPONDS	Bristol	Gloucestershire	Dark blue and grey	168·2	Called "Pennant Stone." Used for Philosophical Institute and Gaol at Bristol. A very durable good stone.
FLETCHER BANK	Shuttleworth	Lancashire.	Grey and blue	...	See remarks, pp. 37, 38. Numerous quarries.
FOREST OF DEAN	Sydney and Coleford	Gloucestershire			
GAZEBY	Bradford	Yorkshire	Pale grey	179	Hard; good for steps. Used for steps at British Museum, etc.
GIFFNEUK, or GIFFNOCK	Giffnock	Renfrewshire		143·9	Soft sandstone, fine grit. Much used for all the principal buildings in Glasgow and the neighbourhood. Can be obtained in large blocks fit for the finest ashlar.
GIPTON WOOD	Leeds	Yorkshire	Light yellow	...	Fine grit; soft stone. Locally called Harehill stone. Used for Yorkshire Bank, and Stock Exchange, Leeds, etc. Sawn into slabs.
GRANGE	Burntisland	Fifeshire.			
GREAT FINSDALE	Morley	Yorkshire.			
GREENSMOOR	Manchester	Lancashire.			
GROSBY	Whitby	Yorkshire.			
GUNNERTON	Barrasford	Northumberland			
HAILES	Slateford	Edinburgh	Greyish white	141·6	Used extensively in Edinburgh : St. Mary's Cathedral, Observatory, etc.
HAREHILLS	Leeds	Yorkshire.			
HARTFORD BRIDGE	Bedlington	Northumberland		...	Hartford Bridge ; Bridge at Windsor ; Lady Burdett Coutts's Church, Westminster.
HAWKSWORTH	Leeds	Yorkshire	Light brown	137·9	Close grained stone, similar to that from Spinkwell and Horsforth.
HIPPERHOLME	Hipperholme	Do.	Light grey and grey	153·8 to 158	
HONLESS HILL	Coleford	Gloucestershire			Used for the late Prince Consort's mausoleum ; Tortworth Court ; Cardiff and Newport Docks ; Usk and Lands End Lighthouses ; fortification at Popton Point. Some of the stone is blue.
HONLEY	Huddersfield	Yorkshire	Bluish grey	...	A sound, good, hard stone ; supplied in blocks and slabs.
HOPTON	Burnley	Lancashire	...		Very hard laminated flag-rock. Used for paving in Liverpool and docks.
HORNCLIFFE	Horncliffe	Do.	Light brown.		
HORSLEY CASTLE	Coxbench	Derbyshire			
HOWLEY'S HILL	Coleford	Gloucestershire			
HOWLEY PARK	Morley, near Leeds	Yorkshire	Light brown	160	Can be had in large blocks ; not hard, but very durable ; good for dressings, stairs, etc.
HUMBIE, dark	Edinburgh	Haddingtonshire	Grey	135·8	The top bed is a dark stone ; the next bed white ; and the next grey ; stone works freely ; good for fine ashlar. Used at Dundas Castle ; spire, Tron Church, and several other buildings in Edinburgh ; Royal Exchange and Bank, Glasgow, etc.
Do. light	Do.	Do.	White	140·2	Used for ashlar work ; church and new buildings in Belper.
HUNGER HILL	Belper	Derbyshire	Warm light brown	136·0	A soft sandstone, fine grit. Much used for all the principal buildings in Glasgow and the neighbourhood. Can be obtained in large blocks fit for the finest ashlar.
HUNTER'S HILL	Bishopbriggs	Lanarkshire	Pale grey	143·9	
IDLE	Bradford	Yorkshire	Light blue, light brown, light yellow	166	Hard and durable blocks ; flags, landings. Several quarries. Stone used all over England, and also exported. Paving at Guildhall, and in many towns. The light yellow is similar to Bolton and Spinkwell stone, and will work in with them.

TABLE OF SANDSTONES—Continued.

NAME OF QUARRY OR LOCALITY.	SITUATION.	COUNTY.	COLOUR OF STONE.	WEIGHT PER FOOT CUBE IN LBS.	WHERE USED, AND REMARKS.
British Sandstones—(CONTINUED). *Coal Measure Sandstones and Grits.*					
KENTON	Newcastle-on-Tyne	Northumberland	Light ferruginous brown	145·1	Particularly good for fine work, carving, etc.; millstones, flagging, grindstones, St. Thomas' Church, New Town Hall, and nearly all the new buildings in Newcastle-on-Tyne.
KIPPENMUIR	Kippen	Stirlingshire.	...	...	Several quarries.
LEE	Bacup	Lancashire	Light grey	142	Used for Commercial Docks, London; railway bridge over Ouse; County Hospital, York; Victoria Docks; works on N.E.R.
LINGERFIELD	Selby	Yorkshire			
LONGANNET	Perth	Perthshire	Brown	131·75	Quartz grains; siliceous cement; beds five inches thick. Used in Exchange, Edinburgh; Tulliallan Castle, Perthshire.
LONGRIDGE	Delworth, near Preston	Lancashire	Warm dark yellow	146·1	Used for Royal Bank, Liverpool; most of the buildings in Preston.
LONGWOOD EDGE	Huddersfield	Yorkshire	Warm light grey brown	153·4	
LUDGEN'S HILL	Bradford	Do.			
MARSH	South Owram	Do.			
MERRYFIELDS	Pateley	Do.	Whitish grey	...	Resembles Craigleith in colour.
MILKING HILL	South Owram	Do.			
MINERA	Minera	Denbighshire	Greenish brown	139·7	Fine stone for ashlar, etc.; grindstones. Used for building in Liverpool, Chester, etc., and other places. Is a good fire-resisting stone, and much used in furnaces; also in the Patent Safe Company's Office, City.
MORLEY	Leeds	Yorkshire	Bluish grey		Numerous quarries.
MORLEY MOOR	Derby	Derbyshire	Warm brownish grey, often greenish	130·5	Bank at Derby; Berniston House. Used for grindstones.
NEW FARM	South Owram	Yorkshire	Yellowish brown	...	The stone from the North and South Owram quarries is the best for steps, landings, etc. Most of the stone known as "York stone" in the London market comes from these quarries.
NORTH OWRAM	Halifax	Do.			
PILLOUGH	Stanton	Derbyshire	Yellow.		
POOL BANK	Otley	Yorkshire			
PRUDHAM	Fourstones	Northumberland	Light olive	145	Central stations, Newcastle and Carlisle. Army and Navy Hotel, Victoria Street.
QUARELLA	Bridgend	Glamorgan	Grey, green, white	...	Green and white, good for dressings; grey for landings. Colonial Institute; Llandaff Cathedral; All Saints' Church, Torquay.
QUEENSBERRY	Halifax	Yorkshire	Yellowish brown	...	Fine grit.

Name	Locality	County	Colour	Sp. gr.	Remarks
REDGATE	Wolsingham	Durham	Light ferruginous brown	139·6	Fine grained and sound stone. Used in Manchester, and at Todmorden Town Hall and Church.
RINGBY	North Owram	Yorkshire	Light yellow	150	Good for landings, steps, sawn slabs.
ROBIN HOOD	Wakefield	Do.	Greenish grey	...	For particulars, see p. 37. Stone mine.
SCOTGATE ASH	Pateley Bridge	Do.	Brownish yellow, grey	153	
SCOTT HALL and Potter Newton	Leeds	Do.			
SHIPLEY	Bradford	Do.	Yellow.		
SHORROCK	Blackburn	Lancashire.			
SOUTH OWRAM	Halifax	Yorkshire	Yellowish brown	149·5	See remarks above on North Owram stone.
SPINKWELL, GREENCLIFF CLIFFWOOD	Bradford	Do.	Warm brownish yellow		Well-known quarries. Hard, fine grained. Contains 88½ per cent silica; no carbonate of lime or soluble salts. Stone used for new Town Halls, Manchester and Bradford; Liverpool Custom House and Docks.
STAINTON, or Stenton	Barnard Castle	Durham	Ferruginous light brown	142·5	Round Keep, Barnard Castle; Joint-Stock Bank and Market House, Barnard Castle; new station, Darlington; sinks and grindstones.
STANNINGLEY	Leeds	Yorkshire	Light brown	...	Fine grained; suitable for steps, landings, etc.; also for engine beds.
STANTON	Stanton	Derbyshire	Reddish yellow		Several quarries.
SWALES MOOR	North Owram	Yorkshire.		...	
TALACRE	Holywell	Flintshire	Brown olive	150·3	Good blocks obtainable. Used at Bodelwyddan Church. Several quarries.
THORNTONS	Bradford	Yorkshire	Yellowish brown.		
TYNE	Newcastle-on-Tyne	Northumberland	Light brown	...	Hospital and Workhouse, Newcastle.
UPPER MOOR	Pudsey	Yorkshire	Do.	154	Fine grained; flagstones; suitable also for "par-point" work.
VICTORIA	Greenmore	Do.	Do.		
VINEY HILL	Coleford	Gloucestershire	Light purplish grey, with occasional light greenish spots	155·7	Forest of Dean stone; red and silver grey beds — red rather streaky; grey in beds 4 inches thick. Used at Cardiff Pier; buildings in Gloucester.
WEST END	Haworth	Yorkshire.	Light yellow to dark brown	126·7	Whitby Abbey; Lewisham Church; New Library, Cambridge, etc.
WHITBY	Whitby	Do.	Brown		Glass and blast furnaces; building stones; grindstones.
WIDEOPEN	Newcastle-on-Tyne	Northumberland		...	
WIMBERRY	Coleford	Gloucestershire	...	...	Used in Newport and Cardiff for buildings; St. John's and Exeter Colleges; Oxford for interior work; and in many other buildings and works in the neighbourhood.
WINDY NOOK	Windy Nook	Durham.			
WOODBURN	Woodburn	Northumberland.			
WOODEND	Shipley	Yorkshire.			

TABLE OF SANDSTONES—*Continued.*

NAME OF QUARRY OR LOCALITY.	SITUATION.	COUNTY.	COLOUR OF STONE.	WEIGHT PER FOOT CUBE IN LBS.	WHERE USED, AND REMARKS.
British Sandstones—(CONTINUED). *Old Red Sandstone.*					
ACHSCRABSTER CASTLEHILL HOLBURN HEAD SPITTAL	Thurso	Caithness-shire	Greenish brown	...	Weathers very well; good for external paving; very great transverse strength—being nearly twice that of Arbroath, and three times that of Craigleith.
ARBROATH	Arbroath	Forfarshire	Greenish grey	...	Very good for internal steps and paving; several quarries.
AUCHRAY	Dundee	Do.	Purplish grey	158·9	Buildings in Dundee and neighbourhood.
BURGHEAD	Inverness	Inverness-shire	Light drab and buff	...	Very durable; easy to work.
DACRE BANK	Leeds	Yorkshire	Light brown	...	Hull and Barnsley, Railway and Docks; new Railway Bridge, Blackfriars.
DUNTRUNE	Dundee	Forfarshire.	Purple grey	161·1	Glammis and Inverquharity and Cortachy Castles; Lendertis House, etc. etc.
GLAMMIS	Forfar	Do.			
KNOCKLEY	Coleford	Gloucestershire	Grey	159·3	Used for Cardiff Pier; troughs, grindstones, etc.
LEOCH	Dundee	Forfarshire	Light purplish grey	159·2	All principal buildings in vicinity.
LEYSMILL	Arbroath	Do.			
LOCHEE	Dundee	Do.	Bluish grey	158·7	Principal buildings of neighbourhood.
LONGANNET	Perth	Perthshire	Light ferruginous brown	131·7	Stadt House, Amsterdam; Exchange, Edinburgh; Talle View Castle, Perthshire.
MUNLOCHY	Munlochy	Ross-shire	Red and variegated	160·6	Cathedral Church of Ross at Fortrose; Inverness old bridge; Cromwell Court; also for canals, etc.
MYLNEFIELD or Ringoodie	Dundee	Forfarshire	Purplish grey	160·0	Good ashlar or paving; old steeple and principal buildings of Dundee; Dundee Docks; Bell Rock Lighthouse; Royal Asylum, Perth; Kinfauns, Huntly, Pitfour Castles; Rossie Priory, etc.
NAIRN	Nairn	Nairnshire	Whitish	...	Durable; easily worked. Used for most buildings at Nairn.
PYOTDYKES	Dundee	Forfarshire	Purplish grey	162·5	Used for Dundee Waterworks.
SLADE	Arbroath	Do.			
TARRADALE	Urray	Ross-shire	Pink	...	Very durable; hard to work. Inverness Castle.
WELLBANK	Monifieth	Forfarshire.			
WESTHALL	Dundee	Do.			
WEYDALE	Thurso	Caithness-shire.			
WILDERNESS	Micheldean	Gloucestershire	Red	...	Speech House, Harrow; Workhouse, Cheltenham; Quay Wall, Gloucester, etc.

Cambrian.					
SWITHLAND	Groby	Lancashire	Dark blue	...	A slate, but contains grit; good for paving, steps, etc. Used for pavement, Prince Consort's Memorial; also for kerbs at St. Pancras Station.
Irish Sandstones.					
ALTMOVER	Dungiven	Londonderry	...	...	Buildings in Londonderry.
BALLYVOY	Ballycastle	Antrim.			
BLOOMHILL	Dungannon	Tyrone.			
CAHERBARNA	Liscannor	Clare.			
CARLAND	Crievagh	Tyrone.			
COURLEIGH	Shankill	Kilkenny.			
CRONOGORT	West Clare	Clare.			See Watsonstone.
DOONAGORE	Doolin	Do.	Whitish	159·4	Buildings in Nenagh; free working stone. May be procured in large blocks, but scarce. Used in new Barracks, Tipperary.
DRUMBANE	Thurles	Tipperary			Works well; wears well. All dimensions obtainable.
DRUMKEELAN	Mount Charles	Donegal	Yellowish white		Works well. Much used in North of Ireland.
DUNGANNON	Dungannon	Tyrone	Light cream	136·5	
FARROW-BAR	East Barrs	Leitrim.			
GLANDORE	Glandore	Cork	Greenish	...	Hard grit; very durable. Ballymoney Castle; Kilcolunan House.
GLUSHA	Ballintrain	Tyrone.			
KILLACOLLA	Killacolla	Limerick.			
KNOCKDONAGH	Knockdonagh	Clare.			
KNOCKERRA	Kilrush	Do.			
LACKAGH	Lackagh	Tyrone.			
LINHILL	Cookstown	Do.	Brownish	...	Fine grained; several quarries near Lismore.
LISMORE	Lismore	Waterford	Yellowish	...	Large blocks; soft; yellowish when raised, but become hard and white; used for millstones.
LONGFORD	Longford	Longford.			
MALLOW	Mallow	Cork	Brown, some reddish	...	Good flags and blocks; several quarries near Mallow. Stone used in the town; the reddish fine grained, ferruginous.
MONEY POINT	Kilrush	Clare	Dark brown	...	Like Carlow flags, but darker, and covered with wormlike impressions.
MOUNTMELLICK	...	Queen's County	Light grey.		
NEWTOWNARDS	Newtownards	Down	...	...	Several quarries.
ROARK'S	Cluneave	Westmeath.			
SCRABO	Newtownards	Down	Light grey	...	Used for buildings at Newtownards, and for flags; easily worked; geological formation doubtful.

TABLE OF SANDSTONES—*Continued.*

NAME OF QUARRY OR LOCALITY.	SITUATION.	COUNTY.	COLOUR OF STONE.	WEIGHT PER FOOT CUBE IN LBS.	WHERE USED, AND REMARKS.
Irish Sandstones— (CONTINUED).					
SEE FINN	Annalong	Down.			
SHANKILL and Kellymount	Shankill	Kilkenny	Dark greyish brown	...	Known as "Carlow flags." Thin-bedded siliceous grit; fine grained; very hard; very durable. Beds do not require dressing. Used for flagging and roofs.
STURGEONS	Newry	Down.			
WATSONSTONE (Doonagore)	Liscannor	Clare	Grey	...	"Shamrock Stone," for paving, kerbing, edgings, channelling, steps, landings. Used in Keyham Dockyard Extension, etc.
WHITE	Doon North	Limerick.	Reddish grey.		
YOUGHAL	Youghal	Cork			
Coal Measures— *Sandstone.*					
MOUNT CHARLES	Mount Charles	Donegal	...	161	Sligo Bank; Science and Art Museum, Dublin; Letterkenny Cathedral.

LIMESTONES.

The term limestone is applied to any stone the greater proportion of which consists of carbonate of lime; but the members of the class differ greatly in chemical composition, texture, hardness, and other physical characteristics.

Composition.—Chalk, Portland stone, marble, and several other varieties of limestone, consist of nearly pure carbonate of lime, though they are very dissimilar in texture, hardness, and weathering qualities.

Other limestones, such as the dolomites, contain a very large proportion of carbonate of magnesia. Some contain clay, a large proportion of which converts them into marls, and makes them useless for building purposes. Many limestones contain a considerable proportion of silica, some contain iron, others bitumen.

The carbonate of lime in stones of this class is, of course, liable to attack from the carbonic acid dissolved in the moisture of ordinary air, and is in time destroyed by the more violent acids and vapours generally found in the atmosphere of large towns.

Texture.—A great deal depends therefore upon the texture of the stone.

The best weathering limestones are dense, uniform, and homogeneous in structure and composition, with fine even small grains, and of a crystalline texture.

Some limestones consist of a mass of fossils, either entire, or broken up and united by cementing matter. Others are entirely made up of round grains of carbonate of lime, generally held together by cement of the same material. (See p. 56.)

The Royal Commissioners gave a preference to limestones as a class, " on account of their more general uniformity of tint, their comparatively homogeneous structure, and the facility and economy of their conversion to building purposes;" and of this class they preferred " those which are most crystalline."

Many of the most easily worked limestones are very soft when first quarried, but harden upon exposure to the atmosphere.

" This is said to arise from a slight decomposition taking place, which will remove most of the softer particles and leave the hardest and most durable to act as a protection to the remainder."[1]
By others it is attributed to the escape of the " quarry damp."

[1] *Guide to Museum of Practical Geology*, by R. Hunt, F.R.S.

Classification.—Limestones are classed :—1st, Scientifically from a geological point of view ; or, 2d, Practically, according to their physical characteristics.

SCIENTIFIC CLASSIFICATION.—Limestones are known as Carboniferous, Lias, etc., according to the formation from which they are obtained.   These forma-tions are shown in the Tables, pp. 67 to 73, but they need not be further referred to.

PRACTICAL CLASSIFICATION.—The terms *Liver rock, Freestone, Flagstone*, are applied to limestones in the same way as to sandstones (see p. 35).

The difference in the physical characteristics of limestones leads to their classification by the engineer as follows :—

> Marbles.
> Compact limestones.
> Granular   ,.
> Shelly   ,,
> Magnesian   ,,

These will now be described in turn.

MARBLES.

Marble is the name practically given to any limestone which is hard and compact enough to take a fine polish.

The name is frequently, however, erroneously applied to other stones, such as serpentine, merely because they are capable of being polished.

Some marbles—such, for example, as those from Devonshire—will retain their polish indoors, but lose it when exposed to the weather.

Marble is found in all great limestone formations.   It con-sists generally of pure carbonate of lime.   The texture, degree of crystallisation, hardness, and durability, of different varieties vary considerably.

Marble can generally be raised in large blocks.   The hand-somer kinds are too expensive for use, except for chimney-pieces table slabs, inlaid work, etc.

The less handsome varieties are used for building in the neigh-bourhood of the quarries.

The appearance of the ornamental marbles differs greatly. Some are wholly of one colour, others derive their beauty from a mixture of accidental substances—metallic oxides, etc., which give them a veined or clouded appearance.   Others receive a varied and beautiful " figure " from shells, corals, stems of en-crinites, etc.. embedded in them.

Uses.—Marble is used in connection with building chiefly for columns, pilasters, mantelpieces, and for decoration.

The weight of marble makes it suitable for sea-walls, break-waters, etc., when it is cheaply obtainable, but some varieties are liable to the attacks of boring molluscs. (See p. 10.)

In the absence of better material marble may be used for road metal and paving setts, but it is brittle and not adapted to with-stand a heavy traffic. Roads made with it are greasy in wet weather and dusty when dry.

Different forms of Marble. — *Encrinal and Shell Marbles* are those which derive their figure from embedded fossils, encrinites (resembling lily-shaped plants with jointed stems), or fossils of ordinary shells.

Madrepore Marbles are made up entirely of fossil corals.

Ancient Marbles.—Many of the marbles used by the ancients, and handed down to us in the shape of works of art, are not now known in their natural state.

Their markings and tints are frequently imitated in artificial marbles, and the ancient names are applied to the imitations.

Varieties.—A good deal of the marble used in this country comes from the Continent.

Of the varieties found in England, the best known are those of various colours from Devonshire ; black and grey marbles from Derbyshire ; the Purbeck marble from Dorsetshire ; Mona marble from Anglesea.

Purbeck marble was formerly much used in churches, especially in the detached shafting of clustered columns in Early English thirteenth century work, and has also been used in modern restorations. For a fine example of the latter see the Chapter House, Westminster Abbey. The colour is usually, when polished, dark grey, the sections of shells showing up in lighter shades, but the tint is variable, and a reddish brown is also found. This marble is obtained from the Upper Purbeck series, and consists of a mass of *Paludina.* An outcrop in sharply folded beds can easily be traced at Peverel Point, near Swanage, and also inland in some disused quarries between Swanage and Corfe, on the margin of the Wealden series.

There are many varieties in *Scotland*, but they are chiefly used locally, and burnt for lime.

Ireland supplies marbles of all colours. Black from Galway, Kilkenny, and other counties ; dark grey (black birdseye) from Kilkenny ; white from Donegal ; red, pink, and dove from Midleton, County Cork ; Irish green from Connemara (a serpentine, see p. 34).

TABLE OF SOME OF THE PRINCIPAL MARBLE QUARRIES IN GREAT BRITAIN AND IRELAND.

NAME OF QUARRY OR LOCALITY.	SITUATION.	COUNTY.	COLOUR OF STONE.	WEIGHT PER FOOT CUBE IN LBS.	GEOLOGICAL FORMATION.	WHERE USED, AND REMARKS.
English Marbles.						
ASHFORD	Rowsley	Derbyshire	Black	...	Mountain limestone	These marbles are highly figured, as they contain embedded fossils of all kinds.
BAKEWELL	Bakewell	Do.	Brown	...	Do.	
BERRY POMEROY	...	Devonshire	Dove coloured, and red and white			
BRIDSTOW	...	Do.				
CHUDLEIGH	...	Do.	Black			Chimney-pieces.
COLE HILL	Matlock	Derbyshire	Do.	...	Mountain limestone.	
DENT	...	Yorkshire	...	...	Granite.	Full of fossil encrinites or "stone lilies."
DREWSLEIGHTON	Drewsleighton	Devonshire	...	...	Devonian	
IPPLEPEN	Newton	Do.	Purple clouded with red and white	163·4	Devonian	Blocks 18 feet square sent to London. Used for shafts of columns, National Provincial Bank, Bishopgate Street.
KILLEY PARK ONE ASH	Matlock	Do.	Green; rose coloured.	...	Mountain limestone.	
ORESTON	Plymouth	Devonshire	...	...	Devonian	Used for Plymouth Breakwater. Attacked by *Pholades.* (See p. 10.)
MARBLE HOLE	Ashford	Derbyshire	...			Black marble mine.
PETIT TOR	St. Marychurch	Devonshire	Grey	...	...	Known as Babbicomb marble. Some blocks consist entirely of fossil corals, and are known as *Madrepore marble.*
PETWORTH	Arundel	Sussex	Do.	...	Wealden	Very similar to Purbeck. The Episcopal chair of Canterbury is made of this marble. Found also near Horsham, Eastbourne, Lewes.
PLYMOUTH	Plymouth	Devonshire	...	...	...	Plymouth Forts and Barracks.
POCOMBE	Exeter	Do.	Purple; white veins	...	Devonian.	

PURBECK, Swanage	Swanage	Dorsetshire	Grey	...	Purbeck beds	A mass of shells; occurs in beds from 6 in. to 9 in. thick; much used for shafts in thirteenth century work. See p. 51.
SHELDON	Matlock	Derbyshire	...	...	Mountain limestone.	
STAVERTON	...	Do.	Dove-coloured, and red and white.	...		
S. TAUTON WILTON	Totness ...	Devonshire Staffordshire.	...	...	Devonian.	
ANGLESEA ISLE OF MAN.	...	Isle of Man	Black.	...		
MONA	...	Anglesea	Dark green, light green, and sometimes purple, with white	...	...	Apparently formed by a combination of crystalline limestone and serpentine.
PORTWASH	...	Isle of Man	Black and grey	...	Carboniferous	Steps of St. Paul's Cathedral; full of encrinites and shells.
PORT ST. MARY RAMSLEY	East Ogwell	Do. Devonshire	Black Red, veined white and pink.	...	Do.	Takes a good polish.
SCARLETT	Castletown	Isle of Man	Pale	...	Do.	Castle Rushen built with this marble.
Scotch Marbles.						
ABOYNE	Aboyne	Aberdeenshire	Grey and white.			
ASSYNT	...	Sutherlandshire	Grey, blue, and dove-colour.			
BALLACHULISH	Glencoe	Argyleshire	White and grey.			
BLAIRGOWRIE	Blairgowrie	Perthshire	White.			
GLENTILT	Blair-Athol	Do.	White and grey	...		Occasional yellow spots.
IONA	Oban	Argyleshire	Grey and white	...		Sometimes has yellowish spots or rims of steatite; does not take a high polish.
SKYE	...	...	Greyish.			
TIREE	Tiree	Hebrides	Pink, with dark green crystals.	172·3	...	Block for bust of Sir W. de la Beche, Museum of Practical Geology.

TABLE OF MARBLES—*Continued.*

NAME OF QUARRY OR LOCALITY.	SITUATION.	COUNTY.	COLOUR OF STONE.	WEIGHT PER FOOT CUBE IN LBS.	GEOLOGICAL FORMATION.	WHERE USED, AND REMARKS.
Irish Marbles.						
ANGLIHAM	Galway	Galway	Black.			
BALLYKILABOY	Ballykilaboy	Kilkenny	Grey.			
BALLYMORE	Dunfanaghy	Donegal	Creamy white.	...	...	Polishes well; highly crystalline; difficult to work; chimney-pieces.
BOSTON	Rathangen	Kildare	Black			
CARLOW	Carlow	Carlow	Do.			
CHURCHTOWN	Churchtown	Cork	Red, black, brownish red	...	...	Museum of Irish Industry, Dublin.
CLIFDEN	Clifden	Galway	White	...	...	Stones of considerable area but small depth; no cubical blocks of any size.
CLONMACNOISE	Clonmacnoise	King's County	Sienna.			
DONERAILE	Doneraile	Cork	Black			
DUNKERROW	...	...	Black, white, purple, veined with green.	...	...	Museum of Irish Industry, Dublin.
GLANNAN	Glasslough	Monaghan	Black	171·4	...	Takes a good polish.
KILKENNY	Kilkenny	Kilkenny	Do.		...	When dressed the figures of shells appear.
KINTAIL	Rathmullin	Donegal	Do.		...	Rathmullin Pier.
LETTERNAPHY	Clifden	Galway	Green		...	Rather coarse; works any size; exported.
LIMERICK	Limerick	Limerick	Black		...	Used in London. Several quarries.
LISSOUGHTER	Recess	Galway	...		...	Serpentine marble.
LITTLE ISLAND	Ballytrasna	Cork	...		...	"Victoria Red" marble.
MENTO	Galway	Galway	Do.	...	...	There are three beds. The middle is called the "London bed," and exported. Blocks 5 feet to 10 feet long and 4 feet to 5 feet wide. Sawn on the spot into slabs, etc.
MIDLETON	Midleton	Cork	Red, pink, dove.			
PALLAS KENRY	Limerick	Limerick	Red	...	...	Spotted from embedded fossils; hard and laminated; takes and retains a good polish.
STREAMSTOWN	Clifden	Galway.				
TULLAMORE	Tullamore	King's County	Grey	...	...	Some varieties clouded. Used for polished work, chimney-pieces, etc.

TABLE OF SOME OF THE PRINCIPAL CONTINENTAL MARBLE QUARRIES.

NAME OF QUARRY OR LOCALITY.	SITUATION.	COUNTRY.	COLOUR OF STONE.	WEIGHT PER FOOT CUBE IN LBS.	GEOLOGICAL FORMATION.	WHERE USED, AND REMARKS.
Continental Marbles.						
BARDIGLIA	Montalto	Italy	Bluish grey, black veins and cloudings	...	...	Called "Bardilla"; sometimes contains veinings resembling flowers.
BROCATELLA	...	France		...		Porphyry.
CARRARA	Carrara	Italy	White	168·6	Jurassic, oolitic, lias, oolitic	Used chiefly for statuary; composed almost entirely of carbonate of lime; sometimes contains particles of quartz and pyrites; weathers badly in a smoky atmosphere, also under changes of temperature. Used for chimney-pieces.
CARYSTIAN ⎱ CIPPOLINO ⎰	Island of Euboea	Greece.				
EMPEROR'S RED	...	Portugal	Red.			
*GIALLO ANTICO	...	...	Yellow	...	...	*Ancient marbles, used for works of art; not now found in quarries, but imitated in artificial marbles.
*NOIR ANTICO	...	Greece	Black.			
PARIAN	Island of Parr	Greece	Cream white	...	...	Preferred to Carrara for statuary; same composition; brighter fracture.
*ROSSO ANTICO	Near first cataract of Nile	Egypt	Reddish brown; deep blood red, or veined with white	...	...	Consists chiefly of felspar.
ROUGE ROYAL	...	France and Belgium	Red and brown, veined	...	...	Chiefly used in this country for chimney-pieces.
SICILIAN, also called Ravaccione.	Near Carrara	Italy	White	169·1	...	Somewhat same composition as Carrara. Used for Marble Arch. Marbles from Sicily are not much used in this country.
SIENNA	Sienna	Do.	Yellow, with veins and cloudings.			
*VERDE ANTIQUE	Near Genoa and Florence	Do.	Mottled with light green, dark green, and black.	...		Generally classed by builders among marbles; really a serpentine. (See p. 34.)

Compact Limestone.—*Composition and Structure.*—Compact limestone consists of carbonate of lime either pure or in combination with sand or clay.

It is generally devoid of crystalline structure, of a dull earthy appearance, and of a dark blue, grey, black, or mottled colour.

In some cases, however, it is crystalline and full of organic remains. It is then properly known as a crystalline limestone.

Some of the Carboniferous limestones are of the compact class, also the Lias limestone, which contains a considerable amount of clay, and is used for making hydraulic lime ; also Kentish Rag from the Cretaceous system, which is more fully described at page 64.

Uses.—The compact limestones are good for building purposes, where their dull colour and the difficulty of working them are not objections.

They are useful for paving sets and road metal under a light traffic.

They are chiefly used however as flux in blast furnaces ; for agriculture, bleaching, tanning, and other industrial purposes.

Weight.—They weigh from 153 to 172 lbs. per cubic foot, and absorb very little water, taking up generally less than 1 per cent of its weight in twenty-four hours.

Granular Limestones. — *Composition and Structure.* — These limestones consist of grains of carbonate of lime cemented together by the same substance, or by some mixture of carbonate of lime with silica or alumina.

Size of Grains.—They are generally found in the Oolitic (or eggstone) formation. The grains vary greatly in size. In some cases they are very small and uniform, very few being of a larger size, as in Caen stone. When the whole of the grains are somewhat larger, as in Ketton stone, they constitute what are called "*Roestones*," the structure resembling that of the roe of a fish. When the grains are still larger, as big as peas, the stones are known as *Pisolites,* or pea stones.

These stones nearly all contain fossil shells. In some cases the shelly matter occurs in larger quantity than the grains. They are then called *shelly granular limestones.*

Colour.—The colour of these stones is very variable, being sometimes white, light yellow, light brown, or cream-coloured.

Weathering Qualities.—The granular limestones are generally soft and somewhat absorbent. They are therefore liable to the attacks of acid atmospheres, and of frost. but otherwise are fairly durable.

Natural Bed.—This stone "is generally obtainable in large blocks, and it is often difficult when the stone has been sawn to detect its natural bed. This may be sometimes done by directing a jet of water on the side of the block, and it is well to do this as it is of great importance with some of the less durable sorts that they should be set upon their natural bed." [1]

Weight and Absorption.—The weight of this class of stone varies from 116 to 151 lbs., the lighter and more absorbent stones being, as might be expected, less durable than those of a more compact order.

Their absorption of water in twenty-four hours is hardly ever less than 4 per cent of their weight, while it is sometimes as much as 12 per cent.[1]

Varieties.—This class affords some of the principal building stones of this country, many of which will hereafter be described more in detail.

The very fine grained stones may be represented by Chilmark (see page 63); those with larger grains by Portland, Ancaster, and Painswick; and those with large spherical grains by **Ketton** and Casterton; while Bath stone has large egg-shaped grains.

Uses.—Some of these stones—such, for example, as certain varieties of Portland—are well adapted for outdoor work; others —such as Bath, Caen, Painswick—for internal work, carving, etc., while some of the harder kinds—such as Seacombe, Painswick, and some of the beds of Chilmark and Portland—are adapted for internal staircases where there is not likely to be much wear.

Shelly Limestone.—There may be said to be two classes of this stone.

Structure.—The first consists almost entirely of small shells cemented together, but shows no crystals on fracture.

Purbeck is an example of this class.

Stones of the second class consist chiefly of shells, but break with a highly crystalline fracture.

Of this class Hopton Wood stone is an example.

Colour.—This is given in the Table, pp. 67-73.

Uses.—Stone of this class is chiefly used for paving.

Weight and Absorption.—The weight of this class of stone is from about 157 to 169 lbs. per cubic foot, and its absorption is very small, generally much less than 2 per cent of its weight.

Magnesian Limestones.—*Composition.*—Magnesian limestones

[1] Wray.

are composed of carbonate of lime and carbonate of magnesia in variable proportions, together with a small quantity of silica, iron, and alumina.

Many limestones contain carbonate of magnesia, but those with less than 15 per cent do not come into the class now under consideration.

The better varieties of magnesian limestone are those in which there is at least 40 per cent of carbonate of magnesia, with 4 or 5 per cent of silica.

When the magnesia is present in the proportion of one molecule of carbonate of magnesia to one molecule of carbonate of lime (*i.e.* 54·18 carb. lime and 45·82 carb. magnesia), the stone is called a *Dolomite*.[1]

Professor Daniel states that the nearer a magnesian limestone approaches dolomite in composition, the more durable it is likely to be.

Structure.—It is not merely the nature of the constituents or their mechanical mixture that gives dolomite its good qualities: there is some peculiarity in the crystallisation which is all important.

Mr. C. Smith says, " In the formation of dolomite, some peculiar combination takes place between the molecules of each substance; they possess some inherent power, by which the invisible or minutest particles intermix and unite with each other so intimately as to be inseparable by mechanical means. On examining with a highly magnifying power a specimen of genuine magnesian limestone, such as that of Bolsover Moor, it will be found not composed of two sorts of crystals, some formed of carbonate of lime others of carbonate of magnesia, but the entire mass of stone is made up of rhomboids each of which contains both the earths homogeneously crystallised together. When this is the case, we know by practical observation that the stone is extremely durable."[2]

Some magnesian limestones contain sand, in which case their weathering qualities are greatly injured.

Some are peculiarly subject to the attacks of sulphuric acid, which forms a soluble sulphate of magnesia easily washed away.

Analyses.—The following Table gives analyses of some of the principal magnesian limestones. The red and white Mansfield contain a large proportion of silica and are generally classed among the sandstones (see p. 38).

---

[1] After a French geologist Dolomieu, who was the first discoverer of this mineral in the Alps.   [2] Smith's *Lithology.*

	1 Dolo- mites.	2 Bols- over Moor.	3 Hud- dleston.	4 Roach Abbey.	5 Park Nook.	6 Mans- field, Red.	7 Mans- field, White.	8 Mansfield Wood- house.	9 Bulwell	10 North Auston.	11 Steetley
Carbonate of Magnesia	45·82	40·2	41·37	39·4	41·6	16·10	7·30	42·60	32·75	43·07	43·78
Carbonate of Lime	54·18	51·1	54·19	57·5	55·7	26·50	41·30	51·65	62·80	54·89	53·95
Silica	...	3·6	2·53	0·8	0·0	49·40	50·00	3·70	Trace.	0·56	0·44
Iron and Alumina	...	1·8	0·30	0·7	0·4	3·20			2·30	0·73	0·64
Water and Loss	...	3·3	1·61	1·6	2·3	4·80	1·40	2·05	2·15	0·75	1·19
	100·0	100·0	100·0	100·0	100·0	100·00	100·00	100·00	100·00	100·0	100·0

Cols. 2 3 4 5 from the *Report of the Royal Commission.*
1 6 7 8 Smith's *Lithology.*
9 Page's *Economic Geology.*
10 11 *Builder,* 20th November 1886.

Principal Varieties in common use.—A few of the most noted varieties of limestone used in this country will now be described, after which, at p. 67, will be given a list of some of the principal quarries in Great Britain and Ireland.

Bath Stone is one of the best known and most extensively used building stones in this country.

Geological Position.—This stone is obtained from that division of the Oolitic formation which is known as the Great or Bath Oolitic group. Geologically speaking it lies below the Portland stone, being separated from it by the Kimmeridge clay, coral rag, and Oxford clay.

The building stone lies between beds of ragstone : dark veins run at right angles to the beds.

Most of it is of a fine even grain, composed chiefly of carbonate of lime— sometimes interspersed with shelly fragments.

Some of the varieties of this stone contain sand-cracks, vents, clay-balls, etc. ; these should of course be avoided.

Colour.—The colour varies from white to light cream colour and yellow.

Quarrying.—The quarries are worked by tunnelling, and the stone is pro-duced in blocks up to 5 or 6 feet deep, and weighing as much as 10 or 12 tons.

It is important that Bath stone should be quarried in summer when it is freed from the ground moisture or quarry sap. If quarried in winter it is very likely to fall to pieces with the first frost.

Seasoning and Weathering.—The stone is very soft when first quarried, but hardens upon exposure to the air (see p. 49). It is important that it should be placed on, or parallel to its natural bed (see p. 9).

The stone varies greatly in quality : some varieties weather badly, while others are fit for external work in ordinary atmospheres.

Size and Uses.—As it is obtainable in large blocks, and is easily worked, it is particularly valuable for mouldings and carved work.

QUARRIES.—There are several quarries in the neighbourhood of Bath, now incorporated in the "Bath Stone Firms, Limited," among which may be mentioned the following :—

Box Ground.—Found in beds from 10 inches to 4½ feet thick. A coarse but sound stone, which weathers well except in a smoky atmosphere and by the sea. It is harder than Combe Down, but freer from vents. The stone is not now so easily worked as it used to be.

Combe Down stone differs in quality according to locality. Many large abandoned quarries exist.

Farleigh Down stone is soft and very fine grained. It occurs in different beds, from 10 inches to 3½ feet thick, some of a yellow, and others of a red colour. The former does not weather well, and is used for tracery and internal work.

Westwood Ground is produced from stone mines of great size, and is stated to be a very superior stone, free from vents and defects, and procurable in large and sound blocks.

Corsham Down.—This quarry consists of three principal varieties in several beds. The "fine upper," in beds about 3 feet thick, is found about 90 feet below the surface. The stone from these beds is of fine grain, and suitable for sculpture and mouldings. Next below is the "corngrit," from 2 to 3 feet thick, a harder stone, full of little pieces of flint ; of a good colour, sound, and durable, but unable to resist frost. It is difficult to work, but good for carrying weight, and used chiefly for engine beds, columns, landings, steps, etc. Below this again is the "bottom bed," an excellent soft stone, about 4½ feet thick, but occasionally stained with blue patches.

Corsham Ridge is also a recent quarry, and supplies good hard stone. It was used for the face work and carved pediments of the Royal Aquarium at Westminster.

Stoke Ground stone is obtained from mines in the neighbourhood of Limpley Stoke. It consists of one bed from 5¾ to 8 feet thick, yielding blocks up to 6 tons weight. The stone is of a light brown colour, soft, easy to work, fit for carving, and when seasoned for external work.

Other important quarries are *Monks Park, Park Lane, Clift, Spring, Lower Hill, Huddeswell,* etc.

Portland Stone is obtained from the upper parts of the Oolitic series.

It has already been mentioned that there are four distinct varieties of Portland stone used for building, of which, however, three only are generally sent into the market.

The section of a quarry given at page 7 is here reproduced for convenience, but it will not be necessary again to describe the order in which the different varieties occur.

Beginning at the top of the quarry and working downwards, we find (interspersed with cap, flinty tiers, and other beds useless for building purposes four varieties of stone all more or less useful to the engineer or builder These are *True Roach, Whitbed, Bastard Roach* or *Curf,* and *Basebed.*

As these four varieties of stone differ greatly in their characteristics and in the uses to which they may be applied, it will be well to describe them separately.

Chemical Composition.—The chemical composition of the different varieties is almost the same, and it may therefore be given at once for the whole.

The following is the analysis made by Professors Daniel and Wheatstone for the commissioners who selected the stone for the Houses of Parliament :—

Silica	1·20
Carbonate of lime . . .	95·16
Carbonate of magnesia . . .	1·20

Iron and alumina	.	.	.	.	0·50
Water and loss .	.	.	.	.	1·94
Bitumen	.	.	.	.	A trace.

It will be seen that the stone consists almost entirely of carbonate of lime. " The most durable stone has its cementing matter in a solid and half crystalline state ; in the least durable stone it is in an earthy and powdery state." [1]

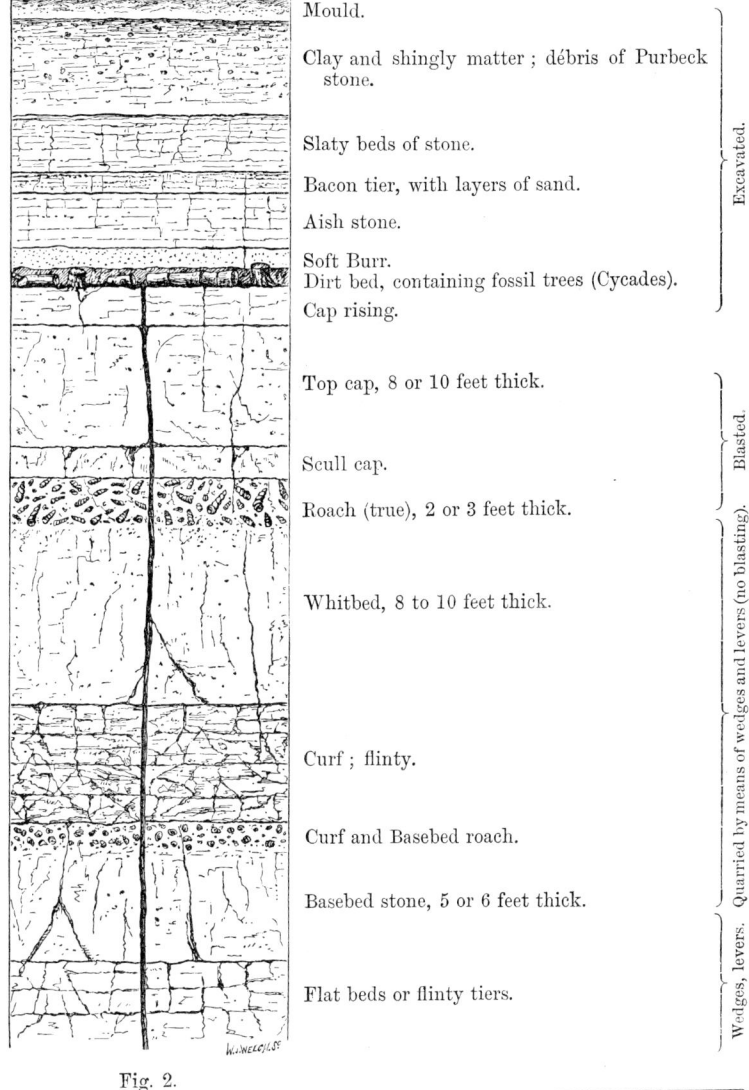

Mould.

Clay and shingly matter ; débris of Purbeck stone.

Slaty beds of stone.

Bacon tier, with layers of sand.

Aish stone.

Soft Burr.
Dirt bed, containing fossil trees (Cycades).

Cap rising.

⎫
⎪
⎬ Excavated.
⎪
⎭

Top cap, 8 or 10 feet thick.

Scull cap.

Roach (true), 2 or 3 feet thick.

Whitbed, 8 to 10 feet thick.

⎫
⎬ Blasted.
⎭

Curf ; flinty.

Curf and Basebed roach.

Basebed stone, 5 or 6 feet thick.

⎫
⎬ Quarried by means of wedges and levers (no blasting).
⎭

Flat beds or flinty tiers.

⎫
⎬ Wedges, levers.
⎭

Fig. 2.

[1] Report of Commissioners respecting stone to be used in building the new Houses of Parliament.

Roach, or *True Roach* as it is sometimes called, is a mass of fossils united by a cement composed of carbonate of lime.

The stone also contains a great number of cavities, large and small, being the moulds left by fossils that have dropped out.

Most of the fossils are merely casts, but in some cases portions of the shell are left.

The true roach may be distinguished from the other as it contains the peculiar fossil shown in Fig. 3, and known as the "Portland screw," which is never found in the bastard roach. This is an important distinction, as the true roach weathers far better than the bastard roach.

Cast of the shell known as the "Portland Screw," *Cerithium Portlandicum.*

Fig. 3.[1]

True roach is one of the best stones that can be used for heavy engineering works.

It is remarkably tough and strong, weathers admirably, and resists the action of water particularly well.

It has been much used for fortifications, breakwaters, dock and sea walls, and is suitable for massive plinths or other ashlar work where a rough face is appropriate ; but the numerous cavities it contains render it unsuitable for fine work, and for positions where smooth faces or sharp clear arrises are required.

The colour of true roach is a very light brown.

Whitbed.—This is the most valuable bed of the Portland series. It immediately underlies the true roach stone, which is firmly attached to its upper surface.

The stone consists of fine oolitic grains, well cemented together, and interspersed occasionally with a small amount of shelly matter. The cementing material is hard and crystalline.

Good whitbed stone resists the weather admirably. It is easily dressed to a smooth surface, and will take a very fine arris. It is suitable for the finest class of ashlar work. Some of it is too hard and not sufficiently uniform in texture for carving ; other blocks are quite fit for the most intricate work. When examined through a microscope the grains of whitbed will be found to have a more oolitic or roe-like appearance than those of basebed (see p. 63). The grain is also more open, and the cementing material is stronger.

The colour is generally white, or nearly so, but some of the best stone has a decidedly brown tint.

It is unfortunate that there is no more marked distinction between whitbed and basebed, as the weathering qualities of the latter are greatly inferior. Basebed is fit only for internal work, and great disappointment is caused when it is used, mistaking it for whitbed, in external work exposed to trying atmospheres. The carver, however, prefers basebed, though it is not so durable, because it looks better and is more easily worked.

Bastard Roach, Basebed Roach, or *Curf.*[2]—This stone resembles true roach in appearance, being a mass of fossils and cavities. The cementing material is, however, inferior, the stone does not weather well, and it is not used for building or engineering works, except in the immediate locality.

The thickness of basebed roach varies considerably in different quarries. In

---

[1] From Lyell's *Geology.*   [2] Sometimes written "Kerf."

some there is scarcely any, in others the bed has a thickness of from 12 to 24 inches, or even more.

These beds are sometimes interspersed with thin layers of flint.

The basebed roach is suitable for foundations, for backing walls, and for internal work where it will not be subjected to blows or traffic.

Basebed is not easily distinguished from whitbed. The external appearance of the stone from both beds is almost exactly the same.

The basebed is, however, more uniform in structure, and freer from shelly matter. Its weathering qualities are not so good as those of whitbed, but as it is softer to work, it is often preferred by masons, and is known in the market as *best-bed*.

If basebed be required for external work, it should be seasoned for a year before use, in order that it may have every chance of weathering well.

The stone from this bed is well adapted for internal work and carving of the highest class.

QUARRIES.—The Portland stone quarries are worked from open faces. Blocks of great size, such as 10 or 12 tons (140 or 170 cubic feet), can easily be procured.

After experimenting upon stone from the different beds, Professor Abel reported that "on the whole the evidence may be considered a little in favour of the opinion that an improvement in the strength of the stone is effected, to some extent, by seasoning." [1]

The whole island of Portland is full of quarries, each of which produces the different beds of stone above described.

The commissioners reported that "the best stone is in the N.E. part of the island, the worst in the S.W. part."

Several of the quarries belong to the Government, but some of the best are in private hands, and the stone is worked in great quantities for the market.

The names of the principal quarries are Waycroft, Wide Street, Maggot, Weston Independent, Inmosthay, Tout, Bowers, etc.

Buildings in which used.—Portland stone, chiefly whitbed, was used for all buildings of importance erected in London from about 1600 to 1800. It was also used for the west front of St. Paul's, for the Horseguards, Somerset House, the General Post Office St. Martins-le-Grand, the India House and Foreign Offices in Downing Street, the Reform Club, and many other important buildings.

Chilmark Stone.—This stone is procured from the Portland and Purbeck series of the oolitic formation as developed near Tisbury, Wardour Castle, in Wiltshire.

It is known also as Wardour stone, and in London as Tisbury stone.

The siliciferous nature of the cement which binds the particles (carbonate of lime) of the stone gives it excellent weathering qualities, while the softness and even grain of some of the beds renders them capable of being elaborately worked.

There are four distinct varieties of the stone.

---

[1] *Professional Papers, Royal Engineers,* vol. xii.

The Trough or *Hard Bed* is of a close even texture, of yellowish-brown colour.

It has an average thickness of 2 feet 6 inches, but stones may be obtained 3 feet 6 inches thick and of any reasonable length and breadth—the random blocks averaging 16 cubic feet.

It is used principally for steps, also for cornices, copings, sills, plinths, chimney-pieces, paving, road metal, heavy engineering works, and in any position exposed to wet and hard wear.

The Scott or *Brown Bed* is of warmer colour than the hard bed. Average thickness of bed 3 feet, maximum 4 feet, random blocks average 16 cubic feet.

Principally used for ashlar mouldings, carvings, random rubble, and for building purposes generally.

The General Bed, from the Garden quarry, is of a rich yellow tint and fine texture. It is capable of being elaborately carved, and is chiefly used for that purpose, also for ashlar, mouldings, etc.

The average thickness of bed is 4 feet, maximum 5 feet.

Strength.

	Resistance to crushing per foot sup.	Tensile strength per square inch.
Hard bed . . .	196 tons.	500 lbs.
Scott bed . . .	104 ,,	206 ,,
General bed . .	100 ,,	355 ,,

Chemical Analysis.

Silica	10·4
Carbonate of lime . .	79·0
,, magnesia . .	3·7
Iron alumina . . .	2·0
Water and loss . .	4·2

Working.—The stone has to be cut with a wet saw, and the relative cost of working the beds compared with Portland is stated by the proprietors to be—

Portland and Hard bed	1·0
Scott and Garden bed	0·6

Buildings in which used.—Salisbury Cathedral, Tisbury Church, Wardour Castle, Fonthill Abbey, Priory Church, Christ Church ; Post Office, Westminster Road, London ; Post Office, Exeter ; Sorting Post Office, Hampstead ; London and County Banks, Hastings and Banbury ; restoration of Chichester and Rochester Cathedrals, and of Chapter House, Westminster Abbey ; Longford Castle, Wilts ; Crewe Hall, near Chester, etc. etc.

Kentish Rag[1] is found in the Greensand formation, in a district running through the central part of Kent, about thirty miles long and from four to ten miles broad, including the towns of Sevenoaks, Maidstone, Lenham, etc.

Beds.—The *Ragstone* is found in beds varying from 6 inches to 3 feet in thickness, alternating with fine sand known as *Hassock*, which is frequently so consolidated as to form a stone that can be used for building.

---

[1] Taken chiefly from *Observations on Kentish Ragstone*, by J. Whichcord

The hassock is generally found adhering to the ragstone, and at the bed of junction organic remains often occur.

The ragstone itself is a very compact, heavy stone, which absorbs very little water, and resists the weather well.

The hassock, attached to it, is a calcareous sandstone, soft, porous, and very perishable under atmospheric influences.

There are several beds in a Kentish ragstone quarry ; many of them are worthless. It may be interesting to mention a few of the most useful.

After a top layer of mould and loam there are two or three beds of hassock and ferruginous sand, after which come the more useful beds, the best of which are mentioned below in succession.

Land Rag.—About 8 or 10 inches deep ; dark grey ; free working. Below this is a bed of fine hassock.

Header laying.—Thin dark stone used for headers.

Green Rag.—10 inches thick ; greenish colour ; free working ; not very sound. Fossils generally found on top bed. Below this is a layer of workable hassock.

Yellow Rag.—Broken up into headers for pitching.

Pelsea yields large hard blocks 12 inches thick ; difficult to quarry.

Next come two inferior and flinty beds interspersed with hassock.

Great Rag is a layer sometimes 3 feet deep, but split into two thicknesses full of cross fissures ; no large stones from it. Broken up for headers, or makes the best description of lime. A very superior layer of hassock (often containing fossils) is found below this bed.

Newington Cleaves.—A flinty bed ; produces some large blocks. Then a flinty bed between two layers of hassock.

Whiteland Bridge produces blocks 12 feet long, any width, and 14 inches thick ; stone very free working ; bluish colour.

Main Bridge.—Like the last bed, but of small scantling. Used for paving kerbs. After the last bed comes some inferior hassock.

Garl yields hard blocks of considerable side, used for headstones. Upper and lower surfaces of the bed show a red colour.

Horse Bridge yields blocks of good stone, 15 feet long and 16 inches thick.

Headstone laying yields blocks about 7 inches thick, used for headstones. Then a deep bed of soft hassock.

White Rag, which is of no use for building, as it tumbles to pieces upon exposure to the air.

THE RAGSTONE is used chiefly for rubble work, being very difficult to dress It does not gain in beauty by being tooled, because even the best kinds are full of small hassocky spots, which show themselves upon a smooth face, turn rusty upon exposure to the weather, and facilitate the decay of the stone.

The ragstone makes very good paving sets and curbs. It is also used for road metal, but yields a good deal of dust in dry weather.

If used as ashlar, great care must be taken to place it on its natural bed, otherwise it will decay.

The ragstone is not suitable for internal work, for, as it is non-absorbent, the moisture of the air condenses upon its surface, causing what is known as sweating.

All ragstone used for external work should have the hassock carefully knocked off.

It is important also to see that the small "pockets" containing iron, which often occur in the stone, are not exposed upon the face, otherwise the iron will oxidise upon exposure to the atmosphere, and cause ugly rust stains.

THE HASSOCK is totally unfit for external work, but it is frequently used as a lining to walls built of ragstone, by which the sweating above mentioned is avoided.

QUARRIES.—There are several quarries, among which may be mentioned the Allington, Castle, and Coombe near Maidstone, and Chilmington near Ashford. Also quarries at Aylesford and at Hythe.

Composition.—The following are analyses of the Kentish Rag and Hassock respectively :—

Kentish Rag.		*Hassock.*	
Carbonate of lime with a little magnesia	92·6	Carbonate of lime . . .	26·2
Earthy matter	6·5	Earthy matter	72·0
Oxide of iron	0·5	Oxide of iron	1·8
Carbonaceous matter . . .	0·4		100·0
	100·0		

Yellow Mansfield is obtained from quarries at *Mansfield Woodhouse*, two miles from Mansfield. It is crystalline, and has a warm yellow colour.

This stone almost exactly resembles the Bolsover Moor stone, which was selected by the Royal Commissioners for the Houses of Parliament.

The only difference is "that its colour is rather deeper, partly owing to its having a greater number of minute black specks, which is a peculiarity more or less to be found in all varieties of the magnesian limestone rocks."

The chemical composition of Mansfield stone, and the characteristics which it shares with other magnesian limestones, are given at pages 58, 59.

Uses.—It is useful for ashlar, mouldings, columns, etc., and is eminently adapted for highly carved work.

Where used.—Amicable Life Assurance Office, Fleet Street. Martyrs' Memorial, Oxford.

Purbeck Stone is obtained from stone mines in the Middle Purbeck series near Swanage. Numerous abandoned mines are to be seen on the hillside, but a few are still worked. The principal veins are, reckoning upwards from the *Cinder Bed*, the *Downs Vein*, *Freestone*, *Freestone Rag*, and *Lane-and-end.* Below the *Cinder Bed* are the *Button*, *Feather*, *Cap*, and *New Vein*, with the *Brassy Bed.* From the above veins kerbing, paving, building stone, and tile stone are obtained.

Caen and Aubigny Stones are Oolitic limestones, which may be mentioned here, as they are a good deal used in this country, though they are found in Normandy.

Caen Stone is of a pale cream-yellow colour. It is very soft when first quarried, but hardens upon exposure ; is easily worked and carved, but weathers very badly ; weighs 120 lbs. per foot cube. Used in Henry VII. Chapel, Westminster Abbey ; the Tower, Buckingham Palace, and many other buildings.

Aubigny Stone is similar to Caen, but more crystalline, harder, and heavier. It also weathers badly. Used at St. Mary's, Stoke Newington, and other buildings. *Maninghen*, quarried near Boulogne. *Chateau Gaillard*, *Tercé*, *Lavoux*, and *Chauvigny* are produced from quarries near Poitiers.

Several other limestones of considerable importance will be found in the following Tables :—

TABLE OF THE PRINCIPAL LIMESTONE QUARRIES IN GREAT BRITAIN AND IRELAND.

NAME OF QUARRY OR LOCALITY.	SITUATION.	COUNTY.	COLOUR OF STONE	WEIGHT PER FOOT CUBE IN LBS.	WHERE USED, AND REMARKS.
English Limestones.					
Cretaceous.					
BEER	Beer Seaton	Devonshire	Light brown	131·7	Cathedral and St. Peter's, Exeter. Soft when first quarried, hard when seasoned.
KENTISH RAG	Maidstone, etc.	Kent	Blue grey, or greenish grey	166·6	Numerous quarries. See Remarks, p. 64.
TOTTERNHOE	Dunstable	Bedfordshire	Greenish white	...	Dunstable Priory; Luton, and many other churches; Fonthill House; Woburn Abbey.
OOLITIC. *Purbeck Beds.*					
PURBECK	Swanage	Dorsetshire	Brownish grey	169·2	Much used for paving. Very durable, but wears rather slippery.
Portland Oolite. PURBECK	Langton and Worth	Do.	Whitish brown	150·0	"Purbeck Portland" is a superior stone, fit for fine ashlar, of good colour, strong and durable. The best beds are considered to be little, if at all, inferior to Portland stone. Used in public buildings in Swanage, etc. Old quarries in the sea cliffs at Tillywhim, Seacombe, Dancing Ledge, Winspit, etc.
Do.	Do.	Do.	Very light brown	151·0	
Portland Oolite. CHILMARK	Salisbury	Wiltshire	Light greenish grey	153·5	See Remarks, p. 63.
PORTLAND QUARRIES	Portland	Dorsetshire	Whitish brown	Roach 155 / Whitbed 145 / Curf 144 / Basebed 137	See Remarks, p. 60.
TISBURY WARDOUR	Salisbury / Do.	Wiltshire / Do.	Light greenish grey. Do.	...	
Coralline Oolite. HEADINGTON	Oxford	Oxfordshire	Blue and brown tints	...	Generally too soft and friable for building. Contains from 91 to 96 per cent carbonate of lime. The worst beds have been used for colleges and churches in Oxford. Has weathered badly, but stood best at Wadham College.

TABLE OF LIMESTONES—*Continued.*

NAME OF QUARRY OR LOCALITY.	SITUATION.	COUNTY.	COLOUR OF STONE.	WEIGHT PER FOOT CUBE IN LBS.	WHERE USED, AND REMARKS.
English Limestones—(CONTINUED).					
HILDENLEY	Malton	Yorkshire	Whitish cream	130·6	Resembles indurated chalk. Used for paving. Kirkham Priory; columns of chapel, Castle Howard.
Great and Upper Oolite.					
ANCASTER	Sleaford	Lincolnshire	White, yellow, pink	160·	Compact, fine grained. Works like Bath stone. Becomes harder after quarried. Weathers well. Used for Belvoir Castle; Wollaton Hall. At St. Pancras Station for dressings.
BALL'S GREEN	Minchinhampton	Gloucestershire	White	…	Known as Painswick stone. Very uniform in grain. Used for New Houses of Parliament, etc. Good for staircases.
BARNAC MILL	Stamford	Northampton	Light brown	136·7	Shelly limestone. Much used for churches in Lincolnshire, Cambridgeshire, and North Suffolk. Peterloo College; Croyland Abbey.
BATH STONE	Corsham, etc.	Wiltshire	Cream	123·	See Remarks, p. 60. Several quarries.
CAMPDEN HILL	…	Gloucestershire	Rich orange cream	140	Various churches in locality. New Town Hall, Leamington.
CASTERTON	Stamford	Rutland	Light brown	…	Known as Stamford marble. Ely Cathedral; Buildings in Stamford; New Record Office, Fetter Lane.
CLIPSHAM	Charlbury	Do.	…	113·	Buildings in Stamford; Ely and Peterborough Cathedrals; Exton, Uppingham, Melton Mowbray, Astley, and other churches.
COLLY WESTON	Colly Weston	Northamptonshire.			
COMBE DOWN	Bath	Somersetshire	Cream	116·	See Remarks, p. 60.
DOULTING (Bramble Ditch, Chelynch)	Shepton Mallet	Do.	Light brown	130·3	A shelly granular limestone. Very regular and uniform in texture; free working, and very durable. Wells Cathedral; Glastonbury Abbey; Chatham Naval Hospital.
DUNDRY	Bristol	Gloucestershire	…	…	St. Mary's, Redcliffe. Restoration of Bristol Cathedral. Llandaff Cathedral.
FARLEIGH DOWN	Corsham	Wiltshire	Cream	122·7	See Remarks, p. 60.
HAM HILL	Stoke	Somersetshire	Light orange brown	141·6	Shelly granules. Beds from 6 inches to 3 feet thick. The best are very durable and strong. The stone is sawn with a tooth saw.
HAYDOR	Grantham	Lincolnshire	Brownish cream colour.	133·7	Essential that this stone should be set upon its natural bed. Lincoln Cathedral; Boston, Grantham, Newark, and other churches; Culverthorpe House; Belvoir Castle.
LINDLEY'S	Ancaster	Do.			

	Locality	County	Colour		Remarks
PAINSWICK .	Stroud	Gloucestershire	Cream colour	...	Known as Painswick stone. Very uniform in grain. Used for New Houses of Parliament, etc.
PUDDLECOTE .	Charlbury	Oxfordshire	Light brown	...	Largely used in Oxford in the twelfth century. Used at Chatford, Dean, Chilson, and Burton-on-the-Water. Several quarries.
SHEPTON MALLET and others	....	Somersetshire	Cream coloured	...	
ST. GILES and others	Lincoln	Lincolnshire	...	...	Lincoln Cathedral. There are two or three quarries round Lincoln for building stone.
TAINTON .	Burford	Oxfordshire	Streaky brown	136·	A shelly oolite. Grains soft and chalky. Absorbs water freely. Blenheim House; churches in Oxford; interior of St. Paul's Cathedral; Byland Abbey.
TYTHERINGTON .	Tytherington	Gloucestershire.			
WELDON .	Corby, Kettering	Northampton-shire	Creamy white	140	New University Library, Cambridge; Chapter House, Lincoln, restoration; Geddington Cross.
WINDRUSH, Hard .	Windrush	Gloucestershire	Cream	135·9	Some beds shelly. Some of even grain. Windrush House; Barrington House.
Do. Soft	Do.	Do.	Do.	118·1	
YEOVIL .	Yeovil	Somersetshire	Greenish grey	...	Used for bridges over Worcester, Shropshire, and Wolverhampton Railway.
Do.	Do.	Do.	Brownish yellow		do.
Do.	Do.	Do.	do.		Do.
Lias.					
KEINTON .	Somerton	Somerset	Blue grey	...	Paving, building, and lime.
KETTON .	Ketton	Rutlandshire	Cream and pink	128·3	Very good durable stone, oolitic grains well defined; fit for stairs, plinths, mullions, and other dressings. St. Dunstan's Church, Fleet Street; the modern part of Peterborough and Ely Cathedrals; Sandringham Hall, etc. etc.; St. Pancras Railway Station; York Minster restoration.
KETTON RAG .	Do.	Do.	White	155·7	The rag beds are white, cemented with highly crystallised carbonate of lime. The crash is of a dark brown colour, very coarse, full of shells, distinct ova, and very ferruginous.
LOAD BRIDGE .	Langport	Somerset	Blue grey	...	Langport and Somerset churches.
RIMPTON .	....	Do.	Light slate colour	...	Used for flags and landings.
SPARKFORD .	Castle Cary	Dorset	Some white, some blue	...	Used for bridges on Worcester, Shropshire, and Wolverhampton Railway.
SUTTON .	Bridgend	Glamorganshire	Very light	136·0	Used for Llandaff Cathedral, Neath Abbey, Corby Castle, etc.
WILMCOTE .	Stratford-on-Avon	Warwickshire	Some white, some blue		Several beds; some polish like marble. Used for paving.
Magnesian Limestone. ANSTON—					
Norfall Quarry	South Anston	Yorkshire	Cream brown	144·0	Used in Houses of Parliament for the structure generally.
Stone Ends .	Do.	Do.	Do.	144·2	Used in Houses of Parliament for plinths of building towards river.

TABLE OF LIMESTONES—*Continued.*

NAME OF QUARRY OR LOCALITY.	SITUATION.	COUNTY.	COLOUR OF STONE.	WEIGHT PER FOOT CUBE IN LBS.	WHERE USED, AND REMARKS.
English Limestones—(CONTINUED).					
Magnesian Limestone.					
AYCLIFFE	Darlington	Durham	Grey	165	Used for building in the district.
BOLSOVER MOOR	Chesterfield	Derbyshire	Light yellowish brown	151·7	Used for Southwell Church, and for paving. This stone was first chosen for the Houses of Parliament, but the quantity required was more than the quarries could supply.
BRAMHAM MOOR	Tadcaster	Yorkshire	Warm brownish yellow		
CADEBY	Doncaster	Do.	Cream	126·6	A friable stone; central beds the best. Used at Day and Martin's, High Holborn; Almshouses, Edgeware.
CALVERLEY WOOD	...	Do.	White		Colne Town Hall, etc.
HUDDLESTONE	Sherburne	Do.	Whitish cream	137·8	Hard veins occur in the stone. Used for York Minster; Selby Cathedral; Huddlestone Hall; Sherburne Church, Swindon; Westminster Hall; the Cross opposite Charing Cross Station.
JACKDAW CRAIG	Tadcaster	Do.	Dark cream		York Minster, and probably most of the churches in York. A good stone, but requires careful selection. Lowest bed is the best stone.
MANSFIELD, Yellow	Mansfield Woodhouse	Nottinghamshire	Yellow	145·8	See Remarks, page 66. Several quarries.
MICKLEFIELD	South Milford	Yorkshire	Cream		Very good stone; much used in York and Leeds.
PARK NOOK	Do.	Do.	Do.	137·2	Subject to pot-holes containing calcareous spar. Used for Pontefract old Church, Campsall Lodge, etc.
ROCHE ABBEY	Bawtry	Do.	Whitish cream	139·1	Stone weathers black, and in lines according to the beds. Roche Abbey; Deanery, York; Tickhill Church and Castle; Blyth Church; Bawtry Church; several other churches in Yorkshire and Lincolnshire; Toxteth Church and Hall; Osberton and Witton Churches, Nottinghamshire.
SMAWSE	Tadcaster	Do.	Light yellowish brown	127·5	Stone crisp and brittle; not fit for steps. Ripon Minster; repairs York and Beverley Minsters, Hull old Church; St. Mary's Church, Beverley.
STEETLEY	Worksop	Nottinghamshire	White	128·2	Used for Houses of Parliament; Steetley Church; Abbey Church, Scarborough; Doncaster.
Do.	Do.	Do.	Yellow	130·6	
WOODHOUSE	Mansfield	Nottinghamshire	Cream brown	145·8	Used for lower part of river front of Houses of Parliament; Southwell Church; Martyrs' Memorial at Oxford.
Carboniferous.					
HOPTON WOOD	Wirksworth	Derbyshire	Grey	158·4	A sort of marble, fine-grained; compact; will take a polish; in large blocks. Good for steps, paving, etc.; weathers well. Used at Chatsworth; Belvoir Castle.
LITTLE ORME'S HEAD	Llandudno	Carnarvon.			
PENMON	Llangoed	Anglesey	Do.	167·3	Britannia Tubular Bridge, etc.

			Colour		Remarks
PRESTHOPE	Much Wenlock	Shropshire	Grey.	...	
TRAETH-BYCHAN	Llangefni	Anglesey			
VALLIS VALE	Frome	Somersetshire.			
Devonian.					
BRIXHAM	Brixham	...	...	...	Breakwater, Torbay; Dawlish sea wall; paving, etc. See *Marbles.*
POMPHLET	Plymouth	...			
STONEY COOMBE	Newton Abbott	Devonshire.			
Scotch Limestones	...	...	...	...	Very few of the Scotch limestones are used for building.
Irish Limestones.					
Carboniferous.					
ANNAHOLE	Castleblaney	Monaghan.	Light grey	...	Very crystalline; works well. Cullen Church; Ardbracken House.
ARDBRACCAN	Navan	Meath			
ARMAGH, NAVAN	Armagh	Armagh	Light pinkish grey	160	Very fossiliferous; takes a good polish; when polished is yellow, when tooled white; large blocks obtainable. Used for houses in Armagh, and chimney-pieces.
BALLINTEMPLE	Cork	Cork	Light grey	173·4	Close grained and compact; works freely and very well; good for carving; large blocks. Point House, Savings Bank, public buildings in Cork.
BALLINTOGHER	Lixnaw	Kerry.	Dark greyish blue	...	Compact texture; works well. Victoria Bridge, Sligo.
BALLISODARE	Ballisodare	Sligo.			
BALLYBEGGAN	Tralee	Kerry.			
BALLYCASTLE	Ballycastle	Antrim.			
BALLYCONNELL	Ballyconnell	Cavan	Bluish grey	...	Hard. Ballyconnell Bridge, lock, mill, etc.
BALLYDUFF	Tullamore	King's County.			
BALLYSHANNON	Ballyshannon	Donegal	Brownish grey	...	Compact slightly crystalline; works freely. R. C. Chapel, barracks, etc., at Ballyshannon.
BALLYVADDY	Carnlough	Antrim.	Dark blue.		
BARLEY HILL	Carrickmacross	Monaghan	Grey.		
BRACKERNAGH	Ballinasloe	Galway	Dark blue		
BROWNS HILL	Carlow	Carlow		...	Full of pearl spar and crystals of quartz; unfit for fine work. Buildings in Carlow.
CARRICK SLAIM	Carrick-on-Shannon	Leitrim	Light blue-grey	...	Carrick-on-Shannon Bridge.
CARRIGACRUMP	Cloyne	Cork	Dark grey	...	Laminated; easily worked; large blocks. Bank of Ireland, Cork.
CASHEL	Lanesborough	Longford	Whitish grey		Athlone Bridge; Longford Chapel.
CASTLEMORE	Castlemore	Cork.			

TABLE OF LIMESTONES—*Continued.*

NAME OF QUARRY OR LOCALITY.	SITUATION.	COUNTY.	COLOUR OF STONE.	WEIGHT PER FOOT CUBE IN LBS.	WHERE USED, AND REMARKS.
Irish Limestones (CONTINUED).					
CHURCHTOWN	Newcastle	Limerick	Light grey	...	Irregular beds. Courthouse and Workhouse, Newcastle.
CLARE CASTLE	Ennis	Clare	Grey	...	Very close grained; tough; easily worked; large blocks. Cork Harbour forts
CLARE HILL	Moira	Down			
COLIN WELL	Ballycolin	Antrim			
CORNAGRADE	Cornagrade	Fermanagh			
CROSSDRUM	Oldcastle	Meath	Whitish grey	...	Works freely; can be obtained in large blocks.
DEVLIN'S	Carnmoin	Londonderry			
DUBLIN, Rathgar, Kimmage, and several quarries	Dublin	Dublin	Light grey to black	...	Calp limestone; used chiefly for building purposes, and the spawls for metal; alternates with beds of hard and soft slate. A good stone; weathers well. Used for old St. Patrick's Cathedral, Trinity College, Portobello Barracks, and public buildings.
DUNGANNON	Dungannon	Tyrone	Blue	...	Used for railway bridges, etc.
DUNKIT	Waterford	Kilkenny			
FALLOW VEE	Carnlough	Antrim			
FARN	Killorglin	Kerry	Grey	...	Killarney Cathedral and Lunatic Asylum.
FERMOY	Fermoy	Cork	Light bluish grey	171·1	
FIRIES	Castleisland	Kerry			
FOYNES	Foynes	Limerick	Blue	...	Railway Station, Foynes; Kilrush, Kildegrat piers. A good stone for engineering work.
GILLOGUE	Limerick	Do.	Dark when quarried, afterwards white	...	Yields a good hydraulic lime. Used at Floating Basin, Limerick Docks.
GOLDEN BRIDGE	Dublin	Dublin			
GORESTOWN	Moy	Tyrone			
GORTON	Carnlough	Antrim			
HOOLE	Fethard	Wexford			Used for building; makes hydraulic lime.
HOWTH	Howth	Dublin	Grey and dark blue	...	Used for rubble; makes good hydraulic cement.

KENMARE	Kenmare	Kerry	Grey	...	Slaty; hard to work. Used for Suspension Bridge.
KILLARNEY	Killarney	Do.	Do.	...	Hard; even grained; difficult to work. Several quarries near Killarney, some yielding flags used for Muckross Abbey.
KILWAUGHTER	Kilwaughter	Antrim.			
KINTOGHER	Kintogher	Sligo.			
LANESBOROUGH	Lanesborough	Longford	Dark grey	...	Works freely; polishes well. Shannon works; Lanesborough Bridge.
LERSCLIP	...	Kildare	Black	165	Good for dressings.
LIMERICK	Limerick	Limerick	Blue	157 to 167	Easily worked; compact. Docks, bridges, etc.
LISBURY	...	...	Dark		Compact; easy to work. Nenagh Gaol.
LISSATONAY	Nenagh	Tipperary.			
LISTOWEL	Listowel	Kerry	Dark bluish grey	172·1	Barracks and Bridewell, Listowel. Dense, close grained, free working, with fossils and crystalline fragments.
LITTLE ISLAND	Cork	Cork	Light grey	...	Very fossiliferous; works freely. Used for forts, Spike Island.
MAGHERAMORE	Magheramore	Antrim.			
MEELICK	Killaloe	Mayo	Dark grey	...	Irregular fracture; large blocks. Moyne and Rossock Abbeys. Has weathered well.
MEELIN	Meelin	Cork.			
MILVERTON	Skerries	Dublin.			
MOUNT ARGUS	Harold's Cross	Do.			
NEWTOWN	Finglas	Do.			
PARISHA	Tickmacriven	Antrim.			
QUARRY LODGE	Rathmore	Kerry.			
RATHLIN	Island of Rathlin	Antrim.			
RED COW, and other quarries	Dublin	Dublin	Grey to black	...	Good serviceable stone for general building purposes.
SHANDON	Dungarvan	Waterford	Light grey.	...	A good building stone; dresses well.
SKERRIES	Dublin	Dublin	Grey		
TOWN	Glenarm	Antrim.	Bluish	...	Good for dressings and monumental work.
TULLAMORE	Tullamore	King's County			
WHITE GATE	Do.	Do.	Bluish grey	...	Very close grained; large blocks; brittle. Used for Queen's College, Lough Corrib Docks, etc.

ARTIFICIAL STONE.

In consequence of the difficulty which exists in many localities of obtaining durable natural stone at a moderate cost, many processes have been invented for the manufacture of artificial stone.

Some of these processes are successful in producing artificial stones which compare favourably in all their qualities with natural stones having a high character.

The expense of artificial stone is a bar to its extensive use for ordinary blocks, but the facility with which it can be moulded to the most intricate forms makes it very economical when it is required to take the place of carvings or other enrichments in natural stone.

A few of the best known artificial stones will now be described. Some of them are merely forms of concrete, and will be mentioned in the chapter devoted to that material.

Artificial stone, prepared by the process of the late Mr. F. Ransome, is made by mixing artificially-dried sand with silicate of soda (dissolved flint) and a small proportion of powdered stone or chalk. These are thoroughly incorporated in a pug or mortar mill, and forced by hand into moulds.

The blocks turned out have a cold solution of chloride of calcium poured over them, and are then immersed in a boiling solution of the same, sometimes under pressure, so that the pores of the material are entirely filled with the solution, after which it is found to be as hard as most building stones. The excess of chloride of sodium is then washed off, otherwise it is apt to cause efflorescence.

It will be seen that the above process depends upon the double decomposition of the silicate of soda and chloride of calcium. The chlorine and soda combine to form chloride of sodium, which is washed out, and the silica attacking the calcium forms silicate of lime, a strong and durable cement which binds the particles of the stone together.

Characteristics.—This stone has a fine homogeneous structure, so that it can, if necessary, be worked and carved like the best building stones.

The great advantage that it possesses is the facility with which it may be moulded into any form required.

Several experiments have been made upon this material.

It absorbs about 6·5 per cent of water.

Its tensile strength is about 360 lbs. per inch.

Its resistance to crushing about 2 tons per inch.

It weighs about 120 lbs. per cubic foot.

Of course these figures vary according to the nature of the material used in making the stone, the age of the specimen, etc.

The composition of this stone indicates that it will weather well, and some experiments made by Professor Frankland show that its resistance to acids was fully equal to Portland, Anston, Parkspring, and other of the best building stones.

Details of the experiments made by different observers will be found collected in Gwilt's *Encyclopædia of Architecture*, page 485.

Uses.—This stone is well adapted for all purposes for which natural sand-stones and limestones are used.   It can, however, be most economically employed for dressings (especially for those of an ornamental character), and for imitation carved work, though its use for this purpose has been condemned from an artistic point of view.

This stone is also used for caissons or hollow blocks for foundations, for grindstones, filters, etc. ; and by substituting grains of corundum and oxide of iron for the sand, a substance called solid emery is produced, which is formed into wheels for sharpening tools, polishing metal surfaces, etc.

Ransome's stone has been used at St. Thomas's Hospital, the India Office, the London Docks, the Brighton Aquarium, the Albert Bridge, and in several other buildings both at home and abroad.

Apœnite is a variety of Ransome's stone, made with 5 parts of sand, 1 of Farnham rock, $1\frac{1}{4}$ of Portland cement, with the same proportion of silicate of soda.

It can be made more quickly, and is considered superior to the other.[1]

Moreover, it has the great advantage that it can be made on the works where it is to be placed in position.

It is used for steps, balustrades, cylinder foundations, etc.

It weighs about 137 lbs. per cubic foot, and absorbs in 24 hours about $5\frac{1}{2}$ per cent of its weight of water.

Victoria Stone consists of washed, finely-powdered granite, bound together with the strongest Portland cement, and then hardened by immersion in silicate of soda.

The silicate is formed by boiling ground Farnham stone in cream caustic soda.

A mixture of four parts of crushed granite with one of Portland cement is allowed to set for three days or more into a hard block moulded to the required shape.   It is then immersed in the silicate of soda for some seven or eight weeks.

The lime in the cement combines with the silicate, the whole mass being indurated by the silicate of lime thus formed.

Characteristics and Uses.—This artificial stone is used chiefly for paving, which is said to be more durable, to be cheaper, and to stand a greater crushing force, than Yorkshire flags.   It is used also for window sills, coping stones, caps for piers, stairs, landings, troughs, tanks, sinks, etc.

It weighs from 140 to 160 lbs. per cubic foot, and absorbs from 2 to 6 per cent of its weight of water in 24 hours.   The thinner flags are less compact and more absorptive than the thicker ones.[2]

" The white colour, semi-transparency, and extreme hardness of this oxychloride, as well as the small quantity which is required for binding together a considerable mass of any material, facilitate the production of imitations of any description of stone, and render it highly probable that it will play an important part in the future history of artificial stone." [1]

Where used.—This stone has been used for the whole of the external stone-work, except the cornice, at Fresh Wharf, London Bridge.

Also for the panels in the tower, and for the chimney shafts at Messrs. Peek and Frean's biscuit manufactory at Bermondsey, and for paving in many parts of London.

Silicated Stone is made in the same way as Victoria stone, and is used for paving slabs and drain pipes.

Sorel Stone is so called after M. Sorel, a French chemist.

---

[1] Dent.   [2] Wray.

Native carbonate of magnesia, or magnesite, is calcined and mixed with sand or powdered marble. It is then wetted with waste liquor from salt works containing a large proportion of magnesium chloride, pugged, and then rammed or stamped into iron, wooden, or plaster moulds.

It hardens rapidly, setting throughout its mass like ordinary hydraulic cement. In 24 hours it is hard enough to remove from the moulds, and the blocks will bear handling in three or four days.

The proportion of magnesia to the inert material bound together varies from 3 to 15 per cent.

This stone has been found to resist an enormous compression. The resistance of 2-inch cubes varied in different experiments from 4923 to 21,562 lbs. per square inch.

Chance's Artificial Stone is made by melting the Rowley Rag, a basaltic rock found in Staffordshire, and then casting it into the shapes required for different architectural ornaments.

Greenstone, whinstone, or any similar rock, may be treated in the same way. The moulds are of sand in iron boxes, and are at a red heat when they receive the melted stone. They should cool slowly, in order to obtain a hard material like the original stone ; if allowed to cool too quickly the material becomes brittle and glassy.[1]

Rust's Vitrified Marble is produced by fusing together a mixture of glass and sand. " The soft pasty mass is taken out of the pot on the end of an iron rod, and placed in a small metal mould of any required shape or design. The large proportion of sand used prevents the mass, when thus suddenly cooled, from acquiring such a high state of tension as to be liable to fly to pieces, which would be the case with glass alone. The material, when cool, is either used in the form in which it is cast, or it is broken up into small pieces by the stroke of a light hammer, to be used in the construction of mosaics for pavements, or other purposes.

" Any colour can be given to the mass when in a semi-fluid state by mixing with it the oxides of iron, chromium, cobalt, or such other colouring materials as are usually employed for fired ware. This vitrified marble has been used for the bosses and coloured portions of the string course which extends round the Home and Colonial Offices, and also at the Albert Memorial in Hyde Park." [2]

Other artificial marbles are made, which partake of the character of plasters, and will be noticed in Chapter III.

Artificial Paving Slabs and Paving Stones of many kinds are in the market. They are often composed of Portland cement concrete (see p.215),very carefully made, with hard aggregates and the very best cement. It is said that the very finely ground German cement (see p.169) is used for this purpose. Silicates are sometimes added to give hardness to the mass.

PRESERVATION OF STONE.

In consequence of the rapid decay of some of our public buildings, the question of the preservation of stone has of late years attracted much attention.

Several methods have been proposed — a great number of different solutions and preparations have been tried ; but none of them combine efficiency and cheapness to such an extent as to have come into very general use.

---

[1] *Descriptive Catalogue,* Museum of Practical Geology, Jermyn Street.   [2] Dent.

It is unnecessary to give a description of these preparations in detail, but they naturally range themselves under two distinct classes which may be noticed.

The first of these classes consists of preparations containing dissolved organic substances; these fill the pores of the stone, and preserve it for a time, but they are themselves subject to decay, and therefore can afford only a temporary protection.

The preparations of the second class are solutions of substances which act either upon the constituents of the stone to which they are applied, or upon one another (when more than one is applied) so as to form insoluble compounds which fill the pores and harden the structure of the stone, at the same time making it also denser, more impervious, and abler to resist atmospheric influences.

Many processes are successful in the laboratory of the chemist; but none is likely to be of use in the practical execution of engineering or building works, which is not economically applicable on a large scale.

It has been recommended that stones should be placed in vacuum chambers so as to introduce solutions more readily—also that stones should be heated, or immersed in solutions. All these methods are impracticable in dealing with large blocks, on account of the expense and inconvenience attending the manipulation.

Any preservative solution, to be of practical value, must be capable of application to the surface to be protected by means of a brush.

Preparations containing Organic Substances.—Filling the Pores with Organic Matter.—*Paint.*—One of the most common methods of preserving the surface of stone is to paint it. This is effectual for a time, but the paint is destroyed by atmospheric influence in the course of a few years. " In London the time hardly amounts to three years even under favourable circumstances." [1] Moreover, it cannot well be used in important buildings where appearance has to be considered.

Oil has also been used as a coating ; it fills the pores of the stone and keeps out the air for a time, but it discolours the stone to which it is applied.

Paraffin is more lasting than oil, but is open to the same objection as regards discoloration of the stone.

Softsoap dissolved in water ($\frac{3}{4}$ lb. soap per gallon), followed by a solution of alum ($\frac{1}{2}$ lb. alum per gallon), has been frequently employed. [1]

Paraffin dissolved in Naphtha.—" $1\frac{1}{2}$ lb. paraffin to a gallon of coal tar naphtha, and applied warm, is perhaps superior to both the former for this special purpose."

" There is, however, no evidence to show that any methods such as these are likely to be successful in affording permanent protection to stone." [2]

---

[1] Ansted.　　　　　[2] Dent.

Beeswax dissolved in coal tar Naphtha has also been proposed,[1] or, when the natural colour of the stone is to be preserved, *white wax dissolved in double distilled Camphine.*

Wax varnish to preserve statues and marble exposed to the air.—The following is given in Spons' *Workshop Receipts:*—" Melt 2 parts of wax in 8 parts of pure essence of turpentine."

The surface should be cleaned with water dashed with hydrochloric acid, but should be perfectly dry, the solution applied hot and thin.

Preparations not containing Organic Substances.—*Soluble silica.*—There is a large class of preparations whose preservative influences depend upon the presence of soluble silica, which combines with substances contained in, or added to the stone under treatment.

By this means insoluble silicates are formed, which not only preserve the stone from the attacks of the atmosphere but also add considerably to its hardness.

Unfortunately the use of these substances sometimes causes efflorescence on the face of the wall to which they are applied. The soluble alkaline salts left in the pores of the stone are drawn to the surface ; these crystallise in the form of white powder, and disfigure, or in some cases injure, the wall.

The soluble silica is sometimes found in the natural state.

A large proportion may be obtained from the Farnham rock, or from the lower chalk beds of Surrey and Hampshire by merely boiling with an alkali in an open vessel.

ALKALINE SILICATES.—Ordinary silica in the form of flints may, however, be dissolved by being digested with caustic soda, or potash, under pressure.

If a piece of porous limestone or chalk be dipped into this solution, part of the silica in solution separates from the alkali in which it was dissolved and combines with the lime, forming a hard insoluble silicate of lime ; part of it remains in the pores and becomes hard.

KUHLMANN'S PROCESS consists in coating the surface of stone to be preserved with a solution of silicate of potash or silicate of soda.

The hardening of the surface is due to the decomposition of the silicate of potash. If the material operated upon be a limestone, carbonate of potash, silicio-carbonate of lime and silica will be deposited ; besides which the carbonic acid in the air will combine with some of the potash, causing an efflorescence on the surface, which will eventually disappear.[1]

When applied to sulphate of lime, crystallisation takes place which disintegrates the surface.

In order to correct the discoloration of stone sometimes produced by the application of preservative solutions, M. Kuhlmann proposed that the surfaces should be coloured.

Surfaces that are too light may be darkened by treatment with a durable silicate of manganese and potash.

Those that are too dark may be made lighter by adding sulphate of baryta to the siliceous solutions.

By introducing the sulphates of iron, copper, and manganese, he obtained reddish-brown, green, and brown colours.

RANSOME'S INDURATING SOLUTIONS consist of silicate of soda or potash, and chloride of calcium or barium.

The surface of the stone is made thoroughly clean and dry, all decayed parts being cut out and replaced by good.

---

[1] Gillmore.

The silicate is then diluted with from 1 to 3 parts of soft water until it is thin enough to be absorbed by the stone freely. The less water that is used the better, so long as the stone is thoroughly penetrated by the solution. The solution is applied with an ordinary whitewash brush. "After say a dozen brushings over the silicate will be found to enter very slowly. When it ceases to go in, but remains on the surface glistening, although dry to the touch, it is a sign that the brick or stone is sufficiently charged ; the brushing on should just stop short of this appearance." . . . "No excess must on any account be allowed to remain upon the face." After the silicate has become *perfectly dry*, the solution of chloride of calcium is "applied freely (but brushed on lightly without making it froth) so as to be absorbed with the silicate into the structure of the stone." [1]

The effect of using these two solutions in succession is that a double decomposition takes place, and insoluble silicate of lime is formed which fills the pores of the stone and binds its particles together, thus increasing both its strength and weathering qualities.

In some cases it may be desirable to repeat the operation, and as the silicate of lime is white or colourless, "in the *second dressing* the prepared chloride of calcium may be *tinted* so as to produce a colour harmonising with the natural colour of the stone."

"Before applying this second process the stone should be well washed with rain water and allowed to dry again."

The following cautions are given in Messrs. Bateman's circular :—

" 1. The stone must be clean and dry.

" 2. The *silicate* should be applied till the stone is fully charged, but *no excess* must upon any account be allowed to remain *upon the face*.

" 3. The *calcium* must not be applied until *after the silicate is dry ;* a clear day or so should intervene when convenient.

" 4. Special care must be taken not to allow either of the solutions to be splashed upon the windows or upon painted work, as they cannot afterwards be removed therefrom.

" 5. Upon no account use any brush or jet for the calcium that has previously been used for the silicate, or *vice versâ.*"

The bottles or drums of silicate have a *black seal,* those of calcium a *red seal.*

Under ordinary circumstances about four gallons of each solution will be required for every hundred yards of surface, but this will depend upon the porosity of the material coated.

This material has been used with success not only for the preservation of stone from decay, but also to keep out damp.

It has been used at St. George's Hall, Liverpool, for preserving the sculpture ; at Trinity College, Dublin ; Cardiff Town Hall ; Greenock Custom House, and for other buildings both in this country and in India.

It is applicable not only to stone and brick surfaces, but also to those rendered with cement or lime plaster.

SZERELMEY'S STONE LIQUID is stated by Professor Ansted to be " a combination of Kuhlmann's process with a temporary wash of some bituminous substance."

The wall being made perfectly dry and clean, the liquid is applied in two or three coats with a painter's brush until a slight glaze appears upon the surface.

This composition was used with some success in arresting for a time the decay of the stone in the Houses of Parliament.

---

[1] Patentee's circular.

The stone liquid is transparent and colourless, but Szerelmey's iron paint is opaque and of different colours, and is applied like ordinary paint (see p. 425).

The *Petrifying Liquid* of the Silicate Paint Company is stated in their circular to be "a solution of silica," thinned with warm water, and applied to clean wall surfaces, which must be warmed if they are not already dry. 1 cwt. will cover from 120 to 150 square yards.

OTHER PROCESSES.—Among other processes which have been tried are—

Solution of Baryta followed by solution of *Ferro-silicic Acid* so as to fill the pores of the stone with an insoluble ferro-silicate of baryta.

Solution of Baryta followed by solution of *Superphosphate of Lime* producing an insoluble phosphate of lime and phosphate of baryta.

Soluble Oxalate of Alumina applied to *limestones* produces insoluble oxalate of lime and alumina.

"These three processes last alluded to all possess the advantage of producing by the changes they undergo within the structure of the stone an insoluble substance, without at the same time giving rise to the formation of any soluble salt likely to cause efflorescence, which necessarily attends the use of alkaline silicates." [1]

Fluate.—Introduced by the "Bath Stone Firms, Limited," for the hardening, preservation, and making waterproof of stone, bricks, marble, etc. It is stated to harden and permanently preserve all stones, bricks, and marble that will imbibe the liquid ; is applied with a brush, and has been used in various public buildings.

Temporary Protection of Stone Surfaces.—During the erection of large buildings the surface of the masonry built in the earlier stages of the work is smeared over with a sort of thin mortar, so as to preserve it from atmospheric influence, and to make it easier to clean down.

Tables illustrating the Properties of Different Stones.—The following Tables give a selection from the results of a great number of experiments upon stone made by various authorities. In many cases the figures given are not directly comparable with each other, inasmuch as the experiments have been made by different observers and under different conditions. They afford, however, a useful indication of what may be expected in dealing with stones of different descriptions.

The Tables on pages 81-83 show the weight required to crush stones of different kinds.

Experiments upon the resistance of stones to crushing have generally been made upon cubes. Professor Rankine says that these experiments indicate "somewhat more than the real strength of the material." The reason for this is that the fracture of stones under compression generally takes place by their shearing on a plane inclined at a slope having $1\frac{1}{2}$ rise to 1 of base. In order to ascertain the strength of any stone for a special purpose, experiments should be made on prisms whose heights are about $1\frac{1}{2}$ times their diameters.

The hardest stones—such as basalts, primary limestones, slates, etc.—give way suddenly. Other stones begin to crack under from $\frac{1}{2}$ to $\frac{2}{3}$ the crushing load.

It should be noticed that the size of cubes experimented upon varies considerably. With test blocks of geometrically similar form, experiment appears to show that the crushing strength varies as the area of the surface on which the crushing pressure acts.

The working stress allowed in practice upon ashlar blocks should not exceed $\frac{1}{20}$ of the crushing weight given above.[2]

---

[1] Dent.   [2] Stoney *On Strains.*

TABLE OF THE RESISTANCE TO CRUSHING, WEIGHT, MODULUS OF ELASTICITY, AND ABSORPTION OF BUILDING STONES.

STONE.	Crushing Load Per Square Foot in Tons.	Length of Side of Cube.	Weight Per Cubic Foot.	Modulus of Elasticity in Direct Compression. Tons per Square Foot.	Number of Specimens Tested.	Absorption of Water. Percentage of Dry Weight absorbed.	Authority.
Granites.							
Aberdeen (grey) .	1412	2″	..	..	..	0·81	U
Do. do. .	1162	⎧ cylinders ⎫	..	..	..	..	U
Do. (red) .	1357	⎨ 2⅜″ diam. ⎬	..	..	..	..	
Do. do. .	1614	⎩ 3¼″ high ⎭ 2″	..	..	..	0·70	U
Aberdeenshire—							
Corennie . .	1284	2¼″	161	540,700	5	0·40	Be
Cove . . .	987	2¼″	169	..	3	0·55	Be
Kemnay . .	1149	2¼″	161	..	6	0·31	Be
Craigton . .	1282	2¼″	..	..	3	..	Be
Peterhead .	1208	2¼″	158	657,000	3	0·29	Be
Dyce . .	1106	2¼″	165	621,400	3	0·19	Be
Hill of Fare .	1360	2¼″	159	460,400	3	0·40	Be
Sclattie . .	850	2¼″	161	321,700	3	0·10	Be
Persley (grey) .	943	2¼″	162	522,300	3	0·19	Be
Rubislaw . .	1194	2¼″	164	525,200	6	..	Be
Ben Cruachan .	877	2¼″	171	..	3	0·29	Be
Penrhyn . .	1060	2¼″	165	..	5	0·12	Be
Cornish (grey) .	956	2¼″	162	538,300	3	..	Be
Mount Sorrel .	820	2″	..	..	..	..	F
Do. do. Syenite	760	2″	..	..	..	..	F
Argyleshire . .	701	1½″	..	..	..	..	F
Killiney . .	692	1″	..	..	..	..	W
Ballyknocken .	203	1″	..	..	..	..	W
Irish (various) .	144 to 864	1″	..	..	..	..	W
Bavarian. Metten	1230	..	..	244,000	..	..	Br
Do. do.	1210	Pressure parallel to bed	..	..	..	..	Br
Do. Passau .	1230	Pressure square to bed	..	199,000	..	..	Br
Do. do.	1290	Pressure parallel to bed	..	..	..	..	Br
St. Gotthard .	720	Pressure square to bed	..	..	..	..	Br
Do. do. .	810	Pressure parallel to bed	..	..	..	..	Br
Worcester, U.S. .	1622	Specimens 6 sides high (24″ × 6″ × ·4″)	..	..	..	..	Wn
Quincy, U.S. .	1239	do. do.	..	..	..	..	Wn
Cape Ann, U.S. .	1555	do. do.	..	571,939	..	..	Wn
Basalts, etc.							
Whinstone . .	769	..	..	..	..	..	F
Grauwacke ⎫	1086	2″	..	..	..	..	F
Penmaenmawr ⎭							
Felspathic greenstone . .	1100	..	..	..	..	..	W
Hornblendic greenstone .	1580	..	..	..	..	..	W
Irish (various)	504 to 2088	1″	..	..	..	..	W
North River Bluestone, U.S. .	1649	..	..	..	..	..	Wn
Slates.							
Compact Welsh ⎫	733 to 1052	3″	..	..	..	0·27	U
Clayslate ⎭							
Slate . . .	1205	..	..	..	..	..	M
Valencia . .	720	1″	Pressure square to bed	..	..	..	W

U, Unwin. Be, Beare. F, Fairbairn. W, Wilkinson. Br, Bauschinger. Wn, Watertown Arsenal.
M, Mallet. G, Grant. R, Rennie. K, Kirkaldy. C, Commissioners on Stone.

TABLE OF THE RESISTANCE TO CRUSHING, ETC.—*Continued.*

STONE.	Crushing Load Per Square Foot in Tons.	Length of Side of Cube.	Weight Per Cubic Foot.	Modulus of Elasticity in Direct Compression. Tons per Square Foot.	Number of Specimens Tested.	Absorption of Water. Percentage of Dry Weight absorbed.	Authority.
Slates—Continued.							
Valencia . .	658	1″	Pressure parallel to bed	..	..	..	W
Killaloe . .	1974	1″	..	..	..	..	W
Glanmore . .	1372	1″	..	..	..	..	W
Monson, U.S. .	1254	Specimens (24″×6″×4″)	..	..	..	..	Wn
Do. U.S. .	534	do. do.	..	803,520	..	..	Wn
Sandstones.							
Bramley Fall .	156	(3″×3″×1½″) on edge	..	..	..	..	G
Do. .	338	(3″×3″×1½″) flat	..	..	..	..	G
Do. .	389	1½″	..	..	..	..	R
Do. .	259	6″	..	..	..	..	K
Do. .	238	2¼″	132	114,900	3	3·70	Be
Prudham Four Stones Quarry	455	2¼″	142	117,250	8	4·16	Be
Corncockle .	384	2¼″	132	156,300	3	4·57	Be
Gunnerton . .	354	2¼″	131	111,950	3	5·16	Be
Cragg . . .	574	2¼″	136	153,300	3	4·13	Be
Corsehill . .	445	2¼″	130	121,000	8	7·94	Be
Red, do. . .	504	6″	..	..	..	..	K
Polmaise . .	551	2¼″	142	138,100	5	4·58	Be
White Piean .	613	2¼″	138	157,400	3	4·25	Be
Arbroath . .	558	2¼″	151	164,800	3	2.32	Be
Auchinlee . .	203	2¼″	129	84,080	3	6·90	Be
Craigleith . .	862	2¼″	138	..	3	3·61	Be
Do. . .	353	1½″	..	..	..	..	R
Do. . .	504	2″	..	..	..	..	C
Do. . .	778	6″	..	..	..	..	K
White Hailes .	662	2¼″	144	105,100	3	3·71	Be
Blue do. .	459	2¼″	143	86,960	3	4·70	Be
Dean Forest .	530	2¼″	151	172,000	3	2·71	Be
Gatelaw Bridge .	496	2¼″	129	168,100	3	5·84	Be
Binnie . . .	569	2¼″	135	152,300	3	5·22	Be
Do. . . .	322	2″	..	..	..	..	C
Darley Dale .	455	2″	..	..	..	..	C
Darley Top . .	517	2¼″	139	160,200	3	3·40	Be
Hermand . .	457	2¼″	142	191,900	3	4·70	Be
Howley Park .	467	2¼″	140	116,500	3	4·90	Be
White Grinshill .	209	2¼″	122	115,500	3	7·80	Be
Grinshill . .	331	6″	..	..	..	..	K
Hercules Ridge .	336	2¼″	138	171,200	3	3·60	Be
Ackworth . .	389	2¼″	141	90,960	3	5·00	Be
Robin Hood .	574	2¼″	145	132,200	3	3·90	Be
Aspatria . .	234	2¼″	123	80,100	3	8·50	Be
Scotgate . .	734	6″	..	..	..	..	K
Wilderness . .	475	6″	..	..	..	..	K
Cunliffe . .	677	6″	..	..	..	..	K
Quarella (white) .	547	6″	..	..	..	..	K
Do. (green) .	432	6″	..	..	..	..	K
Giffneuk . .	310	2″	..	..	..	..	C
Kenton . .	318	2″	..	..	..	..	C
Morley Moor .	318	2″	..	..	..	..	C
Park Spring .	487	2″	..	..	..	..	C
Stanley . .	383	2″	..	..	..	..	C
Runcorn . .	267-443	4″	..	..	16	..	U
Do. . .	427-514	4″	..	..	6	..	U
Do. . .	295-416	3″	..	..	6	..	U
York Grit . .	712	3″	..	..	..	..	U
Bunter (Kronach)	300	..	..	..	..	..	Br
Do. (Wertheim)	720	..	..	..	..	..	Br
Keuper (Wurtemburg) . .	580	..	..	..	..	..	Br
Keuper (Coburg) .	290	..	..	..	..	..	Br
Molasse (Allgau) .	465	..	..	..	..	..	Br
Do. (Kempten)	870	..	..	..	..	..	Br
Do. (Kempten)	1850	..	..	..	..	..	Br
Potomac Red, U.S.	1223	..	..	..	..	..	Wn

TABLE OF THE RESISTANCE TO CRUSHING, ETC.—*Continued.*

STONE.	Crushing Load Per Square Foot in Tons.	Length of Side of Cube.	Weight Per Cubic Foot.	Modulus of Elasticity in Direct Compression. Tons Per Square Foot.	Number of Specimens Tested.	Absorption of Water. Percentage of Dry Weight absorbed.	Authority.
Sandstones—Continued.							
Portland Red, U.S.	685	Specimens 6 sides high (24″×6″×4″)	..	..	..	..	Wn
Do. do., U.S.	397	do. do.	..	173,760	..	..	Wn
Ohio, U.S. . .	556	do. do.	..	..	..	..	Wn
Do. U.S. . .	375	do. do.	..	127,280	..	..	Wn
Dolomites.							
Mansfield (white)	462	2¼″	140	235,600	4	5·01	Be
Do. do.	337	2″	..	..	..	..	C
Do. (red) .	592	2¼″	143	198,700	4	4·58	Be
Do. do. .	609	2″	..	..	..	..	C
Do. yellow magnesian	577	2¼″	145	406,100	5	4·62	Be
Anstow . .	302	2¼″	132	350,000	3	7·50	Be
Do. . .	196	..	..	..	..	..	F
Limestones.							
Ancaster(brown)) Weather bed)	552	2¼″	156	499,800	4	2·42	Be
Do. freestone	184	2¼″	140	130,500	4	6·27	Be
Portland, base bed	287	2¼″	137	..	3	6·84	Be
Do. do.	147	2¼″	124	..	3	11·10	Be
Do. whitbed	205	2¼″	132	236,900	3	7·51	Be
Ketton . . .	102	2¼″	128	156,900	3	8·10	Be
Do. . . .	164	2″	..	\.	..	..	C
Do. . . .	312	3″	..	..	..	..	U
Do. (rag) .	577	2″	..	..	..	..	C
Banchory limestone . .	956	2¼″	175	..	..	..	Be
Hopton Wood .	806	6″	..	..	..	..	K
Corsham Down .	70 to 128	2¼″	129	116,000	5	11·12	Be
Do. "fine upper"	84	12″	..	..	..	..	K
Do. " bottom bed " .	103	12″	..	..	..	..	K
Farleigh Down .	62	2¼″	120	..	5	12·88	Be
Monk's Park .	139	2¼″	137	166,100	5	8·02	Be
Box Ground .	82 to 126	2¼″	128	129,200	5	7·80	Be
Do. do. .	95	2″	..	..	..	..	C
Coombe Down .	93 to 151	2¼″	128	..	5	6·00	Be
Corngrit . .	107 to 172	2¼″	134	243,700	5	8·88	Be
Do. .	116	12″	..	..	..	..	K
Stoke Ground .	70 to 107	2¼″	126	109,300	5	10·84	Be
Winsley Ground .	85 to 115	2¼″	133	164,000	5	7·74	Be
Westwood Ground	90 to 140	2¼″	130	128,900	5	8·85	Be
Doulting) Fine Beds) .	81 to 117	2¼″	125	121,000	5	10·87	Be
Doulting) . Chelynch) . .	135 to 211	2¼″	150	379,000	5	3·36	Be
Hamhill .	166	2¼″	136	164,500	3	..	Be
Do. . .	259	2″	..	..	..	..	C
Ancaster freestone	184	2¼″	140	130,500	4	6·27	Be
Huddlestone .	278	2″	..	..	..	..	C
Park Nook .	278	2″	..	..	..	..	C
Roche Abbey .	250	2″	..	..	..	..	C
Totternhoe . .	124	2″	..	..	..	..	O
Anstow .	196	..	..	..	..	..	F
Caen . . .	198	3″	..	..	..	12·56	U
Hoosier Buff, U.S.	456	Specimens 6 sides high (24″×6″×4″)	..	..	..	..	Wn
Do. do. U.S.	264	do. do.	..	200,880	..	..	Wn
Indiana, U.S. .	473	do. do.	..	..	..	..	Wn
Do. U.S. .	321	do. do.	..	335,800	..	..	Wn
Vermont marble, U.S. . .	770	do. do.	..	451,150	..	..	Wn
Lee marble, U.S. .	1160	do. do.	..	829,440	..	..	Wn

Tensile Strength of Stone.—Stone is rarely employed so as to be subject to a tensile stress. The following TABLE is chiefly from Mr. Stoney's work on *Strains*, and he remarks that it would be well to have the figures corroborated.

MATERIAL.	Tearing Weight per Square Inch in lbs.	Authority.
Arbroath paving	1261	Buchanan.
Caithness do.	1054	Do.
Chilmark	500	Do.
Craigleith stone	453	Do.
Hailes	336	Do.
Humbie	283	Do.
Binnie	279	Do.
Whinstone	1469	Do.
Marble, white	722	Hopkinson.
Do.	551	Do.
Slate	9600 to 12,880	Rankine.

TRANSVERSE AND SHEARING STRENGTH OF STONE.

Description of Stone.	Dimensions.			Modulus of Rupture.	Shearing Strength.	Authority.
	Length.	Breadth.	Depth.	lbs. per sq. in.	lbs. per sq. in.	
Hoosier Buff Limestone .	19″	4″	6″	1601	1066	Wn
Indiana Limestone	Do.	Do.	Do.	943	{ 608 1154 }	Do.
Vermont Marble .	Do.	Do.	Do.	1531	{ 690 1847 }	Do.
Lee Marble . .	Do.	Do	Do.	1586	2052	Do.
Potomac Red Sandstone .	Do.	Do.	Do.	3009	2293	Do.
Portland Red Sandstone .	Do.	Do.	Do.	1576	1392	Wn
Ohio Sandstone .	Do.	Do.	Do.	774	906	Do.
North River Bluestone . .	Do.	Do.	Do.	5618	3107	Do.
Worcester Granite	Do.	Do.	Do.	2393	2359	Do.
Quincy Granite .	Do.	Do.	Do.	2131	2355	Do.
Sandstone . .	...	...	...	1100 to 2360	...	Rankine
Slate . . .	...	...	...	5000	...	Do.

Wn, Watertown Arsenal.

Absorption.—TABLE showing the bulk of Water absorbed in twenty-four hours by various stones :—

Nature of Stone.	Bulk of Water absorbed as compared with bulk of stone per cent.	Authority.
Several specimens of good granite and syenite	½ per cent	W
Do. do. indifferent specimens	1 ,,	W
Do. do. very bad . . .	3 ,,	W
Trap and basalt	A trace.	W
Do. do.	$\frac{1}{10}$ to $\frac{1}{5}$ p. c.	
Sandstones.		
Craigleith—Very durable . .	8 per cent	C
Park Spring Do. . .	8 ,,	C
Giffneuk—Moderately durable .	10 ,,	C
Heddon Do. .	10·4 ,,	C
Kenton Do. .	9·9 ,,	C
Mansfield Do. .	10·4 ,,	C
Hassock—Very bad . . .	20·0 ,,	W
Limestones.		
Marble	A trace.	
Portland—Very durable . .	13·5 per cent	C
Ancaster—Durable . . .	16·6 ,,	C
Bath (Boxground) . . .	17 ,,	C
Ketton—Durable . . .	15·1 ,,	
Chilmark	8·6 ,,	C
Roche Abbey—Durable . .	17·2 ,,	C
Kent Rag Do. . .	1⅓ ,,	W
Ransome's stone (artificial) . .	12 ,,	W
Victoria Do. do. . .	7·6 ,,	W
Apœnite Do. do. . .	12 ,,	W

W, Deduced from Experiments detailed in Wray *On Stone.*
C, Royal Commission on Stone for Houses of Parliament.

The following Table shows approximately the weights of different classes of stone, and may be useful.

TABLE giving the Weight and Bulkiness of Different Varieties of Stone.

	Weight per cubic ft. in lbs.		Weight per cubic ft. in lbs.
Granites and Syenites	. 162 to 187	Limestones, Shelly . .	157 to 167
Trap and Basalt . .	. 164 to 187	,, Magnesian .	126 to 153
Slate	. 165 to 181	,, Kent Rag . .	166
Sandstones . .	. 116 to 170	,, Lias . . .	127 to 156
Marble . . .	. 168 to 172	,, Chalk . .	117 to 174
Limestones, Compact .	. 155 to 172	Quartz	165
,, Granular .	. 116 to 151	Felspar	162
,, Shelly granular	130 to 140		

RESISTANCE TO WEAR.—Mr. Walker exposed the undermentioned descriptions of granite and whinstone to very heavy waggon traffic for seventeen months, and found their vertical wear to be as follows :—

	Inch.		Inch.
Guernsey granite	. ·060	Heytor granite	. ·141
Herm ,,	. ·075	Aberdeen red granite .	·159
Baltic whinstone	. ·082	Dartmoor granite	. ·207
Peterhead blue granite .	·131	Aberdeen blue granite .	·225

Mr. Newton's experiments on the flags used in Liverpool showed Kilrush flags to be most durable, Caithness flags next. The flags found to be least durable were those from Llangollen and Yorkshire.[1]

_____

[1] D. Kinnear Clarke *On Roads and Streets.*

Chapter II.

BRICKS, TILES, TERRACOTTA, ETC.

THERE are many different forms in which clay after it is
burnt or baked is used by the builder and engineer.
Some of the more important of these will now be described
under the following classification :—
Bricks.
Fireclay and Fire-bricks.
Terracotta.
Stoneware.
Miscellaneous clay goods of Earthenware, Fireclay, Stoneware,
Terracotta.

BRICKS.

Ordinary building bricks are made of clay or other earths
subjected to several processes (which somewhat vary according to
local practice, influenced by the nature of the material), formed to
the required shape in moulds, and burnt.

BRICK EARTHS.

Constituents of Brick Earth. —The earths used for making
ordinary bricks generally consist of alumina and silica, either
alone or in combination with other substances, such as lime,
magnesia, iron, etc.

It is beyond the province of these Notes to go into the chemistry
of the subject, but it will be useful just to glance at the part
played by each of these constituents, and the effect that it has
upon earth considered as a material for brickmaking.

It may here be remarked that mere inspection or even chemical analysis of a clay gives very little information as to its suitableness for brickmaking. No test is satisfactory but that of actually trying the clay by making a few bricks with it.

In the absence of facilities for burning full-sized bricks, a fair indication of the quality of a clay for brickmaking may be arrived at by making it into a small brick about 3 inches long by 1½ inch wide by 1 inch thick. This small brick may be burnt in a common house fire, being protected from contact with the fuel by placing it inside a shield made by roughly rolling a piece of sheet-iron round into a hollow cylinder of about 2 inches diameter.

Alumina is a principal constituent in nearly every kind of clay. It gives the material its plastic qualities, but it shrinks and cracks in drying, warps, and becomes very hard under the influence of heat.

Silica exists to a greater or less extent in all clay, in a state of chemical combination with the alumina, forming Silicate of alumina.

It is found also in nearly all clays in an uncombined state—as sand.

Silica is infusible alone, or in the presence of alumina only, except at very high temperatures.

If, however, the silica and alumina be in nearly equal proportions, the presence of a small quantity of oxide of iron will render them fusible at a comparatively low temperature.

Pure silicate of alumina is plastic, but shrinks when drying, and warps with heat.

The action of sand is to prevent cracking, shrinking, and warping, and to provide the silica necessary for a partial vitrification of the materials, which is generally desirable.

The larger the proportion of sand present, the more shapely and equable in texture will the brick be.

An excess of sand in clay renders the brick made from it too brittle.

The difference between the silica which is in a state of chemical combination and that which exists merely as sand is not shown in ordinary chemical analyses, and this is one reason why they are not so useful as they might be in determining the value of a clay for brickmaking. [1]

Lime has a twofold effect upon the clay containing it.

It diminishes the contraction of the raw bricks in drying, and it acts as a flux in burning, causing the grains of silica to melt, and thus binding the particles of the brick together.

An excess of lime causes the brick to melt and lose its shape.

Again, whatever lime is present must be in a very finely divided state. Lumps of limestone are fatal to a clay for brickmaking. When a brick containing a lump of limestone is burnt, the carbonic acid is driven off, the lump is formed into "quicklime," and is liable to slake directly the brick is wetted or exposed to the weather. Pieces of quicklime not larger than pin-heads have been known to detach portions of a brick and to split it to pieces.

The presence of lime may be detected by heating the clay with a little dilute sulphuric acid. If there is lime present an effervescence will take place.

Bricks containing lumps of quicklime should be immersed for several hours

---

[1] Silica sometimes exists in clay in a soluble condition combined with lime ; it is then injurious, as it may absorb moisture which has been known to destroy walls by making the bricks swell.

before use, so as to kill the lime and prevent it from slaking after the bricks are made, or even built into the work.

Iron Pyrites often occurs in clays, and should be carefully removed. If not, the pyrites is partially decomposed in the kiln, will oxidise in the brick, crystallise, and split it to pieces.

Carbonaceous matter, when it exists naturally in clays to any considerable extent, is objectionable. " When not burnt completely out in the kiln, which is sometimes with the denser clays difficult, the bricks are of a different colour in the interior and exterior, and will not bear cutting for face work without spoiling the appearance of the brickwork.

" But, worse than this, such bricks when worked in the wall occasionally pass out soluble compounds like those absorbed from soot by the bricks or flues, and like those (when used again in new work) discolour plastering or stucco work." [1]

Alkalies, when they exist in clay to a considerable extent, make it unfit for the manufacture of bricks. They act as a flux, and cause the clay to melt and become shapeless.

Salt.—" Common salt is nearly always present in minute quantity in clays ; but when these are taken from the sea-shore, or without or beneath the sea-washes, or from localities in and about the salt formations (Trias), they frequently, though in all other respects excellent clays, are unfit for burning into good bricks.

" Chloride of sodium [2] is not only a powerful flux when mixed even in very small proportion with clays, but it possesses the property of being volatilised by the heat of the brick kilns, and in that condition it carries with it in a volatile state various metallic compounds, as those of iron, which exist in all clays, and act as fluxes. The result is that bricks made of such clays warp, twist, and agglutinate together upon the surfaces long before they have been exposed to a sufficient and sufficiently prolonged heat to burn them *to the core* into good hard brick. Place bricks can be made of such clay, but nothing more, and these are nearly always bad, because never after free from hygrometric moisture." [1]

Oxide of Iron in clay influences the colour of the bricks to be produced (see p. 90). The tint resulting after burning depends upon the proportion of iron in the clay and the temperature to which it has been raised.

When in the presence of silica and alumina whose proportions are nearly equal, iron renders them fusible.

Practical Classification of Brick Earths.—Brick earths are generally divided into three classes.

1. *Plastic* or *Strong Clays* (called by the brickmaker "*foul clays*"), which are composed of silica and alumina, with but a small proportion of lime, magnesia, soda, or other salts. These are also known as *pure clays*.

2. *Loams* or *Mild Clays*, consisting of clay and sand, and sometimes called *sandy clays*.

3. *Marls* or *Calcareous Clays*, which contain a large proportion of carbonate of lime.

---

[1] Mallet *On Brickmaking.*   [2] Common salt.

Malm is an artificial imitation of natural marl, and is made by mixing clay and chalk in a wash mill. It is sometimes called *washed clay.*

It generally happens that a clay as found in nature is unfit for brickmaking by itself.

It will probably turn out to be deficient in some necessary quality which has to be supplied by mixing it with other clays, or by adding the constituent required, such as sand or lime.

A good Brick Earth should contain sufficient flux to fuse its constituents at a furnace heat, but not so much as to make the bricks run together and become *vitrified.*

Such earths contain from $\frac{1}{5}$ to $\frac{1}{3}$ alumina, and from $\frac{1}{2}$ to $\frac{3}{5}$ silica, the remainder consisting of carbonate of lime, carbonate of magnesia, oxide of iron, etc.

The bricks made from such clays are a silicate of alumina and lime or other fluxes.

The following Table gives the analysis of some brick clays :—

	Burham Clay.[1]	London Brick Clay.[1]	Loam.[1]	Marl.[2]
Silica	42·92	49·5	66·7	} 43·00
Alumina .	20·42	34·3	27·0	
Oxide of iron .	5·00	7·7	1·3	3·00
Carbonate of lime	18·91	1·4	0·5	46·50
Carbonate of magnesia	·12	5·1	...	3·50
Potash and soda	·33	...	...	} 4·00
Water . .	6·68	...	...	
Organic matter .	5·01	1·9	5·0	...

Characteristics of different kinds of Brick Earth.—The quality of the bricks produced depends to a very great extent on the selection and mixing of the clay.

PURE or FOUL CLAYS are sometimes used for bricks without the addition of other substances. In such a case any sand they contain acts merely to prevent excessive contraction. For want of a flux it does not become fused so as to bind the particles of the brick together.

Bricks made from such clays are rather *baked* than *burned.* They are not so well able to resist the action of the weather as those which are partly vitrified through the aid of a flux.

Pure clays are therefore very much improved by the addition of sand or loam, by adding lime to act as a flux, or ashes to provide alkalies for the same purpose.

---

[1] Abney.　　　　　　　[2] Knapp.

LOAMS are so loose and sandy that they require a flux to fuse and bind the particles together, and to take up the excess of sand that would otherwise remain in an uncombined state.

MARLS are, of all the clays, the best suited for making bricks without mixture with other substances, though they are often mixed with chalk or lime when there is any deficiency in that constituent.

The Colour of Bricks depends upon the composition of the clay, upon the kind of sand used for moulding, on the state of dryness of the bricks before burning, on the temperature at which they are burnt, and upon the amount of air admitted to the kiln.

Pure clay, free from iron, will burn *white,* but the colour of white bricks is generally produced by adding *chalk* to the clay.

The presence of *iron* produces a tint which varies from *light yellow* to *orange* and *red ;* the colour increases in intensity according to the proportion of iron contained in the clay.

To obtain a clear *bright red* brick the clay should be free from impurities, and should contain a large proportion of *oxide of iron,* which is converted by burning into the red oxide, but not fused.

When there is from 8 to 10 per cent of *oxide of iron,* and the brick is raised to an intense heat, the red oxide of iron is converted into the black oxide, combines with the silica, and fuses, producing a *dark blue* or *purple* colour.

When a small quantity of *manganese* is present, with a large proportion of iron, the brick becomes darker still, *blue* or even *black.*

A little *lime* in the presence of a small quantity of *iron* produces a *cream colour ;* an increase of the iron changes the colour to *red,* and an increase of the lime produces a *brown* colour.

Magnesia in the presence of *iron* makes the brick *yellow.*

A clay containing *alkalies* and burnt at a high temperature becomes a *bluish green.*

BRICKMAKING.

The operations involved in brickmaking are very numerous, though not intricate; they differ in several particulars in different localities, according to local custom, generally influenced by the nature of the clay.

To describe these operations in sufficient detail to be of any practical value would require a separate treatise of considerable bulk and profusely illustrated.

Such descriptions would be beyond the province of these Notes, and would be unnecessary, for the practices in the brickfields of different localities are fully described in Dobson *On Brick and Tile Making*, one of Weale's series of very valuable technical works.[1]

It may be useful, however, to give a general sketch of the operations involved in brickmaking—not such details as would be of practical use to the brickmaker, but just so much as will enable any one using bricks to understand and appreciate more clearly the qualities and peculiarities of different varieties, many of the characteristics of which are caused by differences in the processes of manufacture.

With this object the various operations will now be rapidly and shortly sketched in succession.

Preparation of Brick Earth.—*Unsoiling.*—The surface of the site from which the clay is to be obtained is first stripped of its turf and mould, which is removed to a spoil bank and kept to be respread over the site after the clay has been dug out.

The mould is sometimes called *encallow*, and the process of removing it *encallowing*.

Clay-digging and Weathering.—In the autumn the clay is dug, and the various descriptions which it is intended to mix, together with the ashes which are to be incorporated in the mixture, are wheeled to heaps, in some places called *kerfs*, in which they remain during the winter, sometimes during two or three winters, so that they may be thoroughly disintegrated by the action of frost.

This mellowing of the clay renders the bricks made from it less liable to warp.

Clearing from Stones.—If the clay contain pebbles or pieces of ironstone they must be carefully picked out by hand; or if they are found in large numbers, the clay must be washed in small quantities and strained through a grating so as to separate all the stones from the mass.

Grinding.—When the clay is of a hard marly character and full of lumps, or contains fragments of limestone, known by brickmakers as *race*, it requires to be ground between cast-iron rollers, which must be set sufficiently close to reduce the hard particles to powder.

Tempering.—This is done after the winter's frosts, generally in March or April, before the brickmaking begins.

---

[1] Now published by Messrs. Crosby Lockwood and Company, Stationers' Hall Court, E.C.

It consists in digging and turning over the kerfs or heaps of clay; sometimes the clay is also well trodden under foot, in some places it is passed through a pug-mill, and occasionally, for the very best bricks it is kept damp in cellars for a year or two to ripen.

PREPARATION OF MALM.—The clay is dug in the autumn and at once tipped, together with a proportion of ground chalk in pulp, into a wash mill. This consists of a brick-lined circular tank in which are revolving harrows, knives, or implements of some kind to disintegrate and mix up the clay and chalk.

The exact proportion of the chalk differs according to the composition of the clay, but in some cases the chalk is about $\frac{1}{16}$ of the bulk of the clay.

The mixture having been reduced to a creamy consistence is strained off through fine gratings into large shallow tanks called *backs*, and there left till it is nearly solid.

After that it is *soiled* in layers from 1 to 3 feet deep, *i.e.* covered about $\frac{1}{6}$ its bulk with screened cinders, and allowed to remain during the winter.

In the spring the *backs* are dug out, the layers of clay and ashes being thoroughly incorporated in a pug-mill.

In some places the preparation of malm is known as *washing*.

Malm bricks are made with the mixture of clay and chalk described above.

Washed Bricks.—These contain a certain proportion of malm, and are made in two ways.

In some parts of the country—in Essex for example—they are composed wholly of an inferior malm, made like the malm described above, except that the proportion of chalk is only one half of that in the ordinary malm, and the cinders are unscreened.

In other brickfields, including those near London, a certain proportion of ordinary liquid malm is poured over unwashed clay, and mixed with it so that the whole becomes an inferior malmed clay.

Quantity of Clay required.—The quantity of clay used for making bricks is very variable, depending upon the nature of the clay and the processes to which it is subjected.

The quantity required for 1000 bricks of ordinary size ranges from $1\frac{3}{4}$ to $3\frac{1}{2}$ cubic yards, measured before digging. The stronger the clay, the more of it is required.

Hand-Moulding.—The moulds used are rectangular boxes without top or bottom, about 10 inches long, 5 inches wide, and 3 inches deep, the exact size depending upon that of the brick required, and upon the contraction of the clay in burning, which may be about $\frac{1}{10}$ or $\frac{1}{12}$ of the linear dimensions.

The moulds are sometimes made of wood edged or lined with iron, or of sheet-iron strengthened at the sides with wood, or, as in the best works, with sides and ends of brass protected by wood.

The mould stands either upon the table at which the moulder works—in which case the bottom of the brick is flat—or it rests upon a *stock board*, or bottom made to fit the mould, and upon which is a raised projection which forms an indentation or *frog*[1] in the bottom of the brick.

The process of hand-moulding consists in dashing a clot of clay into the mould, and pressing it home so as thoroughly to fill every corner.

When a stock board is used the lower side of the brick rests upon it. The superfluous clay protruding above the top is swept or scraped off by a *strike* or straight-edge of wood or steel.

Thus the lower surface of the brick is indented by the frog on the stock board, but the upper surface is struck smooth.

When there is no stock board the bottom of the brick rests upon the moulding-table, and the top surface is formed by means of a *plane*, which is a piece of board about 9 inches by 3 inches with a short vertical handle near one end.

Slop-moulding is the term used when the mould is dipped frequently in water to prevent the wet bricks from sticking to it.

Sand-moulding is when the mould is sprinkled with sand or fine ashes for the same purpose, and is considered to produce cleaner and sharper bricks than slop-moulding.

BEARING OFF.—As each brick is moulded it is disposed of in one of two ways—

1. It may be carried by a boy in the mould to the drying floor or ground, and there deposited, the mould being taken off and returned to the moulder. Or,

2. It may be deposited upon a *pallet* (a piece of board $\frac{3}{8}$ inch thick, the same width as the mould but longer), and placed by the boy upon a bearing-off or *hack barrow* for removal to the drying ground. These barrows are made with springs, and run upon smooth wrought-iron *wheeling plates*, so as to shake the brick as little as possible.

Drying.—The raw moulded bricks may be dried either in sheds under cover or out of doors.

DRYING IN SHEDS.—Drying sheds are extensively used in

---

[1] Sometimes called a *kick*.

Nottinghamshire and the Midland counties, and they insure the great advantage of being independent of weather in drying the bricks.

Where coals are cheap the sheds may be warmed by flues run- ning under the floor. This secures the raw bricks against the effects of frost, and enables the brickmaking to be continued throughout the winter.

DRYING OUT OF DOORS.—*Hacking.*—Bricks to be dried out of doors are placed upon *hacks*, which are long parallel banks raised about 6 inches from the ground. They have a slight in- clination toward the kiln to facilitate drainage and transport of the bricks, and are sometimes made of brick rubbish and ashes so that they may be quite dry.

The bricks are placed upon the hacks, sometimes laid square in plan, sometimes diagonally, and piled up. They should not be more than eight courses deep, or the lowest course would be liable to be crushed.

Scintling.—When the raw bricks are half dry they are *scintled,* that is, placed diagonally and a little distance apart, so that the wind may pass between them. They therefore take up more room in plan, but as the bricks are drier, about fourteen courses may be piled up instead of eight as before.

The bricks on the hacks are generally protected from the wea- ther by light coverings or roofs, made of straw, called *lews,* or by boards or tiles.

In permanent brickyards the hacks are sometimes covered by sheds, with sliding roofs so that they may be uncovered in favour- able weather and closed in case of rain, or when the sun is so strong as to warp the bricks.

The time of drying varies according to the weather. The raw bricks generally take about ten days before being *scintled,* and about three to six weeks for the whole process of drying.

Machine-Moulding.—When there is a large number of bricks to be made at the same spot, it pays to set up machines for moulding, and in cases where the clay is very hard, stony, or in any way refractory, machines become a necessity.

There are several varieties of machines for brick-moulding, but they may be divided into two general classes. These will be just referred to, but to describe even the most common brickmaking machines in detail would be beyond the province of these Notes.

Plastic Clay Machines.—In these the clay is first pugged in the machine. next forced through an opening of about 10 inches by 5 inches in a plastic band, from which the bricks are cut off by means of wires, and then dried or burnt as usual.

Machines of this class are suitable for any plastic clay free from hard lumps or stones, which, if they exist, must be taken out or crushed.

Dry Clay Machines in which the clay is first reduced to powder, then filled by the machine dry into a mould, and subjected to enormous pressure which consolidates it, and forms a well-shaped brick with hard even surfaces.

These machines are well adapted for stony and marly clays, and they are economical inasmuch as they save the expense and time employed in drying, which process is, of course, unnecessary.

Some machines receive the clay in a *semi-dry* condition, and deal with it in the same way as the dry clay machines just mentioned.

PRESSED BRICKS are made by placing ordinary raw bricks when nearly dry in a metal mould or die, and subjecting them to powerful compression under a piston.

This may be done in hand presses, or in larger machines worked by steam power.

Such treatment gives the bricks good faces and arrises, but they require more care in drying and burning, and the oil necessarily used, gives them a glazed surface, which peels off upon exposure to the weather.

Dressed Bricks after being moulded are beaten with a dresser shaped like a small cricket bat, and sometimes tipped with iron. This toughens the clay, corrects its warping, and improves the arrises of the brick.

" *Polished Bricks,* as they are called, are rubbed upon a bench plated with iron, to make their surfaces perfectly even, and are also dressed with a dresser as before described.

" This process is only gone through with the very best bricks, and its cost is such that it is not employed to any very great extent." [1]

Frog.—Most hand-made bricks have a hollow on one of the larger surfaces called the " frog " or " kick."

The object of this is to afford a key for the mortar. The reason why there is not a hollow on both sides in hand-moulded bricks is that the top of the brick is struck off to a flush surface by the moulder.

Bricks should be laid with the hollow uppermost.[2]

Wire-cut bricks (see above) have, of course, no frogs.

In some machine bricks made by pressure there is a frog on each side.

Burning.—Bricks are burnt in " *clamps,*" or in " *kilns,*" according to the practice of the locality.

---

[1] Dobson.   [2] Except in over-sailing courses or when they are to receive a layer of asphalte for a damp course.

CLAMP BURNING.—Clamps are stacks of dried raw bricks dexterously built up over a system of flues roughly formed with burnt bricks, and leading from *live-holes* or *eyes* at which the fire is introduced.

Clamps in a permanent brickfield are made in a very elaborate manner, which is fully described in Dobson *On Brick and Tile Making*.

The following brief notes refer merely to the rough kind of clamp used in a temporary brickfield.

Bricks intended to be burnt in clamps should always be made from clay mixed with ashes or breeze, so that they contain fuel which enables the fire to seize upon them and burn them throughout more readily.

Building the Clamp.—Before building a clamp the ground is first made firm, thoroughly drained, and sometimes formed to a dish-shaped section.

On the ground is placed a foundation of burnt bricks on edge close together to keep out the moisture from the earth, so that the bricks may lean inwards towards the centre of the clamp, and not tend to fall outwards.

[This foundation course is sometimes omitted, especially if the ground has been the site of a recently burning clamp, so that it is pretty dry.]

The clamp itself is commenced by laying two courses of burnt bricks on edge, in parallel lines. In the lowermost course of these two the parallel lines run diagonally across the clamp, with spaces about 2 inches wide between the lines. The uppermost course is laid crossways upon the lower, the lines of bricks being parallel to the end of the clamp, and the bricks close together. The spaces between the bricks are filled in with breeze.

In laying these courses the live-holes[1] about 9 inches square are left right across the clamp, filled up with faggots, and the whole covered with breeze from 5 to 7 inches thick (marked *p* in Fig. 6).

Upon the layer of breeze, *p*, is placed the first course of raw bricks as headers; along the clamp, above this course, another layer of breeze, *o*, from 4 inches to 7 inches thick; and then alternate heading and stretching courses of raw bricks on edge,

---

[1] In many cases there is only one live-hole—near the end of the clamp, as in Fig. 4. If, however, the burning does not keep pace with the building of the clamp live-holes are formed at intervals, so as to start the burning at other points

up to a height of 10 or 12 feet.[1] On the uppermost course of raw bricks a 3-inch layer of breeze is sometimes spread, and the top of the clamp is covered with a casing of burnt bricks two or three courses deep.

In some cases a thin layer of breeze (about $\frac{3}{8}$ inch thick) is spread over the top of each course of raw bricks throughout the clamp. In other cases a little breeze is inserted only at the edges of the courses.

In rough clamps used for temporary works, the ends and sides are smeared with clay. In permanent brickfields they are cased with burnt bricks.

The bricks nearest the outside of the clamp are tilted a little upward by the insertion of bats, so that they may not have a tendency to fall outwards.

Figs. 4 to 7 are taken from measurements of a clamp in one of the Kentish brickfields.

Fig. 4 is a plan of one end of the clamp, showing the live-hole and the arrangement of the bricks in the first course.

Fig. 5 is a plan of part of the second course.

Fig. 6 is an enlarged section of one of the sides of the clamp, showing at *q* how the bricks are packed so as to give a batter to the outer wall of burnt brick.

o is a layer of breeze 4 to 7 inches thick, *p* a layer 5 to 7 inches thick.[1]

Fig. 7 is a longitudinal section of the clamp—which may be of any length.

Time of Burning.—The operation of burning takes from two to six weeks. A good deal depends, however, upon the situation and direction of wind. The nearer the live-holes are together the quicker the burning.

The completion of the burning process is indicated by the settling down of the top of the clamp, caused by the shrinking of the bricks as they become burnt.

Quality of Bricks.—The bricks taken from a clamp will be found very unequal in quality. Those from near the eyes are often fused together, or misshapen, forming *burrs.* Those near the outside are underburnt and soft, and are called *Place bricks.*

Again, much depends upon the proportion of breeze used in the clamp. Too much will cause the bricks to be weak and porous.

---

[1] In some clamps a third layer of breeze some 2 or 3 inches thick is introduced between *o* and the mass of raw bricks above it.

ILLUSTRATIONS OF PARTS OF A CLAMP.

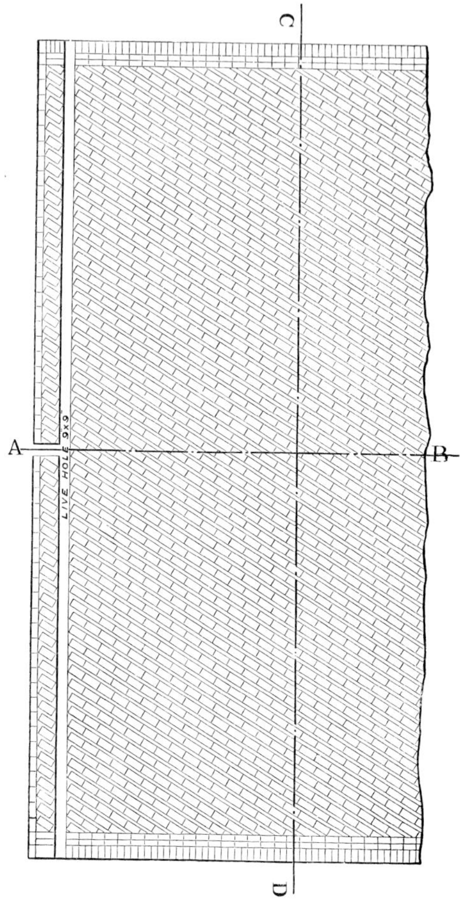

C

A———————————————————B

LIVE HOLE 9x9

D

Fig. 4.

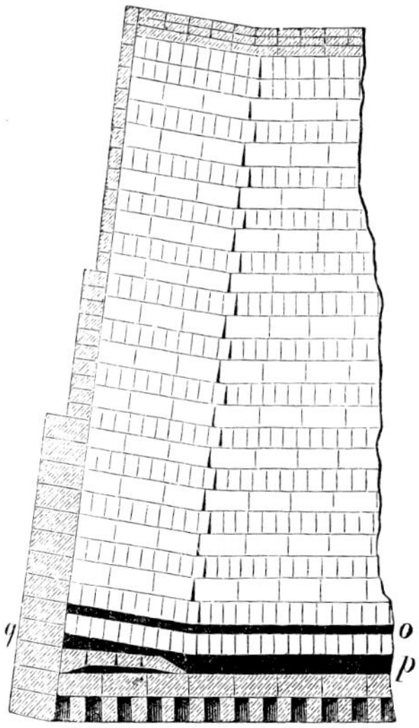

Section of part of Clamp on line C D
Fig. 6.

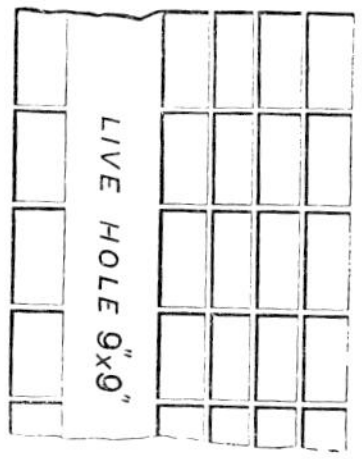

q   o
p

LIVE HOLE 9x9"

Plan of arrangement of
Second Course.
Fig. 5.

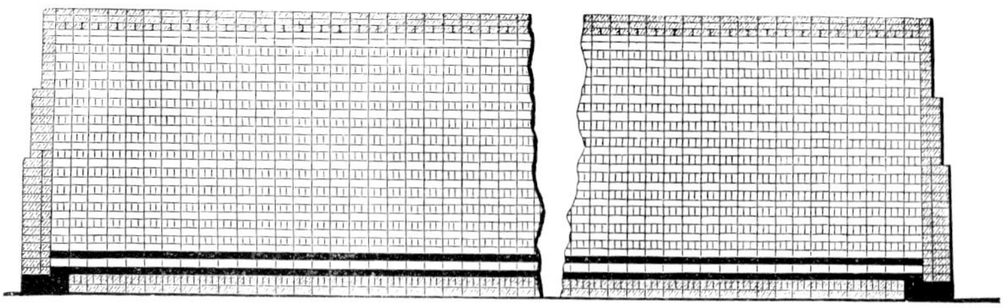

Broken Section on line A B.

Burnt Brick shown thus

Fig. 7.

The quantity of breeze required will vary a good deal according to the quality of the clay, but the following may be taken as an approximation per 1000 bricks :—

Mixed with the clay—$\frac{1}{2}$ chaldron ; in the clamp—$\frac{1}{4}$ chaldron. Besides about 2 to 3 cwts. of coal in each fire-hole, that is about $\frac{1}{2}$ cwt. per 1000 bricks.

KILN BURNING.—There are several descriptions of kilns used for burning bricks, but it will only be necessary to refer to those that are likely to be used by the engineer or builder in establishing a temporary brickfield to supply bricks for special works in progress.

Several forms of kiln, used chiefly in permanent brickmaking yards, may be excluded, or very lightly touched upon, as being interesting to the brick-manufacturer rather than to the engineer.

SCOTCH KILN.[1]—The form of kiln most commonly used in the

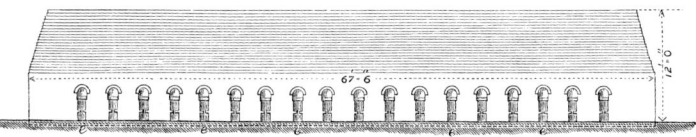

Fig. 8. *Elevation.*

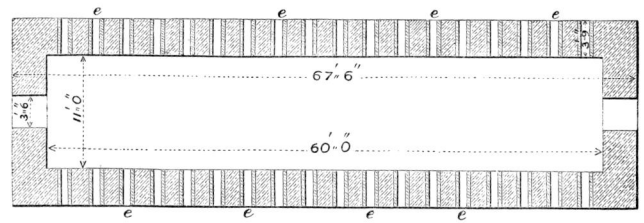

Fig. 9. *Ground-plan.*

United Kingdom for making a moderate supply of bricks is known as the *Scotch kiln.*

The Scotch kiln is a rough rectangular building, open at the

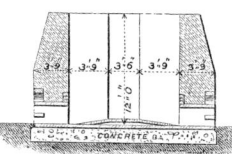

Fig. 10. *Cross Section.*

top, and having wide doorways at the ends. The side walls are built of old bricks set in clay, and in them are several openings called fire-holes, or " eyes," (*e e*, Fig. 9), built in firebricks and fireclay, opposite one another.

The dried raw bricks are arranged in the kilns so as to form flues connecting the fire-holes or eyes, and

[1] Sometimes called the *Dutch Kiln.*

they are packed so as to leave spaces between the bricks from bottom to top, through which the fire can find its way to and around every brick.

After the dried bricks are " *crowded,*" *i.e.* filled into the kiln, the ends are built up, and plastered over with clay.

At first the fires are kept low, simply to drive off the moisture.

After about three days the steam ceases to rise ; the fires are allowed to burn up briskly ; the draught is regulated by partially stopping the fire-holes with clay, and by covering the top of the kiln with old bricks, boards, or earth, so as to keep in the heat.

In from 48 to 60 hours the bricks will be sufficiently burnt, and they will be found to have settled down.

The fire-holes are then completely stopped with clay, all air excluded, and the kiln is allowed to cool very gradually.

Fuel.—About a half-ton of soft coal is required for burning 1000 bricks. The exact quantity depends upon the nature of the clay, the quality of the fuel, and the skill in setting the kiln.

Size of Kiln.—A convenient size for a kiln is about 60 feet by 11 feet internal dimensions, and 12 feet high. This will contain about 80,000 bricks. The fire-holes are 3 feet apart. These kilns are often made 12 feet wide, but 11 feet is enough to burn through properly.

Time of Burning and Produce.—A kiln takes on an average a week to burn, and, including the time required for crowding and emptying, it may be burnt about once every three weeks, or ten times in an average season. This will produce about 800,000 bricks, that is about as many as would be turned out by two moulders in full work.

The bricks in the centre of the kiln are generally well burnt. Those at the bottom are likely to be very hard, some clinkered. Those at the top are often badly burnt, soft, and unfit for exterior work.

A modification of the Scotch Kiln is used in Essex and Suffolk. The floor is made with openings like lattice work, through which the heat ascends from arched furnaces underneath.[1]

Comparative Advantages of Kiln and Clamp Burning.—The following advantages are claimed for kiln-burning over clamp-burning :—

[1] Dobson.

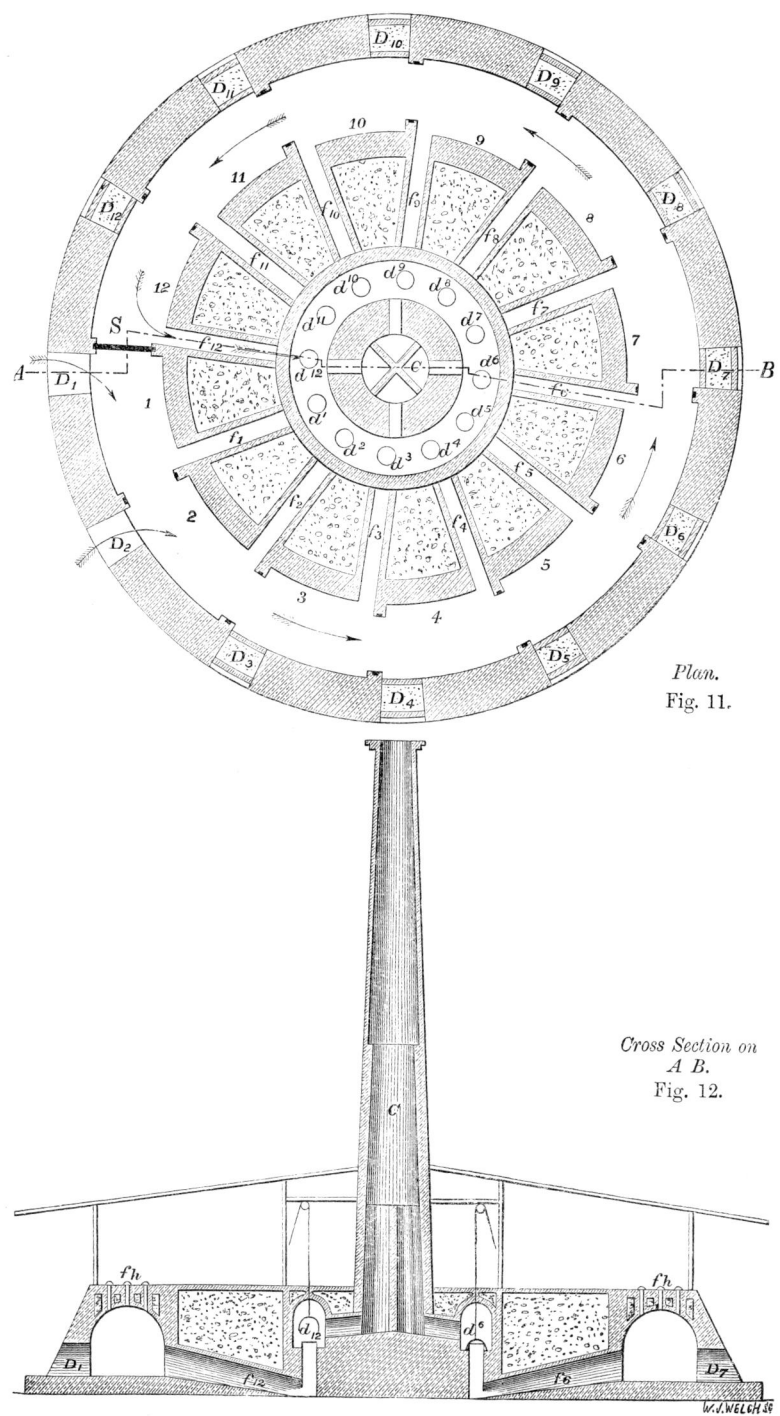

Plan.
Fig. 11.

Cross Section on
A B.
Fig. 12.

W.J.WELCH SC

1. In kilns the bricks are nearly all turned out of the same quality, being equally burnt, and are more uniform in colour; whereas the bricks produced from different parts of the same clamp vary greatly in quality, and many of them are almost useless.

2. For kiln-burning the bricks need not stand so long on the hacks to dry, because the fires in the kiln can be regulated so as to drive off the moisture gradually.

This prevents warping, which often occurs with bricks clamped in too moist a condition.

3. Though the kiln-burning requires more fuel, yet the speed with which the crowding, burning, and discharging take place, the absence of waste, and the superior quality of the bricks produced, render it preferable even from an economical point of view.

HOFFMANN'S KILN is used chiefly in brick-manufactories on a large scale, where a great number of bricks is required annually, and a continuous supply has to be kept up ; but it is also employed in making bricks for very extensive works where several millions of bricks are required.

This kiln is circular in plan.

It consists of an annular tunnel-shaped chamber (marked 1 to 12, Fig. 11) of brickwork lined with firebricks.

At certain equidistant points there are grooves formed in the sides of this annular chamber, so that it can be screened across by an iron shutter temporarily inserted (see S, Fig. 11) at any of these points.

The portions of the kiln between these points are called " chambers " or " compartments."

They are marked 1 to 12 in Fig. 11 ; each of them is connected by means of a flue, *f*, with a high central chimney, C.

The number of compartments varies in different kilns from 8 to 24, but a 12-chambered kiln is found in practice to be the most convenient for the purpose of the engineer in providing a temporary brickfield to supply special works.

Each flue can be cut off from the chimney by lowering upon it a cast-iron damper, *d*.

Each compartment has a doorway leading outside the kiln, marked D in the figure. This doorway can be filled up by dry brick walls with sand packed in between them.

Fire-holes with covers are provided at *fh, fh*, by which fuel in the shape of powdered coal may be supplied to the bricks.

The object of these arrangements is to utilise all the heat produced by the fuel, and thus to save expense in firing.

Thus in a 12-chambered kiln on a certain day the chambers might be in use as shown in Fig. 11.

The annular chamber is closed by an iron shutter at S, between compartments 12 and 1.

The flue from No. 12 is placed in communication with the chimney by raising the damper, d_{12} ; all the other dampers are closed

The state of things is then as follows ---

Chamber No. 1 is being filled with raw bricks.

,, No. 2 is being emptied of cold burnt bricks.

,, Nos. 3, 4, 5, 6 contain bricks which have been burnt and are cooling.

,, Nos. 7, 8 contain bricks which are being burnt, fuel being supplied to them through the fire-holes, fh, fh.

,, Nos. 9, 10, 11, 12 are drying and becoming very hot under the influence of the heat from Nos. 7, 8.

The cold air is entering through the open doors of 1 and 2, and proceeds in the direction shown by the arrows. Becoming partly heated by passing over the cooling bricks in 3, 4, 5, 6, it enters 7 and 8, whence it goes on in a highly heated state to dry and heat the raw bricks in Nos. 9, 10, 11, 12. Meeting the screen between 12 and 1, it passes through the flue f_{12}, goes up at d_{12} into the chimney.

The next day

No. 2 would be filled with raw bricks.

No. 3 would be emptied of cold burnt bricks.

Nos. 4, 5, 6, 7 would contain burnt bricks cooling.

Nos. 8, 9 would contain bricks burning.

Nos. 10, 11, 12, 1 would be drying.

The screen would be between 1 and 2, and the smoke, etc., would escape up the chimney through the flue f_1, the damper d_1 being raised, and all the other dampers down. The doors D_2 D_3 would be open ; all the other doors shut.

A similar change is made each day, so that the kiln burns continuously, never being allowed to go out except for repairs.

Size and Produce of Kiln.—Each chamber, if made about 36 feet long, 15 feet mean width, and 8 feet high, will hold 25,000 bricks.

$12 \times 25,000 = 300,000$ bricks may therefore be burnt in the whole kiln every twelve days, or (as the bricks are not filled in or unloaded on Sundays) say once a fortnight.

Such a kiln will therefore burn some four or five million bricks per annum.

Disadvantages.—The great drawback to the use of Hoffmann's kiln is the first cost of its construction.

It is necessary to burn some six to ten million bricks before the saving in fuel has compensated for the cost of building the kiln.

It is therefore not adapted for burning bricks for special works unless they are on a very large scale.

It is, however, the most economical form of kiln for permanent brick-making works turning out a large annual supply of bricks.

Advantages.—The advantages of Hoffmann's kiln are—

1. *Economy of Fuel.*—In Scotch and similar kilns a great deal of the heat from the burning fuel, and also all the heat from the bricks when cooling, passes away and is wasted ; by this kiln they are both preserved, and utilised in drying and heating the bricks before burning.

The result is that only 2 to 3 cwt. of *coal dust* and *slack*, costing 4d. or 5d., are required per 1000 bricks, instead of half a ton of *good* coal, costing 4s. or 5s.

The prices given are those quoted by the patentees, and vary of course in different localities.

2. There being no rapid draught, the hot gases fill the chambers, and the bricks in all parts of the kiln are burnt equally well.

3. The bricks can at any time be examined, and the burning regulated, through the fire-holes.

4. As the fuel is thrown into the chambers after they are at a high heat, wood, turf, or coal can be used.

5. The charging and emptying of the kiln goes on continuously and without interruption, so that a regular supply of bricks can be maintained.

6. The height of the chambers, only 8 or 9 feet, is such that there is no danger of crushing the lower courses when the bricks are raw or at a high temperature.

7. The bricks are not liable to injury by sudden changes of temperature.

8. There is no smoke, as the combustion of the fuel is perfect.

Modifications of Hoffmann's Kilns are used in different parts of the country. In many of them the chambers are differently arranged. They are often placed in a straight line, and the waste heat from each is utilised in a somewhat similar manner. Among these may be mentioned Lancaster's, Morand's. Clayton's, Pollock and Mitchell's, and Chamberlain and Wedekind's kilns.

Bull's patent Semi-continuous Kiln is said to utilise the waste heat and thoroughly to consume the fuel, without expense in construction of a very large kiln. The expenditure of coal is stated to be about 3 cwts. per 1000 bricks.

The bricks are packed in a somewhat elaborate manner. The whole construction is fully explained in *Engineering* of the 22d October 1875.

CUPOLAS, or, as they are locally called, *ovens*, are small circular domed kilns. They are used in Staffordshire for the celebrated bricks of that district (see p. 108). They are sometimes used in other localities, and also for burning firebricks.

OTHER FORMS OF KILN.—An immense number of different kilns are in use for burning bricks and tiles of special descriptions. New forms of kiln are invented nearly every week. It would be impossible, for want of space, to describe even a few of these, and such a description, if given, would be interesting to the manufacturer rather than to the engineer or builder.

Classification of Bricks.—Building bricks may, for the purposes of the engineer or architect, be divided into three classes.

1. *Cutters* or *Rubbers, i.e.* bricks intended to be cut or rubbed to some shape different from that in which they were originally moulded.

2. *Ordinary Bricks,* intended to be used without cutting except where required to form the bond.

The best of these are selected for fronts, and are termed facing bricks.

Specially hard varieties are used for coping, also for paving, quoins, and other positions where they will be subjected to unusual wear.

3. *Underburnt and misshapen Bricks,* only fit for inside work.

Of each of these classes there are in most brickfields several varieties, varying in quality according to circumstances. Their general characteristics are, however, as follow :—

CUTTERS or RUBBERS are purposely made sufficiently soft to be cut approximately to the shape required with a trowel, and then rubbed to a smooth face and to an accurate shape.

To ensure this they are made of washed earth carefully freed from lumps of all kinds, and uniform in composition throughout its mass.

The best rubbers are burnt to a point a little short of vitrification.

Inferior kinds are often stinted in firing; the cohesion between the particles is small, and they are easily destroyed by rain or frost.

For the sake of durability it is better to avoid rubbers in all exposed work, and to use " purpose-made" bricks moulded to the shape required and thoroughly well burnt.

This is often done in good work.

The characteristics of good rubbers are mentioned at page 112.

ORDINARY BUILDING BRICKS.—The second class of bricks includes the bulk of those required for building. The qualities and characteristics of these vary, not only in different localities, but also in the same brickyard (see p. 106).

Such bricks are made either from washed earth or malm, from partly washed earth, or from earth which has merely been tempered, not washed at all.

They should be hard and well shaped, those most uniform in colour being selected for facing, and the whole of the remainder being fit to use for good sound work.

UNDERBURNT AND MISSHAPEN BRICKS.—The underburnt bricks of the third class are generally known as *grizzle* or *place* bricks, in some places as *samel* bricks.

They are always soft inside, and sometimes outside also, are very liable to decay, and unfit for good work.

They are, however, often used for the inside of walls.

Names of different Varieties of Bricks.—As before mentioned, the names given to different classes of bricks vary in different districts, and even in different brickfields of the same district.

CLASSIFICATION OF CLAMP-BURNT BRICKS.—The subjoined list of the names for clamp-burnt bricks, adopted in a Kentish brickfield supplying the London market, may be taken as a specimen.

Following it is a description of some of the more important varieties.

The bricks are divided generally into three classes—*Malms, Washed,* and *Common*—according to the manner in which the earth for them is prepared (see p. 92). For the third or *Common* class the earth is not washed at all. All three classes are moulded and burned in exactly the same manner, and are then further sorted into a number of varieties according to the manner in which they have been affected by the fire.

The classes are subdivided as follows :—

Price per Thousand at Brickfield.

Malms	Cutters	140/.
	Best Seconds	70/.
	Mean do.	80/.
	Brown Facing Paviors	55/.
	Hard Paviors	50/.
	Shippers	32/6.
	Bright Stocks	37/6.
	Grizzle	19/.
	Place	16/.
Washed	Shippers	28/6.
	Stocks	20/.
	Hard Stocks	20/.
	Grizzles	17/.
	Place	13/.
Common	Shippers	28/.
	Stocks	24/.
	Grizzles	16/.
	Rough Stocks	16/.
	Place	12/.

The prices above mentioned serve only to show the relative value of the different classes of bricks. The actual market rates vary of course from time to time, and depend upon seasons, etc.

Of the above classes *cutters* have already been described.

Seconds are similar to cutters, but with some slight unevenness of colour.

Bright Fronts are the corresponding quality from " washed " earth.

Facing Paviors are hard-burnt malm bricks of good shape and colour used for facing superior walls.

Hard Paviors are rather more burned, and slightly blemished in colour. They are used for superior paving, coping, etc.

Shippers are sound, hard-burned bricks, not quite perfect in form. They are chiefly exported, ships taking them as ballast.

Stocks are hard-burned bricks, fairly sound, but more blemished than shippers. They are used for the principal mass of ordinary good work.

Hard Stocks are overburnt bricks, sound, but considerably blemished both in form and colour. They are used for ordinary pavings, for footings, and in the body of thick walls.

Grizzle and *Place* bricks are underburnt. They are very weak, and two out of five " common " or unwashed place bricks are allowed to be bats, the stones left in the unwashed earth making them very liable to breakage.

These two last-mentioned descriptions are only used for inferior or temporary work, and are commonly covered with cement rendering to protect them from the weather when intended to be permanent.

Chuffs are bricks upon which rain has fallen while they were hot, making them full of cracks, and perfectly useless.

Burrs are lumps of bricks vitrified and run together. They are used for rough walling, artificial rock-work. etc

Bats are broken bricks.

Of the above varieties those from " common " or unwashed clay are hardly ever quite perfect in form on account of the stones left in the earth, which make them shrink unequally, and become distorted in burning.

Bricks from " washed " clay suffer in the same way to a less degree.

CLASSIFICATION OF KILN-BURNT BRICKS.—Kiln-burnt bricks are generally pretty equally burnt, and are classed chiefly according to the process by which they are made.

Thus in one yard the classification is as follows :—

> Patent bricks.
> Common hand-made.
> Copper moulds.
> Pressed bricks.
> Dressed pressed bricks.

In another yard the classes are

> Best white pressed.
> Second do. do.
> Pink do. do.
> White wire cut.
> Second do. do.
> Pink do. do.

The Burham Company's bricks are thus classified in their circulars :—

> No. 1. Pressed Gault (Facing).
>   2. Do. (Mingled).
>   3. Do. (Paviors).
> No. 1. Wire Cut (Facing).
>   2. Do. (as they rise from kiln).
>   3. Do. (Mingled or discoloured).

Varieties of Bricks in the Market.—The bricks used in ordinary buildings generally are, or should be, the best that are made in the neighbourhood.

Some descriptions of bricks, however, are universally known, and are used even outside the locality in which they are made, either for special purposes, or in buildings of such importance as to justify incurring the expense of carriage.

A few of the more important of these varieties may now be noticed.

White Bricks.—The best materials from which to make white bricks are a refractory clay, which will naturally burn to pale yellow or white, and a fine white or yellow sand, which vitrifies slightly under a strong heat.

In the absence of such material, however, every clay which does not contain more than 6 per cent of iron will burn into a white brick, provided it is strong enough to stand a sufficient quantity of chalk mixed with it. In the case of very refractory clays the mixture with a large proportion of chalk will render the resulting brick friable.

The processes usually gone through in the manufacture of white bricks do not differ very materially from those applied to other bricks.

"The best mode of manufacture is to grind the clay dry, mix it tho roughly with sand while dry, and then through a press." [1]

White bricks are frequently burnt in close kilns, carefully protected from smoky flames and soot, thoroughly burned in a dead heat, and allowed to cool down ; gradually, or the faces will be full of cracks.

The clays from which white bricks are made are generally heavy, and they are in such case lightened by being made hollow or perforated.

Green stains are often noticed on the surface of white bricks if they are underburnt.

These stains can generally be rubbed off when the brick is dry ; if they reappear they can be permanently removed " by mixing up a wash of clay and sand of which the brick was made with sulphate of copper, painting over the brick with it, and leaving it till it is perfectly dry, and then rubbing it off with a brush." [1]

White bricks may be procured from several parts of England. Some of the best come from Suffolk, Essex, Arsley, Ewill in the district of the Medway ; from Dorsetshire (Beaulieu bricks and others) ; from the London brickfields ; from Exbury and Cowes. Others are made in Cambridgeshire, Devonshire, Lincolnshire, and the Midland counties.

A few of the best known varieties will be further noticed.

GAULT BRICKS are made from a band of bluish tenacious clay which lies between the Upper and Lower Greensand formations.

This clay in its natural state contains sufficient chalk to flux the mass, and to give the brick a white colour.

The bricks made from this clay are of very good quality ; extremely hard throughout, very durable, but difficult to cut.

They are generally white, but the lower qualities have a pink tinge caused by irregularities in burning.

Bricks made from Gault clay are generally very heavy. To remedy this a large frog is sometimes formed in the brick, or it is perforated throughout its thickness.

Bricks of this description are manufactured by the Burham Company at Burham, near Rochester, and at Aylesford, near Maidstone ; also at Folkestone, near Hitchin, and at other places.

Suffolk white Bricks are also made from the Gault clay.

They contain a very large proportion of sand which makes them useful for rubbers.

They are rather soft for ordinary building purposes, but harden in time, which is attributed to the silicic acid in the clay acting upon the chalk so as to form some of it into a silicate of lime.

Beaulieu Bricks, of a light straw colour, are made from clay dug upon the Beaulieu river, near Southampton.

Ballingdon Bricks, made by Beart's process near Sudbury, in Suffolk, are much used for facework.

BEART'S PATENT BRICKS are made at Arsley, near Hitchin, from the Gault clay.

There are different classes. " White rubbers, hand-made, moulded, solid brick, equal to the best Suffolks. No. 1, best selected white facing brick (pierced) and ordinary. These two are of uniform colour, hard and well burnt, and used extensively for facings. No. 2, mingled red and pink, vary

---

[1] *Building News*, Sept. and Oct. 1874.

from the above only in colour, and are equal in every respect to the best made stock bricks." [1]

The clay contains lime, and requires to be burnt with great care, or the lime will remain in a quick state, and slake after the brick is in use.

Staffordshire Blue Bricks are made from the clays and marls of that county, which contain from 7 to 10 per cent of oxide of iron.

They are burnt in circular ovens with domed tops, being raised to a very high temperature, which causes the peroxide or red oxide to be converted into the protoxide or black oxide of iron.

These bricks are generally of a dark-blue or nearly black colour, with smooth glassy surfaces. They are very durable, impervious to water, and will resist enormous pressure.

Bricks of this description are extensively used throughout the country for paving, coping, channels, and other special purposes in which great hardness and durability are required.

For building ordinary strong work the second-class Staffordshire bricks are more suitable than the first quality, as the former have rougher surfaces to which the mortar adheres more firmly.

An inferior class of these bricks is made by the use of a surface wash of iron. These look well for a time, but the colour does not wear well.

Dust Bricks are blue bricks, for which coal-dust is used in moulding instead of sand. They have glossy surfaces, are very hard, and are used for paving. [1]

Red and *Drab* coloured bricks are also made in Staffordshire. The former are used for building, and the latter chiefly as a fire-brick, where intense heat is not required. [2]

Tipton Blue Bricks are Staffordshire blue bricks from the neighbourhood of the town after which they are named.

Black Bricks are obtained from Cowbridge in South Wales, from Maidenhead in Berkshire, and from other places.

Some inferior black bricks are made with a mixture of soot, and are weak and almost useless.

FAREHAM RED BRICKS are made from a moderately plastic clay, which is found in very deep beds around the town of Fareham, and in other places in the neighbourhood.

They are dressed or batted (as described at p. 95) when partially dry, and thus brought to a very true surface. They are also carefully burnt in small oven kilns holding from 20,000 to 30,000 each.

These bricks are of a fine deep-red colour, and have been much used in London for superior buildings.

The facework of St. Thomas's Hospital is of Fareham bricks, and many have been used in the new Law Courts.

Sometimes these bricks are rubbed so as to obtain very fine surfaces and thin mortar joints, but this removal of the outer skin is bad, as it tends to make the brick decay quickly under atmospheric influences.

NOTTINGHAM PATENT BRICKS are made by the dry clay process, the clay being ground and subjected to pressure of about 200 tons on the brick in moulding.

They are very close in texture, and have good surfaces and arrises, but they appear to be deficient in toughness, and do not " ring" properly or weather well.

They are of a dull red colour. Many of them are burnt in Hoffmann's

---

[1] Gwilt.   [2] Dobson.

kilns, in which case the ends are generally of a yellowish shade. This is owing to the ends being exposed to the fire, whereas the other parts of the brick are protected.

Sometimes bricks are purposely packed on end, so as to protect the ends from fire, make them red, so as to afford headers of an uniform colour.

These bricks were used for part of the St. Pancras Station.

LANCASHIRE RED PRESSED FACING BRICKS are made by Platt's patent brickmaking machine.

Dutch Clinkers are small bricks, well burnt, very hard, vitrified through out, and sometimes warped.

They are used almost entirely for paving.

Adamantine Clinkers are similar bricks, harder, denser, and heavier, of a fine pink-white colour and smooth surface.

They are sometimes chamfered on the edge so as to give a firmer foothold when used for paving.

Terro-metallic Clinkers are bricks of the same size and shape, made from a clay which is burnt very hard to a dark-brown or nearly black colour.

Enamelled Bricks have a white or light yellow glazed surface like that of china.

This is produced by a thin coating of white material over the brick, which in inferior descriptions is apt to peel off.

Bricks of this kind are much used for the sake of cleanliness in lavatories, urinals, butchers' shops, dairies, etc. ; also in order to obtain reflected light, as in some of the underground railway stations.

Salted Bricks have a thin glaze over their surfaces, produced by throwing salt into the fire during the burning process.

Moulded Bricks are produced in every variety of pattern, from simple sections like those of cornice, plinth, and string-course bricks, already mentioned, up to the most elaborate decorated blocks of different forms, such as voussoirs for arches, diaper patterns for walls, panels, string-courses, etc.

The simpler patterns are made in moulds furnished with shifting sides and ends on which the pattern is raised or sunk. These can be screened up against the soft clay, and then released so as to liberate the moulded brick.

Sometimes the pattern is formed on the stock-board, or on a plaster cast which takes its place.

In the more elaborate patterns iron moulds are used, which are opened and closed by simple machinery.

Pether's Ornamental Bricks are made by the Burham Company from Gault clay forced into a hinged iron mould.

They can be made to almost any design, however elaborate, and afford a cheap and very durable means of decoration.

Pallette Bricks " rebated on edge so as to hold a $1\frac{1}{2}$-inch fillet securely in the wall, splayed from $\frac{7}{8}$ inch at one edge to $\frac{1}{2}$ inch at the other, have been occasionally used of late but are not recommended, as the advantage gained is not to be compared to the extra labour and expense involved." [1]

Concrete Bricks should hardly be noticed in this chapter, as they are not made of clay, and they do not require burning.

Bodmer's Bricks consist of a species of fine concrete, the constituent parts of which vary, some being of about $\frac{1}{6}$ to $\frac{1}{8}$ of its weight of sand with selenitic lime or cement, others of black furnace slag mixed with about $\frac{1}{8}$ of its weight of lime or cement according to quality.

---

[1] Seddon.

The ingredients are filled into moulds, and subjected to considerable pressure which binds the particles together.

The moulded bricks are then left to ripen and harden out of doors for a period which varies with the setting properties of the lime or cement used.

The resulting bricks are hard and dense, with good arrises and surfaces, and they weigh about 58 cwts. a thousand.

The cost of labour for making these bricks is said to be from 3s. to 3s. 6d per thousand.[1]

Wood's Patent Concrete Bricks are similar to those just described. They are made at Middlesborough from slag reduced by agitation in water to the state of sand. The slag sand is ground and mixed with $\frac{1}{7}$ its bulk of lime. The mixture is forced into moulds under a pressure of about half a ton per square inch. The bricks are dried in the air, and are then ready for use.

These bricks may be made with ordinary sand or crushed stone instead of slag sand.

Slag Bricks are made by running molten slag into iron moulds. The blocks are removed while the interior is still molten, and then annealed in ovens.

Characteristics of good Bricks.—*Freedom from Flaws or Lumps.* —Good building bricks should be sound, free from cracks and flaws, also from stones, or lumps of any kind.

Lumps of lime, however small, are specially dangerous; they slake when the brick is exposed to moisture, and split it to pieces.

A small proportion of lime finely divided and disseminated throughout the mass is an advantage, as it affords the flux necessary for the proper vitrification of the brick.

In examining a brick, lumps of any kind should be regarded with suspicion and tested.

Shape and Surface.—In order to ensure good brickwork the bricks must be regular in shape and uniform in size.

Their arrises (or edges) should be square, straight, and sharply defined.

Their surfaces should be even, not hollow; not too smooth, or the mortar will not adhere to them.

Absorption.—The proportion of water that a brick will absorb is a very good indication of its quality.

Insufficiently burnt bricks absorb a large proportion and are sure to decay in a short time.

It is generally stated in books that a good brick should not absorb more than $\frac{1}{15}$ of its weight of water.

The absorption of average bricks is, however, generally about $\frac{1}{6}$ of their weights, and it is only very highly vitrified bricks that take up so little as $\frac{1}{13}$ or $\frac{1}{15}$. (See p. 114.)

---

[1] Spons' *Illustrated Price Book.*

Texture.—Good bricks should be hard, and burnt so thoroughly that there is incipient vitrification all through the brick.

This may be seen by examining a fractured surface, or the surface may be tested with a knife, which will make hardly any impression upon it unless the brick is underburnt.

A brick thoroughly burnt and sound will give out a ringing sound when struck against another. A dull sound indicates a soft or shaky brick.

A well-burnt brick will be very hard, and possesses great power of resistance to compression. (See p. 115.)

CHARACTERISTICS OF GOOD RUBBERS.—A really first-class rubber (see p. 105) will not be easily scored by a knife even in the centre, and the finger will make no impression upon it.

Such a brick will be of uniform texture, compact, regular in colour and size, free from flaws of any description.

Method of distinguishing Clamp-burnt, Kiln-burnt, and Machine-made Bricks.—In clamp-burnt bricks the traces of the breeze mixed with the clay can generally be seen.

Kiln-burnt bricks very often have light and dark stripes upon their sides, caused by their being arranged while burning with intervals between them. Where the brick is exposed it is burnt to a light colour ; where it rests upon or against other bricks it is dark.

In some cases care is taken to prevent this, and the best kiln-burnt bricks are of an uniform colour.

Machine-made bricks may generally easily be distinguished, if wire-cut, by the marks of the wires ; if moulded, by the peculiar form of the mould, letters on the surface, etc., or sometimes by having a frog on both sides.

In many cases the marks made by pronged forks, used for hacking the bricks, may be seen on their sides.

Size and Weight of Bricks.—Before the year 1839 a duty was paid upon bricks; their size was then practically fixed by Act of Parliament, and it has since remained materially unaltered.

Ordinary bricks in the neighbourhood of London are about $8\frac{3}{4}$ inches long, $4\frac{1}{4}$ wide, and $2\frac{1}{2}$ inches thick, and weigh about 7 lbs. each. In different parts of the country the size and weight vary slightly ; in the north of England and in Scotland they are larger and heavier. (See p. 113.) A very large brick is inconvenient for an ordinary man to grasp, and a heavy brick fatigues the bricklayer, who has to lift it when wet and lay it with one hand.

In order to obtain good brickwork it is important that the length of each brick should just exceed twice its breadth by the thickness of a mortar joint.[1]

[1] For the R.I.B.A. standard size of bricks, see the R.I.B.A. *Kalendar*, 1903-04.

The following TABLE gives the SIZE and WEIGHT of some of the best-known varieties of BRICKS in this country. See also Tables, pages 114, 115.

	SIZE IN INCHES.			WEIGHT. LBS.	WEIGHT PER 1000 IN CWTS.
London Stock	$8\frac{3}{4}$	$4\frac{1}{4}$	$2\frac{3}{4}$	6·81	$60\frac{3}{4}$
Red Kiln	$8\frac{3}{4}$	$4\frac{1}{4}$	$2\frac{3}{4}$	7·0	63
Fareham Reds	8·5	4·15	2·6	6·3	56·2 G
Do. Rubbers	10·9	4·8	2·9	8·8	78·5 G
Catty Brook Pressed Brick (near Bristol)	$9\frac{1}{4}$	4	3	9·5	85
Bridgewater Red Brick	8·75	4·3	2·75		
Lancashire Red Pressed Facing Brick	9	$4\frac{1}{2}$	3	8·9	80
Pressed Brick from Leeds	9·5	4·5	3·5	10	89
Scotch Brick from Sandyfauld, near Glasgow	9·5	4·5	3·5	9·7	86·6
Bricks made from Blaize near Glasgow	9	4·3	3·4	8·6	77
Scotch Brick from Elgin (used for partitions)	12	6	3		
Irish Brick from Athy	8	$3\frac{3}{4}$	$2\frac{3}{4}$		
Burham Wire Cut	8·6	4·0	2·6	5·4	58·2 G
Do. Pressed	8·75	4·2	2·7	6·1	54·5 G
Suffolk Brimstone	9	4·6	2·6	6·8	60·7 G
Do. White	9·2	4·3	2·6	6·3	56·2 G
Staffordshire thin Paving	9	$4\frac{1}{2}$	3	8·9	80
Do. do.	9	$4\frac{1}{2}$	2	6·1	55
Staffordshire Brick-on-edge, for edge paving	9	3	$3\frac{1}{2}$	7·8	70
Tipton Blue	9	$4\frac{1}{2}$	3	10	89
Adamantine Clinker	6	$2\frac{1}{2}$	$1\frac{3}{4}$	2	18
Dutch Clinker	$6\frac{1}{4}$	3	$1\frac{1}{2}$	1·55	14

The Figures marked G are from Grant's Experiments, *Proceedings Inst. Civ. Engineers*, vol. xxv. p. 35.

Tests for Bricks.—The best method of testing bricks is to see if they ring when struck together; to ascertain their hardness, by throwing them on to the ground, or by striking them against other bricks.

The fractured surface should also be examined in order to ascertain if it exhibits the characteristics mentioned at page 112.

Brard's test is sometimes used for bricks, but is not of much practical benefit, for the reasons stated at page 11.

The amount of water absorbed by bricks is to a certain extent an indication of their quality, and their resistance to compression, either singly or when built into brickwork, will show whether they are strong enough for the purpose required.

The following Table shows the weight and absorption of several different classes of bricks. The results marked L are from experiments made by Mr. Baldwin Latham.[1] The remainder are the results of experiments made by a friend of the present writer's.

TABLE SHOWING ABSORPTION OF WATER BY DIFFERENT VARIETIES OF BRICKS.

DESCRIPTION OF BRICK.	WEIGHTS WHEN DRY.		PERCENTAGE OF WATER ABSORBED.
	LB.	OZ.	
Malm Cutters . . .	4	15	22
Malm Best Seconds . .	5	$1\frac{1}{2}$	20
Malm Brown Facing Paviors .	5	$0\frac{1}{2}$	17
Do. Hard Paviors . . .	4	13	$9\frac{1}{2}$
Washed Bright Yellow Fronts .	5	1	20
Malm Shippers . . .	5	$1\frac{1}{4}$	$8\frac{1}{3}$
Malm Bright Stocks . .	4	$13\frac{1}{2}$	22
Washed Do. . . .	5	$0\frac{1}{2}$	16
Common Shippers . . .	5	$0\frac{1}{2}$	9
Common Gray Stocks . .	5	0	$10\frac{1}{2}$
Do. Hard Do. . .	5	$0\frac{1}{2}$	$7\frac{1}{2}$
Malm Grizzles . . .	4	$13\frac{1}{2}$	22
Do. Place	5	$0\frac{1}{2}$	21
Common Place . . .	5	$0\frac{1}{2}$	20
Washed Shippers . . .	5	2	10
Do. Hard Stocks . .	4	$15\frac{1}{2}$	$4\frac{1}{4}$
Do. Grizzle . . .	5	0	21
Common Grizzle . . .	5	1	18
Washed Place . . .	5	0	21
Staffordshire Dressed Blue . .	9	5	2·3 L
Do. Pressed Blue . .	8	11	3·7 L
Do. Common Blue .	9	0	6·5 L
Do. Bastard . .	9	8	11·8 L
Machine-made Red . . .	9	14	9·9 L
Do. from Leeds . .	10	0	10·0
Wire-cut White Gault . .	6	3	19·0 L
Pressed Gault . . .	5	12	19·5 L
Brown Glazed Brick . .	8	6	8·6 L

Strength of Bricks.—In practice bricks are subjected to compression, and sometimes to transverse stress, but not to tension.

The compressive stress brought upon evenly-bedded bricks is generally far less than they are able to bear.

In some cases, however, as in arches and retaining walls, the stress may be concentrated upon a small portion of the brick, or the same effect may be produced by the bed of the brick being uneven.

Such concentrated stresses are apt to crack the portion of the brick upon which they act.

[1] Latham's *Sanitary Engineering.*

TABLE OF THE RESISTANCE OF BRICKS TO COMPRESSION.

Description of Brick.	Dimensions of Specimen.			Average load per Sq. Ft. at which Brick first cracked.	Crushing Load Per Square Foot.			Number of Specimens Tested.	Authority.	Remarks.
	Length. Inches.	Breadth. Inches.	Height. Inches.	Tons.	Min. Tons.	Max. Tons.	Average. Tons.			
Gault . . .	8·8	..	..	51	108	179	146	5	G	
Do. . . .	..	..	..	47	113	139	125	5	G	Frogs filled with cement.
Burham Gault .	9·0	4·5	2·7	26	50	71	61	5	G	
Do. do. .	..	..	..	58	96	118	108	5	G	Do. do.
Do. do. } No. 1 pressed } ·	9·0	4·5	2·6	32	128	197	170	10	G	Do. do.
Do. do. .	..	..	..	31	100	151	116	10	G	
Do. do. .	8·7	4·2	2·7	29	121	176	144	10	G	
Do. No. 2 Wire Cut	8·6	4·0	2·6	26	42	208	137	10	G	
Burham Wire Cut	9·0	4·5	2·6	37	163	262	·216	10	G	
Aylesford Gault Pressed	9·0	4·5	2·2	30	91	179	125	10	G	
Do. do.	9·0	4·5	2·2	38	98	170	130	10	G	Do. do.
Gault Wire Cut .	8·7	4·1	3·0	111	173	173	173	1	U	{ Faces made smooth and parallel by plaster of Paris.
Do. do. .	8·7	4·1	2·9	..	169	169	169	1	U	Do. do.
London Stocks .	8·9	4·2	2·5	..	103	129	115	3	U	
Do. do. .	4·6	4·0	2·4	130	177	181	179	2	U	Half Brick.
Sittingbourne Stocks	8·8	4·1	2·5	22	103	182	134	10	G	
Shippers Stocks	9·0	4·5	2·5	17	89	133	116	10	G	
Do. do.	9·0	4·5	2·5	28	134	197	140	10	G	Frogs filled with cement.
Suffolk Brimstone	9·0	4·6	2·7	17	84	125	105	10	G	
Do. Best Whites	9·2	4·6	2·6	18	55	92	67	10	G	
Fareham Best Reds	8·5	4·2	2·6	33	76	148	104	10	G	
Do. Best Rubbers	10·2	4·9	2·9	4	32	61	45	10	G	
Tipton Best Blue	8·7	4·3	2·5	82	198	(380)	··	10	G	(Not crushed).
Staffordshire Blue (common)	4·4	4·2	3·0	184	386	464	425	2	U	Half Bricks.
Do. do. pressed	9·0	4·3	3·1	..	275	275	275	1	U	
Blue Bricks(West Bromwich)	9·0	4·4	2·7	548	988	1129	1064	6	Wd	
Do. do. .	8·7	4·1	2·8	261	586	692	651	6	Wd	
Terra Metallic White Glazed (West Bromwich)	8·7	4·3	3·1	225	239	299	274	6	Wd	Recessed both sides.
Semi-dry Bricks of Slate Debris (South Wales)	8·7	4·2	2·3	556	1010	1099	1056	6	Wd	
Re l Bricks (Saltley, Birmingham)	9·0	4·4	3·2	139	137	202	180	5	Wd	Bedded between ⅜″ pine. Recess one side, filled with cement.
Red Brick "Craven"	8·7	4·2	2·9	354	412	569	496	6	Wd	Recesses both sides, filled with cement. Bedded as above.
American "Face" Bricks	7·7	3·6	2·1	..	768	937	858b	3	WN	Surfaces ground flat, bedded in plaster of Paris.
The same on end	3·6	2·1	7·7	..	366	522	448	6	WN	Tested on end.
"Common" Bricks	7·7	3·3	·2·1	..	803	1220	1016a	5	WN	Surfaces ground flat, bedded in plaster of Paris.
The same on end	3·3	·2·1	7·7	..	462	1014	698	11	WN	Tested on end.
" Face " Bricks (Mass., U.S.)	..	..	..	..	710	1075	895b	3	WN	Surfaces ground flat with emery.
"Common" Bricks (Mass., U.S.)	..	..	..	..	835	1436	1178b	3	WN	Do.
Bay State Bricks (Mass., U.S.)	..	..	..	..	668	817	733b	3	WN	Do.

G, Grant's Experiments, *Proc. Inst. C.E.*, vol. xxv.   U, Unwin, *Testing of Materials of Construction.*
Wd, Ward on Brickmaking, *Proc. Inst. C.E.*, vol. lxxxvi.   WN, *Report of Tests on the Strength of Structural Material made at Watertown Arsenal.* The tests marked "a" are further referred to in Table of Experiments on Brick Piers, p. 118 ; those marked " b," on p. 119.

The resistance of bricks to crushing is much reduced when they are built into work, being greatly influenced by the nature of the mortar used.

A series of experiments on rectangular piers of brickwork in mortar, of varying lengths and sections, both solid and hollow, carried out at Watertown Arsenal, U.S., from 1883 to 1891, constitutes a valuable addition to the experimental data available on this subject. [1]

The leading results are given in a tabulated form on page 118.

The heights of the piers tested range from 1' 4" to 12' 6", and in section from 8" to 16" square, nominal dimensions, the actual size of the bricks used averaging about $7.7 \times 3.2 \times 2.2$ inches.

The tests of single bricks, of which the piers were built, show in most cases an exceptionally high crushing resistance, tested on the flat. The specimens were prepared by carefully grinding the bed surfaces, which were set in thin facings of plaster of Paris to ensure even bearings against the compressing plates of the testing machine.

The end surfaces of the built-up piers were flat and carefully bedded.

The bricks were laid on the bed, and the joints broken at every course. Failure occurred in detail by the development of cracks, longitudinal fissures, and the partial crushing of the bricks.

By plotting the results of the experiments, curves may be obtained, fairly representing the averages of the series, and the ultimate crushing strength of solid square piers in percentages of the crushing resistance of the single brick (when prepared for testing as above described) may be expressed approximately as in the Table following, for varying proportions of length to side of square, and for the particular quality of brick and of mortar used in these experiments, and as described in the Tables on pages 118, 119.

[1] A fully detailed description of the Experiments on the Strength of Brick Piers carried out by the Science Committee of the Royal Institute of British Architects in 1895-97 will be found, accompanied by illustrations showing the mode of failure of the piers, in the Journal of that Institute dated 2nd April 1896, 17th Dec. 1896, 31st Dec. 1896, 18th Dec. 1897, 8th Jan. 1898 ; to which the student is referred.

Proportion of Height of Pier when side of Square = 1.	Ultimate Crushing Strength of Pier, when Ultimate Crushing Strength of Single Brick = 100.	
	8 inch Square Pier.	12 inch Square Pier.
2	18·0	14·0
4	16·4	12·0
6	15·0	10·6
8	14·0	9·7
10	13·5	9·0
12	13·0	8·5
14	12·6	8·1

The influence of length is manifest, while the larger section shows a diminution in strength.

The first visible crack occurs at loads varying from 57 to 87 per cent of the ultimate crushing load.

There appears to be little perceptible difference between the strength of solid and hollow piers per unit of brickwork area, but the hollow space in the piers tested did not exceed in area say 30 per cent of the total section.

It is possible that with bricks of softer quality, and with mortar resembling as closely as possible in its physical properties those of the bricks, a higher percentage of the single brick might have been obtained, though not a higher absolute value of crushing resistance. Further experiments are needed to determine these and other points before a satisfactory theory of the strength of brick piers can be formulated.

The influence of the strength of mortar upon the ultimate resistance is seen from a comparison of individual experiments in the same series, as follows :—

Dimensions of Pier.	Quality of Mortar.	Mean Crushing Strength of Mortar in 6″ Cubes.	Ultimate Strength of Pier.	
			Percentage of Single Brick.	Tons per square foot.
		Lbs. per sq. in.		
1′·4″ × 7″·5 × 7″·5	1 lime, 3 sand	124	18·1	162
1′·4″ × 7″·6 × 7″·6	1 Portland cement, 2 sand	546	27·1	242
6′·8″ × 7″·6 × 7″·6	1 lime, 3 sand	124	13·5	120
6′·8″ × 7″·6 × 7″·6	1 P.C., 2 sand	546	16·2	144
2′·0″ × 11″·5 × 11″·5	1 lime, 3 sand	124	13·9	124
2′·0″ × 11″·5 × 11″·5	1 P.C., 2 sand	546	26·4	235
10′·0″ × 11″·5 × 11″·5	1 lime, 3 sand	124	10·9	97
10′·0″ × 11″·5 × 11″·5	1 P.C., 2 sand	546	16·2	144
6′·0″ × 12″·0 × 12″·0	1 lime, 3 sand	124	10·2	75
6′·0″ × 12″·0 × 12″·0	1 P.C., 2 sand	546	15·7	115
6′·0″ × 12″·0 × 12″·0	Neat P.C.	3482	20·8	153

TABLE SHOWING THE RESULTS OF EXPERIMENTS ON THE RESISTANCE TO COMPRESSION OF SQUARE BRICK PIERS IN MORTAR, ENDS FLAT AND CAREFULLY BEDDED.[1]

Height of Pier in Feet.	Ratio of Height of Piers. To least side Square.	Ratio of Height of Piers. To least Radius of Gyration.	8 Inches Square Solid Pier. Avg. Ult. Crushing Load, Per cent of Crushing Load of Single Brick.	8 in. Tons per square foot.	8 in. Average Load at first Visible Crack. Per Cent of Ultimate Load.	8 in. Number of Piers Tested.	12 in. Per Cent of Crushing Load of Single Brick. Solid	Hollow	12 in. Tons Per Square Foot. Solid	Hollow	12 in. Avg. Load at First Visible Crack. Per Cent Ult. Solid	Hollow	12 in. Number of Piers Tested. Solid	Hollow	16 in. Per Cent of Crushing Load of Single Brick. Solid	Hollow	16 in. Tons Per Square Foot. Solid	Hollow	16 in. Avg. Load at First Visible Crack. Per Cent Ult. Solid	Hollow	16 in. Number of Piers Tested. Solid	Hollow
2'·0"	2·1	6					15·1	15·6	154	158	61	59	2	2								
2'·4"	2·1	7·8					13·2	12·02	134	123·2	64	49·2	1	1 2								
2'·4"	2·5	8·2					11·5	12·1	117	123	65	52	1	2								
2'·0"	2·5	9					11·5	13·5	117	137	78	72	2	1								
2'·0"	3·1	11	17·3	177	57	2									9·2	10·0	94	102	69	54	2	2
3'·0"	3·2	11																				
4'·0"	4·2	13									73	72	2	1								
4'·0"	4·5	14					10·5	9·6	107	98												
4'·3"	4·7	14																				
4'·6"	4·7	16	11·6	119	78	2									6·2	7·9	63	80	68	62	2	2
6'·0"	6·3	20																				
6'·0"	6·4	22						10·7	109		70	61	2	1								
6'·0"	7·8	22																				
4'·0"	8·3	24																				
10'·0"	7·8	27																				
8'·0"	8·4	27	15·0	153	67	2																
10'·0"	9·4	29					9·6	9·2	98	93	81	65	1									
8'·0"	10·5	32					8·5		86													
6'·0"	10·5	34	9·8	100	80	2																
10'·0"	12·6	36	10·4	106	64	2																
8'·0"	16·0	44																				
12'·6"	13·0	55 / 45·3																				
Average							9·3		99·3		7·13		23									

Remar[k]: Piers of "Common" bricks in mortar of 1 part Rosendale cement to 2 parts sand. Six in. cubes of this mortar, which was intended to be of uniform strength throughout, at 17 months gave a mean ultimate compressive strength of 161 lbs. per sq. in. Five bricks of the kind used in the piers, tested on the flat, as described on p. 115, gave an ultimate compressive strength varying from 803 to 1220 tons per sq. ft. The figures in columns of percentage of crushing load of single bricks are based on the mean strength of these 5 specimens, or 1016 tons per sq. ft.— Age of piers when tested in most cases was 21 months. Courses laid with breaking joints. Thickness of mortar joint from 0·20 to 0·30 in. Loads gradually applied.

Piers of "Face" bricks [fn] in mortar of 1 part Rosendale cement to 2 parts sand. Ultimate crushing strength of mortar in 6-inch cubes 161 lbs. per sq. inch. Three bricks of kind used in piers, tested on flat, gave an ultimate compressive strength varying from 768 to 937 tons per sq. ft. Mean compressive strength of single brick = 858 tons per sq. ft.

Piers of "Face" bricks [4] in mortar of 1 of lime to 3 of sand. Crushing strength of 3 specimens, tested on flat, from 710 to 1075 tons per sq. ft. Average 895 tons. Average crushing strength of mortar 124 lbs. per sq. inch.

Piers of "Common" bricks [4] in mortar of 1 of lime to 3 of sand. Crushing strength of 3 specimens, tested on flat, from 835 to 1436 tons per sq. ft. Average 1178 tons. Average crushing strength of mortar 124 lbs. pr sq. inch.

Piers of "Bay State Bricks"[4] in mortar of 1 of lime to 3 of sand. Crushing strength of 3 specimens from 668 to 817 tons. --Average 733, strength of mortar as above.

[1] Report of Tests on the Strength of Structural Material made at Watertown Arsenal.　[2] Soaked in water 16 hours.　[3] Laid with two bond stones 4 ft. apart, 3" thick.　[4] Age when tested from 14 to 24 months.　[5] Pier 12"×8".

With the above may be compared the following experiments made by Dr. Böhme at Berlin.[1]

Dimensions of Pier.	Quality of Mortar.	Crushing Strength of Mortar in Cubes.	Ultimate Strength of Pier.		
			Percentage of Single Brick.	Tons per Square Foot.	
		Lbs. per sq. in.			
10″ × 10″ × 9″·5	1 lime, 2 sand	177	44·1	116·2	Beneckendorfer brick.
Do.	1 cement, 6 sand	1748	53·0	139·4	
Do.	1 cement, 6 sand	1748	53·9 (wet)	141·9	
Do.	1 cement, 3 sand	3000	60·0	158·8	
Do.	1 cement, 3 sand	3000	64·5 (wet)	169·8	
10″ × 10″ × 9″·5	1 lime, 2 sand	177	43·5	70·0	Hertzfelder brick.
Do.	1 cement, 6 sand	1748	57·9	93·2	
Do.	1 cement, 6 sand	1748	54·3 (wet)	87·4	
Do.	1 cement, 3 sand	3000	62·5	100·7	
Do.	1 cement, 3 sand	3000	65·0 (wet)	104·6	

Crushing strength of single Beneckendorfer brick = 263·2 tons per square foot. That of Hertzfelder brick = 160·9 tons per square foot.

Six piers, tested by General Q. A. Gilmore, 22 inches in total height, 12 inches square, composed of North River brick in mortar of 1 of Rosendale cement to 2 sand, with bluestone caps and bases 3 inches thick, gave an ultimate crushing resistance varying from 111 to 130 tons per square foot, with average of 119 tons. Cracks developed at loads varying from 58 to 84 per cent of the ultimate load. The age of the pier when tested was $21\frac{1}{2}$ months.[2]

When the crushing strength of the single brick is taken as the measure of the ultimate strength of the pier, the method of testing and of the preparation of the specimen brick should be taken into account. Curioni has shown that the use of sheet lead as a bedding for the compressed surfaces of bricks leads to a reduction of strength of from $7\frac{1}{2}$ to 50 per cent,[3] while Professors Unwin and Beare have shown that a considerable loss of strength in crushing cubes of stone arises from a similar use of the same material.

The preparation of the surfaces, and the rigidity of the iron compressing planes of the testing machine, resisting by friction

[1] Unwin, *Testing of Materials of Construction.*
[2] *Notes on the Compressive Resistance of Freestone*, etc.
[3] *Min. Proc. Inst. of Civil Engineers*, vol. lxxiii.

the tendency to disintegration, both tend to augment the compressive resistance of the single brick when compared with its strength in the built-up pier.

The following are the results of experiments made by Mr. Kirkcaldy for Blackfriars Bridge :—

BRICKS USED IN PIERS.	Size of Piers Four Courses High	Cementing Material.	TONS PER FOOT SUP.	
			At which failure began.	At which crushing took place.
	Inches			
Common Stocks, recessed one side	14 × 14	Lime Mortar	17	27
Do. do.	Do.	Do.	21	30
Red Bricks (machine made) .	Do.	Do.	20	40
Do. (hand made) . .	Do.	Do.	20	36
Gault Bricks	Do.	Roman Cement	24	59
Do.	9 × 9	Do.	54	72
Clark's Sudbury (machine made)	Do.	Portland Cement	49	76
Uxbridge Red (hand made) .	Do.	Do.	44	53

Transverse Strength of Bricks.—Recent American experiments on different samples of Paving brick give a modulus of rupture on a span of 6″ ranging from about 1000 to 3500 lbs. per sq. in.[1]

Tensile Strength of Brick.—The tenacity of brick is given by the late Professor Rankine as varying from 280 to 300 lbs. per square inch. Experiments on twelve samples of brick tested at the East River Bridge Works showed an ultimate tenacity ranging from 90 to 358 lbs. per square inch. Average, $168\frac{1}{2}$ lbs. per square inch.[2]

Different Forms of Bricks.—The different forms in which bricks are made for special purposes are almost innumerable.

It would not be worth while, even if space were available, to describe them all ; but a few of the principal varieties may be mentioned.

Ordinary Bricks are of rectangular section, both longitudinally and transversely, and solid throughout. They have already been described.

Purpose-made Bricks are those which are specially moulded to shapes suited for particular situations, such, for example, as the voussoirs of arches struck to a quick curve, the corners of obtuse-angled structures, etc.

There are several advantages in having the bricks thus purpose-moulded : cutting is saved, and the surface-skin of the brick is left intact, which enables the brick to resist the weather far better than if the surface were removed by cutting.

Arch Bricks are shaped as voussoirs of arches.

Compass Bricks taper in one direction at least. If they taper in thickness they are suitable for the voussoirs of an arch, and are called *Arch bricks* or *Side-wedge bricks.* If, however, the thickness is constant, and they vary gradually in width, they are useful for steining walls, and are sometimes called *Bullheads.*

The name *Compass bricks* is sometimes applied only to bricks tapering in

[1] *Engineering News*, 18th April 1895.   [2] *Trans. Am. Soc. Civ. Eng.*, vol. vii.

both directions, as in Fig. 13. Such bricks are used for parts of furnaces etc. etc.

Perforated Bricks (Fig. 14) have cylindrical holes through their thickness, which makes them easier to burn (because the fire can penetrate them more thoroughly), and lighter to handle.

Such bricks are often made from the denser and heavier clays.

An objection sometimes stated against them is that they transmit sound readily.

Splits are bricks of the ordinary area, but of reduced thickness, being 9 inches by $4\frac{1}{2}$ inches wide, and 1, $1\frac{1}{2}$, or 2 inches in thickness.

Soaps are bricks 9 inches long, $2\frac{1}{4}$ inches wide and $2\frac{1}{4}$ inches thick.

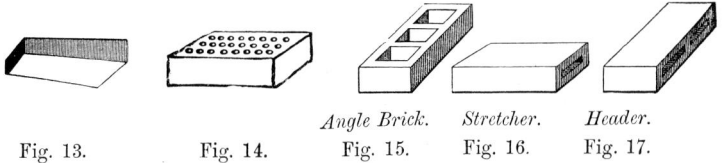

| | | *Angle Brick.* | *Stretcher.* | *Header.* |
| Fig. 13. | Fig. 14. | Fig. 15. | Fig. 16. | Fig. 17. |

Some varieties of these bricks, pierced with elaborate patterns, and used for ventilating purposes, are made in stoneware and terracotta (see p. 139).

Hollow Bricks should be moulded from the best and most homogeneous clay. They may be of large size, as their shape enables them to be thoroughly burnt, and makes them lighter to handle.

There are a great many forms of these bricks used for building hollow walls.

Figs. 15, 16, 17 show hollow bricks made by Messrs. Clayton, Son, and Howlett's machines.

The three figures show an angle brick, stretcher, and header in position as for the angle of a wall, but spread out so as to show their construction. They are so arranged that a solid side or end is always presented on the face of the wall.

In other forms the perforations are somewhat different ; for example, as in Figs. 18, 19—

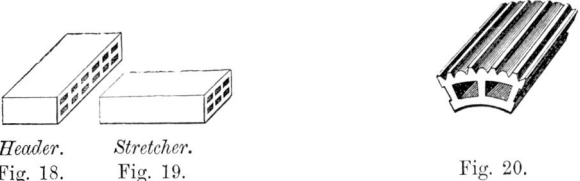

Header.   *Stretcher.*
Fig. 18.   Fig. 19.   Fig. 20.

The form and use of other hollow bricks are shown by the section, Fig. 27.

Tubular Bricks are hollow bricks in which there is one large perforation running through the length of the brick.

Tubular bricks are also made in the form shown in Fig. 20, so that several of them built up together form a pillar.

Somewhat similar bricks, but flat, instead of round, are made for building up pilasters.

Plinth, Cornice, and String-Course Bricks are made of several patterns.

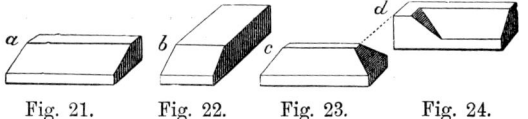

Fig. 21.   Fig. 22.   Fig. 23.   Fig. 24.

They have to be arranged so as to be built in as headers and stretchers, and also for angles.

Thus Figs. 21 to 24 are all plinth bricks : *a* is a stretcher, *b* a header, *c* an external angle, *d* an internal angle.

Those that are intended to project should have a throat on the lower side, as in Fig. 25.

Sometimes several different forms of moulded bricks are combined to form

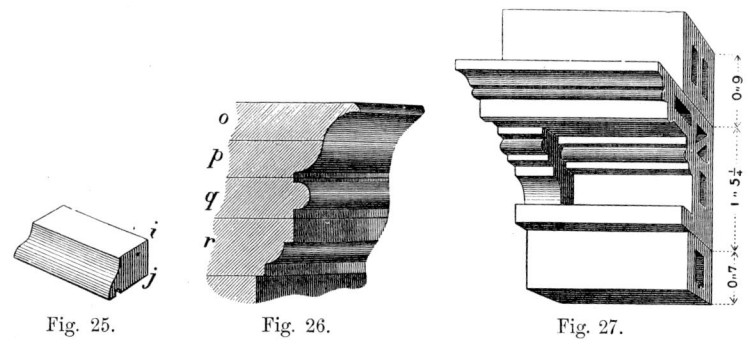

Fig. 25.   Fig. 26.   Fig. 27.

a cornice, as in Figs. 26, 27, which are from an advertisement by the Broom-hall Company.

Bricks shaped like *o*, Fig. 26, are known as *Hollow Cornice;* those of section like *p* are *Full Cornice*, while *q* and *r* are *Moulded Cornice bricks.*

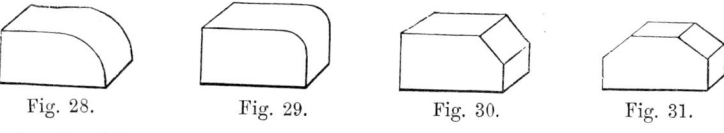

Fig. 28.   Fig. 29.   Fig. 30.   Fig. 31.

Round-ended and Bull-nosed Bricks.—Figs. 28 and 29 are for use at corners where sharp arrises would be liable to damage.

Splay Bricks are bevelled off on one side, like Fig. 30. They are sometimes called *slopes.*

Double Cant Bricks have a splay on both sides, like Fig. 31.

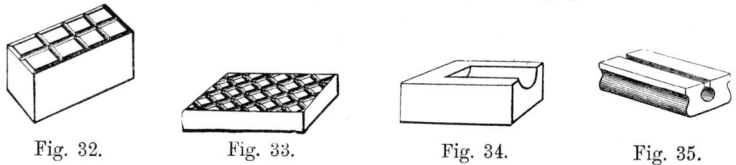

Fig. 32.   Fig. 33.   Fig. 34.   Fig. 35.

Pavings are made generally of dark blue Staffordshire ware, very hard, the surfaces rendered less slippery by being indented with flutes, or with a diamond pattern. See Figs. 32, 33.

Gutter Bricks, called also *Channel and Sough* bricks, are made of various sections, such as that in Fig. 34, which shows a gutter brick with *stop end*.

Drain Bricks are of the form shown in Fig. 35. A number of these placed side by side form a suitable floor for a cattle-shed, or for any building where much liquid falls on the floor, and has to be carried off at once.

Coping Bricks are made of several different sections to suit walls of different thicknesses.

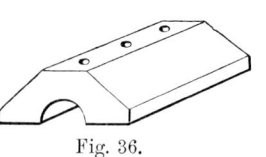

Fig. 36.

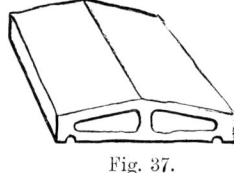

Fig. 37.

When they are to project they should always be throated as in Fig. 37.

They are either prepared to receive palisades, as in Fig. 36, or left plain with a curved or an angular top as in Fig. 37.

Copings for Platforms and Wing Walls are for railway or other platforms, and for retaining and wing walls. They are made either plain, or (for platforms) with indented or fluted surfaces.

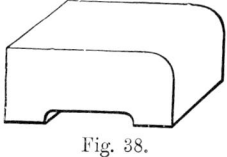

Fig. 38.

Coping bricks are made of considerable size, even as large as 18 inches by 6 inches by 6 inches.

Stopped ends and angles are made for all coping bricks.

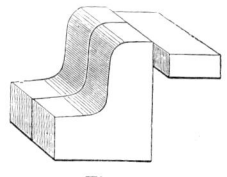

Fig. 39.

Fig. 40.

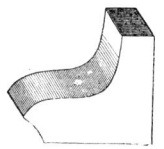

Fig. 41.

Kerb Bricks for footpaths are made of the section shown in Fig. 39, and of other sections.

Tunnel Heads are of the form shown in Fig. 40, and are made generally in fireclay for parts of furnaces.

Boiler Seatings, of the shape shown in Fig. 41, are also made in fireclay.

Besides the forms of bricks above illustrated, there are several which cannot be described, such as *Sink bricks*, made in the form of a dished sink

Manger bricks, which when put together form a manger ; *Sill bricks,* which are shaped like the centre and ends of a stone sill.

Colouring Bricks.[1]—Bricks may be coloured either (1) by mixing substances with the clay which will produce the required colour when burnt ; or (2) by dipping the brick in colouring matter after it is burnt.

The former method may be adopted when the colouring matter is cheap and plentiful ; the latter when it is expensive.

(1.) When the colours are mixed with the clay it should be remembered that red ochre burns yellow.

Yellow ochre burns red.

Iron burns red at low temperature ; black at high temperature.

Manganese burns black.

Light red, ⎫
Indian red, ⎬ retain their colours when exposed even to a white heat.
French ultramarine, ⎭

The above-mentioned colours may be mixed with the clay in different proportions according to the tint required.

(2.) When bricks are to be coloured by dipping, the colouring matter is added to a mixture of linseed oil and turpentine, containing a little litharge as a drier.

The colouring matters used are as follows :—

In dark red bricks, Indian red.

  ,, blue   ,,   French ultramarine.

  ,, black   ,,   manganese.

  ,, grey   ,,   white lead and manganese.

The bricks are heated on an iron plate, and dipped when hot, then slightly washed with cold water, and allowed to dry.

If the brick be burnt after being dipped, it will be covered with a glaze.

The colour penetrates about $\frac{1}{8}$ inch into ordinary porous bricks (not so far into terracotta), and it stands the weather well.

If the bricks cannot conveniently be heated and dipped, the liquid may be heated and laid on.

FIRECLAY AND FIREBRICKS.

Fireclay is the name given to any clay which is capable of standing a high temperature without melting or becoming soft. Such clays are also called *refractory.*

Uses in Building.—Fireclay is required in buildings for setting stoves, ovens, backs of ranges, etc.

It is also used for the manufacture of firebricks, fire lumps, drain pipes, chimney pots, and other similar articles.

Where found.—Clay of this description abounds in the coal-measures, just beneath the several seams of coal.

The following list gives some of the counties in which fireclay

[1] Roorkee *Treatise of Civil Engineering.*

is found, together with the localities producing the best known descriptions :—

Ayrshire . . .	.	Kilmarnock, Dean, Hillhead, Perceton.
Buckinghamshire	.	Hedgerley.
Cornwall . . .	.	St. Austells.
Derbyshire	.	Burton-on-Trent.
Devonshire	.	Plympton.
Dorsetshire	.	Poole.
Fife . . .	.	Lillyhill.
Lanarkshire	.	Garnkirk, Glenboig.
Monmouthshire .	.	Newport.
Northumberland	.	Newcastle-on-Tyne.
South Wales	.	Dowlais, Neath.
Staffordshire	.	Brierly Hill, Wolverhampton.
Worcestershire	. {	Stourbridge, Dudley, Tipton, Hanford, Gornal, etc.
Yorkshire	.	Wortley (near Leeds), Elland, Stannington, etc.

Composition.—A refractory fireclay will contain nearly pure hydrated silicate of alumina.

The more alumina that there is in proportion to the silica, the more infusible will be the clay.

The composition of different fireclays varies, however, considerably.

They contain

From 59 to 96 per cent silica.

    „   2 to 36   „   alumina.

    „   2 to 5   „   oxide of iron.

A very small percentage of lime, magnesia, potash, soda.

The fire-resisting properties of the clay depend chiefly upon the relative proportions of these constituents.

If the oxide of iron or the alkalies are present in large proportion, they act as a flux, and cause fusion ; the clay is no longer fireproof or refractory.

It will not, however, be necessary to enter in detail upon the part played by each of the constituents that are found in fireclay. These constituents are the same as those found in brick earth (though their proportions are different), and the effect they produce upon the clay is the same in both cases (see p. 87).

The presence of an extremely small proportion of lime, potash, or soda, may, however, improve the clay, by soldering the particles firmly together.[1]

When a clay containing iron requires the addition of sand to

[1] Percy's *Metallurgy.*

prevent its cracking, it is a common practice to add burnt clay instead, so as to produce the same beneficial effect without risk of making the clay fusible.

The chemical analysis of a clay is not a very safe criterion of its qualities. The silica shown may be either soluble silica influencing the chemical constitution of the clay, or it may be sand which is chemically inert.

In the analysis there is no distinction made between the two.

" A good fireclay should have an uniform texture, a somewhat greasy feel, and be free from any of the alkaline earths." [1]

The following TABLE shows the ANALYSES of different CLAYS used for the manufacture of FIREBRICKS :—

	SiO_2. Silica.	Al_2O_3. Alumina.	KO. Potassa.	NaO. Soda.	CaO. Lime.	MgO. Magnesia.	FeO. Protoxide Iron.	Fe_2O_3. Peroxide Iron.	Water. Water.	Organic Matter. Organic Matter.
Brierly Hill, Stafford-shire, P	51·80	30·40	...	Trace	...	0·50	4·14	...	13·11	
Burton-on-Trent, G	58·08	36·89	·20	1·88	·55	·14	...	2·26		
Cornwall, P	46·32	39·74	...	...	0·36	0·44	0·27	...	24·75	
Dinas, G	97·62	1·4	·10	·10	·29	...	...	·49		
Dowlais, best, P	67·12	21·18	2·02	...	0·32	0·84	...	1·85	6·21	1·90
Glascote, near Tam-worth, P	50·20	32·59	2·32	...	0·36	0·44	...	3·52	12·69	
Glasgow, P	66·16	22·54	...	...	1·42	Trace	5·31	...	3·14	
Hedgerley, Bucks, G	84·65	8·85	...	...	1·90	·35	...	4·25		
Howth, near Dublin, P	74·44	19·04	2·07	...	0·45	0·27	...	0·61	3·71	
Ireland, P	79·40	12·25	...	...	0·50	...	...	1·30	5·20	
Kilmarnock, Ayrshire, G	58·92	35·65	1·14	1·06	·39	·35	...	2·49		
Newcastle, P	55·50	27·75	2·19	0·44	0·67	0·75	...	2·01	10·53	
Plympton, Devon, G	74·02	21·37	·82	·09	·40	·36	...	1·94		
Poole, Dorset, P	48·99	32·11	3·31	...	0·43	0·22	2·34	...	11·96	
Stourbridge, Worces-tershire, P	63·30	23·30	...	...	0·73	...	1·80	...	10·30	
Teignmouth, Devon, P	52·06	29·38	2·29	...	0·43	0·02	2·37	...	12·83	
Wortley, Leeds, G	65·25	29·71	·43	·12	·40	·61	Titanic Acid ·41	3·07		

P, Percy's *Metallurgy*.   G, Capt. Grover, *R.E. Prof. Papers*, vol. xix.

[1] Page's *Economic Geology*.

Grain.—It should be remarked that the infusibility of fire-clays does not depend altogether upon their chemical composition, but also upon their degree of fineness. A fireclay with a coarse open grain will probably prove more refractory than one with a close even texture.

Firebricks are made from fireclay by processes very similar to those adopted in making ordinary bricks.

The clay is dug, weathered, tempered, ground under rollers, passed through riddles to remove lumps, pugged, moulded, burnt in cupolas or in Hoffmann's kilns at a heat slowly increasing until it attains a very high temperature, and then allowed gradually to cool.

There are several varieties of firebrick in general use, named usually after the locality providing the fireclay from which they are made.

Stourbridge Firebricks are made in a district about twenty miles south-west of Birmingham, which contains several varieties of fireclay.

The material used for these firebricks is a black clay found in a thick seam under the coal-measures.

The bricks produced are generally of a pale brownish colour, sometimes reddish or yellow-grey. They are frequently mottled with dark spots, which are stated by Dr. Percy to be due to the presence of particles of iron pyrites.

" With Stourbridge clay it is customary to mix burnt ordinary clay. For common firebricks the proportions of the latter to the former are often as much as two to one. This gives a brick capable of resisting the action of the heat caused by a house fire, though it would not be sufficiently refractory for resisting a furnace temperature. Fireclay being expensive, the inferior brick is naturally cheaper, and is much used." [1]

Kilmarnock and *Newcastle* furnish firebricks somewhat similar to those from Stourbridge.

Dinas Firebricks are made from a so-called fireclay found in Glamorganshire.

It will be seen from the table of analyses on the previous page that the so-called " clay " consists nearly entirely of silica. It is found in the state of sand. About 1 per cent of lime is added, and enough water to make it cohere. The bricks are then moulded by machinery under pressure, dried, and burnt in a close kiln.

The bricks made from this substance will bear a most intense heat, being the only description that will resist the temperature (4000° to 5000° Fahr.) of a regenerative furnace.[2]

They expand under heat, are porous, and will not stand rough usage.

The fractured surface of a Dinas firebrick " presents the appearance of coarse irregular white particles of quartz, surrounded by a small quantity of light brownish-yellow matter. The lime which is added exerts a flux-

[1] Abney's *Notes on Chemistry of Building Materials.*
[2] Dr. Siemens, *Chemical Society,* 7th May 1868.

ing action on the surface of the fragments of quartz, and so causes them to agglutinate. . . . From their siliceous nature it is obvious that they should not be exposed to the action of slags rich in metallic oxides." [1]

Guismuyda Firebricks, made near Swansea, and *Narberth Firebricks*, from Pembrokeshire, are of the same description as those from Dinas.

Lee Moor Firebricks are made from the refuse of china clay produced by the disintegration of felspathic granite (see p. 16), found chiefly in Cornwall and Devonshire granite.

A well-known variety of these bricks is manufactured at Lee Moor, near Plympton. They have a compact surface of a dull reddish-brown ; are very hard and highly refractory.

Windsor or Hedgerly Firebricks are made from the sandy slate-coloured loam used for the manufacture of rubbers, and are of a red colour when burnt. These bricks consist chiefly of sand ; they contain only 9 per cent of alumina, and a large proportion ($4\frac{1}{4}$ per cent) of oxide of iron. They are unable, therefore, to resist very high temperatures.

The following TABLE shows the CRUSHING STRENGTH, WEIGHT when dry, and ABSORPTIVE POWER of different classes of FIRE-BRICKS :—

RESISTANCE TO COMPRESSION, WEIGHT, AND ABSORPTION OF FIREBRICKS.

DESCRIPTION OF BRICK.	DIMENSIONS OF SPECIMEN, in Inches.			Area exposed to crush-ing.	Average weight under which brick cracked.	Average force required to crush brick.	Weight when dry.	Per-centage of water absorbed.	Autho-rity.[2]
	Length.	Breadth.	Thickness.						
				Sq. In.	Tons.	Tons.	Lbs.		
Stourbridge firebrick .	9·08	4·4	2·47	39·9	25·0	50·9	7·2	9·5	L
Lee Moor do. .	9·12	4·34	2·54	39·5	14·8	54·9	7·7	4·9	L
Newcastle do. .	8·91	4·40	2·44	39·2	27·0	45·6	6·1	9·9	L
Dinas do. .	8·92	4·32	2·44	38·7	28·0	49·0	6·9	9·3	L
Welsh do. .	8·64	4·62	2·55	36·8	14·4	53·3	6·9	6·2	L

TERRA COTTA.

Terra Cotta is a kind of earthenware which is frequently used as a substitute for stone in the ornamental parts of buildings.

[1] Percy's *Metallurgy*, p. 238.
[2] Mr. Baldwin Latham, *Sanitary Engineering*.

Many localities furnish clay from which terra cotta may be made, as, for example, Tamworth, in Staffordshire; Watcombe, in Devonshire; Poole, in Dorsetshire; Everton, in Surrey, and other places in Northamptonshire and Cornwall.

MAKING TERRA COTTA.—The great difficulty to be overcome in making terra cotta is the uncertain shrinkage of the clay.

To obviate this as much as possible, different clays are mixed together, and a large proportion of ground glass, pottery, and, in some cases, of sand, is added.

This mixture is ground into fine powder, thrown into water, finely strained, pugged, kneaded, forced into plaster moulds smeared with soft soap, very carefully dried, gradually baked in a pottery kiln, and slowly cooled.

The drying process requires to be conducted with extreme care. If the blocks are subjected to draughts of cold air, if they are of unequal thickness, or if the operation is conducted too quickly, they will warp, twist, and become useless.

Nature of Clay.—As before mentioned, the red clays contain oxide of iron. If this is in considerable proportion (say from 8 to 10 per cent), it makes them very fusible and difficult to burn successfully.

This fusibility is aggravated by the presence of lime, magnesia, and other impurities, and the resulting terra cotta is not so hard and durable as that from the more refractory white clays.

In some cases the white clay is used with an admixture of oxide of iron just sufficient to make it burn to a good red colour.

Fireclays are used for the manufacture of terra cotta, in some instances with very little preparation.

Terra cotta made from fireclays, when properly burnt, is excellent in texture, colour, and surface, but appears ragged and porous directly the outer skin is removed. It manifestly suffers for want of a small proportion of some flux, such as that afforded by the lime and alkalies in the mixed clays.

The mixed clays used for terra cotta contract from $\frac{1}{10}$ to $\frac{1}{12}$ of their linear dimensions in drying and baking.

The red clays shrink only about $\frac{1}{20}$ lineally, while fireclays shrink as much as $\frac{1}{8}$. More than half of this shrinkage is in drying, and the remainder in burning.

Blocks.—The blocks used for building purposes should average from about 1 to 3 cubic feet in bulk, and no block should contain more than 4 cubic feet.

Such blocks are generally made hollow, the thickness of the shell of terra cotta being from 1 inch to 2 inches.

Large blocks should have a diaphragm, or partition of terra cotta across them, to prevent their warping.

If required to bear considerable weight the blocks should be filled with broken brick bedded in good mortar or cement.

BUILDING TERRA COTTA.—The blocks should be so shaped as to form a good bond with the brickwork, or whatever material is used for the backing.

The blocks are usually made from 12 inches to 18 inches long, 6 inches to 15 inches high, and from $4\frac{1}{2}$ to 9 inches thick on the bed. These dimensions are suitable for bonding into brick backing.

When the blocks are of the thicknesses above mentioned, the joints are made square and flush as in ordinary ashlar work.

ADVANTAGES.—The advantages of terra cotta are as follows :—

Durability.—If properly burnt, it is unaffected by the atmosphere, or by acid fumes of any description.

Lightness.—If solid it weighs 122 lbs. per foot cube ; but if hollow, as generally used, it weighs only from 60 to 70 lbs. per foot cube, or half the weight of the lightest building stones.

Strength.—Its resistance, when solid to compression, is nearly $\frac{1}{3}$ greater than that of Portland stone.

Hardness.—Mr. Page found by experiment that it lost $\frac{1}{16}$ inch in thickness, while York stone lost $\frac{1}{4}$ inch with the same amount of friction. It is, therefore, well adapted for floors.

Cost.—It is cheaper in London than the better descriptions of building stone. It is so easily moulded into any shape, that for intricate work, such as carvings, etc., it is only half the cost of stone.

DISADVANTAGES.—Terra cotta is subject to unequal shrinkage in burning, which sometimes causes the pieces to be twisted.

When this is the case great care must be taken in fixing the blocks, otherwise the long lines of a building, such as those of the string-courses or cornices, which are intended to be straight, are apt to be uneven, and the faces of blocks are often " in winding."

Twisted and warped blocks are sometimes set right by chiselling, but this should be avoided, for if the vitrified skin on the surface be removed, the material will not be able to withstand the attacks of the atmosphere etc.

Another drawback is the uncertainty of getting terra cotta delivered as required, whereas a stone may be taken and fixed at once, the carving being left, if necessary, to be completed afterwards.

COLOUR.—Terra cotta is made in several colours, depending chiefly upon the amount of heat it has gone through.

White, pale grey, pale yellow, or straw colour indicate a want of firing.

Rich yellow, pink, and buff varieties are generally well burnt.

A green hue is a sign of absorption of moisture, and is a sign of bad material.

A glazed surface can be given to terra cotta if required.

POROUS TERRA COTTA is made in America " of a mixture of clay and some combustible material—such as sawdust, charcoal, cut straw, tan bark, etc. When baked the combustible material is consumed, leaving the terra cotta full of small holes. It is fireproof, of little weight, great tenacity, strong, can be cut with edge tools, will hold nails driven in, and gives a good surface for plastering." [1]

INFERIOR TERRA COTTA is " sometimes made by overlaying a coarsely-prepared common body with a thin coating of a finer and more expensive clay."

" Unless these two bodies have been most carefully tested and assimilated in their contraction and expansion, they are sure in course of time to destroy one another; that is, the inequality in their shrinkage will cause hair cracks in the fine outer skin, which will inevitably retain moisture, and cause the surface layer to drop off in scales after the winter frosts."

" Another very reprehensible custom is that of coating over the clay, just before it goes into the kiln, with a thin wash of some ochreish paint, mixed with finely ground clay, which produces a sort of artificial bloom, very pretty looking for the first year or two after the work is executed, but sure to wear off before long." [2]

Common window sills, etc., have been made of concrete covered over with a skin of burnt clay to look like terra cotta—this skin soon breaks away.

Where used.—Terra cotta has been extensively used in Dulwich College, in Messrs. Doulton's warehouses, Lambeth, in the Albert

---

[1] *Proceedings R.I.B.A.* 1886, p. 129.
[2] *Reports on Exhibition,* 1876, p. 14.

Hall, in the new Constitutional Club, in the Natural History Museum, in the Law Courts, Birmingham, and in several of the new buildings near the South Kensington Museum.

Stoneware is the name given to articles made from the plastic clays of the Lias formation, obtained chiefly in the south of England.

The best comes from Poole, in Dorsetshire, or from the vicinity of Teignmouth, in Devonshire. It contains about 76 parts silica and 24 of alumina, with a very small proportion of other ingredients.

This clay contains very little infusible matter. It is generally mixed with a certain proportion of powdered stoneware, ground and calcined flints, ground decomposed Cornish granite (see p. 16), or sand, to prevent excessive shrinkage.

They are burnt in domed kilns like fireclay goods, but at a very much higher temperature.

A fractured surface shows that this ware is thoroughly vitrified throughout. It is intensely hard, dense in texture, close in grain, and rings well when struck.

This material is admirably adapted for all purposes where strength and resistance to atmospheric, chemical, or other destroy ing influences are required.

Stoneware articles are often salt glazed (see p. 134), but the material is in itself non-absorbent, and will resist the effect of moisture even when unglazed.

PIPES AND MISCELLANEOUS CLAY WARES.

Pipes and other articles made in clay are practically divided under four heads.

1. *Unglazed Earthenware*, made from ordinary clays, similar to those used for common bricks and tiles.

Earthenware of this description is weak, porous, liable to the attacks of frost, and is not adapted to resist the atmosphere or other destroying agents.

2. *Fireclay Ware*, made from the fireclays of the coal-measures (see p. 125).

This material has a very open grain, is porous, except where protected by glazing; and is weak when compared with terra cotta or stoneware.

It is much used for common varieties of the different articles about to be described, especially in the localities where the fire-clay is found; but it is inferior to stoneware or to terra cotta for nearly every purpose.

3. *Stoneware* is made, as before stated, from clays of the Lias formation, mixed with sand and ground pottery, to prevent shrinkage.

The characteristics of this material have already been pointed out (see p. 133).

Its strength, durability, imperviousness, and resistance to destructive influences make it an admirable material for sanitary ware, sewer pipes, ornamental works exposed to the atmosphere, and for vessels to contain chemical compounds.

4. *Terra Cotta* is often used for pipes and other miscellaneous articles.

Its composition, and the mode in which it is manufactured, have already been described.

It is inferior to stoneware, inasmuch as it is more absorbent and less dense in grain. It is burnt at the same heat as fireclay goods, but is superior to them in strength and durability.

Glazing.—It is often advisable to glaze the surface of articles made in clay, sometimes for appearance, but more generally in order to protect portions exposed to the action of the atmosphere, .to sewage, or other destroying agents.

These glazes are either *transparent*—merely a film of glass—or *opaque*, like an enamel.

TRANSPARENT GLAZES of several kinds are known in the trade. Two methods, adapted to the somewhat rough articles used by the engineer and builder, will now be described.

Salt Glazing is effected by throwing salt into the furnace when the articles it contains are at a high temperature. The heat of the fire volatilises the salt (chloride of sodium), the vapour being in the presence of oxygen and silica is decomposed, the chlorine goes off through the top of the furnace, the sodium combines with the silica in the clay to form silicate of soda, which again unites with the silicates of alumina, lime, or iron in the clay, to form a surface coating of glass.

This method of glazing has great advantages. The vapour of the volatilised salt gets into every crevice, however small, and coats it with an impenetrable film of glass.

It is used for stoneware, and also for articles made from fireclay.

Lead Glazing is carried out by dipping the article to be glazed (after it has been once burnt) into a bath containing oxide oɪ lead and tin—or borax with kelp, sand, etc., ground to powder, and stirred in water to a creamy consistence.

The particles of these different materials adhere to the surface of the article when it is dipped. It is then withdrawn and re-burnt. The high temperature of the furnace causes the particles to run together and to form a film of glass over the whole surface.

This method of glazing is used for terra cotta, and sometimes for articles made from fireclay.

Lead glazing is also used for earthenware crocks, etc., which are made out of inferior clays such as would not stand the high temperature required for salt glazing. A lead glaze will generally chip off easily.

OPAQUE GLAZES are required in cases where it is wished to give (to the whole, or to any portion of an article) an appearance superior to that presented by the ordinary burnt material.

The article to be glazed is dipped before burning into a *slip* formed of superior clay, very finely worked, dried, etc., and brought to the colour required.

Thus the pans of water-closets are often made white inside, and of a cream colour, or some other tint, outside.

Burning.—Terra cotta, stoneware, and fireclay ware, are all burnt in domed kilns.

The heat is applied gradually, and after it has risen to its height is kept up for a period varying from 24 to 72 hours, according to the size of the kiln and of the articles in it. The kiln is then allowed to cool down gradually.

Terra cotta is burnt at a much lower temperature than stoneware.

In order to protect articles of a delicate nature from direct contact with the fire, which would discolour them, they are placed in large fire-clay jars called *seggars*, or enclosed in a casing of fire-brick formed within the kiln, and known as a *muffle*.

Pipes are made from clay, very finely ground, washed, sieved, tempered, pugged, and forced by machinery through a mould, or *dod* as it is called—dried, and baked in a circular kiln.

AGRICULTURAL DRAIN PIPES are made of various sections, but the circulaɪ and ⌒ shaped are those in most common use.

These pipes are sold in 2-foot lengths, and of diameters varying by half inches from 1 inch to 6 inches.

Collars are short pieces of pipe sometimes used to cover the joints between each pair of lengths of the drain pipes, so as to give the ends of the pipes a firm bed.

They are 3 or 4 inches in length, and about 1 inch greater in diameter than the pipes they unite.

They are, however, generally omitted altogether.

SEWER PIPES should be of a vitreous imperishable material, of sufficient strength to resist fracture, having toughness enough to withstand shocks, tenacious, hard, homogeneous, impervious, uniform in thickness, true in section, perfectly straight longitudinally, or formed to the proper curve, uniformly glazed both inside and outside, free from fire cracks and flaws of all kinds.

When struck they should ring clearly.

Porous substances are not so good as those that are vitreous throughout, and pipes burned at a low temperature are inferior to those that have been subjected to a considerable heat.

Sewage pipes are made both from stoneware and fireclay. The former is the stronger material, and is said better to resist the decomposing effect of sewage and other substances having a chemical action.

Salt-glazed pipes only should be used ; if the glaze can be picked off it is proof that the pipes are made out of a clay that would not stand a high temperature ; in fact, that the pipes are not stoneware.

Fireclay pipes should be made thicker than those of the same diameter in stoneware.

Different Forms of Sewer Pipes.—Several forms of sewer pipes have been devised, but only one or two of the most common need be noticed.

Socket Pipes.—Pipes intended to convey sewage are generally made with sockets. Care should be taken that this socket is in the same piece with the pipe, not formed separately, as is sometimes the case.

Half Socket Pipes have a socket on the lower half of the circular section only, so that a broken length may at any time be taken out and replaced, or a junction inserted.

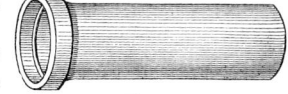

Fig. 44.

The following TABLE gives the dimensions and thickness of stoneware, fireclay, and other clay pipes, as laid down by Mr. Baldwin Latham :—[1]

STONEWARE.			FIRECLAY.		OTHER CLAYS.	ALL PIPES.
Internal Diameter.	Thickness.	Length in work.	Thickness.	Length in work.	Thickness.	Depth of Socket.
Inches.	Inches.	Feet.	Inches.	Feet.	Inches.	Inches.
2	...	...	...	...	$\frac{5}{16}$	
3	$\frac{3}{8}$	2	$\frac{3}{8}$	2	$\frac{3}{8}$	$1\frac{1}{2}$
4	$\frac{1}{2}$	2	$\frac{1}{2}$	2	$\frac{1}{2}$	$1\frac{1}{2}$
6	$\frac{5}{8}$	2	$\frac{3}{4}$	2	$\frac{3}{4}$	$1\frac{3}{4}$
9	$\frac{3}{4}$	2	$\frac{4}{5}$	2	$\frac{7}{8}$	2
10	$\frac{7}{8}$	2	1	2	...	2
12	$1\frac{3}{4}$	2	$1\frac{1}{10}$	2	1	2
15	$1\frac{1}{8}$	2	$1\frac{1}{4}$	2 to 3	$1\frac{1}{4}$	$2\frac{1}{4}$
18	$1\frac{1}{4}$	2 to 3	$1\frac{1}{2}$	2 to 3	$2\frac{1}{2}$	$2\frac{1}{2}$

[1] Specification for Bideford Waterworks.—*Humber.*

Mr. Baldwin Latham states that he " has found in some cases that the thickness given in the above Table for fireclay pipes is not sufficient." [1]

Socket pipes may readily be obtained as small as 2 inches in diameter ; also pipes of 21, 24, and 30 inches in diameter, in 2½ or 3 feet lengths.

BENDS are curved lengths of pipe which are made to varying radii, and of 2, 3, 4, 6, 9, 12, 15, and 18 inches bore. They should always be segments of circles, and should form perfect curves when jointed together.

TAPER PIPES (Fig. 45) are intended to form a connection between two pipes of different diameter.

JUNCTIONS for pipes are made in several different forms. They are usually in 2 feet lengths.

Fig. 45.

Single Junctions are those to form the joint when one pipe enters the side of another. The junction may either be at right angles to the pipe, as in Fig. 46, or joined to it by a gradual bend, as in Fig. 47. The latter is the best construction.

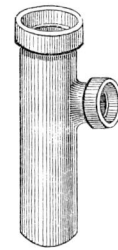
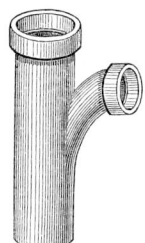

Fig. 46.   Fig. 47.

Double Junctions are to form the joint where two pipes meet a third, either at the sides as in Fig. 48, or at one end as in Fig. 49.

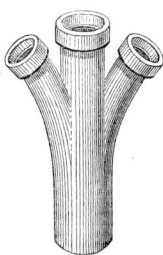
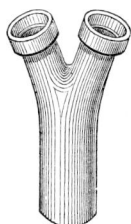

Fig. 48.   Fig. 49.

Mr. Baldwin Latham gives the following directions for forming bends and junctions :—
" The centre from which a branch on a main is struck must be upon a line at right angles to the centre line of the main pipe, the inside of the main pipe meeting the inside of the branch at a tangent on a radius line from which it is struck ; the ends of all curved pipes must be in the direction of the radius of the curves with which they are described." [2]

SADDLES and CHAIRS in earthenware are formed of such a shape as to make a secure junction between the adjacent lengths of a sewer pipe, and yet to enable the sewer to be examined at any time, and any obstruction to be removed without breaking a pipe.

---

[1] Latham's *Sanitary Engineering.*   [2] Specification for Bideford Waterworks.

Figs. 50, 51 show the junctions of Jenning's improved patent drain pipes.

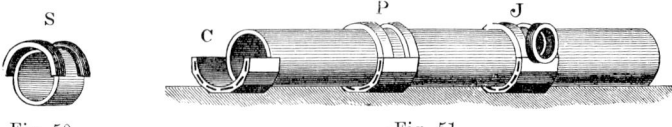

Fig. 50.   Fig. 51.

The chair is shown at C in position supporting the end of a pipe ; another length would be placed in the vacant half of the chair, and then the short piece S placed between the two lengths over the chair.   The bottom of the short piece is flush with that of the lengths of pipe united by it.

Some of the saddles are plain, as at P, which shows one in position.   Others have junctions attached, as at J.

Other saddle and chair junctions introduced by Mr. Jennings have no short piece attached to the saddle.   The chair and saddle are rebated at each end, of a depth equal to the thickness of the adjacent lengths of pipes, which therefore fit into the rebates, and have their inner surfaces flush with those of the saddle and chair.

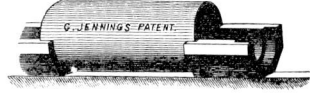

Fig. 52.

Fig. 52 shows one of these chairs.   The saddle is exactly similar in form, being made, however, with or without junctions, as in Fig. 51.

The objection to pipes with half sockets, saddles, etc., is that when the sewer is more than half full they leak or overflow.

OPERCULAR OR LIDDED PIPES were introduced by Messrs. Doulton.

They are similar in form to ordinary socket pipes, but are strengthened by two ribs running lengthways, shown at *r r* in Fig. 53, which is a section of the pipe.

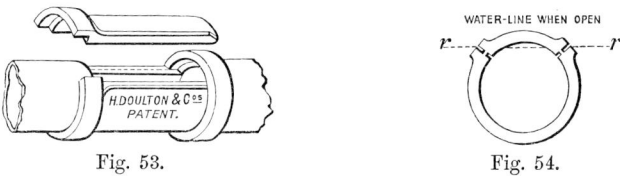

Fig. 53.   Fig. 54.

A longitudinal nick or furrow is made throughout the length of the pipe along these flanges, so that by inserting a chisel the upper portion of the pipe between *r r* may easily be removed.

Thus the whole length of the pipe may be opened up for inspection or for removal of obstructions.

CAPPED PIPES have circular or oval holes in them, with loose covers, so that they can be examined without being broken or taken up.

SYPHON TRAPS in stoneware may be obtained of almost any shape and description, either with or without inlets for examination.

One form is shown in Fig. 55, with the inlet dotted.   This description may be obtained of 2, 3, 4, 6, 9, and 12 inches bore.

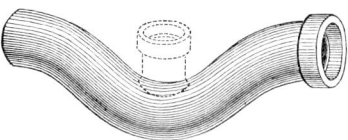

Fig. 55.

The position of inlet shown in Fig. 55 is the usual one, but a trap so formed is liable to choke, and it is better to have the inlet at the upper end of the pipe.

Tests for Sewer Pipes.—" The impermeability of a pipe may be tested by tying a piece of bladder over one end, reversing it, and filling with water. If it is not perfectly impervious, the water will begin to ooze through the pores " of the pipe.

The absorption of water, ascertained by weighing a dry pipe, immersing it for twenty-four hours, wiping dry, and reweighing, was found by Mr. Baldwin Latham to vary from ·429 to 6·89 per cent of the weight of the dry pipe.

The power to resist chemical action may be tested by pulverising a piece of the pipe and boiling it in hydrochloric acid, washing on filter, and noting loss in weight. Dr. Millar has shown that in stoneware pipes there should be no loss.

Stanford's Patent Joint is used in order to get a perfectly close joint between the lengths of socket pipes.

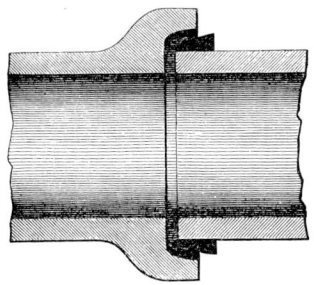

This is ensured by casting upon the spigot and in the socket of each pipe, rings of a durable material (a mixture of coal-tar, sulphur, and ground pipes), which, when put together, fit mechanically, so as to form a water-tight joint without the aid of cement.

Fig. 56.

In putting such a joint together, the surface is sometimes greased with Russian tallow and resin. Among other patent joints in the market may also be mentioned Doulton's, Hassall's, Robin's, Patent Paragon, Sykes, and others.

Miscellaneous Clay Wares.—The variety of articles used by the engineer and builder—which are made from burnt or baked clay—is endless. A few of the more important may now be mentioned.

Perforated Air Bricks are made in stoneware and terra cotta. They are pierced with different patterns, and are most useful for ventilating courses, supplying air to stores, etc.

They are better for this purpose than iron gratings, as they are cheaper, more durable, do not stain the walls with rust, or require painting.

Fig. 57.

The pattern shown is from Mr. Jennings' circular. They are made in all sizes, from 9 × 3 inches up to 18 × 18 inches.

DAMP-PROOF COURSES are made in stoneware (or sometimes in fireclay) pierced with perforations of different patterns.

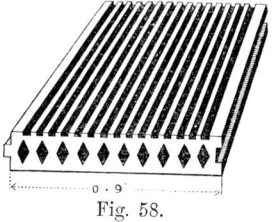

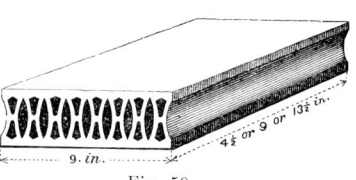

Fig. 58.   Fig. 59.

The slabs are generally 9 inches long. They vary from 9 inches to 18 inches in width, to suit different thicknesses of brick wall, and their own height or thickness varies from $1\frac{1}{2}$ to $2\frac{3}{4}$ inches.

A damp-proof course slab in stoneware, as made by Messrs. Doulton, with ribbed surfaces and tongue and groove joints, is shown in Fig. 58.

A thicker slab, as made by Mr. Jennings, is shown in Fig. 59.

The method in which these damp-proof courses are used is explained at page 6, Part II.

BONDING BRICKS.—These bricks, introduced by Mr. Jennings, are used for uniting the opposite sides of hollow walls.

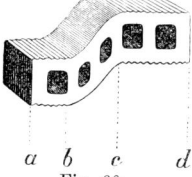

The original bricks were straight. A sketch of one is given in Fig. 61.

The improved bonding bricks are, however, bent (see Fig. 60), so that water endeavouring to pass from the outer to the inner side of the wall would have to go up an incline.

Fig. 61.

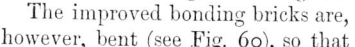

a   b   c   d
Fig. 60.

An illustration of the use of these bricks is given at page 64, Part I.

These bricks are made in four sizes, ranging from No. 4 to No. 7 ; their dimensions being as follows :—

Parts of the brick. See Fig. 60.	No. 4.	No. 5.	No. 6.	No. 7.
a b	$2\frac{1}{4}$	$2\frac{1}{4}$	$4\frac{1}{2}$	$4\frac{1}{2}$
b c	3	3	3	$4\frac{1}{2}$
c d	$2\frac{1}{4}$	$4\frac{1}{2}$	$4\frac{1}{2}$	$4\frac{1}{2}$

WALL FACINGS are made of different patterns, in earthenware and in terra cotta. Those patented by the Broomhall Company are of an L shape, and are used to form a superior facing to walls built of concrete.

SLEEPER BLOCKS, made in stoneware, are useful for carrying floors which are close to the ground in damp situations.

Fig. 62 shows a specimen of one of these, as made by Mr. Jennings. They are made for 9 inch and 4½ inch walls, or wall plates.

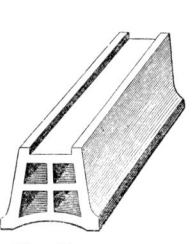

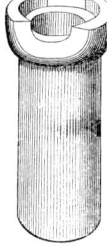

Fig. 62.　　　　　　　　Fig. 63.　　　　　　Fig. 64.

CHIMNEY-FLUE PIPES [1] are made in terra cotta, fireclay, etc.

These pipes are intended for lining chimney flues, instead of pargetting them see Part I. page 85.

They are frequently cylindrical, with plain butt joints, sometimes with ordinary sockets, or with the sides of the sockets cut off, as in Fig. 63.

They are made of 9, 10, or 12 inches in diameter, and generally in 2-feet lengths.

These pipes are sometimes oval, or of a section consisting of a rectangle with the corners rounded, as in Fig. 64.

The oblong sections are manufactured by Messrs. Doulton and Co. in three sizes :—

$$16 \times 10 \text{ inches.}$$
$$14 \times 9 \quad ,,$$
$$10 \times 6 \quad ,,$$

Combined Smoke and Air Flues are made in the form shown in Fig. 65.

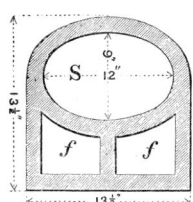

These pipes are intended to be built into chimney breasts. The smoke ascends the flue S, while the foul air is drawn off through the flues marked $f f$.

The blocks containing these flues are made in different forms and sizes. In some patterns the smoke flue is circular, 12 inches in diameter, the whole block occupying 18 × 14 inches. In others there is only one air flue, and the whole block takes up only 14 × 9½ inches in plan.

Fig. 65.

CHIMNEY POTS of every imaginable design are made in terra cotta. Any attempt to illustrate them in detail would be useless.

Billing's Chimney Terminals, with partitions similar to those described in Part I. page 91, are made in terra cotta by Messrs. Doulton and Co.

Invert Blocks of stoneware for sewers have been mentioned in Part II., and their advantages described.

The best of these are provided with a projecting lip, as shown in Fig. 67, which covers the joint between two adjacent blocks.

Sometimes two or three blocks are combined to form an invert, as in Fig. 68.

JUNCTION BLOCKS (Fig. 69) are intended to be built into brick sewers to receive pipe drains.

Sc. *Vent-linings.*

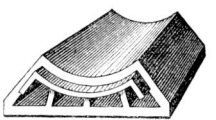

Fig. 66.   Fig. 67.

They are made either direct as at A, or oblique as at B, to suit the position of the drain.

The blocks vary in pattern so as to fit drains of any size, placed at different angles of inclination.

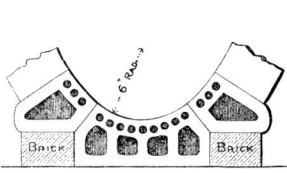
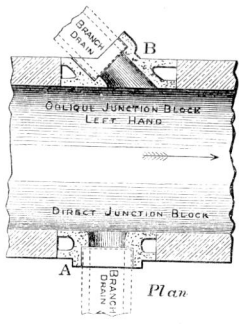

Fig. 68.   Fig. 69.

SEGMENTAL SEWERS are made in stoneware of pieces formed to the shape of segments of the circle, and united by groove and tongue joints.

Gulley Traps for streets and yards, *Sewer Gas Interceptors, Traps, Sluice Valves, Valve Traps, Channel Pipes* for sewage, *Conduits,* and sanitary apparatus of every form and variety, are made in stoneware, but any detailed description of them is necessarily omitted here for want of space.

The same reason makes it necessary to exclude any description of the various ornamental articles executed in terra cotta, such as dental, dog-tooth, and moulded cornices, trusses, medallions, cornices, moulded arch blocks, lintels, jambs, capitals, pier caps, parapet fittings, terminals, etc. etc.

Stoneware is also made of every form and colour for wall decoration, both external and internal.

TILES.

The tiles used in connection with buildings may be divided into two great classes.

1. Common tiles of different shapes used for roofing and paving.
2. Encaustic tiles used for decorative purposes.

Common Tiles are made out of somewhat the same material as ordinary

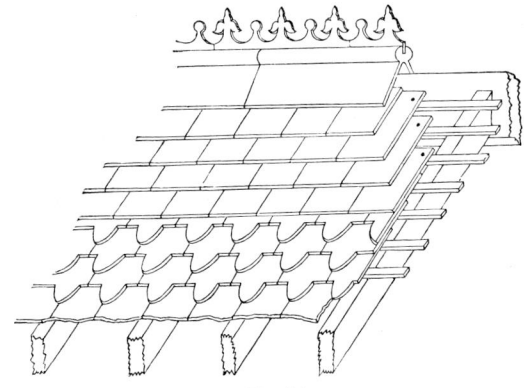

Fig. 70.

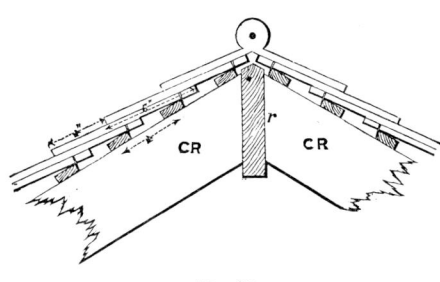

Fig. 71.

bricks, but they should be purer or stronger clays—well worked so as to bear " thwacking," or they will be liable to lose shape in burning.

The clay is weathered either by exposure to frost or sun—allowed to mellow in pits—tempered—pugged, cleared from stones—moulded, trimmed with a knife — thwacked (that is beaten when half dry with a wooden bat to correct warping) and burned in a domed kiln.

Common tiles are made in a great variety of shapes, for roofing, paving, and other purposes.

PAVING TILES—for common purposes—are made in different shapes, such as squares, hexagons, etc., and in sizes varying from 6 × 6 inches, to about 12 × 12 inches, and about 1 inch thick.

Flooring tiles are sometimes known as *Quarries.*

ROOFING TILES.—Of these there are several different kinds, a few of which will now be described.

Tile roof coverings are heavy ; moreover they are apt to absorb water, and to keep the roof wet.

To prevent this they should be glazed, which involves reburning and makes them expensive.

Many descriptions of roof tiles do not fit together very closely, and therefore require pointing to make a tight roof.

Plain Tiles are flat, either rectangular, or cut to various patterns. See Fig. 70.

Each tile is pierced with two holes near its upper edge, through which small oak pegs are driven, by which the tile is hung on to battens or laths, nailed apart at the proper gauge, as described in Part I.

Sometimes the holes are omitted, or two little projections at the back of the tile are provided to take the place of the oak pins.

Pantiles [1] are moulded flat, and afterwards bent to the form shown in Fig. 72, over another mould.

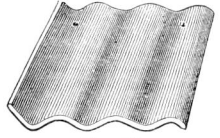

Each tile has a *stub*, projecting about $\frac{3}{4}$-inch from the centre of the back of the upper edge of the tile, by which it is hooked on to the laths.

The method of laying these tiles is described at page 205, Fig. 72. Part I.

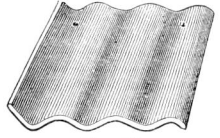

Fig. 73.

Double Roll Tiles are like two pantiles joined together, side by side. They have three stubs on the back.

Corrugated Tiles are similar to pantiles, but each tile contains three or four corrugations, as in Fig. 73.

Improved Corrugated Tiles have flat pieces alternating between the corrugations.

Taylor's Patent Roofing Tiles, now known as the *Broomhall Company's Patent Roofing Tiles*, form a handsome roof covering.

The form of the tiles is shown in Fig. 75. They are laid alternately as

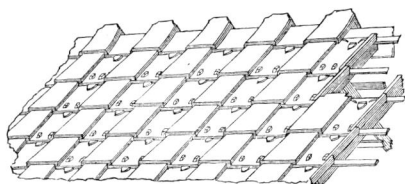

Fig. 74.

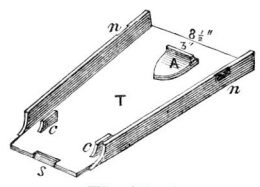

Fig. 75.

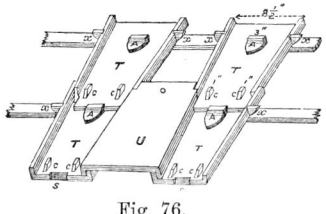

Fig. 76.

capping and channel tiles, as shown in Fig. 76, in which T T are laid as channel tiles, while U, being a tile of the same form as the others, is reversed to fit over them as a capping tile.

A description of the method of laying these tiles has been given at page 207, Part I.

Venetian or Italian Tiles are of the form shown in Fig. 77.

---

[1] Sometimes called *Flemish Tiles.*

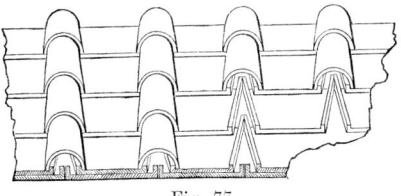

Fig. 77.

The snow is rather apt to lodge upon these tiles, and when it thaws to pass through the roof.

Wade and Cherry's Tiles.—These tiles are each shaped something like the ace of spades, so that their form renders the amount of lap smaller than in ordinary tiles.

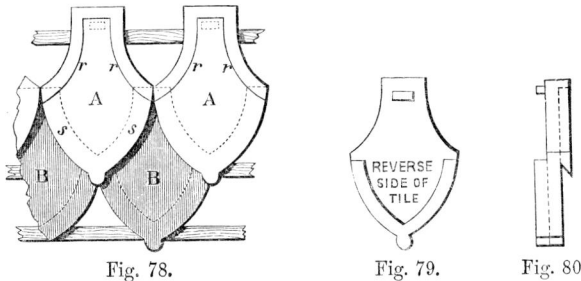

Fig. 78.   Fig. 79.   Fig. 80.

A flange, or raised rim, of dove-tailed or under-cut section is formed on the top half of the uppermost side of each tile (see Figs. 79, 80), and on the lower half of the undermost side (the latter is dotted in Fig. 78). The upper flanges correspond to $r\ r$. Of course B in the figure hooks on to the lower flanges $s\ s$. This holds them firm, and it is said to exclude wind and rain, and to render pointing unnecessary.

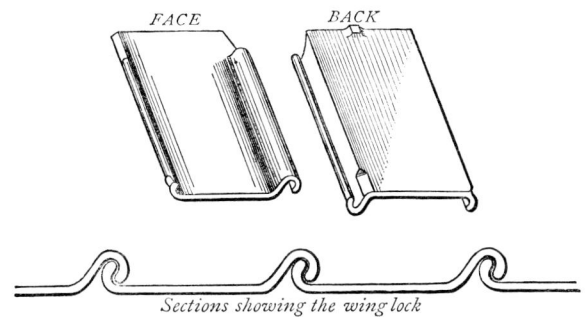

Sections showing the wing lock

Fig. 80*a*.

Foster's Lock Wing Roofing Tiles are illustrated in Fig. 80*a*,[1] which

---

[1] From the Patentee's Circulars.

explains itself. It is claimed for these tiles that they are cheaper than the commonest tiles made, can be hung quickly and without skilled labour, require no pointing, and cannot be blown off the roof, as the stronger the pressure is underneath, the tighter the lock.

Poole's Patent Bonding Roll Roofing Tiles are shewn in Fig. 80*b*,[1] which requires no description.

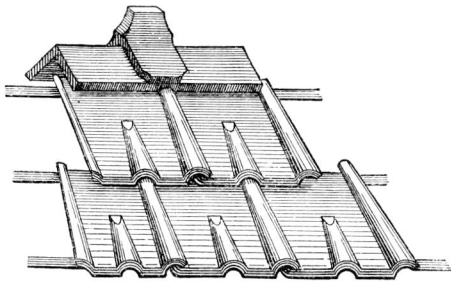

Fig. 80*b*.

Ridge Tiles are made of various forms : Plain, as in Figs. 81, 82 ; with a grooved roll to contain detached *fleurs*, as in Fig. 70 ; or with a plain roll.

The various lengths may be joined by pegs, holes for which are left in the rolls, as in Fig. 71, or they may be made to lap, as in Fig. 82.

Fig. 81.   Fig. 82.

In some varieties fleurs or other ornaments are made in one piece with the tile.

Other kinds, such for example as the Broomhall tiles above mentioned, require special ridges.

Hip and Valley Tiles are made of special shapes, to fit the hips and valleys of tiled roofs. Their form necessarily varies according to the pattern of tile and the pitch of the roof to which they are to be fitted.

Wall Tiles.—*Hall's Hanging Tiles* are glazed of different colours and fixed to walls by a nail in each tile driven into the joints of the brickwork, and are used to cover walls where light is important, as in areas, or for cleanliness, as in dairies.

Encaustic Tiles are those in which the colours are produced by substances mixed in with the clay—not printed on after the tile is made.

Such tiles may be made from ordinary clays and marls carefully prepared—sometimes mixed with finer clays, and also with different colouring substances, such as manganese for black, cobalt for blue.

Those tiles which are ornamented by inlaid patterns of different colours are made in the following manner :—

---

[1] From the Patentee's Circulars.

The clay used is first very carefully prepared—mixed with the colouring matter, and "*slipped*," that is, passed through fine muslin or silk sieves; boiled in the *slip-kiln* until it becomes plastic, *wedged*, that is cut up into pieces, which are dashed against one another to drive out the air and consolidate them ; and *aged*, that is kept for several months, during which fermentation goes on and organic matters disappear. During this time the wedging should be repeated at intervals. After this the clay is *slapped*, that is, cut up by means of a wire into long pieces, which are kept always in the same direction. This consolidates the mass and preserves the grain.

Each tile generally consists of three layers :—The face, which is a slab of very pure clay of the colour required for the ground of the pattern ; the body, which is of coarser clay ; and the back, to prevent warping, which is formed with a thin layer of clay different from the body.

The clay for the face is cut into a pat about $\frac{1}{4}$ inch thick, and as much larger in area as will allow for contraction in burning. It is then placed upon a plaster of Paris slab, upon which the form of the inlaid pattern is left in relief. The face clay pressing upon this receives an indentation corresponding to the form of the pattern required.

It is then backed up with the body of coarser clay, and the thin layer to form the back.

At this stage the maker's name is stamped on the back, and also a few holes to make the cement adhere to the tile when it is set.

Slip clay of the different colours required, according to the design, is then poured into the different parts of the indented pattern on the face.

After this has become hard, the superfluous clay is carefully scraped off, leaving it only in the parts originally indented so as to form the pattern.

The raw clay tiles are then trimmed, carefully dried, baked in ovens, protected from smoke, etc., by being arranged in large fireclay jars called *seggars*.

The burnt tiles may then, if required, be glazed by dipping them into a mixture of powdered glass and water, and reheating.

Inferior Encaustic Tiles.—"A class of pseudo-encaustic tiles is now being largely made, in which the colour, which should be burnt in along with the clay, is merely applied as a transfer printed pattern on the surface.

" Such tiles are frequently coated in the glass oven with a transparent fritted glaze, and serve for flower boxes, wall tiles, and such like purposes.

" To give them the appearance of having true inlaid colours, the edges of these tiles have frequently a little colour applied to them to represent the depth of the insertion of the coloured clay." [1]

Dry Tiles.—These tiles, each of which is of the same colour throughout, are made by the dry process.

The clay having been very carefully prepared is mixed with the colouring matter, " *slipped*," dried, and reduced to fine powder.

It is then placed in a press and subjected to enormous pressure from a steel die. This reduces the powder to a third of its bulk and thoroughly consolidates it; at the same time the pattern, if any, is impressed upon the tile by means of the die.

They are then carefully dried in a hot room, glazed, and fired.

There are several places in which encaustic tiles are made, but the most celebrated manufactory in the country is that of Messrs. Minton, Hollins, and Co., at Stoke-upon-Trent; the founder of which, Mr. Herbert Minton, brought the art to its present state of perfection.

There are other manufacturers at Stoke, Staffordshire, at Poole in Dorsetshire, also at Broseley, near Hereford.

Tesseræ are tiles sometimes made by the dry process just described, and are so accurate in form that they can be laid as mosaic work in pavement without any rubbing or injury of the face.

They are sometimes made out of moist clay, and cut into various shapes by wires.

Majolica Tiles have raised patterns, and their colour " applied in the form of an enamel or coloured opaque glaze. They have not therefore the same amount of durability, and are only used for walls and similar ornamental purposes." [2]

Mosaic Paving Slabs are made by arranging tesseræ in the pattern required. Strips of wood are placed round the whole so as to form a rough frame.

Portland cement is then run in over the backs of the tesseræ, and the whole strengthened and formed into a slab by two layers of common tiles set in cement.[3]

---

[1] Report on International Exhibition, 1871, by Gilbert Redgrave, Esq.
[2] *Ibid.*   [3] Gwilt's *Encyclopædia.*

Uses.—Flat encaustic tiles made by either process may be used for paving or wall decoration, but those with raised patterns must of course be restricted to the latter purpose.

In some cases the tiles for wall decoration are put together in panels before being glazed. A picture is painted upon the panel, the tiles composing it are then separated, burnt at a high temperature and glazed.

Chemical Analysis.—The student who is desirous of pursuing this branch of the subject is referred, for the chemical analysis of a brick earth or brick, to *Notes on the Chemistry of Building Materials*, by Captain Abney, R.E., F.R.S.

CHAPTER III.

LIMES, CEMENTS, MORTAR, CONCRETE, PLASTERS, AND ASPHALTES.

LIMES AND CEMENTS.

THERE are hardly any materials used by the engineer, architect, or builder, on which so much depends as upon mortar and concrete.

There are differences of opinion on many points connected with the preparation and use of these materials, and there is still much prejudice existing in favour of exploded notions and of old-fashioned ideas.

These prejudices are the more difficult to overcome, because the old-fashioned methods of preparing mortar and concrete were, as a rule, less troublesome than those of more recent introduction.

In order to clear the way for a proper understanding of this important subject, it will be well, first, to explain the meaning of some of the commonest terms used in connection with it.

Terms in Use.—The natural *Limes and Cements* used for building are produced by the calcination of limestones or other calcareous minerals, the effect of which is to drive off the carbonic acid and moisture they contain.

Calcination is heating to redness in air.

Quicklime or *Caustic Lime* is the resulting lime as left immediately after calcination.

Slaking is the process of chemical combination of quicklime with water. This gives rise to various phenomena which will be more particularly described hereafter. (See p. 151.)

Slaked Lime is the substance remaining after slaking, and is chemically known as the "*hydrate of lime.*"[1]

---

[1] *Calcium hydrate.* The ordinary chemical nomenclature has been adopted throughout these Notes as being more familiar to readers generally than the new nomenclature. The modern names are given in footnotes.

Setting is the hardening of lime which has been mixed into a paste with water.

This is quite a different thing from mere drying. During drying the water in the paste evaporates, but no setting action takes place.

Hydraulicity.—Lime or cement is said to be more or less hydraulic, according to the extent to which paste or mortar made from it will set under water, or in positions where it is free from access of air.

Limestones and other minerals from which limes and cements are produced differ greatly in their composition, ranging from pure carbonate of lime,[1] such as white chalk or marble, to stones containing from 10 to 30 per cent of clay, in addition to other foreign constituents, such as magnesia, oxide of iron,[2] etc.

As the properties of the resulting lime or cement depend very greatly upon the composition of the stone from which it is prepared, it will be instructive briefly to note the characteristics of the most common constituents of such stones before proceeding further; especially distinguishing those which produce hydraulicity from those which have not that effect.

CONSTITUENTS OF LIMESTONE.

Constituents of Limestones which do not produce Hydraulicity. — CARBONATE OF LIME.[1]—As already noticed, some limestones, such as chalk and marble, consist entirely of this substance, and in all it plays an important part.

When pure carbonate of lime is calcined, the carbonic acid and water contained in it are driven off, and "quicklime" results.[3]

Slaking.—If the quicklime is treated (either by sprinkling or dipping) with as much water as it will easily absorb, it almost immediately cracks, swells, and falls into a bulky powder with a hissing crackling sound, slight explosions, and considerable evolution of heat and steam ;—this is the process of "slaking." By it pure lime is increased in volume from 2 to $3\frac{1}{2}$ times its original bulk,—the variation depending both on the density of the original carbonate and on the manner of conducting the process.

Air-slaking.—If the pure quicklime be exposed to the air, it will gradually absorb moisture, and fall into a powder with increase of volume, but without perceptible heating ; it is then said to be "air-slaked." Some carbonic acid is also absorbed in "air-slaking."

Setting.—If a small pat be made of paste from the slaked lime and placed

---

[1] *Calcium carbonate.*   [2] *Ferric oxide.*

[3] Called also "anhydrous" or "caustic" lime.

under water, it will slowly dissolve, until (if the quantity of water be sufficient, or is changed often enough) it entirely disappears.

In air the surface of the pat will absorb carbonic acid, which reconverts it into a carbonate of lime. This action continually decreases, and practically ceases after forming a surface crust less than half an inch thick—the interior remaining pulpy or friable, according as the situation is damp or dry, and undergoing no further change of any kind.

SAND, of an ordinary description (such as that from flint or grains of quartz) occurring as an impurity in the limestone, has by itself no chemical action with the quicklime, when forming part of a limestone calcined at the temperature ordinarily reached in a kiln, but constitutes with it a mere mechanical mixture ; forming what is called a " Poor Lime," and having the effects described at page 157.

Constituents of Limestone which produce Hydraulicity.—The substances above noticed give the lime no hydraulic properties whatever.

It is most important to understand distinctly what constituents are necessary in a limestone to confer upon it the characteristic of hydraulicity.

These will now be shortly referred to.

CLAY is the most important constituent of those which produce hydraulicity in limestones, indeed the great majority of hydraulic limes owe their properties to the clay they contain.

The effects produced by the presence of clay in a limestone are as follows :—

a. It greatly modifies the slaking action. When a large proportion of clay is present, such action does not take place at all.

b. It confers the power of setting, and remaining insoluble under water, or in other positions where the air has no access.

In order that the clay may properly fulfil its functions, it is necessary—

1. That the amount of clay should be properly proportioned to that of the remaining constituents.

(The effects above mentioned are more marked as the proportion of clay is greater, up to a certain limit when the excess of clay becomes injurious.)

2. That the stone should be calcined at the proper temperature.

(This is a very important and very intricate portion of the subject. The same stone will give very different results according to the degree of calcination to which it is subjected.)

The nature of the changes undergone by the clay, and the evils caused by over-burning or under-burning the stone, are explained at page 235.

These changes are of a somewhat complicated nature, and it will be sufficient at present to note the fact that after proper calcination of a limestone containing clay, the result is a substance containing a proportion of free quicklime together with compounds (formed by the clay and lime) which have the property of becoming hard when formed into a paste, even if secluded from the air or placed under water.

SOLUBLE SILICA.—There are several forms of silica, such as sand, flint, etc., which, as already noticed, are useless in lime, for they are only in a state of mechanical mixture with it. The silica must be in combination with other substances and in a peculiar soluble state, or it will not combine with the lime ; in such a state it is found in clay.

Unfortunately, in most analyses of limestones the soluble or usefully active form of silica is not distinguished from the sand, or silica in an inert state ;

this leads to some confusion, and renders the analyses less useful than they would otherwise be.

CARBONATE OF MAGNESIA [1] combined with lime reduces the energy of the slaking, and increases that of the setting processes ; when other substances are present, its behaviour and combination with them are similar to those of lime.

When carbonate of magnesia is present in sufficient quantity (about 30 per cent), it renders lime hydraulic independently of and in the absence of clay.

ALKALIES AND METALLIC OXIDES.—These if exposed to a great heat become fused and quite inert ; but when subjected only to lower temperatures sometimes tend to produce soluble silicates, and thus to cause hydraulicity.

SULPHATES in small quantities tend to suppress the slaking action, and to increase the rapidity of setting.

The introduction of these is the basis of a class of cements which will be considered presently. (See p. 184.)

CLASSIFICATION OF LIMES AND CEMENTS.

The calcined limestone is divided, according to its action in slaking and setting, into the following classes :—

1. Rich, Fat, or Pure Limes.
2. Poor Limes.
3. Hydraulic Limes.
4. Cements $\begin{cases} \text{Natural.} \\ \text{Artificial.} \end{cases}$

These classes merge gradually the one into the other, without sharp distinctions, the difference between them depending upon the nature and amount of the foreign constituents associated with the lime, and upon the degree of calcination to which the stone has been subjected.

The physical characteristics of the raw stone are no index to the properties of the lime or cement produced from it. These properties may however be inferred from the nature and proportions of the chemical constituents of the stone. A general composition has been assigned to the materials yielding each of the classes above mentioned, but it must be borne in mind that this is only an approximate indication of quality, and that the behaviour when calcined and treated with water is the only safe means of classification.

The following Table shows the composition of a number of limestones and cement stones, chosen as characteristic examples, and intended to give some idea of the varieties actually met with.

In comparing these analyses with others, it must be borne in mind that these show the composition of the raw stone, or raw material from which the lime or cement is produced. Analysis of the burnt lime or cement would in each case have given a higher percentage of clay and sand, and the lime and magnesia would not appear as carbonates. (The carbonic acid would have been expelled during calcination.)

---

[1] *Magnesium carbonate.*

TABLE

GIVING THE COMPOSITION OF VARIOUS LIMESTONES, CEMENT STONES, ETC.,
BEFORE CALCINATION.

		COMPOSITION OF RAW STONE OR RAW MATERIAL.				
Nature of Lime or Cement produced.	Description.	Carbonate of Lime and Carbonate of Magnesia.	Clay, Sand, Iron, etc.	Water and Loss.	Analyst or Authority.	
Rich or Fat Limes.	Carrara Marble (see p. 55)	100 carb. lime	..	..	Vicat.	
	White Chalk . .	98·6 carb. lime ·4 carb. magnesia	2 iron, manganese and phosphates ·8 silica and alumina	..	Schweitzer (Reid).	
		99·0	1·0			
	Bath Oolite (see p. 59)	94·5 carb. lime 2·5 carb. magnesia	1·2 iron and alumina			
		97·0	1·2	1·8	Professors Daniel and Wheatstone; Commission on Stone for Houses of Parliament.	
	Portland Oolite (see p. 60)	95·2 carb. lime 1·2 carb. magnesia	·5 iron and alumina 1·2 silica			
		96·4	1·7	1·9	Do.	
Feebly Hydraulic. Poor Limes.	Siliciferous Oolite, Chilmark Stone (see p. 63)	79·0 carb. lime 3·7 carb. magnesia	2·0 iron and alumina 10·4 silica (nearly all sand)			
		82·7	12·4	4·2	Do.	
	Grey Chalk, Halling (see p. 160)	92 carb. lime .	8 clay	..	Col. Scott, R.E.	
Hydraulic.	Roach Abbey, Dolomite (see p. 59)	57·5 carb. lime 39·4 carb. magnesia	·7 iron and alumina ·8 silica			
		96·9	1·5	1·6	Professors Daniel and Wheatstone; Commission on Stone for Houses of Parliament.	
	Bolsover, Dolomite (see p. 59)	51·1 carb. lime 40·2 carb. magnesia	1·8 iron and alumina 3·6 silica			
		91·3	5·4	3·3	Do.	
	Lower Lias, Aberthaw (see p. 160)	86·2 carb. lime	11·2 clay . . .	2·6	Phillips (Captain Smiths Vicat).	
	Grey Chalk, Sussex (see p. 160)	83 carb. lime .	17 clay . . .	..	Col. Scott, R.E.	

COMPOSITION OF LIMESTONES, ETC.—*Continued.*

		COMPOSITION OF RAW STONE OR RAW MATERIAL.[1]				
Nature of Lime or Cement produced.	Description.	Carbonate of Lime and Carbonate of Magnesia.	Clay, Sand, Iron, etc.	Water and Loss.	Analyst or Authority.	
Eminently Hydraulic.	Blue Lias, Lyme Regis (see p. 160)	79·2 carb. lime	17·3 silica and alumina	3·5	Reid.	Some varieties contain less clay. Average about 12·5 per cent.
	Carboniferous, Holywell, Wales (Halkin Mountain Limestone), see p. 160	71·55 carb. lime 1·35 carb. magnesia — 72·90	3·5 alumina 2·2 iron. ·8 alkalies 20·1 silica — 26·6	·5	Muspratt.	
	Arden, near Glasgow (see p. 160)	68·0 carb. lime ·8 carb. magnesia — 68·8	25·8 clay 2·4 iron ·6 chlorides — 28·8	2·4	Ingram.	
Slow Setting Cements.[1]	Heavy English Portland 3 White Chalk and 1 Clay dried, but *unburnt* (see p. 165)	.. .. 77· carb. lime — 77·0	2·7 alumina 3·5 iron 15·8 silica 1·0 alkalies — 23·0	..	Reid.	
	Portland Cement, good	58 to 63 carb. lime	21 to 24 silica 5 to 9 alumina 3 to 6 oxide iron 0·5 to 1·5 sulphuric acid	..	Specification of l'Administration des Pont et Chaussees.	Currie & Co.'s circular
	Kimmeridge Clay (Boulogne) (Natural Portland) *unburnt*	76·6 carb. lime ·8 carb. magnesia — 77·4	9·2 iron and alumina 13·4 silica — 22·6	..	Gilmore.	
	Native Magnesia (Madras)	99 carb. magnesia .	·5 silica	·5	Dr. Malcomson (Captain Smith's Vicat.)	
Moderately Quick Cements.	Dolomite, Portgyfu, North Wales	21·4 carb. lime 61·15 carb. magnesia — 82·55	5·58 silica 2·07 alumina 8·76 iron — 16·41 . . .	1·1	Professor Cabutt (Lipowitz).	
	Rosendale Cement Stone, Layer No. 9. High Falls, Ulster, New York	43·3 carb. lime 26·0 carb. magnesia — 69·3	20·7 silica and alumina 1·9 iron 2·0 sulphuric acid 4·2 alkaline chlorides — 28·8	1·9	Professor Boynton (Gilmore).	
	Medina Cement Stone (see p. 163)	47·80 carb. lime — 47·80	25·69 iron and alumina. 1·50 sulphuric acid 24·50 silica . . — 51·69	·51	Ingram.	

[1] For analyses of the burnt Portland cement see p. 243.

COMPOSITION OF LIMESTONES, ETC.—*Continued.*

	COMPOSITION OF RAW STONE OR RAW MATERIAL.[1]					
Nature of Lime or Cement produced.	Description.	Carbonate of Lime and Carbonate of Magnesia.	Clay, Sand, Iron, etc.	Water and Loss.	Analyst or Authority.	
Quick Cements.	Roman Cement Stone from Calderwood (Scotland), see p. 164	54·0 carb. lime 14·2 carb. magnesia —	3·4 alumina 13·31 iron 8·8 silica 2·6 phosphates			
		68·2	28·1	3·7	Professor Penny (Gilmore)	
	Medina Cement Stone from Portsmouth, Isle of Wight (see p. 163)	45·32 carb. lime ·50 carb. magnesia	14·15 iron and alumina 1·70 sulphuric acid 37·65 silica			
		45·82	53·50	·68		
	Roman Cement Stone from Boulogne Septaria	61·6 carb. lime	4·8 alumina 15·0 silica 9·0 iron			
		61·6	28·8	9·6	Drapping (Captain Smiths Vicat).	
	Roman Cement Stone from Sheppy Septaria	65·7 carb. lime ·5 carb. magnesia	6·6 alumina 6·8 iron 1·9 manganese 18·0 silica	1·30	Do.	
		66·2	32·5			
	Rosendale Cement Stone, Layer No. 16, High Falls, New York	46·0 carb. lime 17·8 carb. magnesia	30·0 silica and alumina 1·3 iron. ·2 sulphuric acid 4·1 alkaline chlorides	·7	Professor Boynton (Gilmore).	
		63·8	35·6			

Rough Tests.—A few rough tests may be applied to a limestone to see if it is likely to furnish a hydraulic lime or cement.

Such a stone will generally have an earthy texture, and will weather to a brown surface.

Acid will not cause upon it so great an effervescence as upon purer limestones.

When breathed upon or moistened a clayey odour is emitted from the stone.

The best plan, however, is to burn a little of the stone in a small experimental kiln, to judge by the slaking, and by the behaviour of pats made from the paste.

---

[1] For analyses of the burnt Roman cement see p. 243.

LIMES.

Rich or Fat Limes are those calcined from pure, or very nearly pure, carbonate of lime, **not** containing sufficient foreign constituents to have any appreciable effect upon either the slaking or setting actions.

The phenomena attendant upon these actions and the characteristics of the resulting paste exactly resemble those described for pure carbonate of lime (see p. 151), and need not be repeated.

Uses.—The solubility and want of setting power of fat lime render it unsuitable for making mortar, except for the walls of outhouses and for other similar positions. It is nevertheless frequently used for the mortar in structures of a much more imposing character.

It is however better than hydraulic limes for sanitary purposes (being purer), and is very useful for plastering and for whitewashing. It is also extensively employed in the manufacture of artificial hydraulic limes and cements.

Precautions in Using.—Fat lime requires to be mixed with a great deal of sand to prevent excessive shrinkage, but this addition does not materially injure it, as it attains no strength worth mentioning under any circumstances.

The only setting that takes place in it is the formation of a thin surface crust, bearing a small proportion to the whole bulk ; mortar made from such lime may therefore be left and re-worked repeatedly without injury.

Stained Fat Limes.—Some of the lime which finds its way into the London market, under the assumed names of Dorking, Halling, and Merstham, is merely fat lime tinged with iron sufficiently to give it the buff colour characteristic of the hydraulic lime made out of the grey chalk from the localities above mentioned (see p. 160).

Of course, this stained lime makes mortar of the same inferior description as would be obtained from a common fat white lime, and has no hydraulic properties whatever.

Poor Limes are those containing from 60 to 90 per cent of carbonate of lime, together with useless inert impurities, such as sand, which have no chemical action whatever upon the lime, and therefore do not impart to it any degree of hydraulicity.

These limes slake sluggishly and imperfectly, the action only commences after an interval of from a few minutes to more than an hour after they are wetted, less water is required for the process, and it is attended with less heat and increase of volume than in the case of the fat limes.

If they contain a large proportion of impurities, or if they are over-burnt, they cannot be depended upon to slake perfectly unless first reduced to powder.

The resulting slaked lime is seldom completely pulverised—is only partially soluble in water, leaving a residue composed of the useless impurities, and without consistence.

The paste formed from the slaked lime is more incoherent, and shrinks less in drying, but behaves in other respects like that made from fat lime—in fact, it is like a fat lime mortar containing a certain proportion of sand.

Mortar made from poor lime is less economical than that from fat lime, owing to the former increasing less in slaking, bearing less sand (as the lime already contains some in the form of impurities), and requiring a more troublesome manipulation than the latter. It is in no way superior as regards setting, and should therefore only be used when no better can be had.

Hydraulic Limes are those containing, after calcination, enough quicklime to develope more or less the slaking action, together with sufficient of such foreign constituents as combine chemically with lime and water to confer an appreciable power of setting without drying or access of air.

Their powers of setting vary considerably. The best of the class set and attain their full strength when kept immersed in water.

They are produced by the moderate calcination of stones containing from 73 to 92 per cent of calcium carbonate, combined with a mixture of foreign constituents of a nature to produce hydraulicity.

Different substances have this effect, as already mentioned (see p. 152), but in the great majority of natural hydraulic limes commonly used for making mortar, the constituent which confers hydraulicity is *clay*.[1]

The phenomena connected with the slaking of limes varies greatly according to their composition. With none is it so violent as with the pure carbonate of lime (see p. 151), and the more clay the limes contain the less energy do they display, until we arrive at those containing as much as 30 per cent of clay,

---

[1] In some varieties, as before mentioned, a portion of the carbonate of lime is replaced by carbonate of magnesia, which increases the rapidity of setting.

when hardly any effect at all is produced by wetting the calcined lime, unless it is first ground to powder.

The setting properties of hydraulic lime also differ very considerably in proportion to the amount they contain of the clay or other constituent, which gives the lime its power of setting without drying or the access of air.

This led to their being subdivided by Vicat into three classes, as shown in the following Table :—

CLASSIFICATION OF HYDRAULIC LIMES.

Name of Class.	Percentage of Clay, associated with Carbonate of Lime only, or with Carbonate of Lime and Carbonate of Magnesia.	Behaviour in slaking after being wetted.	Behaviour in setting under Water.
Feebly Hydraulic	5 to 12 p. c.	Pauses a few minutes, then slakes with decrepitation, development of heat, cracking, and ebullition of vapour.	Firm in 15 to 20 days. In 12 months as hard as soap—dissolves with great difficulty, and in frequently renewed water.
Ordinarily Hydraulic	15 to 20 p. c.	Shows no sign of slaking for an hour, or perhaps several hours—finally cracks all over, with slight fumes, development of heat, but no decrepitation.	Resists the pressure of the finger in 6 or 8 days, and in 12 months as hard as soft stone.
Eminently Hydraulic	20 to 30 p. c.	Very difficult to slake—commences after long and uncertain periods—very slight development of heat, sensible only to touch—very often no cracking, or powder produced.	Firm in 20 hours—hard in 2 to 4 days—very hard in a month—in 6 months can be worked like a hard limestone, and has a similar fracture.

Varieties of Lime in Common Use.—FAT LIMES.—White chalk, marble, the Oolitic limestones, and shells, when calcined, furnish the fat limes in ordinary use. A great variety of fat limes is found in England, Scotland, and Ireland.

Oysters and other shells require burning at a high temperature. They contain gelatine, which is converted into charcoal, and burns with difficulty ; the result is a tendency to produce a badly slaking lime.

HYDRAULIC LIMES.—*Grey Chalk Lime* (called " stone lime " in London) is of a feebly hydraulic character.

It is obtained from the lower chalk beds in the South of England, the present supplies coming from Halling, Dorking, Lewes, Petersfield, Merstham, etc.

This lime is usually of a light buff colour, and slakes very freely. When used with two parts of sand in brickwork, a good sample should sensibly resist the finger-nail at a month old.

Lias Lime varies greatly in its properties according to the locality of the beds from which it is procured, some being only moderately hydraulic, and others eminently so.

The raw stone is of a dark blue colour (hence the lime is called " blue lias "), and the burnt lime a pale grey.

It slakes very sluggishly, and should set well in wet situations (according to its composition) in from one or two to several days.

This lime is sold both in lump and ground. The latter is, as a rule, the best, as the softer stones, containing more clay, are selected for grinding, but it may be adulterated with sand, or be air-slaked (see p. 207).

The lime is ground to nearly the same fineness as Portland cement (see p. 167), and sold in sacks, or, for export, in casks.

Mr. Reid says that limestones which approach nearest to the analysis given in the Table, p. 155, "should have the preference."

Lias lime is procured chiefly from the Midland and South-western counties— the best known being that from Barrow-on-Soar, in Leicestershire ; from Watchet, in Somersetshire ; Lyme Regis, in Dorset ; Whitby, in Yorkshire ; and Rugby, in Warwickshire. The Aberthaw lime, found near Cardiff, is derived from the *Lower Lias* formation.

The Carboniferous Limestones yield very valuable hydraulic limes, among which may be mentioned the Halkin Mountain limestone, from Holywell, in Flintshire ; lime found near Berwick, in Northumberland, etc.

The Arden lime, found in this formation near Glasgow, is of an eminently hydraulic character, and has been much used for docks and other important work. It partakes rather of the character of a Roman cement, and will not stand a large proportion of sand.

The Milton or Hurlett lime, and the Kilbride lime, from the same neighbourhood, are of a similar description.

Hydraulic lime is found also in Fifeshire, at Dunbar, etc. etc.

The Magnesian Limestones, found in Durham, Yorkshire, Derbyshire, Doncaster, and Notts (see p. 57), also furnish hydraulic limes, which are sometimes of a powerful character.

In Ireland the *calp* limestone yields a hydraulic lime, but it is very variable in quality.[1] A good hydraulic lime is obtained from the Gillogue quarry in the carboniferous formation near Limerick.

The lias has not been met with to any extent in Ireland, and is usually imported.

Artificial Hydraulic Lime may be made by moderately calcining an intimate mixture of fat lime with as much clay as will give the mixture a composition like that of a good natural hydraulic limestone, of which the product should be a successful imitation.

A soft material like chalk may be ground and mixed with

[1] Wilkinson's *Practical Geology of Ireland.*

the clay in the raw state. Compact limestone, on the other hand, is more commonly burnt and slaked in the first instance (as being the most economical way of reducing it to powder), then mixed with the clay and burnt a second time.

Lime so treated is called " twice kilned" lime.

The mixture may be made by violently agitating the materials together in water by machinery, or by grinding them together in a dry state, afterwards adding water to form them into a paste.

The paste in either case is moulded into bricks, which are dried, calcined, and otherwise treated like ordinary lime.

Artificial hydraulic limes are not much manufactured or used in this country.

CEMENTS.

The cements used in building and engineering works are calcareous substances, similar in many respects to the best hydraulic limes, but possessing hydraulic properties to a far greater degree.

They may be divided into two classes—

1. Natural Cements.
2. Artificial Cements.

They are distinguished from limestones by not slaking or breaking up when mixed with water after calcination.

Cements are used chiefly in foundations in wet places; in subaqueous work of all kinds; for important structures, where great strength is required, such as dock walls and lighthouses, also for making concrete and cement mortar.

The more exposed parts of ordinary structures, such as the copings of walls, are frequently built in cement, also the tops of chimneys.

Cements are also used in the walls of cesspits, the joints of drains, etc.; for protecting the outer faces of walls and buildings from the weather; for thin walls where extra strength is required ; for pointing, filleting, and many minor purposes.

NATURAL CEMENTS.

Natural cements are burnt direct from stones containing from 20 to 40 per cent of clay, the remainder consisting chiefly of carbonate of lime alone, or of carbonate of lime mixed with carbonate of magnesia.

CARBONATE OF MAGNESIA by itself, when calcined, yields anhydrous magnesia, which does not slake like quicklime, but if powdered and made into paste sets through its whole mass.

permanently expanding, but not breaking up. It is soluble in water, but not so readily as lime.

CEMENT STONES or NODULES are frequently found in thin strata, amongst those of hydraulic limestone. They are usually brown or fawn-coloured, of compact texture, and with an earthy fracture.

Those met with in this country generally contain a large proportion of clay (about 30 or 32 per cent), are burnt at a low temperature, and yield a quick-setting cement of no great ultimate strength.

These stones will not bear much heat without fusing, as they contain a large proportion of iron (see p. 240).

Stones containing a lower proportion of clay (about 22 per cent) are strongly burnt, and yield a heavy slow-setting cement.

The natural cement found at Boulogne (see Table, p. 156) is of this character, and a similar description has been met with at Rugby ; but slow-setting natural cements are rare in this country.

More than 40 per cent of clay injures the cement. If the stone is half clay, it should be used as a " pozzuolana" (see p. 201) ; if there is more than two-thirds clay, it will not set under water.

Slaking and Setting.—Lumps of burnt cement stones are hardly affected by water ; when ground to powder and wetted, they produce a paste which, without any preliminary slaking action, sets under water in from five minutes to as many hours, and acquires within a year a strength varying from that of soft brick to that of the stronger kinds of stone ; the differences in setting powers and strength depending upon the composition of the stone.

The shrinkage of cements setting in air is very slight, the paste being much denser than that made from lime, in consequence of the absence of the expansion caused by slaking.

Roman Cement (originally called Parker's Cement) is made by calcining nodules found in the London clay. These contain from 30 to 45 per cent of clay ; before being burnt they have a fine close grain, pasty appearance, and greasy surface when broken.

The burning is conducted at a low temperature and requires great care.

The colour of the calcined stone is generally a rich brown, and is no guide to the quality of the cement.

Weight and Strength.—Good Roman cement should not weigh more than 75 lbs. per bushel, and should set very quickly (within about 15 minutes of being gauged into paste), but attains no great ultimate strength (see Table, p. 164).

Specifications should mention a minimum weight for these and similar cements, for a heavy cement is likely to be over-burnt, and moreover a stale cement will have become heavier by absorption of carbonic acid from the air.

The little strength possessed by Roman cement rapidly diminishes on the addition of sand.

1 or $1\frac{1}{2}$ part of sand to one of cement is the greatest proportion that should be added.

Storing.—Roman cement is sold in a ground state, and kept in casks, which must be kept carefully closed and dry, otherwise the cement will absorb carbonic acid and become inert. For the same reason it is important to examine this cement carefully before using it.

Uses.—It should be mixed in very small quantities and used at once, and on no account beaten up again after the setting has commenced.

The properties of Roman cement make it valuable for temporarily pointing joints in work to be done and set between tides, and for other purposes where quick setting is desirable, and no great ultimate strength is required.

It is also used for external rendering or stucco, but is liable to efflorescence on the surface, which presents an unsightly appearance (see p. 242).

Market Forms.—Roman cement is usually sold in casks ; sometimes, if it is to be used at once, in sacks.

The inside dimensions of the casks are 2 feet 4 inches high, 1 foot $4\frac{1}{2}$ inches diameter at middle, 1 foot $3\frac{1}{2}$ inches diameter at ends.

Each cask usually contains $3\frac{1}{2}$ trade bushels [1] of 70 lbs. each—*i.e.* 245 lbs.

The sacks measure 3 feet 7 inches by 2 feet, and contain 3 trade bushels —*i.e.* 210 lbs.

MEDINA CEMENT is made from the septaria found in Hampshire and the Isle of Wight, and from those dredged up out of the bed of the Solent.

It sets very rapidly, is of a light brown colour, and resembles Roman cement in its characteristics, but is stronger for the first three months (see Table, p. 164).

It is sold in casks containing $3\frac{1}{2}$ trade bushels of 68 lbs. each, or sacks containing 3 bushels.

HARWICH AND SHEPPY CEMENTS are similar materials made from nodules found in the London clay at Harwich and Sheppy.

---

[1] There are two kinds of bushel used in connection with cements :—(1) The "*striked bushel*," being a measure containing 1·28 cubic feet, lightly filled, and struck smooth at the top with a straight edge (see p. 173)—21 of these bushels go to a cubic yard ; (2) The *trade bushel*, which is a given weight established by practice, and varying for each cement. The weights of trade bushels of different kinds of cement are given at p. 258.

Unless cement is ordered by weight, there is likely to be some confusion between the two kinds of bushel above mentioned. It is desirable where possible to order cement by the ton net.

WHITBY, MULGRAVE'S, or ATKINSON'S CEMENT is made from the septaria of the Whitby shale beds of the Lias formations in Yorkshire. It is something like Portland cement in colour, takes slightly longer to set than Roman cement, and absorbs more moisture, but resembles it in its characteristics generally.

CALDERWOOD CEMENT is a variety of Roman cement of a dark colour from nodules found in Scotland.

East Kilbride in Lanarkshire furnishes a very similar cement.

The following Table, compiled from different sets of experiments by Mr. Grant,[1] shows the strength of two different samples of Roman cement, and of one of Medina cement, and also the weakening effect of sand when added to one of the former :—

Age and Time immersed in Water.	ROMAN CEMENT.					MEDINA CEMENT. Neat.
	Sample A. Neat.	Sample B. Neat.	1 Cement (B) 1 Sand.	1 Cement (B) 2 Sand.	1 Cement (B) 3 Sand.	
7 Days . .	202	120·5	47·5	7·0	10·0	211·0
14 ,, . .	173	169·9	65·6	42·8	19·2	303·4
21 ,, . .	186·5	155·2	74·2	45·9	17·4	298·0
1 Month .	260·3	358·2	81·2	41·9	. .	306·0
3 Months .	322·5	220·4	121·9	91·75	...	448·8
6 ,, . .	472·7	252·5	314·3	...	..	412·4
9 ,, . .	471·1	251·5	...	...	...	457·2
12 ,, . .	643·1	268·5	...	...	...	476·9
2 Years . .	546·3	...	...	...	...	276·0

N.B.—The sectional area of the briquette was 2¼ square inches.   More recent tests by manufacturers appear to show higher results than those in the above Table.

ARTIFICIAL CEMENTS.

Hydraulic cement is made artificially by a process similar to that already described for artificial hydraulic limes (see p. 160), a higher proportion of clay being added to make the mixture resemble the composition of a natural cement stone.

The twice-kilned lime is not however used, but the raw limestone or chalk is if necessary crushed by machinery before mixing.

The cements usually manufactured are of a heavy slow-setting character, and require to be calcined at a high temperature, which produces incipient vitrification. As it is impossible to maintain a perfectly uniform temperature all through the mass, the result is a mixture of products of different degrees of calcination, including half-raw under-burnt portions of light yellow cement, and dense heavy clinker (see p. 198).

A judicious selection of them for grinding, and more especially the rejection of the under-burnt portions, is essential to the production of good and uniform cement.

[1] *Min. Proceedings Civil Engineers*, vols. xxv. and xxxii.

As the best of the cements are burnt to the state of clinkers the subsequent breaking and grinding are tedious and costly operations. Fine grinding is however most essential to properly develope the strength of the cement when used, as it commonly is, with sand.

Portland Cement is so called from a fancied resemblance in its colour to Portland stone.

It is by far the most valuable of all the cements, and is made by intimately mixing and calcining together substances of different kinds, so as to obtain a material containing, as a general rule, when burnt some 58 to 63 per cent of lime combined with about 22 per cent of soluble silica—7 to 12 per cent of alumina—and small percentages of oxide of iron, magnesia, etc. (see p. 241).[1]

_The materials used may be either chalk and clay—which are mixed by the wet process—or limestone and clay or shale mixed by the dry process.

MANUFACTURE FROM CHALK AND CLAY.—The cement best known in this country is made on the banks of the Thames and Medway, from chalk and clay mixed by the wet process.

The proportion of chalk and clay mixed together depends upon the composition of the chalk before burning. The result required is to obtain a mixture containing before burning some 23 to 26 per cent of clay.[2]

With white chalk (which itself contains no clay) 3 volumes of chalk are mixed with 1 volume of alluvial clay or mud from the lower Thames or Medway.

If the chalk itself contains clay, the proportion of clay added is modified accordingly.

For example, with grey chalk, 4 parts of chalk are used to 1 of clay.

The chalk and clay are mixed in water to the condition of a creamy liquid, which is called " slurry," the fine particles in suspension are allowed to settle in large tanks, reservoirs, or " backs," for several weeks, and when the deposit becomes nearly solid, the water is run off, the residue is dug out, sometimes pugged, dried on iron plates over coking ovens, or over the flues from the kiln, burnt in intermittent kilns (see p. 194), at a very high temperature, and then ground to a fine powder.

This method of manufacture is of course applicable only when the materials to be mixed can easily be liquefied in water.

The above is the wet process as ordinarily practised on the Thames and Medway, but in very modern works modifications have been introduced, some of which may be mentioned.

Under the patents of Messrs. I. C. Johnson & Co. of Greenhithe the undermentioned processes have been adopted at their various works, and some of them have been introduced at other works.

The chalk and clay are mixed with much less water (only about 10 per cent) to the consistency of batter pudding. The slurry thus formed is passed through gratings into a pit, whence it is lifted by buckets fixed to the circumference of a vertical revolving wheel, and passed through millstones which grind it to a minute degree of fineness.

---

[1] For very light quick-setting cements the proportion of lime is considerably less (see p. 241).

[2] 28 or even 30 per cent of clay may be used for light quick-setting cements used for stuccoing.

The mixture is then pumped up and spread over the floor of a large arched chamber which branches out from the kiln at a height of about 15 feet above the fire bars. The top of the kiln is closed, so that the waste heat and gases have to pass through this chamber and over the surface of the slurry, which is thus quickly and thoroughly dried. It is then burnt in the usual manner.

It will be seen that by this system the *backs* are rendered unnecessary. This is a great advantage, for they take up much room ; moreover, when slurry is allowed to sub-side in a deep back the heavier particles have a tendency to separate from the other, so that the resulting material is not uniform in composition ; and lastly, in using this system, any slurry found by analysis to be defective can more easily be dealt with than it can in the large mass contained in a reservoir.

The method of drying the slurry is effective and also economical. It utilises the whole heat from the kiln, a large proportion of which in ordinary kilns escapes at the top.

Ransome's system consisted in burning the dried slurry in a revolving iron chamber lined with fire-brick, and fed with waste-gases on the same system as in a regenerative furnace (see p. 303). Among the advantages claimed for this system were economy in space, fuel, grinding, and time, and improvement in quality by the exclusion of the fuel from the cement.

This system may be regarded as the progenitor of the improved *Rotatory Process* now used extensively in the United States and Germany, and which is in course of adoption in this country.[1]

MANUFACTURE FROM LIMESTONES AND CLAY OR SHALE.—In some parts of the country the denser limestones are used in the absence of chalk for the manufacture of Portland cement : hard shales have also often to be used instead of clay.

Thus the Warwickshire (Rugby and Stockton), Somersetshire (Bridgwater), Dorset (Poole and Wareham), Portland cements are made from Lias limestone and clay, and in Cheshire (Doveholes) the limestones of the Carboniferous formation are used for the same purpose.

When dense limestones are used for the manufacture of Portland cement, they must be crushed by machinery. The shale or clay is roughly burnt to ballast (see p. 198), the two are then mixed in the proper proportion (according to their composition) to give the percentage of clay and lime required—and are ground to a fine powder.

This powder is passed into the pug-mill of a brick-making machine—thoroughly mixed—slightly moistened, and then moulded semi-dry into bricks. These bricks are then dried upon hot plates to drive off any remaining moisture—burnt in kilns as here-inafter described—and then ground to powder.

The process of manufacture just described is adapted for hard limestones and shaly clays, which cannot be reduced to liquid and thus mixed together. The dry process is stated by Mr. Reid to be very efficient and economical. He says, moreover, "The carbonate of lime is in so finely comminuted a state, and so accurately blended with the silica and alumina, that no injurious development from this source can possibly arise, at all events in the direction of the cracking or blowing danger."

The plant required for a Portland cement manufactory is so extensive that it can hardly ever be worth while for an engineer or builder to manufacture for himself. This branch of the subject will, therefore, not be pursued further ; but any one who is interested in it will find full details of the processes of manufacture, with much other useful information, in Mr. Reid's works : *A Practical Treatise on the Manufacture of Portland Cement ; The Practical Manufacture of Portland Cement*, translated from Lipowitz ; and *The Science and Art of the Manufacture of Portland Cement ;* also in the *Minutes of the Proceedings of Civil Engineers*, vol. lxii. p. 67, and vol. cxlv. p. 44 ; and in *Calcareous Cements*, G. R. Redgrave.

PORTLAND CEMENT MADE FROM SLAG.—Ordinary blast furnace slag (see p. 261) contains nearly the same constituents as Portland cement, but not in the same proportions—the proportion of lime being too small. The writer has no experience of this material, but it is said to attain the same strength as ordinary Portland cement, and in a shorter time. The process is, of course, useful only in blast-furnace districts, for it would not pay to transport slag to other places for the purpose.

"Great caution is necessary in adopting a cement of this nature, more especially when it is recollected that blast-furnace slags differ materially in their composition. . . . It would appear, however, that when care is taken to see that the constituents of the cement exist in suitable proportions, a very serviceable article is capable of being produced."[2]

---

[1] See page 197.   [2] Dent's *Cantor Lectures.*

The reader is also referred to the remarks under the heading of Adulterations in Portland Cement on page 192.

Portland cement differs very considerably in its characteristics and action. It can be manufactured more cheaply when under-burnt, because then a greater bulk of cement is produced with a given quantity of material, and it requires less fuel and less grinding; it also sets more quickly, but never arrives at the same ultimate strength as a properly burnt cement. Under-burnt cement contains, moreover, an excess of free quicklime, which is apt to slake in the work and cause great mischief. This may be remedied by exposing the cement, and allowing these particles to become *air-slaked.*

TESTS OF QUALITY.—A very slight difference in the manufacture may make a great difference in the character of the material, and rigid testing is necessary in order to secure the best cement.

Before using Portland cement for important work, the undermentioned points should be inquired into :—

Fineness of Grit.—The cement should be ground to a fine powder.

This can be roughly tested by rubbing it between the fingers, or, accurately, by passing it through a sieve with meshes of known size.

With regard to the exact degree of fineness that is advantageous, there is some difference of opinion.

There seems, however, to be no doubt that properly burnt cement, when ground extremely fine, is, as compared with one coarsely ground, much stronger when used with sand, and also safer, for there are none of the coarse particles which exist in well burnt and coarsely ground cements, especially when they have any tendency to excess of lime.

A heavy well burnt cement is difficult to grind properly, and it will often contain a considerable proportion of coarse particles which ought to be separated and reground.

The experiments by Messrs Grant, Colson, Mann, and others show that when used *neat* (*i.e.* without admixture of sand) a coarse-grained cement is stronger than one finely ground.

When mixed with sand, however, as it generally is in actual use, the finely ground cement makes stronger mortar than the other, the difference in its favour being greater as the proportion of sand in the mortar is greater.

A lightly burnt cement is easily ground fine, and "at 7 or even 28 days may appear to be superior to heavy, which is with difficulty ground as fine as the lightly burnt, but in the long run the heavy, if not too coarsely ground, will surpass the lightly burnt, and if the heavily burnt were as finely ground as the light it would be a great deal stronger from the beginning, the time of setting being of course the same.   Fine cement, as it takes more sand, goes farther than coarse, it is also much safer when it verges on the blowing point from excess of lime." [1]

Grinding is better than sifting, "Heavy clinker ground fine will when tested give higher results than lighter cement of equal fineness obtained by sifting." [1]

Mr. Mann, in his experiments on the adhesive strength of cement, found that with cement sifted through a sieve having 176 meshes to the lineal inch, —i.e. 30,976 to the square inch, the so-called "coarse" grains stopped even by this fine mesh influenced the cement as follows :—

	Adhesive strength after 7 days ; in lbs. per sq. inch.
The fine particles only	91
Ditto with 25 per cent of the coarse grains . .	63
Ditto with 75 per cent of coarse grains . . .	26
Coarse grains only	8

Mr. Mann, says that this fine sieve was found to "afford more definite and reliable results than those having larger meshes."

The experiments of the late Mr. Grant, Mr. Mann, and others, have shown that the larger grains in a coarsely ground cement, besides being in many cases a source of danger, are almost useless, sometimes quite useless, as cement, being more or less inert, so that even if safe they play no other part than that of additional sand.

Mr. Grant found that "coarser cement than would pass through the sieve of 2580 to the square inch, was at least no better than sand, and that when it contained free lime it was a source of weakness if not of danger." [2]

There is no general consensus of opinions or practice among engineers with regard to the degree of fineness which it is best actually to require in specifications for Portland cement.

At first the cement used for the Metropolitan Main Drainage Works was specified to be ground "extremely fine," but the exact size of mesh it should pass is not fully determined.

In 1876 Mr. Mann said, "1-50th inch square (2500 holes per square inch) is as fine a mesh as can be conveniently used in practice, smaller ones clogging very easily ; on the other hand, cement reduced to this fineness has a very appreciable superiority with sand, as compared with even slightly coarser samples." [3]

In the second series of very elaborate and useful experiments made by the late Mr. Grant, the resident engineer of the Metropolitan Main Drainage Works, the cement tested had all been passed through a sieve of only 400 holes to the square inch, the weight of the sifted cement being 110·56 lbs. per bushel.

---

[1] Grant, *M.P.I.C.E.*, vol. lxii. p. 102.
[2] *M.P.I.C.E.*, vol. lxii. p. 243.
[3] Captain Innes, R.E., *R.E. Corps Papers*, vol. xxi. p. 1.

Of late years, however, manufacturers in Germany and Austria have introduced cements ground to a much greater degree of fineness.

Cements are easily procurable which will entirely pass sieves of 2580 meshes to the inch (400 to the square centimetre), leaving only 10 per cent on sieves of 5806 meshes to the inch (900 to the square centimetre), and it has been stated that cement can be procured of which only 3 to 10 per cent is rejected by a sieve of 32,000 meshes to the inch.[1]

These German cements are not much, if at all, used in engineering works in England, but they are used by the manufacturers of patent cement paving and similar materials.

With regard to ordinary English cements, Mr. Mann found, that of the cement received from nine makers, from 38 to 50 (average 45·6) per cent was stopped by a sieve with 30,976 meshes per square inch, and with eight varieties Mr. Grant found the residue to average 49·5 per cent.

There seems to be no reason, except the extra cost, why *all* the cement should not be ground to go through the finer mesh, but at present a percentage to be stopped is allowed in most specifications.

Experiments made by a friend of the writer's showed that a cement which left 10 per cent core on a 2500 mesh sieve, left 20 per cent on the 30,976 mesh sieve. A cement which left only 10 per cent core on the 30,976 mesh sieve cost about 4s. a ton more for grinding than the other. It appears therefore that to obtain 10 per cent more cement cost 4s., and that the extra grinding is not economical, except for cements, which cost more than 40s. a ton delivered on the works, as they do sometimes abroad.

Where fineness of grit is alluded to in specifications, as it always should be, 2500 meshes to the square inch is frequently specified, although various Public Departments and a few engineers specify that not more than 10 per cent by weight shall be rejected by a sieve of 5800 meshes to the square inch, and there seems no doubt that this requirement, which is estimated to add only $\frac{1}{10}$ to the cost of the cement, is a very desirable one to enforce until still finer grinding can be obtained.

Great care must be taken, however, that finely ground cement is not lightly burnt, to prevent which the weight, or better still (see p. 172) the specific gravity, of the cement should be specified too.

Where cement is to be sent abroad, and thus rendered expensive by the addition of freight, or it seems especially desirable to have a material which gives the *maximum* strength with the *minimum* bulk.

The table (page 174) shows the degree of fineness and other particulars specified by various public departments and on various works.

Gauge of Wire of Sieves.—It is a curious thing that though many engineers specify the number of meshes to the square inch in the sieves to be used, very few mention the gauge of the wire of which the sieves are to be made, although it is manifest that the size of the orifice of the mesh in the sieve must depend upon the thickness of the wire ; two sieves with the same number of meshes to the inch, but of different gauges of wire, must pass cements quite different as to fineness.

It is therefore necessary to state the gauge of the wire to be used, and as uniformity in this is desirable, the author has obtained from Messrs. Currie and Co. of Leith a list of gauges appropriate for sieves of the meshes stated below.

These are shown in the second column of the following table—the gauges shown in the third column are those used by Messrs. Adie in making sieves for the late Metropolitan Board of Works.

---

[1] *M.P.I.C.E.* 1880, vol. lxii. p. 242.

Meshes per sq. inch.				Gauge of Wire. BWG.	Gauge of Wire. BWG.
400	.	.	.	29	34
900	.	.	.	37	36
1,600	.	.	.	39	...
2,500	.	.	.	39	37
3,600	.	.	.	39	...
5,800	.	.	.	40	39
14,400	.	.	.	43	41
32,000	.	.	.	48	...

Weight.—This particular is generally carefully ascertained. It used to be considered that a good weight per bushel was a sign of thorough burning, but it is now realised that the weight is greatly influenced by the degree of fineness to which the cement is ground, upon the degree to which it has been aerated, and upon the way in which the measure is filled.

The weight is, however, generally specified in connection with the degree of fineness required.

The weight of the Portland cement in the market varies from 95 lbs. to about 120 lbs. per striked bushel.

The heavier cements are slow-setting, but as a general rule,[1] they ultimately have a greater tensile strength than those of small specific gravity.

A heavy cement is likely to be thoroughly burnt throughout, but care must be taken to ascertain that its weight is not caused by its containing a large proportion of coarse unground particles.

In some cases a heavy cement contains a large proportion of over-burnt particles, and unless these are most carefully ground to a fine powder, they slake very slowly, frequently not till they have been used in the work, in which case they cause serious injury.

In very heavy cements there is some danger of an excess of lime, which, if left in a free state, that is uncombined with the silicic acid of the clay, is liable to cause disintegration in the work. It also renders the cement unfit for the joints of sewers,[2] or for any position where it would be liable to the attacks of chemical agents, which would destroy the carbonate of lime.

As above mentioned, the weight of a given bulk of cement depends to a great extent upon the fineness of its grit ; a coarse cement is heavier than one equally well burnt which is finely ground.

---

[1] The experiments of Messrs. Grant, Colson, and Mann show that this rule does not always hold good.

[2] Mr. Baldwin Latham, *Min. Proc. Inst. Civ. Eng.,* vol. xxxii. p. 63.

The weight depends also upon the amount of aeration the cement has received, sometimes after weighing one bushel it will be found that the next bushel weighs 1 lb. or 1½ lbs. less. In testing the weight of large quantities, therefore, the sample bushels should be taken from different parts of the heap.

Lastly, the weight depends upon the method in which the measure is filled —one sample must not be more tightly packed than another.

To ensure this the cement should be poured into the measure as described at p. 173.

The effect of fine grinding upon weight is shown in the following results, obtained by Messrs. Currie and Co. of Leith. The figures will, however, vary considerably with different cements.

Meshes per square inch of sieve.	Percentage retained by sieve.	Weight of Cement per bushel in lbs.	Weight of Cement per cubic foot in lbs.
2,500	10	115	90
3,600	10	112	87
5,500	10	109	85
14,400	10	104	81
32,000	10	98	76

It is evident from the above that the weight test ought never to be used without the sieve test or it would be a direct incentive to coarse grinding.

On the other hand, to use the sieve test only would lead to being supplied with light easily ground cements of no great tensile strength.

The practical difficulty, however, of accurately comparing the weights of cements makes the weight test unreliable, and engineers therefore sometimes require the cement to be of a given specific gravity, which cannot vary with the different degrees of fineness of grit.

The weight of the Portland cement originally used on the Metropolitan Main Drainage Works was specified to be at least 110 lbs. per striked bushel.

The cement actually supplied averaged 114·15 lbs. per bushel in weight.

The cement for the later series of experiments by Mr. Grant was specified to weigh 112 lbs. per striked bushel. It weighed 113·2 lbs., including the coarser particles, but only that portion of it was used which passed through a mesh of 400 holes to the inch, and weighed 110·56 lbs. per bushel, as above stated.

In a few cases a cement weighing 123 lbs. per bushel was experimented upon. It does not seem desirable, as a rule, to specify a weight for cement of more than from 110 to 115 lbs. per bushel. Mr. Grant recommends that when a weight is specified it should not be more than 112 lbs. a bushel.

The weight is generally stated in lbs. per bushel; 21 bushels (each containing 1·283 cubic foot) make a cubic yard. Sometimes it is stated in lbs. per cubic foot.

A very heavy and strong cement is required in important engineering works; but for ordinary purposes of building great tensile strength is not of the first importance, and in some cases, *e.g.* for rendering walls, a lighter and more quickly-setting cement may be used with advantage.

Specific Gravity.—Mr. Mann found [1] that the specific gravity of cement supplied by the best English manufacturers slightly exceeded 3·0.

The particles rejected by a sieve of 2900 meshes to the square inch had a specific gravity from 3·08 to 3·13, and the fine particles passed by that sieve a specific gravity of from 2·97 to 3·06, but in some cases the coarse and fine particles of the same cement had the same specific gravity.

In an inferior cement he found the specific gravity 2·80, and that of the finely sifted portion only 2·55.

Grant's experiments [2] show the specific gravity of differently burnt cements to be as follows :—

Light burnt	.	.	.	.	3·130
Hard „	.	.	.	.	3·134
Medium „	.	.	.	.	3·131

Mr. Grant's specification for specific gravity is "not less than 3·1."

The apparatus described in the next paragraph is recommended by Mr. Grant for ascertaining the specific gravity of cement.

Keate's Specific Gravity Bottle.—This bottle consists of two bulbs, the lower somewhat exceeding the upper in capacity. The exact capacity of the lower bulb is of no importance. On the neck between the bulbs is a file mark *b*, on the neck of the upper bulb is a similar mark *a*.

The capacity of the upper bulb between the marks *a* and *b* must be accurately determined, and may conveniently be either 500 or 1000 grains in water measure at 60° Fahr.

In ascertaining the specific gravity of a solid in small fragments— small shot, for example—the following is the mode of procedure : fill the bottle with distilled water up to the mark *b*, accurately counterpoise the bottle so filled in a balance, drop the substance (of which the specific gravity is to be taken) carefully and gradually into the bottle until the water rises from *b* to *a*. Ascertain exactly the weight of the material so added. If the capacity of the upper bulb be 1000 grains of water, the weight of the material required to raise the water from *b* to *a* is its specific gravity ; if the capacity of the bulb be 500 grains of water, the weight of the substance added must be multiplied by 2, which will give the specific gravity.

The principle of the apparatus is very simple ; the capacity of the upper bulb is an exact measure of distilled water, and when the water is raised from *b* to *a* by dropping a solid into the bottle, the bulk of that solid equivalent to the given volume of distilled water is ascertained and the relation between the weights of the two is given by the weights of the substances added, which is either the specific gravity direct, if the capacity of the bulb is 1000 grains, or it can be ascertained by multiplying the weight of the solid by the number which represents the part of 1000 represented by the capacity of the bulb, etc.

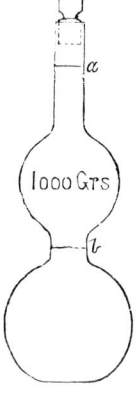

Fig. 82*a.*

If the solid be soluble in water, any convenient liquid can be used in the place of water in making the experiment, the only thing necessary being carefully to counterpoise the bottle filled with the liquid up to *b* in this manner. Petroleum, oil, turpentine, or any liquid suitable to the nature of the material to be tested, may be used, all other things remaining the same.

The only precautions to be observed are that the air, which is apt to cling somewhat to the solid matter when dropped into the liquid, is carefully removed, and that if a very volatile liquid be used in the place of water the bottle should be stopped or corked to prevent evaporation.[3]

Method of Weighing. — In order that the cement may be accurately weighed, great care must be taken in filling the measure.

---

[1] *M.I.C.E.*, vol. lxii. p. 224.   [2] *M.I.C.E.*, vol. lxii. p. 130.
[3] Taken *verbatim* from Appendix iv. p. 129, *M.I.C.E.*, vol. lxii. (Grant).

This may be done by allowing the dry cement to run down a board or shoot, inclined at an angle of 45°, into the measure, any superfluity being carefully struck off with a light straight-edge.

A vessel with holes in it is sometimes used for filling instead of the shoot. An accurate method is to fill the measure through a sieve of about $\frac{1}{16}$ inch mesh held a short distance above it, or the cement may be poured through a hopper placed about two feet above the measure. A drawing of the hopper is sometimes supplied in connection with the specification.

COLOUR.—This point should be examined, though it is not of very great importance. Bad cement may be of a good colour.

Good Portland cement, as received from the manufacturers, should be of a grey or greenish-grey colour.

A brown, or earthy colour, indicates an excess of clay, and shows that the cement is inferior—likely to shrink and disintegrate.

A coarse bluish-grey powder is probably overlimed and likely to blow.

The colour may best be observed by rubbing the cement on the hand or on a piece of white paper.

TEST FOR TENSILE STRENGTH.—This is the most important test in most cases, and it should be made with the aid of a proper machine, as hereinafter described (see p. 187).

Seven Days' Test.—The tensile strength of Portland cement, as required by the original specification for the Metropolitan Main Drainage Works, was 400 lbs. on the briquette area of $2\frac{1}{4}$ inches after six days' immersion.

Shortly after this the specified breaking weight was raised to 500 lbs. per area of $2\frac{1}{4}$ square inches.

The average strength of the cement supplied under this specification during five years was 806·63 lbs. on the briquette area.

The standard breaking weight specified on these works was afterwards raised to 787 lbs. on the briquette area, or 350 lbs. per square inch after seven days' immersion, the specification being as follows :—

"The whole of the cement shall be Portland cement of the very best quality, ground extremely fine, weighing not less than 112 lbs. to the striked bushel, and capable of maintaining a breaking weight of 350 lbs. per square inch seven days after being made in a mould and immersed in water during the interval of seven days."

The rigid testing on these and other engineering works has raised the tensile strength of the best cement manufactured since that date.

The breaking weights specified on the works mentioned below are shown in the following Table (page 174).

The tensile strength of Portland has been increased of late years, besides which improvements in the methods of testing, and the increased care with which they are carried out, cause higher results to be shown with cements similar to those of former years.

Cements can be obtained which will stand a tensile stress of 550 lbs. per square inch, even higher.

PARTICULARS REQUIRED BY VARIOUS PUBLIC DEPARTMENTS AND OTHERS IN THEIR SPECIFICATIONS FOR PORTLAND CEMENT.

DEPARTMENT OR WORK.	FINENESS.		Weight per bushel in lbs.	TENSILE STRENGTH NEAT.		Age when tested.	Rate at which weight is applied.	REMARKS.
	Number of meshes per square inch in sieve.	Residue. Per cent.		On 2¼ sq. inches.	On 1 sq. inch.			
India Office	5800	20	Not less than 116, not more than 120, average 118	lbs. 800	lbs. 355	1 day in air, 6 days in water	100 lbs. in 12 seconds and hang at rest 15 seconds	Sieve of brass wire ·007 inch diameter.
Admiralty	2500 and 5776	0 / 15	Specific gravity 3·15 when fresh burned and ground, or 3·10 when some time has elapsed since grinding	787	350	7 days	100 lbs. in 15 seconds	„ ·005 „ „
War Office	2500	15	110	900	400	7 days	...	For ordinary buildings.
London County Council⁴	5800	10	Specific gravity not less than 3·1	None specified			...	The cement to be gauged with three times its weight of dry sand, which has passed through a sieve of 400 and been retained upon one of 900 meshes to the square inch. Briquettes to be put into water 24 hours after made, and to remain in water 27 days more. Then to have 250 lbs. per square inch applied at the rate of 200 lbs. per minute.
Clyde Trust¹	2500	15	Not less than 115	787	350	7 days	...	
Trinity House²	1600	10	Not less than 116	350 / 500 / 750	155 / 222 / 333	2 days / 4 days / 7 days	"Slowly"	
Peterhead Harbour³	2500	10	Average at least 116 lbs., not less than 114	350 / 500 / 800	156 / 222 / 355	48 hours after gauging / 4 days / 7 days	"Slowly"	

¹ Mr. Deas.   ² Sir J. Douglas (Stoney).   ³ Sir John Coode.   ⁴ Calcareous Cements (Redgrave).

There is, however, great danger in raising this test too high. A cement having great tensile strength after a short interval of setting can only be obtained by using a maximum amount of lime in the manufacture, and an excess of lime in the cement may cause it to expand in the work months or even years after it is used.

It is better, therefore, to require a moderate tensile strength, such as from 300 to 350 lbs. per square inch.

Thirty Days' Test.—It will be noticed that most of the tests above mentioned are applied seven days after the cement is gauged and formed into a briquette.

It has, however, often been suggested that, especially in the case of a heavy slow-setting cement, seven days is too short a period for its properties fully to develop, and a period of thirty days has been suggested as one which would give the cement a fairer trial.

There would be a practical disadvantage in having to keep a consignment of cement thirty days before it is accepted or rejected, otherwise the longer period might be preferable. If it were adopted, of course a far higher tensile test would be necessary.

Mr. Mann found the increase in strength, in samples kept under water thirty days, to be about 20 per cent as compared with those kept only seven days.

Mr. Grant says that cement required to bear 350 lbs. per square inch after seven days should bear 450 lbs. after thirty days.

It has also been suggested that as the cement has generally to be used in the form of mortar, it should be mixed with sand before being tested. Such, however, is not the common practice in this country.

Testing with Sand.—For many years Portland cement was always tested *neat, i.e.* without admixture of sand, and this practice is still almost universal among engineers.

It has, however, frequently been pointed out by Mr. Grant, Mr. Colson, and others, that as Portland cement is nearly always used with a mixture of sand, it would be better to test it, as far as possible, under the condition in which it is used in practice, as it is thus that its probable behaviour when in use can best be ascertained.

The Germans have for some time made their tests on briquettes of 1 Portland cement and 3 sand after twenty-eight days' setting.

It is found that the testing of neat cement forms but little guide to their behaviour when mixed with sand, thus, "coarsely ground cement will, as a rule, give somewhat higher results when tested neat than finely ground, but when mixed with sand, say in the proportion of 1 to 3, the superiority of the finely ground cement becomes apparent."[1]

Sand, however, retards the hardening, and it is found that briquettes formed of 1 Portland cement to 3 sand must be left at least twenty-eight days before being tested.

The length of time required for this test with sand renders it very difficult, indeed almost impossible, to carry out on ordinary works for want of storage room—and it has also other disadvantages.

It is impossible to compare tests made with mixtures of cement and sand, unless the sand is always of exactly uniform composition and quality as regards size, sharpness, and surface of grains, degree of dampness, etc., and sand so uniform in quality would be very difficult to obtain.

The practical difficulties involved in testing cement mixed with sand have prevented it from being universally adopted ; there can be no doubt, however, that for large works where ample storage exists and sand of uniform quality can be ensured, more information about the future behaviour of the cement can be obtained by this test than by testing the cement neat.

TESTING BY COMPRESSION.—Again it has been pointed out that cement when in actual use is generally subjected to compression—very rarely to tension—and that it would be more useful to test its resistance to a compression than to a tensile stress.

The apparatus for testing cement by compression is, however, cumbersome and expensive, and tests of compressive strength are never specified by engineers.

Many experiments have, nevertheless, been made on the resistance of Portland cement to compression. Mr. Grant found the measure of the "compressive strength to be about twenty times that of the tensile strength,"[2] but Herr Bauschinger considers that "there is no fixed relation between the crushing and tensile strength of cement."[2]

[1] Grant, *M.P.I.C.E.* 1880, vol. lxii. p. 104.

[2] *M.P.J.C.E.*, vol. lxii. pp. 108, 208.

ADHESIVE STRENGTH.—Yet another objection is made to the method of testing hitherto and still adopted by most engineers.

It is pointed out that, as the principal function of cement is to produce adherence between portions of the materials used, its capacity to do this should be ascertained, that the strength of its adherence to the materials should be tested ; *i.e.* its adhesive strength, not merely its cohesive strength, that is, the strength with which its own particles cohere.

Mr. Mann has made a great many valuable experiments on the cohesive strength of cement forming it as a joint between two pieces of limestone and finding what weight was necessary to tear them apart.

The following table gives the results of some of his experiments, and he points out that it tends to show that under the ordinary cohesive tensile test a really good cement may be rejected. For example, in experiment No. 9 the cement had a very low cohesive strength after twenty-eight days—only 309 lbs. per square inch—which might have caused its rejection, whereas its adhesive strength—110 lbs. per square inch—was greater than any of those tried with it.

COMPARISON of ADHESIVE and COHESIVE STRENGTH.[1]

No.	DESCRIPTION.	Average strengths in lbs. per square inch.	
		Adhesive.	Cohesive.
1	Ordinary cement. Age 7 days . . .	59	532
2	,, ,, ,, . . .	51	336
3	Fine cement sifted through sieve, } Age 7 days 30,976 meshes to square inch	94	428
4	,, ,, ,,	57	345
5	,, ,, ,,	65	500
6	,, ,, Age 28 days	105	500
7	,, ,, ,,	109	387
8	,, ,, ,,	84	428
9	,, ,, ,,	110	309
10	,, ,, ,,	85	320

He found also that " with one or two exceptions, the quick-setting cements manifested a greater development of adhesive strength than the slow, while, in the case of cohesive strength, quick-setting seems generally to produce an opposite effect." [2]

In the experiments tabulated below, Mr. Mann shows very clearly that the adhesive strength of a very finely ground cement is much greater than that of one coarsely ground.[3]

COMPARATIVE CEMENTITIOUS STRENGTHS of SIFTED and UNSIFTED CEMENTS.[3]

	AVERAGES IN LBS. PER SQUARE INCH.		
	Age 7 days.	Age 28 days.	Age 13 weeks.
Cement with coarse particles removed by sieve with 30,976 meshes to square inch . .	78	93	116
Ordinary cement, as received from the manufacturers . . .	57	78	98

In the next Table Mr. Mann shows the strength of adhesion of Portland cement to various materials.

---

[1] *M.P.I.C.E.*, vol. lxxi. p. 262.   [2] *M.P.I.C.E.*, vol. lxxi. p. 262.
[3] *M.P.I.C.E.*, vol. lxxi. p. 260.

STRENGTH of ADHESION of PORTLAND CEMENT to VARIOUS MATERIALS.[1]
In lbs. per square inch.

MATERIAL.	AVERAGE ADHESIVE STRENGTH.				REMARKS.
	7 days.	28 days.	13 wks.	6 mnths	
Bridgwater brick .	19	...	...	...	Ordinary cement.
,, ,,	24	66	...	...	Sifted through No. 176 sieve.
Slate (sawn) .	49	...	...	...	Ordinary cement.
,, ,,	53	82	...	62	Sifted through No. 176 sieve.
Portland stone .	26	50	...	...	Ordinary. Fragments torn out of surface.
,, ,,	29	62	...	55	Sifted through No. 176 sieve. Fragments torn out of surface.
Ground plate glass	...	102	113	...	Ordinary cement.
,, ,,	...	...	145	...	Sifted through No. 176 sieve.
Plate iron .	23	68	...	...	Ordinary.
,, ,,	44	66	...	...	Sifted through No. 176 sieve.
Sandstone .	...	49	...	...	Ordinary. Fragments torn out of surface.
Polished marble .	38	...	...	...	Ordinary cement.
,, ,,	52	71	...	75	Sifted through No. 176 sieve.
Polished plate glass	47	40	70	...	Ordinary cement.
,, ,,	55	49	51	...	Sifted through No. 176 sieve.
Granite (chiselled)	41	...	...	...	Ordinary.
,, ,,	78	97	153	...	Sifted through No. 176 sieve.
Limestone (sawn) .	57	78	98	...	Ordinary cement.
,, ,,	78	93	116	...	Sifted through No. 176 sieve.

N.B.—No. 176 sieve was made of silk and had 30,976 meshes per square inch.

THE BRIQUETTE.—The tensile stress that a cement will bear depends greatly upon the manner in which the test is made, the form of briquette, the method in which the cement is gauged, the amount of water used, etc. etc.

Method of making Briquettes.—The briquettes, whether of neat cement or of cement and sand, are made in brass moulds of the form shown in Fig. 86.

The following directions are taken chiefly from Mr. Grant's papers, the circulars of Messrs. Currie and Sons, Messrs. Gibbs and Co., and Mr. Faija.

Briquettes of Neat Cement.—Supposing the briquettes are to be made of neat cement, and of the 1-inch square section, the procedure would be as follows :—

The cement from the different casks or sacks should be turned well over, samples from different parts of the heap should be mixed, and if hot should be spread out, especially in hot weather, so as to become thoroughly cool, and water should be carefully added, noting the proportion required to bring the mixture into such a condition that repeated pats with the trowel will bring the moisture up to the surface.

Two cakes of about 2 or 3 inches diameter and ½-inch thickness should be made, and the time noted in minutes that they take to set sufficiently to resist the finger nail.

" If after two hours the cake is soft enough to take the impression of a finger nail, it may be considered slow-setting."[2]

If the cement should be slow-setting, all the briquettes may be made at once, but if quick-setting, only three or four at a time, and if very quick only one should be made at a time.

The moulds should be cleaned with a greasy cloth, and a number of pieces of thin blotting-paper each rather larger than a mould are then placed upon a marble, glass, or slate slab, and on these the moulds are placed.

Then about 4 lbs. cement, or enough for ten briquettes, are weighed and placed upon the non-absorbent slab in a heap ; in the centre of this a hole is made, into which from

[1] *M.P.I.C.E.*, vol. lxxi. p. 266.　　[2] Grant, *M.P.I.C.E.*, vol. lxii. p. 104

a graduated glass is gradually poured the quantity of water previously determined, the mixture being worked with a trowel until it becomes a short, harsh paste, the water is then discontinued, but the working with the trowel continued until the paste becomes pat and smooth.

With this paste the brass moulds are filled as quickly and solidly as possible, a small trowel being used, and the mortar beaten or lightly rammed and gently shaken until all the air has been driven out of it and the mortar has become elastic. The surplus should be cut off level, and the surface left smooth.

The whole operations of making the paste and filling the moulds until the briquettes are placed on one side should not take more than five minutes. The quicker it is done, provided it is done properly, the better, for it is most important that the cement should be at rest before the setting action commences.

When the moulds have been filled they should be numbered and laid aside, in some place where they will be secure against shaking or vibration in a wet damp atmosphere, or covered with a damp cloth till they have set sufficiently to be taken out of the moulds.

This will probably be in less than twenty-four hours, the time varying according to the rate of setting of the cement, but it must be done with great care to avoid flaws, and not too soon, or the briquettes will lose their shape and be difficult to fit into the clips of the machine.

The briquettes should then be placed upon "sheets of glass or on slabs, and laid in a flat box having a cover lined with several layers of linen, woollen, or cotton cloth, kept damp. In this box they are kept until they have hardened sufficiently to be put into water. This will vary from one or two hours to a day or more, but for uniformity, unless in the case of specially slow-setting cements, briquettes of neat cement may be kept for twenty-four hours before being transferred from this box to the shallow tanks in which they are to remain until the moment of testing.

"The numbers on the neat cement briquettes may be made with a sharp point or with a strong pencil.

"The water in the testing room should be kept at a temperature as nearly uniform as possible, say from 60° to 70° Fahr., but if the boxes in which the briquettes be kept are covered, moderate changes of temperature will not materially affect the results."[1]

Briquettes of Cement and Sand are made in a similar manner. About 1 lb. of cement and 3 lbs. carefully washed standard sand will make ten of the 1-inch briquettes. The proportions of water required will be from 8 to 10 per cent, and the mixture must be beaten into the mould with a spatula or light wooden mallet, so as to be as solid as possible.

The briquettes are treated like those made with neat cement, except that they should not be removed from the moulds until at least forty-eight hours after they are made, and should be kept in the boxes another forty-eight hours before they are numbered.

Nature and Proportion of Water in Cement Mortar.—No more water should be used than is necessary to make the cement fit for use, an excess produces porosity and retards the process of hardening.

Grant's experiments show that with 19 per cent of water, making the briquettes into a stiff paste, they stood from 28 to 40 per cent more tensile stress than when 25 per cent of water was used, making the cement of the consistency of stiff grout.[2]

9 oz. of water to 40 oz. of cement, or about 22 per cent, is recommended by Messrs. Gibbs and Co.

With hot or quick-setting cements neat more water will be required than with cool or slow-setting cements.[2]

With mixtures of 1 cement and 3 sand, about 11 to 12 per cent may be used for those which set in less than thirty minutes, and 10 per cent for those that take longer.

Briquettes mixed with salt water are rather stronger than those with fresh water, but salt water should not be used in cement intended for building or rendering the walls of houses to be inhabited, because it tends to keep them damp. Dirty water would of course injure the cement by introducing impurities which would prevent proper adhesion, and hot water should not be used except for experiments to make the cement set more quickly.

Shape of Briquette.—The cement to be tested is formed into a briquette shaped in one of the forms shown in section in Figs. 83 to 86.

---

[1] Grant, *M.P.I.C.E.*, vol. lxii. p. 124.
[2] Grant. *M.P.I.C.E.* 1880, vol. lxii. p. 158.

The briquette is placed in the clips of a testing machine (see p. 187), and broken by slow tension. Each of the figures shows the briquette in the clips ready to be attached to the machines.

There is no doubt that the shape of the briquette has an important influence upon its strength.

The transition from the thicker parts of the briquette to the minimum or breaking section should be gradual—all angles avoided—the shoulders should be so shaped that the bearing of the clips upon them is uniform—the clips hung so that the stress shall pass through their central points.

The form first used in this country is shown in Fig. 83 ; the principal angles were afterwards rounded off as shown in Fig. 84, which is not a good form, for it generally breaks as shown by the dotted line, and not at the minimum section.

Whenever the clips bear upon a considerable part of the surface of the briquette, as in Figs. 83, 84, it is very difficult to prevent them from pressing more at one point than another, and thus causing want of uniformity in the stress.

To avoid this the clips are sometimes done away with, and the briquette is suspended by pins with knife-edges passed through holes in its ends, as in Figs. 85, 86, which represent one of the forms used by Mr. Grant in his experiments.

The last form adopted by the Board of Works is shown in Figs. 86, 86a. The change of form in the briquette is very gradual ; the clips are rounded so

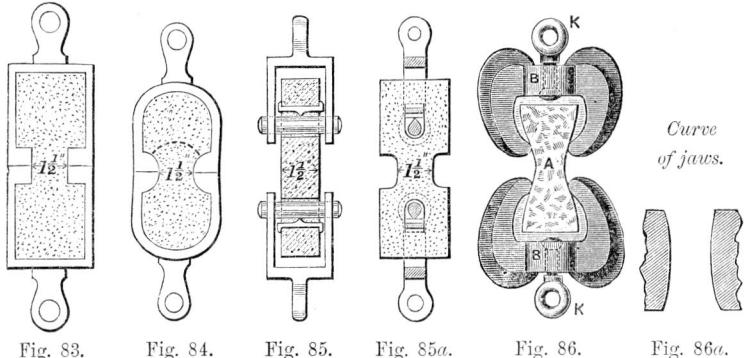

Fig. 83.   Fig. 84.   Fig. 85.   Fig. 85a.   Fig. 86.   Fig. 86a.

as to bear on it at only four points, are hung on knife-edges kk, and have loose joints at BB, so that the stress may pass through their centre points. This form of briquette seems to be the best that has been introduced.

It will be understood that the briquettes shown in the figures are all $1\frac{1}{2}$ inch × $1\frac{1}{2}$ inch = $2\frac{1}{4}$ square inches, at the waist or part intended to be broken. This is clearly seen in Figs. 85, 85a. In many cases the weakest section of the briquette is made only 1×1 square inch (see p. 188), and briquettes of this size are said to give a higher resistance per square inch than the larger ones. In testing, the mean of six briquettes should be taken.

TESTS FOR COOLNESS.—In some cases cement which appears perfectly good in every way has a tendency to crack and swell when placed under water. This action, which is commonly known as "blowing," is caused by the cement being under-burnt, by its containing an excess of lime, or by its not being properly "cool," that is, free from unslaked particles.

In order to detect this tendency to blow, the briquette placed under water should be carefully watched.

If it is inclined to blow, it will show signs of expansion after a day or two under water; in extreme cases the samples will entirely break up, but a few cracks about the edges are the commonest indications.

Pats should also be made about 3 inches in diameter and $\frac{1}{2}$ inch thick, gauged in neat cement with thin edges, and placed upon pieces of glass or other non-porous material.

One is placed under water and watched ; if twenty-four hours after its immersion there are no fine cracks round the edges, the cement may be considered safe.

With slow-setting cements surface cracks commencing at the centre are merely the result of the surface drying too rapidly.[1]

The other pat is left in the air, and should remain of a dark grey colour. If it is yellow or ochrey, the cement contains too much clay, and it is likely to be deficient in tensile strength.

ADDITIONAL TESTS FOR PORTLAND CEMENT.—Besides the ordinary tests above mentioned, the following rough tests will give an indication as to some important qualities of the cement before using it.

1. A bottle is filled with paste made from the neat cement. If, after the cement has been set some days, the bottle remains uncracked, it may be considered that the cement is not too hot.

If the cement has shrunk within the bottle it is probably under-burnt ; the shrinkage can be detected by pouring in a little coloured water.

2. Another test is to fill a piece of glass tubing with neat cement paste, and to note whether there is any shrinkage.

3. A rough method of ascertaining whether the cement is cool enough for use, is by plunging the bare arm into the cement.

If it feels hot the cement has not been sufficiently weathered, and requires further turning over.

HARDENING AND SETTING.—It is important to know how long a cement takes to harden and set. This is generally roughly ascertained by the impression of the finger nail upon the cakes of cement, as described on page 177, but as a rule no means are used for ascertaining this in a more accurate manner.

Time for Setting.—It is extremely difficult to define the time required for the setting of different classes of cements, samples from the same lot may take five minutes or five hours to set, according to its age, temperature, the quantity of water used, etc. As a rough guide, however, the following times for setting may be taken under normal circumstances :—

Quick Setting Cements	15 minutes
Slow ,, ,, ...	2 hours
Very Slow ,, ,, ...	5 hours

[1] Grant *M.P.I.C.E.*, vol. lxii.

Vicat's Needle Apparatus.—Fig. 87 shows an apparatus, invented by M. Vicat, for ascertaining the time of hardening and setting of cements. It is taken from Messrs. Currie's circular, together with the following directions for its use.

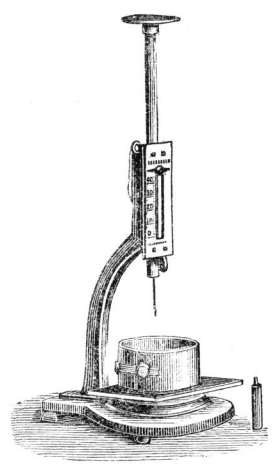

"To use this apparatus, gauge 14 oz. of neat cement with the requisite quantity of water, mix quickly into a stiff paste, and with this fill the circular brass mould resting on the glass plate, and which has a height of 4 centimetres and a diameter of 8 centimetres. The moment at which the needle having 1 sq. millimetre section and 300 grammes' weight is not able to penetrate completely to the bottom of the paste, marks the commencement of hardening. The interval from the time of gauging till the beginning of the hardening process *is the time the cement should be worked and used, if the strength of the work is to correspond to the quality of the cement.* As soon as the paste has become so hard that the needle does not leave an observable impression, it is set, and is the time that should be noted as *setting time.*

"The same apparatus may be used for ascertaining the correct consistency of cement ; only the point of 1 sq. millimetre is replaced by a cylinder of 1 centimetre in diameter. The circular mould is filled as quickly as possible, and the piston immediately let down gently into the paste. The consistency of the paste may be considered correct when the piston sticks at a height of about 6 millimetres from the bottom of the mould. In this manner the exact quantity of water required may be ascertained."

Fig. 87.

STORING.—Portland cement is generally received in sacks in this country, in casks abroad; these should at once be emptied, the cement spread out for a month or so on a wooden or concrete floor, to a depth not exceeding 3 or 4 feet, in a large weather-tight room, and occasionally turned over, so that it may become thoroughly air-slaked and cooled. (See tests for coolness, p. 179.) During the time it is thus exposed the cement if fresh will increase considerably in bulk.[1]

Sometimes cement which if tested when new would crack or "blow," will be found after this cooling to have lost the tendency to do so.

This air-slaking or "cooling" is of the very greatest importance, particularly with cements which, on account of their very high tensile resistance, may be suspected of containing an excess of lime.

STRENGTH.—The strength of Portland cement varies, as has already been mentioned, according to its original composition as regards the percentage of chalk and clay used in its manufacture, with the degree of burning to which it has been subjected, and according to the fineness to which it has been ground.

The strength of gauged Portland cement rapidly increases with age, the breaking weight on a sectional area of $2\frac{1}{4}$ square inches

[1] In some moist climates abroad cement would be deadened by this treatment, so that when sending cement to such climates it should be thoroughly cooled before it is packed into the casks.

being as shown in the second column of the following Table for neat cement weighing 112 lbs. per bushel.[1]

The strength of Portland cement mortar decreases as the proportion of sand is greater, which will be seen by the remaining columns of the Table.

Age and time immersed.	PROPORTION OF CLEAN PIT SAND TO 1 CEMENT.					
	Neat Cement.	1 to 1	2 to 1	3 to 1	4 to 1	5 to 1
1 Week . .	445·0	152·0	64·5	44·5	22·0	
1 Month . .	679·9	326·5	166·5	91·5	71·5	49·0
3 Months .	877·9	549·6	451·9	305·3	153·0	123·5
6 Months .	978·7	639·2	497·9	304·0	275·6	218·8
9 Months .	995·9	718·7	594·4	383·6		
12 Months .	1075·7	795·9	607·5	424·4	317·6	215·6

The following Table[2] shows the strength of Portland cement at different periods of time with different degrees of fineness and with different proportions of sand.

TABLE[2] showing the CEMENT coarsely ground and sifted through a fine sieve neat and mixed with sand.

The German mould used = 5 square centimetres in section.

AGE OF BRIQUETTE.	NEAT.		THREE OF SAND.		FIVE OF SAND.	
	10·2 per cent residue on a sieve of 2580 meshes per square inch.	Sifted so as to pass all through sieve of 32,257 meshes per square inch.	10·2 per cent left on sieve of 2580 meshes per square inch.	All passed through sieve of 32,257 meshes per square inch.	10·2 per cent left on sieve of 2580 meshes per square inch.	All passed through sieve of 32,257 meshes per square inch.
Weeks.	Lbs. per square inch.	Lbs. per square inch.	Lbs. per square inch.	Lbs. per square inch.	Lbs. per square inch.	Lbs. per square inch.
1	353	346	75	252	31	136
4	533	380	171	330	97	208
8	585	469	206	358	118	223
25	701	495	282	397	166	272

Remarks.—This Table shows (1) that the coarsely ground cement broke at a higher point when used neat than when finely ground or sifted, but at a much lower point when mixed with sand ; (2) that at 25 weeks with 3 of sand the gain was equal to 41 per cent, and with 5 of sand equal to 64 per cent ; (3) that the strength of the fine cement with 5 of sand was equal to a greater strain than that of coarse with 3 of sand, especially at the earlier stages. The proportion remaining on a sieve of 5000 per square centimetre (32,257 per sq. inch) was 49·5 per cent.—J. G.

The resistance of Portland cement to crushing after nine months is greater than that of many building materials, being 3 tons per square inch.

---

[1] Grant's Experiments.
[2] From *M.P.I.C.E.*, vol. lxii. p. 149 (Grant).

TABLE[1] showing the TENSILE STRENGTH of various Limes, Selenitic and Portland Cements, with different proportions of Sand at the end of 12 months. Five of the Briquettes kept dry and five in water. Size $1\frac{1}{2}'' \times 1\frac{1}{2}'' = 2\frac{1}{4}$ square inches. Form as in Fig. 84, page 179. Sand 96 lbs. per bushel.

Neat Proportion by Volume.	Grey Lime. L.		Selenitic Grey Lime. L.		Lias Lime. N.		Selenitic Lias Lime. N.		Lias Lime. W.P.S.		Selenitic Lime. C.		Selenitic Rugby Lias.		Selenitic Aberthaw Lime.		Rugby Lias Cement No. 3. Weight 74 lbs.		Portland Cement, 114 lbs. p. bushel. R.		W.P.S. Portland Cement, 120 lbs.		Neat Proportion by Volume.
	Dry.	Wet.	Dry.	Wet.	Dry.	Wet.	Dry.	Wet.	Dry.	Wet.	Dry.	Wet.	Dry.	Wet.	Dry.	Wet.	Dry.	Wet.	Dry.	Wet.	Dry.	Wet.	
Neat																			1059	1050	1243	1460	Neat
1 to 1																			762	751	804	913	1 to 1
2 to 1																			536	583	479	551	2 to 1
3 to 1	113	154	288	318	108	215	178	296	91	182	278	234	206	340	289	459	176	319	445	486	405	439	3 to 1
4 to 1	99	129	147	314	110	134	141	222	62	138	179	290	133	230	188	330	141	248	370	384	362	315	4 to 1
5 to 1	68	102	123	195	73	105	99	162	48	100	163	188	74	174	161	276	91	168	323	321	295	280	5 to 1
6 to 1	47	62	90	146	52	61	117	180	41	76	130	167	65	148		171	45	103	254	278	222	225	6 to 1
8 to 1																			119	229	157	141	8 to 1
10 to 1																			101	138	116	127	10 to 1
12 to 1																			73	129	93	96	12 to 1

TABLE showing the COMPRESSIVE STRENGTH of the same Limes and Cements mixed with different proportions of Sand weighing 127 lbs. to the bushel. All kept dry. Size 6 inch cubes—each number the average of ten cubes.

	Grey Lime. L.	Selenitic Grey Lime. L.	Lias Lime. N.	Selenitic Lias Lime. N.	Lias Lime. W.P.S.	Selenitic Lime. C.	Selenitic Rugby Lias.	Selenitic Aberthaw Lime.	Rugby Lias Cement No. 3. Weight 74 lbs.	Portland Cement, 114 lbs. p. bushel. R.	W.P.S. Portland Cement, 120 lbs.	
	Lbs.	Lbs.	Lbs.	Lbs.	Lbs.	Lbs.	Lbs.	Lbs.	Lbs.	Lbs.	Lbs.	
6 to 1	5712	10,393	6417	9,654	12,913	14,918	20,787	19,107	9676	56,425	48,384	6 to 1
8 to 1	2576	4,278	6204	10,988	6,014	8,579	19,196	12,208	5980	42,806	51,363	8 to 1
10 to 1	2912	4,569	6451	5,745	4,793	7,571	11,849	8,624	3248	29,971	29,232	10 to 1
12 to 1										20,809	16,329	12 to 1

[1] Grant, *M.P.I.C.E.*, vol. lxii. p. 165.

Limit to Increase of Strength with age.—Mr. Grant's experiments with a cement weighing 123 lbs. per bushel led him to the conclusion that it attained its maximum strength after constant immersion for two years, and that there is no reason to fear that a good cement ever deteriorates. The period at which the maximum strength is attained varies, however, with the class of cement—in the case of a light cement it would probably be much shorter.

Market Forms.—Portland cement is sold in casks or in sacks for home consumption, and in casks for export.

The inside dimensions of the casks are sometimes as follows, but they vary. Length, 27½ inches. Diameter at middle, 19 inches. Diameter at ends, 16 inches.

Each cask usually contains 400 lbs. (net).

Those for export should be well hooped and nailed, and lined with stout brown paper.

The sacks measure 22 inches × 38 inches, and each usually contains 2 trade bushels or 200 lbs. of cement, but sometimes is filled so as to contain 2 cwt.

Good Portland cement is slow-setting as compared with the cements made from most natural cement stones, but surpasses them in ultimate strength ; and is more extensively used than any other, for all kinds of work for which cement is suitable. It is particularly well adapted for making concrete.

Scott's Processes.—General Scott's two processes depend upon an intimate admixture with the lime of a small quantity of a sulphate, usually sulphate of lime,[1] before, or at the same time that, the water is added.

All limes are improved by them, and converted into a kind of cement, the slaking action being suppressed, and the lime setting without expansion, thus forming a denser and harder mortar.

The quickness of setting is greatly increased by these processes for all limes, and their ultimate strength is also improved.

Scott's Cement is prepared by passing the fumes of burning sulphur through lumps of quicklime placed on gratings, and raised almost to a red heat, by which about 5 per cent of it is turned into a sulphate.

The calcined stone, if properly burnt, will be found to have lost all power of slaking ; upon being ground it becomes a fine homogeneous powder, of a tint similar to that of the unslaked lime from which it is prepared.

Good Scott's cement should be finely ground, and contain not less than 10 per cent of soluble silica ; it should weigh fully 60 lbs. per striked bushel, and when mixed with two parts of sand should be strong enough to come out of the mould in twenty-four hours. After being left for seven days in a dry place, the weight required to break it should be not less than 66 lbs. per square inch.

This material was coming into considerable use some years ago for making mortar, but especially for plastering. It is not now in the market, having been superseded by selenitic cement, in which the same qualities and characteristics are obtained by a much simpler process of manufacture. It has, however, been described here, as there is sometimes a confusion between the two.

Selenitic Cement,[2] sometimes known as selenitic lime, is also an invention of General Scott's.

This cement, like the other, contains a small proportion of sulphate of lime,

---

[1] *Calcium sulphate.*

[2] So called from *Selenitic,* the scientific name for *gypsum* which is sulphate of lime, and forms, when burnt and ground, *plaster of Paris.*

which is added in the form of plaster of Paris, mechanically mixed and ground with lime. Lime may, however, be selenitised by adding a small proportion of any sulphate, or by mixing it with sulphuric acid.

The sulphate begins to take effect directly water is added. Its presence arrests the slaking action, causes the cement to set much more quickly, and enables it to be used with a much larger proportion of sand than ordinary lime without loss of strength.

Nature of Lime.—This cement may be made from any lime possessing hydraulic properties. The limes from the magnesian limestones are much used for the purpose, also those from the grey chalk. But the best limes for selenitising are those from the Lias formation and grey chalk.

Fineness of Grit.—The cement should be finely ground so as to pass through a sieve of 900 meshes to the inch.

Proportion of Sulphate.—The quantity of sulphate the cement should contain depends upon the quality and description of the lime used for its manufacture, and varies from 4 to 7 per cent, the usual proportion being about 5 per cent.

When more than $7\frac{1}{2}$ per cent of sulphate is required to stop the slaking action, the lime may be considered not suitable for making selenitic cement. In this case, however, the lime may be rendered suitable by mixing it with one containing more clay.

Where used.—This cement has been used at the New Law Courts; Grosvenor Mansions; Chesterfield Mansions, etc. ; and for plastering at the Alexandra Palace, Manchester New Town Hall, and several of the principal new buildings in London, and other large towns.

Strength.—The Table on the opposite page is taken from a circular issued by the Selenitic Cement Company.

It shows the comparative strength of selenitic and Portland cement, with different proportions of sand, and also the increase of strength which accrues to the lime when it is prepared by the selenitic process.

Ordinary lime may be selenitised during the process of mixing it into mortar. The method of doing this is described at p. 212.

SELENITIC CLAY is a preparation of clay and sulphate of lime, which, when added to a pure or nearly pure lime, confers upon it hydraulicity, and also the quick-setting properties of selenitic cement.

Methods of artificially producing Hydraulicity.—In addition to the manufacture of hydraulic limes and cements by the intimate mixture and calcination of the necessary ingredients, hydraulic properties are sometimes conferred upon mortars made from fat lime by adding to them such substances as are known to produce hydraulicity.

POZZUOLANA MORTARS.—These are formed by adding to ordinary fat lime or feebly hydraulic mortars such a proportion of pozzuolana (see p. 201) as will make good their deficiency in clay. This proportion depends upon the composition of the lime in the mortar to be improved.

The success of pozzuolana mortar depends upon the intimate mixture of the ingredients, which should be reduced to a fine powder, and ground in a mill for 20 to 30 minutes.

Good pozzuolana mortar behaves like that made from eminently hydraulic lime.

ALKALINE SILICATES, produced by boiling flints in an alkali, may be added to mortar in the form of a thin syrup.

They are found to greatly quicken the setting of fat lime mortars, making

TABLE showing the FORCE necessary to tear apart common Stock Bricks, bedded one across the other, with various qualities of Cement and Lime, the Joint in each case having a Sectional Area of about Twenty Square Inches.

Nature of Lime or Cement.	Composition.	Age 14 days.	Age 21 days.	Age 28 days.	Age 35 days.	Lbs per sq. inch in 28 days.	REMARKS.
Portland Cement . .	1 cemt. 6 sand	166 lbs.	206 lbs.	313 lbs.	309 lbs.	15½ lbs.	Portland with less than four parts of sand would be out of the question in point of price.
,, ,, . .	1 ,, 5 ,,	304 ,,	336 ,,	325 ,,	433 ,,	16¼ ,,	
,, ,, . .	1 ,, 4 ,,	217 ,,	367 ,,	463 ,,	520 ,,	23 ,,	
Blue Lias lime . . .	1 lime 3 sand	63 ,,					
Blue Lias lime (Selenitic) .	1 ,, 6 ,,	172 ,,	237 ,,	249 ,,	286 ,,	12½ ,,	
,, ,, ,, .	1 ,, 5 ,,	161 ,,	268 ,,	233 ,,	267 ,,	11¾ ,,	
,, ,, ,, .	1 ,, 4 ,,	144 ,,	268 ,,	253 ,,	283 ,,	12½ ,,	
,, ,, ,, .	1 ,, 3 ,,	259 ,,	235 ,,	252 ,,	308 ,,	12½ ,,	
Barrow Lias lime . .	1 ,, 4 ,,	110 ,,	141 ,,	137 ,,	...	6¾ ,,	
,, ,, . .	1 ,, 3 ,,	143 ,,	137 ,,	183 ,,	...	9 ,,	
Barrow Lias lime (Selenitic)	1 ,, 6 ,,	270 ,,	409 ,,	399 ,,	430 ,,	20 ,,	One of the best English limes.
,, ,, ,,	1 ,, 5 ,,	177 ,,	572 ,,	399 ,,	438 ,,	20 ,,	
,, ,, ,,	1 ,, 4 ,,	233 ,,	292 ,,	418 ,,	539 ,,	21 ,,	
,, ,, ,,	1 ,, 3 ,,	255 ,,	388 ,,	541 ,,	435 ,,	27 ,,	
White chalk lime . .	1 ,, 3 ,,	66 ,,	97 ,,	93 ,,	...	4¾ ,,	
White chalk lime (Selenitic)	1 ,, 6 ,,	207 ,,	221 ,,	218 ,,	...	10¾ ,,	
,, ,, ,,	1 ,, 5 ,,	215 ,,	243 ,,	243 ,,	...	12¼ ,,	
,, ,, ,,	1 ,, 4 ,,	179 ,,	197 ,,	197 ,,	...	9¾ ,,	
,, ,, ,,	1 ,, 3 ,,	160 ,,	199 ,,	209 ,,	...	10½ ,,	
Burham lime (Selenitic) .	1 ,, 6 ,,	203 ,,	343 ,,	408 ,,	556 ,,	20½ ,,	A first-class, clayey Medway lime, selected for Selenitic mortar, and now supplied by Messrs. Lee for the purpose.
,, ,, ,, .	1 ,, 5 ,,	291 ,,	424 ,,	368 ,,	490 ,,	18½ ,,	
,, ,, ,, .	1 ,, 4 ,,	336 ,,	385 ,,	454 ,,	430 ,,	22¾ ,,	
,, ,, ,, .	1 ,, 3 ,,	325 ,,	454 ,,	484 ,,	424 ,,	24¼ ,,	

them to resemble hydraulic limes and cements in this respect, according to the quantity used, but they do not materially increase their strength.

Means for testing Tensile Strength of Cement.—It has already been mentioned that the tensile strength of Portland cement for important works should always be tested by direct experiment. There are several different machines by means of which this test can be accurately applied.

A few of these will now be described.

Adie's Testing Machines were among the first adopted for this purpose, and are still widely known and extensively used.

Fig. 88.   *Adie's No. 1 Cement-Testing Machine.*

Adie's No. 1 Machine.—This machine, by means of a straight lever, applies a known strain to a briquette of cement (see p. 177), until the latter breaks across at the narrow central part, the area of which is accurately known. Fig. 88 gives an elevation of the machine.

Fig. 88a is a *split mould* for the briquette. It is arranged to divide longitudinally into two parts, so that the briquette may easily be liberated without the aid of a press. Split moulds are sometimes hinged at one end.

These figures and the following instructions are from the circular of the maker (Mr. Adie, 15 Pall Mall).

Maker's Instructions.—To set up the machine, drop the spindle R into its place in the table, then put the pillar G in position and insert the notched plate in groove of spindle, bolting down G so that the beam when strained [by putting a moulded brick of cement into the clips B and C, and then tightening by means of the wheel R] may take its position freely in the centre of fork H. The wire cord passes twice round pulley at H, and once round that at G, and should never be very tight. When the cement is set, open the mould carefully, undoing both screws simultaneously, and treat the briquette as described at p. 177.

Insert the brick in the clips B and C, then turn the wheel R till the beam at H rises to the pulley [*i.e.* well above the zero line], and roll the weight D gently along until the fracture takes place.

The weight N should not be in its place, except when testing below 300 lbs., when the top row of figures is used.

If the weight D is not sufficient to break the brick, roll it back and hang on the extra weight in a notch in the beam near H.

The speed with which the weight is brought to bear upon the briquette, or at which it is increased, materially affects the result. When the weight is moved rapidly the resulting tensile strength, as shown by the machine, is higher than when the weight is wound slowly.

ADIE'S CEMENT TESTER WITH AUTOMATIC REGISTER.—In order to arrange for a uniform speed, Mr. Adie has added an automatic regulator to his machine, which is shown in Fig. 88*b*.

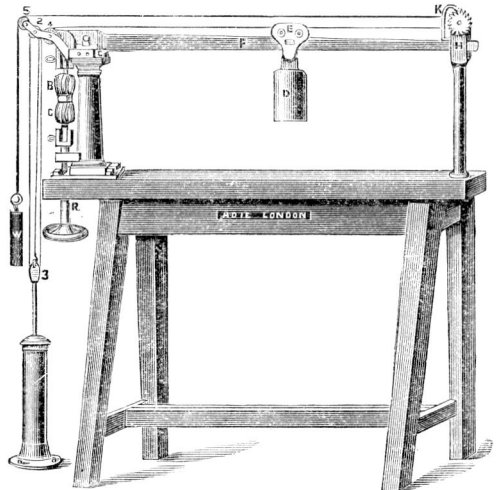

Fig. 88*b.   Cement Tester, with Patent Automatic Regulator.*

Patentee's Instructions.—To make use of this regulator, fix the brass tube to the floor, vertically underneath the pulley marked No. 2 in the drawing above, and fill the tube with clean water. Attach the long cord to the left side of the vernier E, pass it downwards over No. 4 round No. 3 and then upwards round No. 2, and down again to the eye at 3. Attach the short cord to the right side of E, pass it under the pulley at H and twice round it, then along over No. 5 and hang on the weight W.

Put the brick into its place and make all arrangements before actually putting on the strain. Screw up R, lifting the point of the beam well above the zero line on the pillar H, so as to free the check K, and while the weight D is travelling observe that the check is not allowed to touch the rachet wheel ; screw up R more if necessary to allow for any slip or springing of the clips. The catch stops the rolling weight when the fracture takes place, and the result is shown in lbs. on the scale.

To commence another test take off the weight W, before gently letting down the piston, at the same time easing back the weight D by the hand.

ADIE'S NO. 2 TESTING MACHINE.—Fig. 89 is a smaller machine by Mr. Adie for testing briquettes having a central section of 1 inch square. It can also be used as a weighing machine.

This figure, slightly modified, and the following instructions, are from the maker's circular.

Patentee's Instructions.—If the standard A be not in its place when received from the manufacturer, bolt it down to the stand, so that the beam when strained (by putting a moulded brick of cement into the clips B and C and then tightening by means of the wheel underneath at D) may take its position freely in the centre of fork E. Wind the cord once round the pulley at A, and twice round that at E. Mix the cement to be tested with as little water as possible, consistent with perfect homogeneity, and having laid the mould on a flat surface, or on the iron plate supplied for the purpose, fill it with the

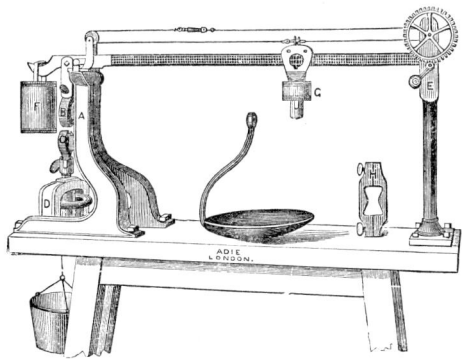

cement as described, and scrape the top flush. When set, take the briquette out of the mould carefully, and place it on the flat plate in water for seven days, it will then be ready for testing. To effect this, place it in the clips, turn the wheel at D till they clasp the brick with sufficient force to raise the end of the beam nearly up to the pulley above.

(*a*) For strains from 0 to 130 lbs., using the bottom row of figures, hang the transferable weight F in the notch at the end of the beam (as shown in the drawing), and roll the vernier weight G along, taking about one minute to travel the length of the beam.

Fig. 89. *Adie's No. 2 Cement Testing Machine.*

(*b*) From 90 to 200 lbs., using the middle row of figures, remove F from the machine, and roll G forward as before.

(*c*) Above 150 lbs., using the top row of figures, hang on F in the notch under G, and roll G (carrying F with it) forward as before.

To use as a weighing machine, remove the sliding block carrying D and C ; take out the top clip B, and hang on the scale pan instead. A hook passing through a hole in the stand can be supplied with one-tenth bushel measure to weigh the cement if required.

Michaelis's Double Lever Cement Testing Apparatus, Figs. 90 and 90*a*.— *Maker's Directions.*—This apparatus consists of a japanned cast iron column, which

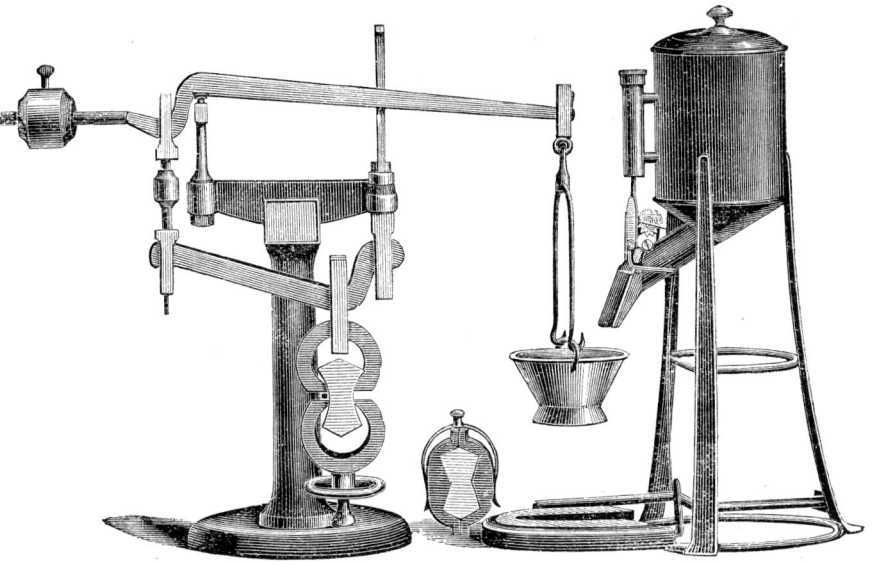

Fig. 90. Fig. 90*a*.
Michaelis's Double Lever Cement Testing Apparatus.

carries two levers, the combined leverage of which is 1 to 50 : that of the longer, being 1 to 10, and that of the shorter, 1 to 5.

Each lever has three hardened steel knife edges acting upon hardened steel concave bearings, so that an extremely accurate balance is obtained.

The short arm of the upper lever is provided with a movable counterpoise, to secure the correct position of the levers, which is indicated by a mark on the upright catch at the top of the column. At the extremity of the long arm is suspended a small brass frame to carry the shot bucket.

On the lower lever, near its fulcrum, is suspended the upper clamp or clip for holding the briquettes. The lower clamp is fixed to the base of the column and adjusted by means of a screw.

To make a test, the cement briquette is taken out of the water, dried and put into the clamps, which must be accurately applied to the sides of the briquette, and the screw applied until the upper edge of the long lever is opposite the mark on the upright catch. Fine shot is then poured from the self-acting shot-run into the bucket suspended from the long lever until the briquette breaks, when the supply of shot is instantly cut off. The breaking strain per square inch is thus exactly fifty times the weight of the bucket and shot ; but to avoid all calculation and possible risk of error, a Salter spring balance with a special dial is supplied, upon which the bucket and shot are weighed and the exact breaking strain of the briquette at once shown.

Michele's Machine.—Fig. 91 is a sketch taken from the illustrated advertisement of the machine.

A briquette of neat cement having been made as before described, and immersed in water for the specified number of days, is placed in the clips V L, as shown.

The handle H is then turned ; it is fitted with a pinion which works in the rack R. The end of the rack being drawn down by the motion of the pinion, draws down the clip L, and brings a stress upon the briquette, which in its turn draws down V and the short arm S of a bent lever.

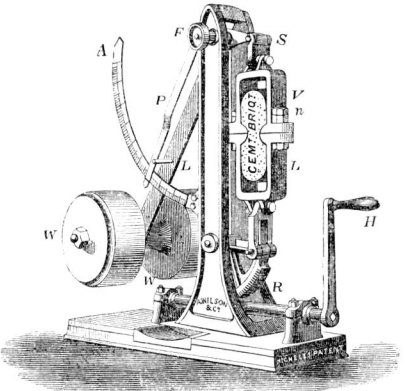

Fig. 91.   *Michele's Cement Testing Machine.*

The long arm L of this lever carries two weights, WW ; as the short arm S is drawn down, these weights are lifted.

While they rise, the leverage with which they act increases with their horizontal distance from the fulcrum F. When the stress produced is sufficient to overcome the resistance of the briquette it breaks across.

The nuts n which secure the clips prevent the weights WW from falling back more than about half an inch.

The stress applied is measured along the graduated arc A. The pointer p is carried up with the long arm of the lever as it rises, but remains when the weights fall, to show the point to which they rise.

These machines are made to test up to 1500 lbs. on the briquette.

Faija's Testing Machine is shown in Fig. 92, from the patentee's circular.
The ordinary-sized machine adapted to briquettes of 1 square inch section will test from 1 to 1000 lbs.

The machine is 14 inches high, 14 inches long, by 3 inches wide, and weighs less than 30 lbs. A special gearing prevents the strain from being put on too quickly.

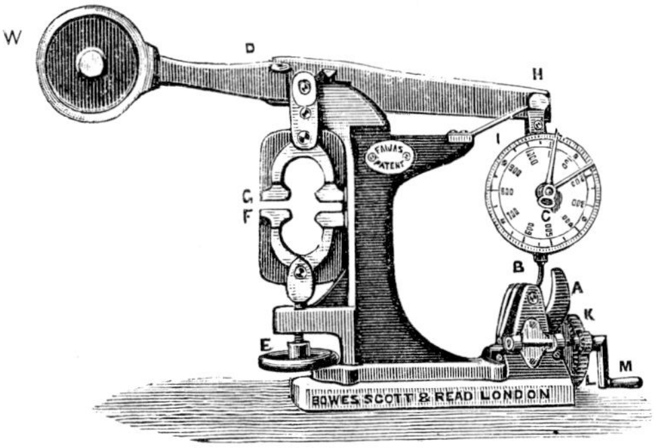

Patentee's Instructions.—On receiving the machine, clean off all old oil and relubricate, attach the balance weight W to the short end of the lever.

To USE THE MACHINE.—See that the quadrant A is in the position shown in sketch, so that the chain B to the dial C is slack, and the lever D free and balanced.

Turn the wheel E from right to left, until the lower clip F can be raised into contact with the upper clip G.

Fig. 92. *Faija's Testing Machine.*

Insert the briquette to be tested in the clips, taking care that it is put in true and evenly, and so that the pull on it and the clips is true and vertical ; then turn the wheel E from left to right, which will bring down the lower clip F, and secure the briquette firmly in the clips. (It is generally advisable to put such a strain on the briquette by turning wheel E that about 100 lbs. is indicated on the dial.) When in this position there should be about half an inch between the under side of knife edge H, and the buffer or recoil spring I.

Having seen that the pinion K is in gear with the wheel L, turn the handle M until the briquette breaks. The loose pointer will show on the dial the strain in lbs. at which the briquette broke.

To RETURN TO ZERO.—Throw the pinion K out of gear with the wheel L by removing the pin and pushing it to the left ; turn the wheel L from left to right until the quadrant A has returned to its normal position with the chain B slack ; put the loose pointer back to zero ; release the lower clip F by turning wheel E from right to left ; remove the broken briquette, and insert the next that is to be broken.

Reid and Bailey's Cement Tester is shown in elevation in Fig. 93, which is taken from the makers' circular.[1]

The briquette having been inserted in the clips c c holds down the short arm of a straight lever. The long arm has a graduated measure attached to its end ; this is gradually weighted by water running from the cistern above.

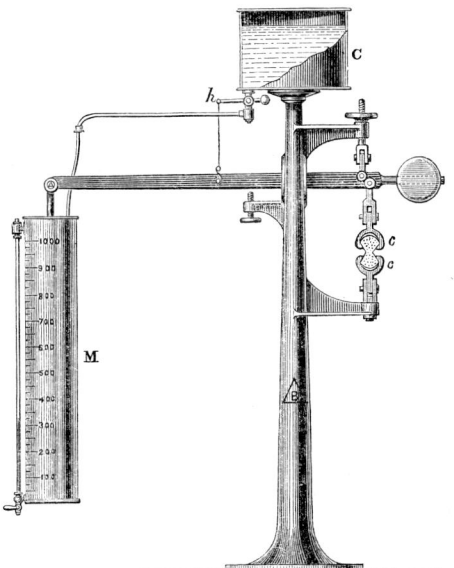

Fig. 93. *Reid and Bailey's Cement Tester.*

When the briquette breaks, the fall of the long arm of the lever draws down *h*, and shuts off the supply cock. The weight required to rupture the briquette is indicated by the

---

[1] Messrs. Bailey and Co., Salford.

amount of water in the measure.   Mr. Reid states that this machine is reliable and accurate.   The weight is applied very gradually and without tremulous vibration, and is recorded automatically by the machine itself.

Bailey's Table Pattern Cement Tester. — Another of Messrs. Bailey and Co.'s cement testers is shown in Fig. 93*a*, which explains itself. It takes sections of 1 inch square, and is sometimes fitted with an automatic arrangement for pouring the shot into the can.

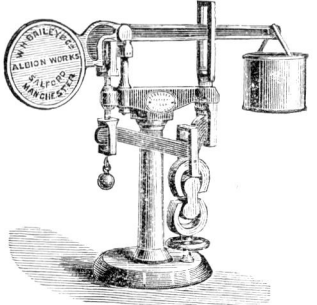

Thurston's Testing Machine.—In order to avoid the difficulty of getting the stress fairly distributed over the area to be fractured, which always occurs in tensile tests, cement has been tested by *twisting.*

Professor Thurston's machine used for testing metals by torsion has, in America, been applied to cements.

This machine has not, however, been adopted in this country, and it need not therefore be described.

Fig. 93*a*.

Simple Tests without Machines.—A tank of water suspended from the specimen may be used as a good simple method of testing the tensile strength.   The weight in lbs. at different depths can be marked inside the tank.

The following specification has been used where there is no testing machine available.

"The cement is to be made into small blocks 1 inch square, and 8 inches long, after being made, these blocks are to be immersed in water for seven days, and then tested by being placed on two supports 6 inches apart, when they must stand the transverse strain produced by a weight of 75 lbs. placed in the centre." [1]

Chemical Tests.—The importance of having a chemical test for Portland cement, in addition to the tests already mentioned, has lately been strongly urged, in consequence of failures arising from an excess of magnesia which has slaked and expanded in the work, causing rupture ; an excess of lime may have the same effect, or cause weakness ; [2] more than 2 per cent of magnesia, or $1\frac{1}{2}$ per cent of sulphuric acid, is said to be injurious to Portland cement.

Adulterations in Portland Cement.—It is stated in the circulars of some cement manufacturers that iron slag is used for the adulteration of Portland cement.   If this is suspected, the only way to avoid it is to refuse to take cement from any manufacturer who has slag on his premises.

Slag has been refused both for cement making and as an aggregate for concrete, for fear that the lime that it contains should disintegrate after use in the work.

This is doubtless a wise precaution.   Slag properly treated by being burnt with lime is, however, sometimes used as the basis of a good Portland cement, as described at p. 166.

---

[1] Messrs. D. B. Stevenson.   *M.P.I.C.E.*, vol. lxxxvii. p. 229.
[2] *M.P.I.C.E.*, vol. lxxxvii. p. 163.

LIME AND CEMENT BURNING.

Limestone is calcined (burnt into lime) in " clamps " or in " kilns " of different forms.

Clamps consist merely of heaps composed of alternate layers of limestone and coal, having a fire-hole below, and covered with clay or sods to prevent the escape of heat.

This is a very wasteful method of burning, and should only be used where limestone and fuel are abundant.

Very similar arrangements for burning lime are in some parts of the country called *Sow Kilns.*

Lime Kilns are divided into two classes, *Tunnel Kilns and Flare Kilns.*

Tunnel Kilns are those in which the fuel and stone are placed in alternate layers.

Flare Kilns have the fuel below, so that the flame only reaches the stone in the kiln above.

Either form of kiln may be worked on the *continuous* or on the *intermittent system.*

The Continuous system is that in which the lime is gradually removed from the bottom of the kiln in small portions, fresh limestone being added at the top to make up for the burnt lime removed at the bottom.

The Intermittent system consists in burning and discharging a whole kilnful at a time. After the stone is well burnt through, the kiln is allowed to cool down, and the burnt lime is removed. The empty kiln is then recharged, and the operation repeated.

The continuous system is most generally applied to tunnel kilns.

The lime so produced is likely to be unequally burnt, but the process is a cheaper one.

By the intermittent system, in which the whole kilnful is burnt at once, the lime is more uniformly calcined throughout.

TUNNEL KILNS, called also CONTINUOUS, " RUNNING," " PERPETUAL," or " DRAW-KILNS."

A kiln of this class is shaped internally either like a cylinder, an inverted cone, or a pair of vertical cones base to base. It is lined with firebrick, and has an opening below, generally protected from the weather by a shed.

At the lower extremity of the cone is a grating, upon which is placed a layer of brushwood, and then alternate layers of coal and moistened stone, reaching to the top, the largest pieces being in the middle, where they will get most heat.

As the lime becomes burnt it is withdrawn through the grating, and fresh stone and fuel are added at the top.

This kiln is economical in fuel, requiring only about $\frac{1}{5}$ the weight of the lime produced, but the lime is not equally well burnt throughout, and it requires great experience to manage the kiln properly.

Fig. 94 is a section of the form of kiln frequently erected as a temporary arrangement to burn lime during the progress of large works.

The kiln may be built of either bricks, stone, or concrete, or sunk into the ground.

When concrete or very rough masonry is used, bonding timbers are built in, or iron bars are fixed externally to bind the structure together.

The interior is lined with firebricks, a hollow space being left behind the lining.

The fuel and broken stone are thrown in at the top of the kiln, and lie in alternate layers, the thickness of each layer of stone being from six to eight times that of the fuel.

At the lower end of the kiln lies the wood for kindling the fire, resting upon a grate of loose bars, which can be drawn out one at a time.

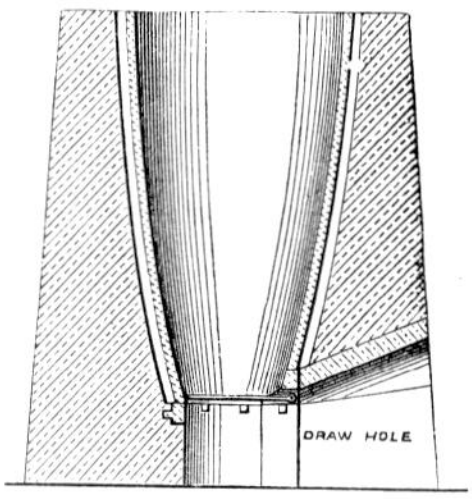

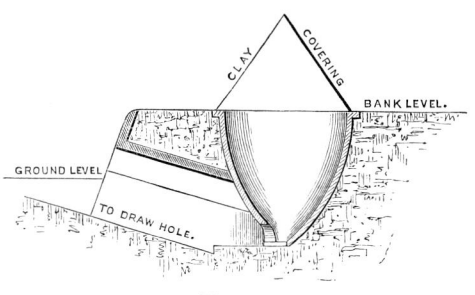

Fig. 94.

The fire having been lighted at the bottom below the grating, the heat passes through the layers ; those nearest the bottom are burnt first, and are withdrawn through the grating by removing one or more of its bars.

As the burnt lime is taken out at the bottom, the bulk of the contents of the kiln slide down, and the space thus left at the top is filled with fresh layers of fuel and stone.

It is convenient to have a shed in front of the drawhole, to secure the freshly burnt lime from the weather.

The size of the kiln varies according to the supply required. A kiln of the form shown in Fig. 94, 16 feet high, 4 feet wide at the bottom, and 9 feet at the top, will hold "about 25 tons of limestone, and will burn sufficient lime to keep twenty bricklayers constantly supplied with mortar."[1]

Fig. 95 shows in section a form of kiln largely used in the Midland counties for burning Lias lime.

The conical mound on the top is composed of layers of fuel and stone, plastered over with clay.

"Care is taken that the clay plastering covering the conical mound does not give too much vent in any one part to the products of combustion, lest too strong a draught should be set up toward such orifice, and cause overburning of the lime in its course.

Fig. 95.

"The fuel is made to burn in a smouldering fashion throughout its operation.

"At the opening of the drawhole, in order to ignite the contents of the kiln, a few large pieces of coal are built up.

"The fuel layers vary from 6 to 3 inches in thickness, those at the bottom being the thickest. The layers of mineral vary from 10 inches at the bottom to 18 inches at the top."[2]

Flare Kilns, called also Intermittent Kilns, are generally in the form of

---

[1] Hill's *Lectures on Machinery.*   [2] Cooke's *Aide Mémoire.*

a cylinder, surmounted in most cases by a conical vault. The broken limestone rests upon arches, roughly formed from large pieces of the same material.

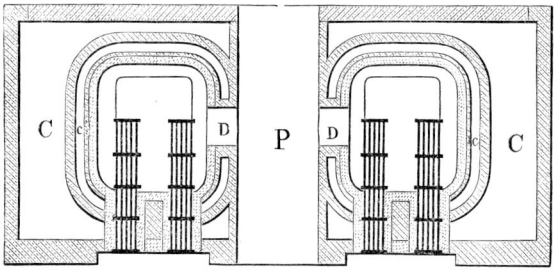

Fig. 96. *Plan.*

Fig. 97. *Section.*

These rough arches must be carefully built, and the heat applied gradually so as not to split the stones.

The fire is lighted below, only the flame being in contact with the stone, thus producing much cleaner lime than that obtained by the methods in which they are mixed together.

Such a kiln is more easily managed than the kinds which are worked continuously, and the lime produced is more uniform in quality. The necessity of letting the fire out after each charge is burnt is a great inconvenience, and also causes waste of fuel.

For the same kind of lime this kiln requires about $\frac{5}{3}$ (*i.e.* nearly double) as much coal as does the tunnel kiln. Moreover, the intermittent kiln requires relining every twelve months, which is a source of great expense.

Fig. 96 is the plan, and Fig. 97 a section of a pair of flare kilns, such as are used for burning grey chalk into lime. L is the hole through which the lower part of the kiln is loaded, l that for the higher levels. The drawholes D D open into a central passage P.

In the section the kiln to the right is shown as loaded, the other as empty. The rough arches of limestone are shown in the former. The fire bars for the fuel are shown in plan and section, the spaces C C are packed with broken chalk, $c\,c$ with chalk dust. The firebrick lining is hatched with broken lines, the ordinary brickwork with continued and broken lines alternately.

Fig. 98 is the section of a simpler flare kiln in common use.

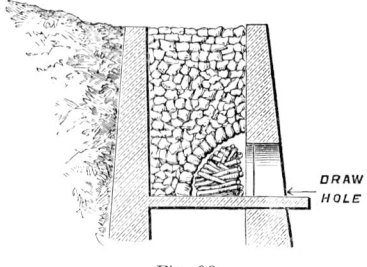

Fig. 98.

Portland Cement Kilns (*Common Form*).—Fig. 99 is the section, and Fig. 100 an elevation, of a form of kiln commonly used for burning Portland cement in the Medway district.

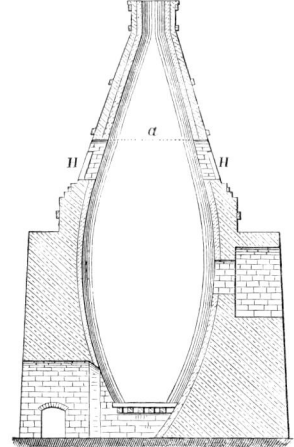

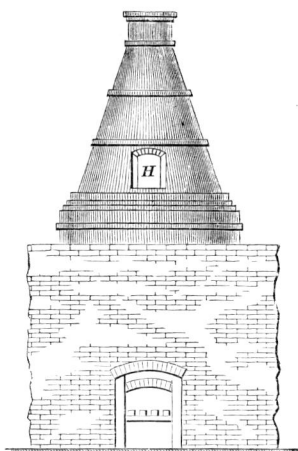

Fig. 99.   *Section.*   Fig. 100.   *Elevation.*

It is worked upon the intermittent system ; the coke and slurry are, however, packed in alternate layers 6 and 4 inches deep.

Such kilns hold about thirty tons. Their contents are burnt in forty-eight hours, and are drawn about once every four days.

The kiln is lined with firebrick, sometimes only up to the line *a*, but better throughout, and loaded at the holes H H H.

The firebrick lining should be detached from the mass of the brickwork, so as to be free to expand and contract under the great changes of temperature to which it is subjected.

The inside should be painted over with wet stuff from the *backs* each time before the kiln is charged. This will greatly increase its durability.

In some forms of this kiln the top has a wider opening, and the short vertical neck or chimney is frequently omitted.

The description of kiln used varies in different places. A modification of Hoffmann's kiln similar to that used for bricks may be economically adopted where a large continuous supply of cement is required year after year.

The contents are burnt at a high temperature, but the amount of firing depends upon the proportions of the mixture. If the lime be in excess it can hardly be overburnt, but if there be too much clay it will fall into dust.

The Michele-Johnson Kiln is a modification of the kiln mentioned at p. 166.

The arched chamber there referred to as branching out from the kiln has a very thin arch over it. Above this the cool slurry is spread for a preliminary drying before it is forced through the openings of the arch and spread over the floor of the chamber ; there it is further dried by the hot air and gases from the kiln, which pass through the chamber on their way to the chimney.

The Rotatory Process of Cement Manufacture.—This system is in extensive use in the United States, and on the Continent, and is being adopted in this country. The principal feature of a rotatory kiln is a cylinder about 60 feet long, and 5 to 6 feet diam., of steel plate, lined with refractory material, slightly inclined to the horizontal, and rotated by gearing. A burner at the lower end injecting powdered coal, oil, or gas, supplies the heat for roasting the raw material fed in at the upper end. The roasted clinker falling from the lower end is then passed through other cylinders, and, emerging at a suitable temperature, after preliminary crushing is finally ground in the usual way. The advantages claimed for this system are. diminished cost of production, greater uniformity in quality of clinker, and the production of an artifically matured cement dispensing with subsequent aeration. The present prudent system of storage and turning over will, however, probably not be abandoned until it has been proved that aeration can safely be dispensed with.

Roman Cement Kilns.—Fig. 101 is a plan and Fig. 102 a cross section

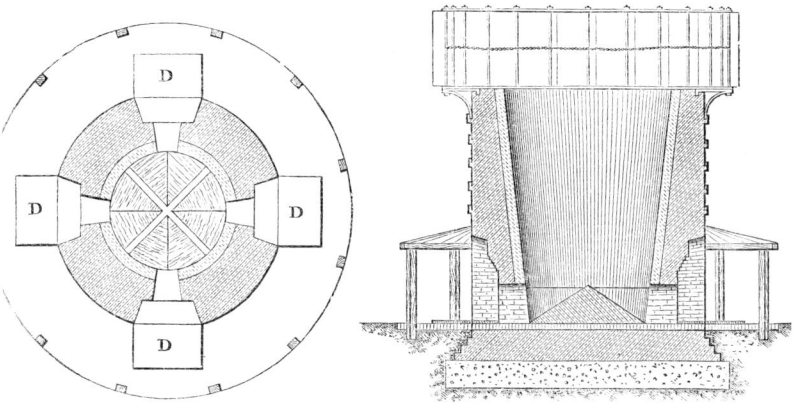

Fig. 101. *Plan.* Fig. 102. *Section.*

of the kiln used for burning Roman cement. It is worked on the constant system. The stone is packed in strata separated at intervals of from 6 to 9 inches by thin layers of fuel. The cone in the centre guides the burnt cement to the drawholes D D where it is taken out.

General Remarks on Burning.—*Gradual heating* is necessary in burning lime or cement stone.   If the heat be suddenly applied, the carbonic acid and moisture will be driven out with such violence as to blow the stone to pieces.

Appearance of the Burning Stone.—As long as the burning is incomplete, and any carbonic acid is left in the stone, it will remain of a dull red colour. When the carbonic acid is all expelled, the stone in the kiln becomes peculiarly bright, which is a sign that the calcination is complete, and that the lime may be withdrawn.

The Temperature at which a lime or cement should be burnt depends upon its composition.

A pure or fat lime requires only heat enough to drive off the carbonic acid and moisture.

Limes containing clay require a somewhat greater heat, in order that the silicates and aluminates may be formed which give the hydraulic properties required.

A great deal depends, however, upon the composition of the clay.

A large proportion of iron and alumina (especially of iron) as compared with the silicic acid, greatly facilitates the action which takes place in calcination, and the prepared mortar also sets more quickly.

Great care must be taken, however, that the heat is not sufficient to fuse the particles of the lime or cement.

Thus Roman cements, in which the quantity of iron and alumina together nearly equals the silicic acid, are burnt with little fuel at a low temperature.

Portland cement, on the other hand, in which the iron and alumina are less than half the silicic acid, is burnt at very high temperatures.   There is very little danger of fusing the particles, and the heat may with advantage be raised to a point just short of vitrification.[1]

The Size of the Lumps into which the lime or cement stone is broken greatly influences the burning operation.

The denser the stone and the higher the temperature at which it is to be burnt, the smaller must be the pieces into which it is broken.

Pure or fat limestones are broken into pieces containing from one to two cubic feet.

Hydraulic limestones into pieces containing about a quarter of a cubic foot.

Roman cement stones and others of the same quick-setting class are broken into pieces containing one or two cubic inches.

The Quantity of Fuel is of course influenced partly by the form of kiln, but chiefly by the nature of the stone and by the temperature at which it is to be burnt.

Thus, for the calcination of pure dense limestones about $\frac{1}{3}$ to $\frac{1}{4}$ their weight of coal is required.

For hydraulic limestones about $\frac{1}{5}$ to $\frac{1}{7}$ their weight.

For Roman and other quick-setting cements about $\frac{1}{8}$ to $\frac{1}{12}$ of the weight of stone.

For Portland cement about $\frac{1}{3}$ the weight of the dried slurry.

PORTLAND CEMENT CLINKER.—The clinker of good Portland cement, when properly burnt, is of a dark *greenish-black* colour, differing in density according to the amount of fuel used.

---

[1] General Scott in *R.E. Papers*, vol. xi.

It is almost impossible to burn the contents of any kiln quite uniformly throughout, and the clinker will be found differing in colour accordingly.

It should not be clinkered into large masses—should rattle well as it comes out of the kiln—should be honeycombed in texture and nearly free from dust.

Some will be found of a *bright yellow* colour, and of light specific gravity. This will set very quickly, and it should be picked out.

Some, again, will be of a *pink* or *dirty white* colour, but more or less heavy according to the heat to which it has been subjected. Clinker of this kind has been imperfectly burnt, and must be again passed through the kiln. It is a dangerous substance to use.

Dense *glazed black* clinker indicates excess of lime, and will also yield dangerous cement inclined to blow ; *dark blue* clinker, a sluggish cement ; and a *brown* clinker with much dust, a weak cement.[1]

Dangerous Limes and Cements.—Sometimes, from defects in the process of calcination of a stone which should produce an eminently hydraulic lime or cement, compounds result which are of a most dangerous character.

These are caused either by over-burning or under-burning.

Over-burnt.—In the former case, a hard and heavy substance is produced, burnt almost to a clinker, which slakes with very great difficulty, and after a considerable lapse of time.

This can only be remedied by screening out the hard portions and grinding them to a fine powder ; otherwise any larger particles that may be left will slake after the mortar has been laid in the work, and may do great damage to the masonry.

Under-burnt.—When, on the other hand, the stone has been under-burnt, a somewhat similar result occurs, but from a different cause.

The substance produced consists partly of a perfect cement or hydraulic lime, and partly of free quicklime. The latter is prevented, by the setting action of the cement, from slaking at once, but does so eventually, and with the same disastrous consequences as occur with over-burnt lime.

This dangerous action may, however, be got rid of by free exposure of the lime or cement, so as to air-slake the caustic portions, or by frequently reworking the mortar, or by adding a proportion of soluble silicates, which anticipate and prevent the slaking action. The latter is, however, seldom if ever done in practice.

Dead-burnt Lime is lime that has been imperfectly calcined and will not slake with water.

This may be caused by under burning, so that only part of the carbonic acid is expelled, the resulting substance being a compound of quicklime and carbonate of lime, which refuses to slake.

Hydraulic lime may be rendered " dead " by over-burning ; the silicates are partially fused and coat the stone, so that the evolution of the carbonic acid is prevented [2] (see p. 239).

Flare-burnt Lime is lime burnt in flare kilns, in which it is kept cleaner than in tunnel kilns owing to its not being in contact with the fuel.

---

[1] Certain varieties of overclayed cements yield a deep bronze-coloured clinker which, as it cools on coming from the kiln, disintegrates spontaneously into a fine flakey greyish powder which produces an inert cement.—*M.P.I.C.E.*, vol. lxii. p. 81.

[2] Dent.

SAND AND SUBSTITUTES FOR SAND.

Sand is known as " argillaceous," " siliceous," or " calcareous," according to its composition.

It is procured from pits, shores of rivers, sea-shores, or by grinding sandstones.

It is chiefly used for mortar, concrete, and plaster. The qualities it should possess for those purposes are pointed out at page 204.

PIT SAND has an angular grain, and a porous, rough surface, which makes it good for mortar, but it often contains clay and similar impurities.

RIVER SAND is not so sharp or angular in its grit, the grains having been rounded and polished by attrition.

It is commonly fine and white, and therefore suited for plastering.

SEA SAND also is deficient in sharpness and grit from the same cause. It contains alkaline salts, which attract moisture and cause permanent damp and efflorescence.

SCREENING.—When sand contains lumps or stones it should be " screened," or, if required of great fineness, passed through a sieve.

WASHING.—Sand found to contain impurities, such as clay, loam, etc., which unfit it for almost every purpose, should be washed by being well stirred in a wooden trough having a current of water flowing through it which carries off the impurities. It is sometimes washed by machinery, such as an Archimedean screw revolving and carrying up the sand, while a stream of water flows down through it.

EXAMINATION OF SAND.—Clean sand should leave no stain when rubbed between the moist hands. Salts can be detected by the taste, and the size and sharpness of the grains can be judged of by the eye.

SIZE OF GRIT.—Where this is specified, as it is in connection with the cement and sand test for Portland cement (see p. 175), it is generally required that the sand should pass through a sieve of 400 meshes to the square inch and be retained by one of 900 meshes.

Substitutes for Sand.—BURNT CLAY is sometimes used as a substitute for sand in mortar.

It is prepared by piling moistened clay over a bonfire of coals and wood. As the clay becomes burnt and the fire breaks through, fresh layers of clay and coal, " breeze," or ashes, are piled on, and the heap may be kept burning until a sufficient supply has been obtained.

The clay should be stiff. Care must be taken that it is thoroughly burnt. Raw or half-burnt pieces would seriously injure mortar.

CRUSHED STONE.—Sand is sometimes very economically obtained by grinding the refuse "spalls" left after working the stones for walling. It is generally clean if carefully collected, but the sharpness of its grit depends upon the nature of the stone from which it is procured.[1]

---

[1] Mr. Kinniple's experiments show that mortar of 1 Portland cement and 1 crushed sandstone is 55 per cent stronger than that of 1 Portland cement 1 pit sand.— *M. P. I. C. E.*, vol. lxiv. p. 330. Experiments by a friend of the writer's showed

SCORIÆ from ironworks, old bricks, *Clinker* from brick kilns, and *Cinders* from coal, make capital substitutes for sand when they are quite clean and properly used. *Wood Cinders* are too alkaline. *Crushed Slag* from furnaces may be dangerous if it contains lime.

POZZUOLANAS, ETC.

POZZUOLANA is a name given to several substances which somewhat resemble each other ; including the Pozzuolana proper, also Trass, Arènes, Psammites, etc.

These are clayey earths containing 80 to 90 per cent of clay, with a little lime, and small quantities of magnesia, potash, soda, oxide of iron,[1] or manganese.

When finely powdered in their raw state without being calcined, they may with great advantage be added to fat lime paste.

In consequence of the amount of clay they contain they confer hydraulic properties upon the lime to a very considerable degree.

The Italian pozzuolana may with advantage be used with fat lime and sand in the following proportions :—[2]

> 12 Pozzuolana well pulverised.
> 6 Quartzose sand well washed.
> 9 Rich lime recently slaked.

Natural Pozzuolana is a naturally-burnt earth of volcanic origin, found at Pozzuoli, near Vesuvius, and in other parts of southern Europe.

It is found in the form of powder more or less coarse in grain, of a brown colour, sometimes passing into red, grey, yellow, and white.

TRASS is also a naturally-burnt argillaceous earth, found on the sites of extinct volcanoes, chiefly near Andernach on the Rhine.

It occurs in lumps of a greyish colour and earthy appearance, is used in the same way as pozzuolana, and confers hydraulic properties upon fat limes.

ARÈNES are natural mixtures of sand and clay. They appear not to have been subjected to heat, but they confer hydraulic properties upon fat lime, probably because they contain a large proportion of soluble silica.

PSAMMITES may be considered as " very feeble pozzuolanas in the crude state, and acquire but a slight increase of hydraulic energy by any degree of calcination.

." Even their feeble powers, however, confer upon them this advantage, that for mortars not absolutely immersed in water when green, and when there is ample time for their properties to develop themselves before submersion, they can be employed in larger proportions than any species of sand holly inert would admit of."[2]

DISINTEGRATED GRANITE, SCHISTS, and BASALT furnish sand having the same characteristics as the Psammites.

ARTIFICIAL POZZUOLANAS are prepared from clays of suitable composition by a slight calcination.

Pounded bricks or tiles possess the properties of pozzuolana in some degree

---

mortar of 1 Portland cement and 2 crushed limestone to be 50 per cent stronger than that of 1 Portland cement and 2 sand.

" Briquettes made from crushed syenite from which the impalpable powder had been removed were 18 per cent stronger than those in which it had not been removed."—*M.P.I.C.E.*, vol. lxxxvii. p. 203.

[1] *Ferric oxide.*   [2] Gillmore.

MORTAR.

Ordinary Mortar is composed of lime and sand mixed into a paste with water.

When cement is substituted for the lime, the mixture is called *Cement Mortar.*

Uses.—The use of mortar in brickwork or masonry is to bind together the bricks or stones, to afford them a soft resting-place, which prevents their inequalities from bearing upon one another, and thus to cause an equal distribution of pressure over the beds. It also fills up the spaces between the bricks or stones and renders the wall weather-tight.

It is also used in concrete (see page 215) as a matrix for broken stones or other bodies to be amalgamated into one solid mass; for plastering, and other purposes.

The quality of mortar depends upon the description of materials used in its manufacture, their treatment, proportions, and method of mixing. These particulars will now be considered.

Description of Lime or Cement to be used in Mortar.—FAT LIMES should only be allowed for inferior or temporary work.

On account of their being cheap and easy to manipulate, they are often used in positions for which they are entirely unfit.

Mortar made from fat lime is not suitable for damp situations or for thick walls. In either case it remains constantly moist; when placed in positions where it is able to dry it becomes friable, and in any case is miserably weak.

Even the economy of fat lime mortar is in many cases doubtful; for walls built with it are injured by frost, require constant repointing, and perhaps before many years rebuilding.

M. Vicat says of fat limes:—"Their use ought for ever to be prohibited, at least in works of any importance."

Sir Charles Pasley adds with regard to fat lime mortar that "when wet it is a pulp or paste, and when dry it is a little better than dust."

Evils of Fat Lime Mortar.—If a pure or feebly hydraulic lime mortar is used in massive brickwork or masonry, it is only the outer edges of the joints that are affected by the carbonic acid in the air. A small portion of the exterior of the joints sets, but the mortar in the inside of the wall remains soft. The result of this is that a heavy pressure is thrown upon the outer edges of the bricks or stones, and they become " flushed," that is, chipped off. In some cases, from the same cause, the headers of brickwork are broken, so that the face of the wall becomes detached, and liable to fall away.

Again, these weak mortars retain or imbibe moisture, which, when it freezes, throws off the outer crust. Pointing is then resorted to. If this is done with the same sort of mortar, the same result ensues, and in an aggravated degree, for as the operation is repeated, the joint becomes wider. In the end it will often be found that more has been expended in patching up work done with bad mortar than would have sufficed to provide good mortar at the first.

HYDRAULIC LIME or CEMENT should, therefore, always be used in mortar for work of any importance. In subaqueous constructions it is, of course, absolutely necessary.

If there is any choice, the class of hydraulic lime used will depend upon the situation and nature of the work to be done.

For ordinary buildings, not very much exposed, slightly hydraulic limes will suffice to form a moderately strong joint, and to withstand the weather.

For damp situations, such as foundations in moist earth, a more powerful hydraulic lime should be prepared.

For masonry under water an eminently hydraulic lime or cement mortar will be necessary. If the work be required to set very quickly, Roman cement, or a cement of that class, would be used; whereas, if quick setting be not necessary, but great ultimate strength is required, a heavy Portland cement should be adopted.

Cement is also generally used for copings, plinths, arches, and other important parts in ordinary house-building.

Description of Sand to be used in Mortar.—Sand is used in mortar to save expense and to prevent excessive shrinkage.

Ordinary sands are not in any way chemically acted upon by the lime, but are simply in a state of mechanical mixture with it; with hydraulic limes and cements the effect of sand is to weaken the mortar.

When fat lime is used, however, the porous structure, caused by the sand, enables the carbonic acid of the air to penetrate farther, and to act upon a larger portion of the joint.

Moreover, the particles of fat lime adhere better to the surfaces of the grains of sand than they do to one another; therefore the sand is in two ways a source of strength in fat lime mortar.

It is of the utmost importance that the sand used for mortar should be perfectly clean, free from clay or other impurities which will prevent the lime from adhering to it.

Sand for this purpose should have a sharp angular grit, the grains not being rounded, their surfaces should not be polished, but rough, so that the lime may adhere to them.

It has been found that, speaking generally, the size of the grains of sand does not influence the strength of the mortar.

Mr. Mann's experiments tend, however, to show that in samples four weeks old Portland cement mortar made with fine sand was weaker than that made with coarse sand.

Very fine sand is objectionable for mortar, as it prevents the air from penetrating, which is necessary in order that the mortar may set.

Although coarse irregular-grained sand may make the best mortar, when very thin joints are used finer sand is sometimes necessary.

Calcareous sands, on the whole, give stronger mortars than siliceous ones.

Sea sand contains salts, which are apt, by attracting moisture, to cause permanent damp and efflorescence.

This moisture will effectually prevent a fat lime from setting, or rather drying, but would tend to increase the strength of a hydraulic lime or cement (see page 240).

Great care must be taken to exclude all organic animal matter from the sand, or substitutes for sand, that may be used in mortar for building or plastering the walls of dwellings, otherwise they will putrefy, and render the walls and ceilings sources of unwholesome emanations.

SUBSTITUTES FOR SAND IN MORTAR.—Any of the substances mentioned at page 200 may be used as substitutes for sand in mortar, some of them with advantage, as there pointed out.

Smiths' ashes and coal dust are used to make the *black mortar* used for pointing, slating, and for some kinds of rubble masonry.

The Description of Water to be used in Mortar.—The water used for mixing mortar should be free from mud, clay, or other impurities.

Salt Water is objectionable in some situations, as it causes damp and efflorescence.

The salts it contains attract moisture, which improves the strength of hydraulic limes and cements by preventing them from drying too quickly, but is fatal to a pure lime for the reasons given above.

Dirty water, and water containing organic matter, are of course objectionable for the same reasons as dirty sand.

Mr. Dyce Cay gives a table of experiments made with 8 oz. fresh water to 36 oz. neat Portland cement, and 7 oz. sea water to 36 oz. Portland cement, which seems to show, as far as experiments with neat cement could show it, that " roughly speaking the salt water briquettes are as strong in a week as the fresh water ones are in a fortnight, and as strong in a fortnight as the fresh water ones are in a month." [1]

Strength of Mortar as compared with Bricks in a Wall.—Lime is much more expensive than sand. It is, therefore, a source of economy to add as much sand as is possible without unduly deteriorating the strength of the mortar.

So long as the joints of masonry or brickwork are weaker than the stones or bricks, the strength of the wall will increase in proportion as the strength of the mortar increases, until they are nearly equal in power of resistance.

The mortar need not be quite equal in strength to the bricks, because in a bonded wall the fracture is constrained to follow a longer path than when the work is put together without breaking joint.

The object, then, is to produce such an equality of resistance as will compel the fracture to follow a straight line, *i.e.* to break the material of the wall straight across rather than to follow the joints.

This cannot always be done, with a due regard to economy, where the wall is built with very hard stone, but it can be done with the generality of bricks.

In some cases a stronger mortar, no doubt, adds to the strength of the wall. For example, when the bricks are very bad, they will sometimes weather out on the face, leaving a honeycomb of mortar joints.

Again, unusually strong mortar is required sometimes for the voussoirs of arches—to prevent sliding—for the lower joints of chimneys and walls, etc. etc.

As a rule, however, it can hardly be economical to make the strength of the mortar joints greater than that of the bricks or stones they unite.

Proportion of Ingredients.—In considering the proportion of sand to be mixed with different limes and cements it is necessary to bear in mind that the strength of the joint formed by the mortar will have an influence upon that of the wall.

The following Table shows how different limes and cements are weakened by the addition of various proportions of sand :—

[1] *M.P.I.C.E.*, vol. lxii. p. 212.

TABLE

Showing the effect of different Proportions of SAND in MORTARS
made from various CEMENTS.

NATURE OF MATERIAL.	Age when tried.	PROPORTION OF CEMENT OR LIME AND SAND.					
		1 c. Neat.	1 c. 1 s.	1 c. 2 s.	1 c. 3 s.	1 c. 4 s.	1 c. 5 s.[1]
		BREAKING WEIGHT IN LBS. UPON AREA OF 10 INCHES.					
Portland Cement . .	11 days	Bricks broke first	504	433	303	420	238
Medina	,,	400	352	278	201	149	83
Roman	,,	400	279	178	154	149	73
Atkinson's . . .	,,	...	385	175	79	49	
Scott's Cement . .	,,	292	286	308	328	281	194
Lias Lime . . .	,,	119	80	124	29	37	42

The above figures are from experiments made for General Scott by tearing asunder bricks united by the different kinds of mortar, and set in air. The sectional area torn asunder being $4 \times 2\frac{1}{2} = 10$ inches in each case.

The Table at page 183 gives fuller particulars as to the loss of strength caused by adding sand to Portland cement.

The proportion of the ingredients in mortar is generally specified thus :—" 1 quicklime to 2 (or more) of sand," meaning that 1 measure of quicklime in lump[2] is to be mixed with 2 measures (or more) of sand.

Now, the quantities of sand put at different times into a measure vary a little, according to the amount of moisture the material contains ; but so little that practically it makes no difference, and this mode of measuring sand is very convenient and sufficiently accurate.

With the lime, however, many conditions have to be fulfilled in order to make it certain that the same quantity always fills the same measure.

The specific gravity of the calcined stone, the size of the lumps, the nature of the burning, the freshness of the lime, all

[1] Portland cement mortar made with 8 parts of sand to 1 of cement may advantageously be used in preference to lime mortar (see p. 213).

[2] The pieces of calcined stone are called " lump-lime," or in the North " lime-shells."

cause the actual quantity contained in a given measure to differ considerably.

In order to avoid this uncertainty it has been proposed that the *weight* of lime for a given quantity of sand should be specified.

Practically, however, this has not been carried out to any great extent, and the bulk of lime to be used is generally specified as well as that of the sand.

The following proportions are given by General Scott for mortar in brickwork built with ordinary London stock bricks.

	PARTS BY MEASURE.	
	Quicklime.	Sand.
Fat limes . . .	1	3
Feebly hydraulic limes . .	1	$2\frac{1}{2}$
Hydraulic limes (such as Lias) .	1	2
Roman cement . .	1	1 or $1\frac{1}{2}$
Medina „ .	1	2
Atkinson's „ .	1	2
Portland „ . .	1	5
Scott's „ . . .	1	4
Selenitic „ . . .	(see p. 184).	

" The proportions here recommended apply only to works above the surface of the ground, or free from the action of a body of water."

" For hydraulic purposes and foundations 1 sand to 1 quicklime is as much as should be admitted. With cement mortar 2 sand may be used with 1 cement, unless actually in contact with water, when 1 part of sand should be the limit allowed." [1]

Preparation and Mixing.—The quicklime and sand having been procured, and their proportions decided, the preparation of the ingredients commences.

SLAKING.—A convenient quantity of the quicklime is measured out on to a wooden or stone floor under cover, and water enough to slake it is sprinkled over it.

The heap of lime is then covered over with the exact quantity of sand required to be mixed with the mortar; this keeps in the heat and moisture, and renders the slaking more rapid and thorough.

In a short time—varying according to the nature of the lime —it will be found thoroughly slaked to a dry powder.

In nearly all limes, however, there will be found overburnt

[1] General Scott in *R. E. Corps Papers*, vol. xi.

refractory particles, and these should be carefully removed by screening—especially in the case of hydraulic limes; for if they get into the mortar and are used, they may slake at some future time, and by their expansion destroy the work.

Quantity slaked and Time required.—The fat limes may be slaked in any convenient quantity, whether required for immediate use or not. Plenty of water may be used in slaking without fear of injuring them, and they will be found ready for use in two or three hours.

Hydraulic limes should be left (after being wetted and covered up) for a period varying from twelve to forty-eight hours, according to the extent of the hydraulic properties they possess; the greater these are, the longer will they be in slaking. Care should be taken not to use too much water, as it absorbs the heat and checks the slaking process. Only so much should be slaked at once as can be worked off within the next eight or ten days.

With strong hydraulic limes, or with others that are known to contain overburnt particles, it is advisable to slake the lime separately, and to screen out all dangerous lumps, etc., before adding the sand, or the safest plan is to have the lime ground before using it.

Ground Lime.—When lime is purchased ready ground there is sometimes danger of its having become *air-slaked*, by which wear and tear of machinery in grinding is saved at the expense of loss of energy on the part of the lime.

At the same time, if unadulterated and fresh, ground lime is likely to be of good quality for the reasons stated at p. 160.

Quantity of Water used.—The quantity of water required for slaking varies with the pureness and freshness of the lime, and is generally between one-third and one-half of its bulk.

A pure lime requires more water than one with hydraulic properties, as it evolves more heat and expands more in slaking.

A recently-burnt lime requires more water than one that has been allowed to get stale.

Mixing.—The great object in mixing is to thoroughly incorporate the ingredients, so that no two grains of dry sand should lie together without an intervening layer or film of lime or cement.

On extensive works a mortar-mill is universally adopted for mixing the ingredients, and, indeed, is absolutely necessary for the intimate incorporation of large quantities.

A few different forms of mortar-mill are shown and described at page 228 *et seq.*

The heap of slaked lime covered with sand, above described, (p. 207) is roughly turned over and shovelled into the revolving pan of the mortar-mill, enough water being added to bring the mixture to the consistency of thick honey.

When the ingredients are thoroughly mixed and ground together, the mortar is shovelled out of the pan on to a "banker" or platform to keep it from the dirty ground, whence it is taken away by the labourers in their hods.

A good deal has been said regarding the number of revolutions that should be given to the pan. Nothing seems to have been settled upon this point except that the mortar should be thoroughly mixed, yet not kept so long in the mill as to be ground to pap. About twenty minutes is a good time for running each charge of about $\frac{2}{3}$ of a cubic yard.

On very small works the mixing is effected by hand or in a pug-mill. It is evident, however, that such a mixture must be very incomplete unless a great deal of time is devoted to it.

Before hydraulic lime is mixed in this manner it is absolutely necessary that it should first be ground to a fine powder, and with any description of lime the smallest refractory unslaked particles should be carefully screened out.

Mortar, when made with cement, should be mixed dry, the ingredients being carefully turned over together two or three times before the water is added. By this process a very thorough incorporation of the materials can be effected.

Quantities mixed.—If a hydraulic mortar is allowed to commence to set and is then disturbed, it is greatly injured. Care should be taken, therefore, to mix it only so long as is required for thorough reduction and incorporation of the ingredients, and only to prepare so much as can be used within a few hours. With fat limes it matters little whether large or small quantities of mortar are made at once, because they set very slowly.

Very quick-setting cements must be used immediately they are mixed.

Bulk of Mortar produced.—The bulk of mortar produced in proportion to that of the ingredients differs greatly according to the nature of the lime or cement and the quantity and description of the sand added to it.

The more hydraulic limes produce a smaller amount of mortar because they expand less in slaking.

The following Table shows the bulk of mortar found by experiment to be produced from a few of the most common ingredients in ordinary use. It must be regarded only as a guide to the approximate quantities. The actual bulk would vary according to the freshness of the lime and the coarseness of the sand.

MORTAR made from given Quantities of LIME and CEMENT and SAND.

Description.	Quick-lime or Cement.	Sand.	Water.	Mortar made.	Remarks.
	Cub. ft.	Cub. ft.	Gallons.	Cub. ft.	
White chalk lime in lump	1	2	$7\frac{1}{4}$	$2\frac{5}{12}$	The quantity of water mentioned includes that required for both slaking and mixing.
Do. do. . .	1	3	8	$3\frac{1}{12}$	
Portland stone lime in lump	1	3	$7\frac{3}{4}$	$3\frac{2}{3}$	
Grey chalk lime in lump	1	2	$7\frac{1}{2}$	$2\frac{1}{2}$	
Do. do. .	1	3	$8\frac{1}{2}$	3	
Stone lime (Plymouth,[1] in lump	1	3	$9\frac{1}{3}$	$3\frac{1}{9}$	
Lias (Keynsham)[1] in lump	1	3	$7\frac{1}{2}$	$2\frac{2}{3}$	
Lias (Warwickshire) in lump	1	2	3	$2\frac{1}{4}$	
Do. do. ground	1	2	3	$2\frac{1}{2}$	
Lias (Keynsham)[1] do.	1	2	3	2	
Do. do. do.	1	3	$4\frac{1}{4}$	$2\frac{1}{5}$	
Lias (Lyme Regis) do.	1	2	$4\frac{2}{3}$	$2\frac{1}{3}$	
Arden lime ground .	1	1	$3\frac{5}{8}$	$1\frac{7}{12}$	
Roman cement[1] . .	1	1	5	$1\frac{2}{3}$	
Portland cement[2] . .	1	1	2	$1\frac{2}{3}$	
Do.[2] . .	1	2	$1\frac{2}{3}$	$2\frac{2}{3}$	
Do.[2] . .	1	3	$1\frac{1}{2}$	$3\frac{1}{3}$	
Do.[2] . .	1	4	$1\frac{2}{3}$	$4\frac{1}{8}$	
Do.[2] . .	1	5	$1\frac{2}{3}$	$5\frac{1}{8}$	

[1] Deduced from Cooke's *Aide Mémoire.*

[2] Deduced from Grant's Experiments, *M.I.P.C.E.*, vol. xxv. The quantities varied according to the amount of water used ; the Table shows the average.

Where authorities are not given, the quantities stated have been derived from experiments made for this work.

The results of further experiments, giving the same kind of information, will be found in Hurst's *Surveyor's Handbook.*

The use of Sugar in Mortar.—It was pointed out many years ago [1] that the bad qualities of rich limes "may be in some degree corrected by the use of a comparatively small quantity of the coarsest sugar dissolved in the water with which they are worked up," and that sugar was extensively used in the East for common mortars made of calcined shells, which when well prepared "resist the action of the weather for centuries." A recent discussion on the subject has led to experiments being made to ascertain the effect of sugar on Portland cement; and it was found that the addition of from $\frac{1}{8}$ per cent to 2 per cent of pure sugar to Dyckerhoff's German Portland cement increased its strength after three months considerably. The sugar is said to "retard the setting," and thus permit the chemical changes in the cement to take place more perfectly. More than 2 per cent of sugar made the cement useless. [2]

Selenitic Mortar is generally made by mixing selenitic cement and sand. It was at one time made by mixing a small proportion of calcium sulphate with ordinary lime and sand.

The licenses issued by the patentees render it necessary that selenitic cement should be used. The proportion of sulphate required to develop the characteristics of the material is added to the cement before it is sold, and the process of mixing the mortar is carried on under the following rules, which are taken from the circular of the patentees :—

SELENITIC MORTAR MADE WITH SELENITISED LIME OR SELENITIC CEMENT.

N.B.—One bushel [3] of prepared selenitic lime requires about six gallons of water (two full-sized pails).

If prepared in a Mortar Mill.—1*st*, Pour into the pan of the edge-runner four full-sized pails of water.

2*d*, Gradually add to the water in the pan 2 bushels of prepared selenitic lime, and grind to the consistency of creamy paste, and in no case should it be thinner.

3*d*, Throw into the pan 10 or 12 bushels of clean sharp sand, burnt clay, ballast, or broken bricks, which must be well ground till thoroughly incorporated. If necessary, water can be added to this in grinding, which is preferable to adding an excess of water to the prepared lime before adding the sand.

When the mortar-mill cannot be used, an ordinary plasterer's tub (containing about 30 or 40 gallons) or trough, with outlet or sluice, may be substituted.

---

[1] Vicat on *Cements* (Smith), published in 1837.   [2] *Engineering*, 1888, p. 102.
[3] A striked bushel = 1.28 cubic foot (see page 163).

If prepared in a Plasterer's Tub.—1*st*, Pour into the tub 4 full-sized pails of water.

2*d*, Gradually add to the water in the tub 2 bushels of prepared selenitic lime, which must be kept well-stirred until thoroughly mixed with the water to the consistency of creamy paste, and in no case should it be thinner.

3*d*, Measure out 10 or 12 bushels of clean sharp sand or burnt clay ballast, and form a ring, into which pour the selenitic lime from the tub, adding water as necessary. This should be turned over two or three times, and well mixed with the larry or mortar hook.

Both the above mixtures are suitable for bricklayers' mortar or for first coat of plastering on brickwork (see p. 248).

N.B.—The Selenitic Cement Company recommend that the workman intrusted with the making up of the selenitic mortar be supplied with suitable measures for his lime and sand, to ensure that the proportions stated in the circulars be adhered to. The want of this frequently leads to unsatisfactory results.

A box measuring inside $13\frac{1}{8}$ inches by $13\frac{1}{8}$ inches by $13\frac{1}{8}$ inches would contain about 1 bushel, and would be useful for measuring the lime, and should be kept dry for that purpose ; and a box without a bottom, measuring inside 36 inches by 18 inches by 18 inches would contain about $5\frac{1}{4}$ bushels, and would be very useful for measuring the sand.

Increase or decrease the quantities given proportionately with the requirements. The prepared selenitic lime must be kept perfectly dry until made into mortar for use.

N.B.—It is of the utmost importance that the mode here indicated of preparing the mortar, concrete, etc., should be observed—viz. First well stirring the prepared selenitic cement in the water before mixing it with the sand, ballast, or other ingredient, otherwise the cement will slake and spoil.

SELENITIC MORTAR MADE WITH ORDINARY LIME.—A few years ago persons using selenitic mortar were permitted to add the sulphate of lime for themselves, and where selenitic cement is not procurable the process might still be useful.

It is conducted as follows :

Three pints of plaster of Paris are stirred in 2 gallons of water. After the mixture is complete it is poured into the pan of a mortar mill ; then 4 gallons of water are added, and the mill revolved three or four times, so as to ensure thorough mixing.

A bushel of finely-ground unslaked lime is now added ; the mixing is continued till the whole becomes a creamy paste, and then 5 bushels of sand are gradually introduced, the whole being thoroughly mixed.

No more is mixed than will be required during the day.

If the mortar gets heated or sets too slowly, a little more plaster of Paris should be added, but not more than $\frac{1}{2}$ pint extra per bushel of lime.

When the lime used in this last-described process is deficient in hydraulic properties, a proportion of selenitic clay should be added so as to bring the total amount of clay in the prepared lime up to about 20 per cent. Any lime requiring more than $7\frac{1}{2}$ per cent of plaster of Paris added to stop slaking with heat will require selenitic clay.

It will be seen that the addition of the plaster of Paris, clay, etc., requires considerable skill and judgment, and the simpler process is to use the selenitic cement, in which the necessary additions have already been carefully made by the patentees.

The following Table, from the patentees' circular, shows the strength of selenitic cement mortar with different proportions of sand as compared with mortars made with other cements.

TABLE

Showing the relative BREAKING WEIGHTS in lbs. of BRIQUETTES having a sectional area at the neck of two and a quarter square inches.

Nature of Lime or Cement.	Age in Days when fractured.	COMPOSITION OF MORTAR.				REMARKS.
		3 sand to 1 cement or lime.	4 sand to 1 cement or lime.	5 sand to 1 cement or lime.	6 sand to 1 cement or lime.	
		BREAKING TENSILE STRESS ON 2¼ SQUARE INCHES.				
Portland cement .	167	...	206	149	113·5	
White chalk lime .	164	67·5				
Do. (Selenitic) .	161	63	58	78	72·3	
Burham lime (Selenitic)	165	...	...	170	210	Good Medway grey lime, sold by Messrs. Lee.
Do. do.	234	...	340	...	...	
Do. do.	161	255	...	...	250	
Halkin lime (Selenitic)	76	128·5	197	99	111	Good hydraulic lime.
Dolgoch lime (Selenitic)	62	155	156·5	157	206·5	Very hydraulic lime.

Mixture of Lime and Cement.—Bad lime is much improved by mixing Portland cement with it.

General Gilmore says :—" Lime paste may be added to a cement paste in much larger quantities than is usually practised in important works without any considerable loss of tensile strength or hardness.

" There is no material diminution of strength until the volume of lime paste becomes nearly equal to that of the cement paste, and it may be used within that limit without apprehension under the most unfavourable circumstances in which mortars can be placed."

Portland Cement Mortar with large proportion of Sand.—Mortar composed of 1 Portland cement, 8 sand, and 1 of slaked fat lime is much better and generally cheaper than 1 of grey lime to 2 sand—the slaked lime slightly weakens the mortar, but is necessary to prevent it from working " short." Loam is sometimes used instead of the slaked lime, but it weakens the mortar still more. This mortar is greatly preferable to that made from lime when frost is to be feared.

The following was used in the outer wall of the Albert Hall :—

1 Portland cement.
1 grey lime (Burham).
6 clean pit sand.

The lime was slaked for twenty-four hours, then mixed with sand for ten minutes. The cement was then added, and the whole ground for one minute. Such a mixture must be used at once.

Grout is a very thin liquid mortar sometimes poured over courses of masonry or brickwork in order that it may penetrate into empty joints left in consequence of bad workmanship.

It may also be necessary in deep and narrow joints between large stones.

It is deficient in strength, and should not be used where it can be avoided.

Precautions in using Mortar.—Fat lime mortars, unless improved by adding pozzuolana and similar substances, are so wanting in strength that any precautions in using them are of but little avail.

In using hydraulic limes and cements it should be remembered that the presence of moisture favours the continuance of the formation of the silicates, etc., commenced in the kiln, and that the setting action of mortars so composed is prematurely stopped if they are allowed to dry too quickly.

It is, therefore, of the utmost importance, especially in hot weather, that the bricks or stones to be imbedded in the mortar should be thoroughly soaked, so that they cannot absorb the moisture from the mortar ; and also in order to remove the dust on their surfaces, which would otherwise prevent the mortar from adhering.

Mortar should be used as stiff as it can be spread ; the joints should be all well filled ; grout should never be used except with large blocks or in other cases where from the position or form of the joint it cannot be filled by mortar of proper consistency.

In frosty weather the freezing and expansion of the water in the mortar disintegrates it and destroys any work in which it may be laid.

Mortar should always be placed for the use of the builder on a small platform or " banker," or in a tub, to keep it from the dirt.

Cement mortars have, of course, peculiarities depending upon the nature of the different cements. These have been noticed in treating of those substances.

CONCRETE.

Concrete is an artificial compound, generally made by mixing lime or cement with sand, water, and some hard material, such as broken stone, gravel, burnt clay, bits of brick, slag, etc. etc.

These ingredients should be thoroughly mixed so as to form a close conglomerate free from voids.

The lime, or cement, sand, and water, combine to form a lime or cement mortar in which the hard material is imbedded, so that the result is a species of very rough rubble masonry.

The broken material is sometimes for convenience called the *aggregate*, and the mortar in which it is encased the *matrix*.

The strength and other qualities of concrete depend chiefly upon the matrix. They are, however, influenced also by the aggregate, and it will be well to make a few remarks upon these two parts of the material separately before proceeding further.

The Matrix, as before stated, is the lime or cement mortar in which the hard broken material, or aggregate, is imbedded.

The lime, or cement, sand, and water, should be so proportioned that the mortar resulting from their mixture is the best that can be made from the materials available. As a rule it should be better than the mortar used for walling, especially if the concrete is to be used in important positions. The reason for this is, that in concrete, the mortar receives less assistance, from the form and arrangement of the bodies it cements together, than it does in masonry or brickwork.

In some cases the mortar is mixed separately, just as if it were to be used in building brickwork or masonry, and then added to the hard material.

More generally, however, the ingredients are mixed together in a dry state, and sprinkled while they are being mixed.

For further remarks on the subject of mixing, see p. 219.

The Aggregate is generally composed of any hard material that can be procured near at hand, or in the most economical manner.

Almost any hard substance may be used when broken up. Among these may be mentioned broken stone, bits of brick, of earthenware, burnt clay, breeze, and shingle. If there is any choice, preference should be given to fragments of a somewhat porous nature, such as pieces of brick or limestone, rather than to those with smooth surfaces, such as flints or shingle, as the for-mer offer rough surfaces to which the cementing material will readily adhere. When weight in the concrete is undesirable, a light porous material such as breeze [1] may be used, but when great weight is an advantage, as in the works of a breakwater or sea wall, the aggregate may be of the heaviest material that can be procured.

Any aggregate of a very absorbent nature should be thoroughly wetted, especially if it is used in connection with a slow-setting lime or cement, otherwise the aggregate will suck all the moisture out of the matrix, and greatly reduce its strength.

Shape.—Many engineers prefer aggregates composed of angular fragments rather than those consisting of rounded pieces, *e.g.* broken stone rather than shingle. The reason for this is that the angular fragments are supposed to fit into one another, and slightly aid the coherence of the mortar or cement by forming a sort of *bond*, while the round stones of the shingle are simply held together by the tenacity of the matrix. Moreover, the angular stones are cemented together by their sides, the rounded stones only at the spots where they touch one another, and angular stones are as a rule rougher and the cement adheres better to their surface.

Size.—The aggregate is generally broken so as to pass through a $1\frac{1}{2}$ or 2 inch mesh. Very large blocks cause straight joints in the mass of the material, which should be avoided if the cement is to bear a transverse stress or to carry any considerable weight.

Of the aggregates in common use, *broken brick, breeze,* or *coke* from gasworks if clean, and *burnt clay* if almost vitrified throughout, all make very good concrete. *Gravel* and *ballast* are also good if angular and clean. *Shingle* is too round and smooth to be a perfect aggregate. *Broken stone* varies ; some kinds are harder, rougher on the surface, and therefore better, than others. *Flints* are generally too round, or, when broken, smooth and splintery. *Chalk* is sometimes used; and the harder varieties make good concrete in positions where they are safe from moisture and frost.

Slag from iron furnaces is sometimes too glassy to make good concrete, but when the surface is porous it is one of the best aggregates that can be used. It is hard, strong, and heavy, and the iron in it combines chemically with the matrix, making it much harder than it would otherwise be. Some slag, however, contains lime which may be dangerous (see p. 166).

The results of experiments as to the relative value of some of these aggregates are given at p. 227.

The materials for concrete may be broken by hand, except when large quantities are required, in which case a Blake's stone-crusher is generally employed.

The size of the pieces of which the aggregate is formed influences the content of the void spaces between them, and therefore the quantity of lime and sand that must be used. Unless the mortar is of such a description that it will attain a greater hardness than the aggregate, the object should be for the concrete to contain as much broken material and as little mortar as possible.

---

[1] Concrete made of breeze is also used when it is required to receive nails, as in lintels, or to be proof against fire.

The following Table shows the amount of voids in a cubic yard of stone broken to different sizes, and in other materials :—

						1 Cubic Yard contains Voids amounting to
Stone broken to 2½-inch gauge	.	.	10 cubic feet.
Do.	2	do.	.	.	.	10⅔ do.
Do.	1½	do.	.	.	.	11⅓ do.
Shingle	.	.	.	.	9 do.
Sand	.	.	.	.	6 do.
Thames ballast (which contains the necessary sand)	.	.	.	.	4½ do.

A mixture of stones of different sizes reduces the amount of voids, and is often desirable.

The contents of the voids in any aggregrate may be ascertained by filling a water-tight box of known dimensions, with the material thoroughly wetted so as not to absorb, and measuring the quantity of water poured in so as to fill up all the interstices ; or by weighing a cubic foot of the aggregate and comparing its weight with that of a cubic foot of the solid stone from which it is broken.

Packing.—In building walls, or other masses of concrete, large pieces of stone, old bricks, chalk, etc., are often packed in for the sake of economy.

Care should be taken that the lumps thus inserted do not touch one another. They should be so far apart, and clear of the face, that the concrete may be well rammed around them.

Where chalk or lumps of absorbent material are used, care must be taken that they are not exposed so as to absorb wet or moisture, otherwise they will be liable to the attacks of frost, and may become a source of destruction to the wall.

Proportion of Ingredients.—The materials to form concrete for ordinary work are generally mixed together in a dry state, the proportion of each being determined by custom, rule of thumb, or experience.

In former days, when lime concrete was more used, a common mixture was

1 quicklime.	**Or**	1 quicklime.
2 sand.		7 Thames ballast (which contains
5 or 6 gravel, broken stone, or brick.		sand and shingle).

The same proportions were for some time blindly adhered to, irrespectively of the nature of the materials used.

The best proportions for the ingredients of a cubic yard of

concrete to be made with any given materials may, however, always be arrived at by ascertaining the contents of the voids in a cubic yard of the aggregate (without sand), and adding to the latter such materials as will make mortar of the best quality and in sufficient quantity to perfectly fill those voids. Where the concrete is not required to be of the best quality, as for example in the backing of heavy walls, the mortar may be made poorer accordingly.

If the aggregate contain sand (as in the case of gravel or ballast), the sand should be screened out of the sample before the voids are measured, and the amount of sand thus screened out will be deducted from that required for the mortar which is to form the matrix of the concrete.

In practice a little more mortar than is actually required to fill the voids should be provided, in order to compensate for imperfect mixing and waste.

Thus, supposing the aggregate available for making concrete to be clean shingle containing 9 cubic feet of voids per cubic yard, a first-rate concrete can be made by adding to each cubic yard of aggregate 4 cubic feet of Portland cement and 8 cubic feet of sand, which will make $10\frac{2}{3}$ cubic feet of 2 to 1 Portland cement mortar (see p. 210), or a little more than sufficient to fill the 9 cubic feet of voids in the shingle.

Again, if the aggregate were ballast, itself containing $4\frac{1}{2}$ cubic feet of sand in each cubic yard, and $4\frac{1}{2}$ cubic feet of voids besides, it would be necessary to add to each cubic yard 4 cubic feet of Portland cement as before, but only $3\frac{1}{2}$ cubic feet of sand, because there are already $4\frac{1}{2}$ cubic feet of sand in the aggregate, making 8 cubic feet of sand altogether, which, with the 4 cubic feet of Portland cement, will make $10\frac{2}{3}$ cubic feet of 2 to 1 Portland cement mortar, or more than sufficient to fill the 9 cubic feet of voids that there are in the ballast without the sand.

If the concrete is not required to be of the first quality, as for example in the backing of heavy walls, the mortar may be made poorer accordingly.

Thus, to make a poorer concrete, with clean shingle for the aggregate, to each cubic yard may be added 12 cubic feet of sand and only 2 cubic feet of Portland cement, making 6 to 1 mortar (or mortar of 6 sand to 1 cement) in more than sufficient quantity to fill the voids.

The chief point to be considered is the quality of the mortar in the concrete. This should be arranged as above described so as to be good enough for the work in which it is to be used, and sufficient in quantity to thoroughly fill the voids of the aggregate, with a little to spare in case of imperfect mixing.

It is a curious thing that engineers have not agreed upon any short way of describing concrete so as to indicate at once its proportions and quality.

In November 1886 Mr. Hayter said at the Institute of Civil Engineers :—[1]

---

[1] *M.P.I.C.E.*, vol. lxxxvii. p. 161.

"In describing concretes it was customary to say that they were mixtures consisting of so many parts of gravel or shingle and sand to 1 part of cement. But in Mr. Hayter's experience two concretes so described might mean admixtures of two different strengths. Thus, assuming a concrete that might be called a 6 to 1 mixture. In specifiying such one engineer might say the concrete was to consist of 1 part Portland cement and 6 parts of gravel or sand of approved quality. Another engineer might say that the concrete was to consist of 1 part Portland cement, 4 parts of gravel or shingle without any sand, and 2 parts of sand."

He pointed out that these two concretes, though both called 6 to 1 concretes, were very different, for in the first there is 1 part of Portland cement to 6 of gravel containing sand, whereas in the second, after the sand is mixed with the shingle, it merely fills the interstices, and the concrete is composed of 1 part Portland cement to 4 of shingle containing sand in its interstices (see p. 218).

The present practice as to briefly describing concrete differs and is often very misleading. It is necessary, therefore, to be very careful in specifications to state exactly how much of each ingredient, shingle or stone, sand and cement is required.

It has been stated that concrete can be made equally good without sand, but sand is a necessary ingredient in all cases where the concrete is required to be waterproof, and it is also desirable on account of strength. Recent experiments have shown that with different aggregates—the proportion of cement, etc., being the same—the concrete made with sand was far stronger both as regards transverse and tensile stress and crushing than that without sand.

Concrete is much used for paving, being made with the very best Portland cement into slabs, and then laid like ordinary stone flags.

For this purpose it is preferable to use an aggregate, such as shingle or granite, much harder than the matrix, and to use very little sand in the latter.

As the matrix becomes worn away, the pebbles of the aggregate project slightly, making the surface slightly rough, and therefore less slippery, and at the same time the matrix is protected from further wear.

Mixing.—As before mentioned, the materials are generally mixed in a dry state, not upon the bare ground, but upon a clean timber or stone platform. The proportions decided upon are measured out either roughly by barrow-loads, or in a more precise manner by means of boxes made of sizes to suit the relative proportions of the ingredients to be used.

Such boxes, in which the quantities to be mixed together can be accurately gauged, should always be used in mixing cement or other concretes intended for important work.

Table showing the Proportions of the Concrete used in various works.

WHERE USED.	PROPORTIONS.	FOR WHAT USED.
1. Peterhead Breakwater	1 Portland cement . . 6 sand, shingle, and broken stone, with granite rubble incorporated therein	Concrete blocks.
Do.	1 Portland cement . . 5 sand, shingle, and broken stones	Cement in bags.
Do.	1 Portland cement . . 4 sand and shingle	Concrete joggles.
2. Newhaven Harbour .	1 Portland cement . . 2 sand 5 shingle	Western sea-wall.
3. Wicklow Harbour .	1 Portland cement . . 7 gravel and sand	In breakwater.
4. Colombo Breakwater	1 Portland cement . . 3 stone 2 sand	In ordinary rings of cylinder foundations.
Do.	1 Portland cement . . 4 stone and sand	In cutting rings of ditto.
Do.	1 Portland cement . . 2 sharp sea sand 4 hand-broken stone ($3\frac{1}{2}''$) 2 machine-crushed ($1\frac{1}{2}''$) screened stone	Blocks.
5. Greenock Harbour .	2 Portland cement . . 7 sand and ballast	Facing to quay wall. *Plastic concrete* behind sheet piling.
Do.	1 Portland cement . . 6 sand and ballast 3 granite chips	Backing to quay wall. *Plastic concrete.*
6. Docks and Dockworks	1 Portland cement . . $1\frac{1}{2}$ sand 5 broken stone	Dock floors and walls to low water.
Do.	1 Portland cement . . 2 sand 6 broken stone	Dock and basin walls between high and low water.
Do.	1 Portland cement . . 3 sand 8 broken stone	Dock and basin walls above high water.
Do.	1 Portland cement . . 4 sand 9 broken stone	Foundations for machinery, etc.
7. Metropolitan Main Drainage Works	1 Portland cement . . $5\frac{1}{2}$ ballast	For sewers.
Do.	1 Portland cement . . 6 gravel	For roofs, floors, etc.
8. For ordinary buildings	1 Portland cement . . 8 gravel	For walls.
Do.	1 Portland cement . . 6 gravel	For floors, roofs, etc.

1. Sir John Coode.   2. *M.P.I.C.E.*, vol. lxxxvii. p. 99.   3. P. 118.   4. *M.P.I.C.E.*, vol. lxxxvii. p. 186.   5. *M.P.I.C.E.*, vol. lxxxvii. pp. 66, 67.   7. *M.P.I.C.E.*, vol. xxv.   8. *Building News.*

The measured materials are then heaped up together, and turned over at least twice, better three times, so as to be most thoroughly incorporated.

The dry mixture should then be sprinkled, not drenched, the water being added gradually through a *rose*, no more being used than is necessary to mix the whole very thoroughly. If too much water be added, it is apt to wash the lime or cement away; at the same time due allowance must be made where the water is liable to soak away or to evaporate quickly.

The moist mixture should then again be turned over twice or three times.

When lime is used it should be in a fine powder.

If a fat lime (which is almost useless for concrete in most positions), it should be slaked and screened.

If a hydraulic lime, it should be finely ground, or, in the absence of machinery for grinding, it should be carefully slaked, and all unslaked particles carefully removed by passing it through a sieve or fine screen.

The lime is often used fresh from the kiln, piled on to the other ingredients during the mixing. This is apt to leave unslaked portions in the lime, and is a dangerous practice.

When Portland cement is used for concrete, it must be thoroughly cooled before mixing. Cements of the Roman class should be fresh.

Laying.—Concrete should, after thorough mixing, be rapidly wheeled to the place where it is to be laid, gently tipped (not from a height) into position, and carefully and steadily rammed in layers about 12 inches thick.

For large masses a somewhat slow-setting cement should be used, and the layers should follow one another so that each is laid before the last has had time to set. This leads to a thorough key being formed between the layers, by which horizontal joints are avoided.

It is essential that the layers should be horizontal; if not, the water trickling off will carry the cement with it.

When circumstances require that each layer should be allowed to set separately, it should be carefully prepared to receive the one that is to rest upon it.

A common practice, which in former years was much insisted upon, is to tip the concrete, after mixing, from a height of 10 feet, or more, into the trench where it is to be deposited.

This process is now considered objectionable, on the ground that the heavy and light portions separate while falling, and that the concrete is therefore not uniform throughout its mass.

Wooden shoots or steeply-inclined troughs are therefore sometimes used, down which the concrete is shot from the place where it is mixed to the site where it is to be used. Such shoots are also objectionable, because the larger stones have a tendency to separate from the soft portions of the concrete.

Its surface should be carefully swept clean, made rough by

means of a pick, washed and covered with a thin coating of cement.

This is especially necessary if it has been rammed, for in that case the finer stuff in the concrete works to the top, and also a thin milky exudation, which will, unless removed, prevent the next layer from adhering.

The joints between the layers are the most important points to be attended to in concrete. When the proper precautions have not been taken, they are found to be sources of weakness, like veins in rocks, and the mass can easily be split with wedges.[1]

When there is not time to allow each layer to set before the concreting is continued, it is better to ram it as quickly as possible, and, before it is set, to add the layers above it.

Anything is better than to allow the layers to be disturbed by ramming, by walking over them, or in any other way, after they have commenced to set.

Concrete made with a very quick-setting cement should therefore not be used for large masses, and if used, not rammed at all.

When concrete has to be laid under water, care must be taken that it is protected during its passage down to the site of deposit, so that the water does not reach it until it is laid.

This protection is afforded sometimes by shoots, by boxes, or by specially contrived iron "skips," which can be opened from above when they have reached the spot where the concrete is to be deposited, so as to leave it there. Sometimes the concrete is filled into bags and deposited without removing the bags.

Concrete is also made into blocks varying in size from 2 to 200 tons. These are allowed to set on shore, and are deposited, the smaller ones in the same way as blocks of stone, those of enormous size by special arrangements which cannot here be described.

Plastic Concrete[2] is a name that has been given to concrete that has been mixed with a very small proportion of water, allowed to set for from 2 to 5 hours, according to the state of the weather, and a little quick-setting cement—such as Roman, Medina, or Orchard—added to it just before it is placed in skips and deposited under water. Concrete deposited in this condition is said to resist the action of the sea and to unite with that previously in position better than concrete deposited in the ordinary liquid condition. On the other hand, it is said that the disturbance of the concrete after it has commenced setting prevents it from ever attaining a proper hardness. The material has not at present been sufficiently used for any decided opinion to be given with regard to its merits.

The Cementing Material to be used for Concrete.—It is hardly necessary to say that when there is a choice the strength and quality of the cementing material should be in proportion to the importance of the part the concrete has to play.

Thus fat lime concretes would be objectionable almost anywhere except as filling in the spandrils of arches.

---

[1] *R.E. Corps Papers*, vol. xxii.   [2] *M.P.I.C.E.*, vol. lxxxvii. p. 66.

Hydraulic lime, or cement, is advisable for concrete in nearly all situations.

Eminently hydraulic limes should be used for concrete foundations in damp ground, and in the absence of cement for subaqueous work of any kind.

Portland cement concretes are adapted for all positions, especially for work under water, or where great strength is required ; also in situations where the concrete has to take the place of stone, as in facing to walls, copings, etc. etc.

For work to be executed between tides, where the concrete is required to set quickly but not to attain any great ultimate strength, Roman or Medina cement may be used with advantage.

When, for the sake of its strength, Portland cement concrete is necessarily used under water, it must be protected by canvas covering or other means from any action which would wash it away before it had time to set.

When concrete is likely to be exposed to great heat, as in fire-proof floors, gypsum has been used as a matrix (see p. 251).

Bulk of Concrete produced.—The bulk of concrete obtained from a mixture of proper proportions of cement, sand, and aggregate, varies considerably according to the nature and proportions of the materials and method of treatment ; but it should in general be a little more than the cubic content of the aggregate before mixing, as the other substances, if in proper proportion, should nearly fit into and disappear in its voids.

The following examples show how the bulk of concrete produced varies according to circumstances :—

Concrete of 1 Portland cement[1] to 6 shingle (or broken stone) and 2 sand.

27 cubic feet shingle or broken stone,
9 „ sand,
$4\frac{1}{2}$,, Portland cement ($3\frac{1}{2}$ bushels),
25 gallons water,

} Make one cubic yard of concrete.

Concrete of 1 Portland cement to 6 broken brick and 2 sand.

30 cubic feet broken brick 2″ mesh,
10 sand,
5 Portland cement,
12 gallons water,

} Made one cubic yard of concrete.

Concrete of 1 Portland cement to 7 Thames ballast[1] (consisting of 2 stone 1 sand).

33 cubic feet ballast,
$4\frac{1}{2}$ cubic feet Portland cement ($3\frac{1}{2}$ bushels),
30 gallons water,

} Make one cubic yard of concrete.

Concrete of 1 Portland cement to 12 gravel, used at Chatham dockyard.

$32\frac{2}{5}$ cubic feet gravel (before shrinkage),
$2\frac{3}{5}$ „ Portland cement,
50 gallons water,

} Made one cubic yard of concrete *in situ.*

[1] Hurst.

Concrete of 1 Portland cement to 8 stone and sand, used at
Cork Harbour works.

27 cubic feet stone broken to 1½-inch gauge, ⎫ Made one
9   ,,   sand,   ⎬ cubic yard of
4½ ,,   Portland cement,   ⎭ concrete *in situ.*

In some concrete landings made with breeze from gasworks and
Portland cement.

29 cubic feet breeze broken to ¾ gauge,   ⎧ Made one
8   ,,   Portland cement,   ⎨ cubic yard of
                                                     ⎩ concrete *in situ.*

Concrete used at Portland Breakwater Fort, stone used in two sizes and
mortar mixed separately.

14 cubic feet stones broken to 3½-inch gauge,
14   ,,   do.   1½   ,,   ⎫ Make one
10   ,,   sand,   ⎬ cubic yard
5   ,,   Portland cement,   ⎪ of concrete
23½ gallons water,   ⎭ *in situ.*

After being rammed the concrete is compressed into about
nine-tenths of the volume it occupies when first made.

Selenitic Concrete.—Concrete may be made with selenitic cement mortar
as the matrix.

Portland cement is sometimes added in small quantities to the selenitic
cement.

"From a series of experiments made on behalf of the patentees, it appears
that a mixture of one part of Portland, four parts of selenitic cement, and
twenty-five parts of sand, was if anything superior to the same Portland
used with four parts of sand."[1]

The patentees' directions for preparing the concrete are as follow :—

For Concrete.—4 full-sized pails of water ; 2 bushels of prepared selenitic lime ; 2
bushels of clean sand.

These ingredients are to be mixed as before in the edge-runner or tub, and then turned
over two or three times on the gauging-floor, to ensure thorough mixing with 12 or 14
bushels of ballast. When the tub is used the sand will be first mixed dry with the ballast,
and the lime poured into it from the tub and thoroughly mixed on the gauging-floor.
An addition of one-sixth of best Portland cement will be found to improve the setting.

EXPANSION OF CONCRETE.—Concrete, when made with hot lime or
cement, swells to an extent amounting to from one-eighth to three-eighths
of an inch per foot of its linear dimensions.

This is owing to the imperfect slaking or cooling of the lime or cement.

It is probable that when such expansion takes place there is a slight disintegration throughout the mass of concrete, and that its coherence is destroyed.

It has been ascertained by experiment that when lime or cement is carefully slaked the concrete practically does not expand at all, and concrete
should be so carefully prepared that no expansion will take place.

In masses of concrete, thin in proportion to their area—such as concrete

---

[1] *Building News*, 30th January 1874.

laid *in situ* instead of paving—cracks are sure to occur unless the area is divided into portions by the introduction of laths so as to break up the surface by dry open joints at intervals.

The expansion which occurs in concrete made with hot lime or cement has been taken advantage of in *underpinning* walls that have settled in parts ; hot concrete forced tightly into openings made below the faulty portions, expands and separates, filling the opening, and lifting the superincumbent work into its proper position.

Uses of Concrete.—Concrete has long been used for the foundations of structures of all kinds, and for filling in the spandrils of arches or the hearting and backs of walls.

Of late years, as the material has improved, it has been employed for many other purposes, a few only of which can now be mentioned.

The walls of ordinary houses, as well as the more massive walls of engineering structures, are now frequently built in concrete, either in continuous mass or in blocks.

Concrete is also used for walls in the form of slabs fitted into timber quartering ; and in hollow blocks, something like those of terra cotta (see p. 131), filled in with inferior material.

This material is also adapted for arches, for stairs, for flooring of different kinds (see Chap. viii., Part II.), and even for roofs.

It can easily be made in slabs well fitted for paving (see page 76), and by the use of wooden moulds can readily be cast in the form of window sills, lintels, dressings of all kinds, steps, etc., and can even be used for troughs and cisterns.

Drain pipes and segments of sewers are also sometimes made of concrete. It was thought that the acids in sewers might act upon the cement, but this has been found practically not to be the case.

The different methods of building monolithic walls, of making blocks, and of casting concrete into different forms, cannot here be entered upon.

Béton is a name given by some writers to any concrete made with hydraulic lime or cement.

By others a distinction is made between the two, *concrete* being the name given when the materials are all mixed together at once, and *béton* when the mortar is made separately.

Practically, however, the word " concrete " covers any form of artificial conglomerate, except artificial stones, which receive distinct names under various patents (see p. 74).

Coignet's Béton Agglomeré is a description of concrete made from a mixture of Portland cement and lime, to which is added a large proportion of sand, no gravel or broken stone being used.

The ingredients are moistened with a minimum quantity of water and pugged in a special mill ; after which the mixture is thrown into a framework of the shape the concrete is intended to assume, and rammed in layers about 6 inches deep.

This material has been largely used in making the Paris sewers, and also occasionally in this country.

Some experiments made to contrast Coignet's Béton with Portland cement concrete showed the former to be a weaker material than the other.[1]

---

[1] *M.P.I.C.E.*, vol. xxxii. Grant's Experiments.

Rock Concrete Tubes with rebated end joints are made at the Bourne Valley Works from the best Portland cement with carefully selected aggregates. These are filled into iron moulds by machinery under heavy percussive action. They are used chiefly as a substitute for brick sewers of from 21 to 36 inches diameter. and are found superior to them in every way.[1]

EXPERIMENTS ON THE RESISTANCE OF CONCRETE TO COMPRESSION.

The following particulars are extracted from the accounts of the well-known experiments by Mr. J. Grant.[2]

Strength of Concrete.—Concrete blocks 12 inches cube, made of Portland cement, weighing 110·56 lbs. per bushel. This cement (neat) broke under a tensile stress of 427 lbs. per square inch after seven days' immersion in water.

The blocks were made in layers 1 inch thick, and compressed by ramming, or in a hydraulic press.

They were kept twelve months before being tested—half of them in air, the others in water.

Composition of Concrete.		CRUSHED AT TONS.	
		Blocks kept in Air.	Blocks kept in Water.
Cement.	Ballast.	Tons.	Tons.
1	to 1	107*	170*
1	,, 2	149	160
1	,, 3	113	115
1	,, 4	103	108
1	,, 5	89	99
1	,, 6	80	91
1	,, 7	75	80
1	,, 8	61	76
1	,, 9	54	68
1	,, 10	48	48

* Exceptional.

These experiments showed that the blocks made with the larger proportions of cement are the stronger, the strength being nearly in proportion to the quantity of cement.

Further experiments showed that compressed blocks were " apparently

---

[1] Manufacturers' Circular.

[2] *Proceedings Institute Civil Engineers,* vol. xxxii. Table 1, Appendix.

stronger than uncompressed blocks in larger proportion than their difference in density."

The relative strength of the concrete cubes made with different kinds of aggregate is shown in the following Table.

Several different proportions between the aggregate and cement were tried, but the following relate to cubes containing eight parts of the aggregate to one of cement.

BLOCKS 12 inches cube (compressed), 8 Aggregate to 1 Cement.[1]

Material for Aggregate.	CRUSHED AT TONS.	
	Blocks in Air.	Blocks in Water.
Ballast . . .	61	76
Portland stone .	110	126
Gravel . . .	74	85
Pottery . . .	97	118
Slag . . .	85	70
Flints . . .	103	117
Glass . . .	65	94

These experiments showed that the concrete of pottery or broken stone was stronger than that of gravel, probably because in the latter case a good deal of the cement is taken up in binding the particles of sand together; partly because the gravel was wanting in angularity.

Tar Concrete is made of broken stones and tar.

About 12 gallons tar are used per cubic yard concrete.

If the tar is too thin, pitch is added to bring it to the proper consistency.

Adding ¼ to 1 bushel of dried and pounded chalk, or dead lime, dried clay, brick dust, or pounded cinders, etc., to every 12 gallons tar, tends to harden the mass.

The materials should be heated, or, at all events, be made perfectly dry, before admixture with the tar.[2]

Mineral tar or bitumen is better for the purpose than coal tar. The former contains an oil which in coal tar is very volatile—escapes, and leaves the tar brittle.

Iron Concrete is composed of cast-iron turnings, asphalte, bitumen, and pitch.

Gas tar is sometimes substituted for the asphalte.

This material has been tried as a backing for armour plates in iron fortifications.

Concrete consisting of 1 part iron borings to 34 of gravel (by bulk) was used with success at the Stranraer Pier.[3]

Lead Concrete, made of broken bricks immersed in lead, has also been used in iron forts.

---

[1] *M.P.I.C.E.* vol. xxxii. Table 5, Appendix.   [2] **Hurst.**   [3] Stevenson *On Harbours.*

MORTAR-MIXING AND CONCRETE-MIXING MACHINERY.

Mortar-mixing Machines.—*Mortar-Mill driven by Steam Power.*—A full description of the different machines in use for mixing mortar would be out of place in these Notes, but a glance at one or two of the commonest forms may be useful.

The mortar-mill in ordinary use on large works is shown in Fig. 103.

A cast-iron pan, P, about 6 or 7 feet in diameter is made to revolve by

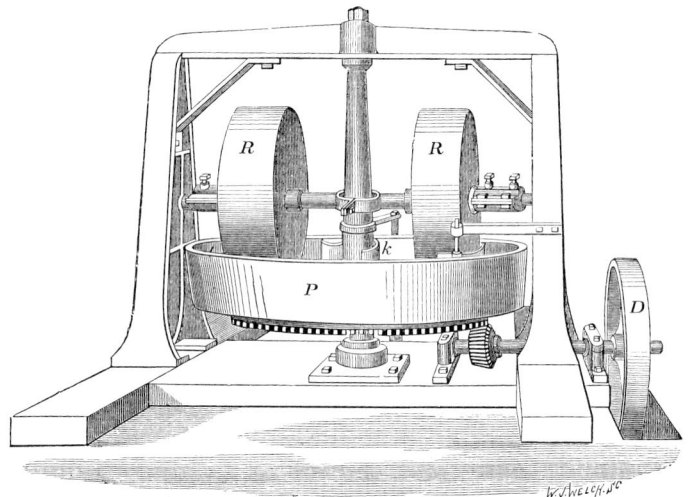

Fig. 103. *Mortar-Mill.*

machinery under a pair of heavy cast-iron rollers filled with concrete, and weighing from 1 to 3 tons the pair.

The ingredients of the mortar are thrown into the pan while it is revolving; plates of iron, marked *k* in the figure, are fixed in suitable positions to guide the material, so that it may all come under the rollers.

The pan has a loose bottom of cast iron, formed in segments, which can be removed and replaced as they wear out.

Machines of this description are generally driven by a small portable engine, a 4-horse power engine being required for a 6-feet pan, and in proportion for other sizes. The band from the engine is passed over the driving-wheel D, and thus turns the spur-gearing which moves the pan.

These mills are made in different sizes, the pans varying in diameter from 5 to 10 feet ; the rollers from 2 feet 8 inches to 3 feet 6 inches.

A mill with a 7-feet pan will turn out about 1½ cubic yard of ordinary lime and sand mortar per hour ; if, however, the mortar is made with burnt ballast, or brick rubbish, which requires grinding as well as mixing, only about 1 cubic yard per hour will be turned out.

Portable Mortar-Mill.—For smaller works, and those which are scattered —as, for instance, along a line of railway—a portable mortar-mill may be used (see Fig. 104).

This machine somewhat resembles the one last described, but is mounted on wheels, and carries a small three horse-power engine with it.

The pan of this machine is sometimes 5 feet, sometimes 6 feet in diameter ; the rollers 2 feet 8 inches or 3 feet in diameter.

Fig. 104.   *Portable Mortar-Mill.*

Such a machine will mix enough mortar to keep ten or twelve bricklayers at work.

Fig. 105.   *Horse Mortar-Mill.*

Horse Mortar-Mill.—A special mill, made by Messrs. Huxham and Brown of Exeter, to be worked by horse-power, is shown on Fig. 105.

Hand Mortar-Mill.—For still smaller works hand mortar-mills may be used of the forms shown in Fig. 106.

The ingredients of the mortar are poured into the hopper H, and find their way into the cylinder C, which contains a series of blades fixed on a central shaft, and made to revolve by means of the handle.

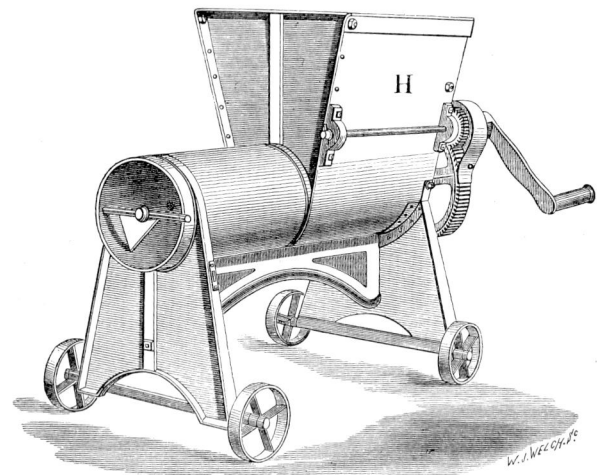

Fig. 106.   *Hand Mortar-Mill.*

It is stated that by the aid of this machine one boy can keep eight men at work, and that one man using it can keep twenty men at work.

Concrete-mixing Machines.—Concrete can be thoroughly well mixed by hand in small quantities ; but when large quantities have to be dealt with, it is difficult, without good organisation, discipline, and very close superintendence, to ensure the thorough incorporation upon which the quality of the material so much depends.

In most cases the use of machinery is a cheap as well as an efficient way of mixing large quantities.

Several arrangements have been devised at different times, suited to the peculiar circumstances of particular works, but it is proposed to describe in these Notes only two or three forms that are commonly used, and one or other of which would be applicable in ordinary cases.

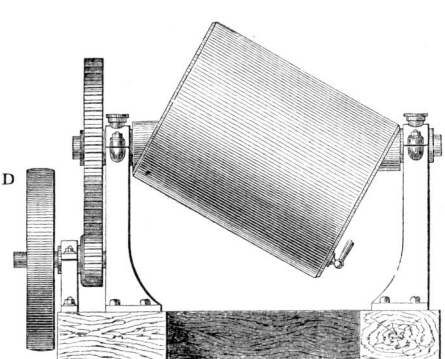

Fig. 107.   *Inclined Cylinder Concrete-Mixer.*

Inclined Cylinder Machine.—A simple form of concrete-mixer consists in an inclined hollow iron cylinder mounted as shown in Fig. 107.

The ingredients of the concrete are filled in by the aid of a hopper through a door at either end, and the cylinder is made to rotate, the band of the engine being passed round the driving-wheel D.

The eccentric motion of the cylinder causes its contents to be rolled over and over, thrown from side to side, and end to end of the cylinder, and thus thoroughly mixed.

A modification of this machine was used at the old Dover Harbour works.

This machine is made in four sizes, containing respectively $\frac{1}{3}$, $\frac{1}{2}$, $\frac{3}{4}$, and 1 cubic yard.

Fig. 107 is taken from the circular of a manufacturer, Mr. H. Sykes, of 66 Bankside, London.

Messent's Patent Concrete-Mixer.—The following description of this machine and the illustration Fig. 108, are taken from the circular of the makers, Messrs. Stothert and Pitt of Bath.

" It consists of a closed box or chamber, A, revolving on an axle, and of such a form as, when half filled with the materials for making concrete, to cause them to be turned over sideways, as well as endways, four times in each revolution

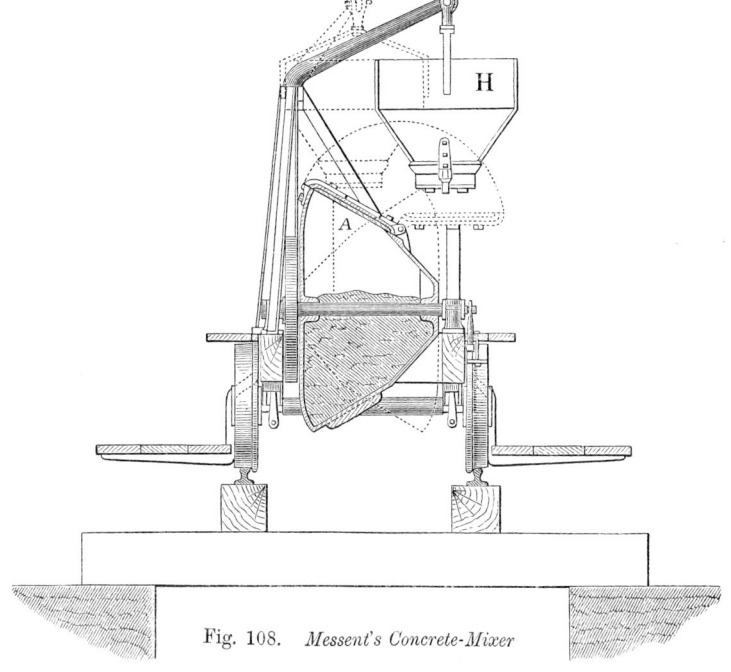

Fig. 108.   *Messent's Concrete-Mixer*

The dotted lines show different positions of the hopper, and also the mixer after a quarter revolution.

of the chamber, so that, in from six to twelve revolutions (the number necessary being varied according to the weight and nature of the materials), a more perfect mixture is effected than can possibly be produced by hand, or (except in a much longer time) by any other machine."

"The mixer is worked by hand or steam power, and is mounted on a trolly of the ordinary railway gauge, and travelled by the same handles that are used for turning it. The travelling gear can, however, be disengaged when the machine has to be taken a long distance by horse or locomotive.

"For filling concrete into a trench, or the hearting of a pier, the machine is supported over the opening, on two balks of timber ; a waggon containing the gravel (and cement in bags) follows on the same line. The hopper H, shown in the figure, suspended from a davit, is made to contain the proper measure of gravel for a charge, whilst the bags contain the proper quantity of cement, and a cistern near at hand (filled by a flexible hose) the proper quantity of water. Two men standing on the waggon (the sides of which are generally raised so that it contains about twice the quantity of an ordinary earth waggon) are able to fill the hopper, in the time employed by four men to give the mixer the requisite number of turns. For counting these, a tell-tale is provided, which indicates when the proper number of turns is completed ; the mixer is then stopped with the door downwards. The door fastening is released, and the charge of concrete falls into its place, the discharge being instantaneous. The opening of the mixer is then turned upwards as in the figure, the door is opened (through the dotted arc, as shown), the hopper, suspended from the davit, is brought over the opening and at once discharged into it, and the water is run in from the cistern at the same time. The door, which closes water-tight, is then shut, and the mixing resumed, the hopper being meantime refilled for the next charge.

"For making concrete blocks, the hand mixer is mounted on a light travelling frame, capable of being moved from one mould to another, and the materials filled into a large tray, holding from 10 to 15 tons, are lifted on to a raised portion of the travelling frame by the steam travelling crane provided for lifting the concrete blocks.

"With the hand-mixer above described, a gang of six men, with a boy for attending to the water cistern, can make from 30 to 40 cubic yards of concrete blocks, and a larger quantity of concrete in bulk in a trench in a day, of better quality and at a cheaper rate than can be done by shovel mixing ; and when the mixers are turned by steam, as at Aberdeen, etc., twice the above quantities are made.

"The mode of applying steam-power varies with the locality and the quantity of work to be done.

"The great advantages of this mixer over others, are its portable shape and self-contained arrangements, which enable it to be easily moved and used in different parts of a work, dispensing with mixing platform and measures ; its economy, and above all, the rapid and perfect amalgamation of materials effected by it, producing, for a certainty, with moderate supervision, concrete of superior strength and quality."

The machine just described was invented by Mr. Messent, the engineer of the Tyne Pier works, Tynemouth.

It has been extensively used on the Tyne at the new breakwaters ; at the harbour works, Aberdeen ; at the Surrey Commercial Docks, London ; for the Sulina works at the mouth of the Danube ; for the Alexandria Dock works at Kurrachee, etc. etc.

Taylor's Patent Concrete Mixer is shown in elevation in Fig. 109. This figure and the following description of the machine are from the catalogue of the makers, Messrs. Stothert and Pitt, Bath.

Fig. 109.

"The above illustration, Fig. 109, shows a Taylor's Patent Concrete Mixer, of 1 cubic yard nominal capacity. This mixer combines the advantages both of the continuous and the intermittent systems. The steel mixer body is formed by two truncated cones bolted, or riveted, together and connected to the hopper in such a way that the charge can be admitted without stopping the machine.

"The method of working is as follows :—The hopper is filled up to the gauge mark with ballast, sand, and cement tipped in from measured bags. The sliding door at the bottom of the hopper is then opened, and the charge passes into the mixer body. After a few revolutions to ensure the thorough incorporation of the dry materials, water is admitted—without stopping the machine—through a hollow shaft to the end of which is attached a pipe fitted with a perforated nozzle, by which means the water is sprayed over the dry materials. When the mixer has made a further fifteen to twenty revolutions, four doors in the mixer body are opened simultaneously, allowing the concrete to fall through a shoot into skips placed underneath the machine, the mixing body continuing to rotate during the whole operation. Whilst one charge is being mixed in the mixing body, the charging hopper is being refilled, thus making the whole process practically continuous.

"These mixers are made with steel framing, as shown above, and if required can be fitted with wheels for any gauge, making the machine portable. They can also be provided without the steel framing, ready to fasten to timber beams in a mixing house. The machines can be fitted with elevators for conveying materials to the hoppers, engines, boilers, and self-propelling gear, forming a self-contained port-able plant."

This machine has been extensively used in H.M. Dockyard Extension Works at Keyham, and elsewhere, and has been found to give excellent results in practice.

The Carey-Latham Concrete-Mixing Machine consists of an arrangement of buckets like those of a dredger. These deliver the sand and ballast into

Fig. 110.

a mixing cylinder, where they are met by a continuous supply of cement— the whole are mixed first dry by revolving blocks until they reach the middle of the cylinder, where the water is added through a perforated shaft, and the mixing is completed with the materials in a wet state.

The quantity of sand in proportion to the ballast is regulated by the arrangement of the buckets, that of cement by an archimedean screw. The machine thus measures as well as mixes the materials. It is made in various sizes to deliver from 5 to 70 cubic yards per hour.

The fig. and part of the above description is from the makers' circular.

Ridley's Concrete-Mixer has a fixed inclined cylinder, with a central shaft carrying longitudinal shelves, which lift the materials as the shaft revolves, and mix them together.

Stoney's Concrete-Mixer is an inclined open iron trough having a shaft passing through its centre, with projecting blades which revolve and mix the materials.

American Concrete-Mixer.—This machine consists of a long box or shoot divided vertically into compartments separated from one another by doors.

The ingredients are placed in the uppermost compartment, and the doors being opened by long levers worked from the top of the shoot, the materials fall gradually from one compartment to the other until they reach the bottom of the shoot thoroughly mixed.

The advantage of this arrangement is that the lower end of the shoot may be placed at the point where the concrete is to be deposited, so that any further handling of the concrete after mixing it is unnecessary.

This machine is not used in England, and therefore no illustration of it is given.

ON THE ACTION OF FOREIGN CONSTITUENTS IN LIMESTONES AND CEMENTS.

The following is an attempt to convey some information with regard to the peculiarities connected with the burning, setting, etc., of limes, cements, and mortars of different classes.

The subject is one almost too intricate for a treatise of such an elementary character as this, in which much chemical knowledge cannot be presupposed. Nevertheless it is touched upon, for without some idea of the principles involved, all dealing with these important materials must be conducted entirely by rule of thumb, or guess work.

A very slight acquaintance, however, with the changes that take place during the different operations will enable the student more easily to remember, and more intelligently to avail himself of, the several characteristics of different limes and cements.

Pure or Fat Limes.—From the experiments made with limes of which the composition is accurately known, it is evident that the differences in their slaking and setting properties are due to the nature and proportion of the foreign constituents they contain. These are chiefly clay and magnesia. Pure or fat limes contain none of these foreign constituents.

CALCINATION.—Pure carbonate of lime[1] contains nothing but lime, carbonic acid, and water. When it is calcined the carbonic acid and water are driven off by the heat, and pure quicklime remains.

SLAKING.—Such a quicklime, when slaked, shows very violent action, great heat is evolved, the mass is greatly swollen, and thoroughly disintegrated.

SETTING.—The residue left after slaking is soluble in water, and has within itself no constituent which will enable it to solidify, except to a very slight extent. It is therefore constantly soft when in a moist situation, and will dissolve under water.

Such portions of its surface, however, as are exposed will imbibe carbonic acid from the air, and will be reconverted into a crust of carbonate of lime, as before described (p. 152).

MORTAR MADE FROM FAT LIME.—It has before been pointed out that the addition of sand improves the setting of fat limes—

1. Because the porous structure caused by the presence of the sand enables the carbonic acid of the air to penetrate farther, and thus to reconvert a greater depth of the lime into carbonate.

2. Because the particles of lime adhere more firmly to particles of sand than to one another.

It has been stated also that pure silica in the shape of sand acts merely mechanically, and enters into no chemical combination with the lime.

For all practical purposes this is true, but experiments have shown that in the course of several years some such action does take place to a very slight degree.

In Petzholt's experiments (described by General Gilmore) he found—

---

[1] *Calcium carbonate.*

1. That in mortar one hundred years old there was more soluble silica than in the original lime.

2. That in mortar three hundred years old there was three times as much soluble silica as in the mortar one hundred years old.

Now, it is a well-known chemical fact that silica does dissolve in alkaline water, though with extreme slowness ; and in this case, no doubt, in the course of a hundred years, a small portion of silica had been so dissolved, and thus enabled to attack the lime.

On the other hand, General Scott mentions a case in which fat lime mortar from a wall five years old was found to be set only on the exterior, and to be in a friable pulpy state inside. Also another case of fat lime mortar fifty years old, which was so soft that it could be beaten up with a trowel.

On the whole we may allow that in fat lime mortar made with siliceous sand a minute proportion of silicate is formed in the course of many years. However, the time required to develop this action is so very long that the fact is of no practical importance to the engineer or builder.

The hardening of lime does not depend merely upon the chemical effect of the combinations which result in the formation of the carbonate, and to a slight degree of the silicate.

It is also caused partly by the crystallisation of the hydrate of lime. The water in fresh mortar contains lime in solution. As the mortar dries the water evaporates, and leaves crystals of lime deposited upon the adjacent particles of lime or sand. These crystals attach themselves firmly to the particles, and will withstand a considerable tensile force.

In the same way, wherever the air can penetrate, the carbonic acid contained in it combines with the lime, and deposits crystals of carbonate of lime.

As before mentioned, the formation of silicates in a fat lime mortar, if it ever occurs, is so slow as to be of no practical value.

The uselessness of fat lime mortar for good work is shown by the following extracts from some of the greatest authorities on the subject :—

Sir Charles Pasley says that " chalk-lime mortar when wet is a pulp or paste, and when dry it is little better than dust."

Vicat says, " Their use should for ever be prohibited in works of any importance."

General Freussart says, " Where good hydraulic lime is to be had, no other kind should be used for any purpose whatever."

General Scott, in a paper on the subject, says, " In the foregoing remarks the worthlessness of pure or fat lime mortars for all constructions, and especially for such as involve the use of heavy masonry, or which will remain damp for any length of time, has been insisted on ; and it has been explained that their unfitness for thick and damp walls results from their not containing within themselves any property by which solidification can be brought about."

Hydraulic Limes and Cements containing Clay.—With a lime containing clay the action is different from that of a pure lime, and not quite so simple.

Before attempting to explain this action it will clear the ground to make a few remarks regarding the nature and composition of clay. Some of the information now about to be given has, however, been anticipated in the chapter on Bricks.

Clay is a compound of silica and alumina with water, chemically known as "hydrated silicate of alumina."

Silica and alumina alone, or in the presence of each other, are infusible, except at extremely high temperatures.

The presence of iron, however, causes the mixture (silica, alumina, and iron) to fuse at a comparatively low temperature.

The same effect is produced to a still greater degree by potash, soda, and chlorides of potassium and sodium.

Many clays naturally contain iron and also the alkalies above mentioned.

Lime is also an infusible substance. When burnt with clay the lime is attacked by the alumina as well as by the silica of the clay, and both silicate of lime and aluminate of lime are formed.

CALCINATION.—When a limestone containing clay is burnt, the carbonic acid from the carbonate of lime and the water from the clay are partially or wholly driven off, and the ingredients are re-arranged in a new set of compounds, the exact nature of which varies both with the original composition of the stone, and with the degree to which it is burnt. In general terms it may be said that these compounds consist of quicklime mixed with silicate of lime and aluminate of lime.

The silicate of lime is formed at a comparatively early stage in the burning, but it is only at the higher temperatures that the alumina and lime enter into combination to form aluminate of lime.

When the burning has been carried to the proper point, these substances are left in a condition in which they will combine with one another and with a proportion of water (when made into a paste with the latter). During this combination they form a new set of compounds, and eventually yield a hard substance insoluble in water.

"When Portland cement is thoroughly well made, hardly any causticity can be detected by the tests owing to the silicates which have formed round the particles of quicklime." [1]

The limit of temperature to which the burning should be carried varies for different stones, and can only be found out experimentally.

The proportion of clay contained by the limestone, also the composition of the clay, both affect the question of the degree of burning, and it will be well to consider these points separately.

PROPORTION OF CLAY.—*Effect in Stones burnt at a moderate temperature.*—If the stone contain a *large proportion* of clay and is burnt at a *moderate temperature,* the silicic acid in the clay attacks the lime, forming calcium silicate. The alumina in the clay does not combine with the lime as long as the temperature is moderate.

If there be sufficient clay present, the whole of the lime is so converted into silicate of lime, and the result is a quick-setting cement like those of the Roman class.

If there is only a *small amount* of clay (that is, not sufficient to provide the necessary amount of silica for the conversion of all the lime into silicate), some of the lime will be left uncombined—*i.e.* in a quick state. This uncombined lime will slake upon the addition of water.

The slaking action will, however, be sluggish as regards the mass of the stone, for it is impeded by the presence of the clay.

Such a state of things exists in *hydraulic limes ;* in these the greater part of the compound body has been converted into a silicate of lime ; but there is sufficient uncombined quicklime remaining to develop a slaking action. This action, however, is in most cases feeble, and sometimes almost suppressed, in consequence of the bulk of quicklime being so small in comparison with that of the mass.

Effect in Cements burnt at a high temperature.—When, however, the calcination is carried to a further stage, and the stone is burnt at a *very high temperature,* not only is the carbonic acid driven off and some of the lime converted into *silicate of lime,* but a further combination takes place—the alumina of the clay combines with the lime, forming *aluminate of lime,* and at the same time a further *silicate of lime* is formed. In addition to these combinations, there are others of an intricate character which result in the formation of double silicates of lime and alumina.

The aluminate of lime was found by M. Frémy to set readily, when powdered and wetted, without access of air, and also to be capable of cementing together inert particles such as those of sand.

The double silicates of lime and alumina also have the property of setting when hydrated—*i.e.* when mixed with as much water as they will take up.

When Portland cement is raised to the very high temperature necessary for the *proper burning* of that material, the whole of the lime in the mixture is converted into either silicate or aluminate—the entire mass is composed of either one or the other of these compounds, and the result is great strength.

In an *underburnt* Portland cement, however, the *aluminate* is not formed, some of the lime is left free ; the resulting cement is quick-setting, but weak and apt to

---

[1] Scott and Redgrave ; *M.I.C.E.,* vol. lxii. p. 78.

"*blow*," the uncombined particles of lime slaking either when they are wetted, or after a considerable lapse of time.

COMPOSITION OF CLAY.—If the clay contains a large proportion of iron and alumina (especially of iron) as compared with the silica, the calcination must be at a comparatively low temperature, or the particles will be fused.

In the Roman, Medina, and Atkinson's cements the quantity of iron and alumina together nearly equals the silica. These are therefore burnt at a low temperature.

When, however, the iron and alumina are in comparatively small proportion compared with the silica, the mixture can be burnt at a very high temperature without danger of fusion. This is the case with Portland cement.

The presence of potash or soda in conjunction with the alumina produces the same effect as the presence of iron, but to a greater degree. If material containing them be exposed to a high degree of calcination, it will fuse into glass or slag.

The same relation holds good between the composition of the clay in hydraulic limestones and the temperature at which they are burnt.

"The larger the amount of the iron and alumina present, the more readily will the lime and the clay, when the limestone is raised to a red heat, pass successively from that condition in which the lime retains all its own proper energy for water to that in which the lime and clay prefer, to partnership as it were, to enter into combination with it in a gradual and quiet manner, and to that in which the formation of the silicates is completed without the intervention of water, and the resulting vitrified compounds show themselves quite indifferent to it, or are only affected by it after having been submitted to its action for some time." [1]

EFFECTS CAUSED BY DIFFERENT DEGREES OF CALCINATION.—It has already been pointed out that the temperature at which the calcination is effected greatly influences the nature of the hydraulic lime or cement produced.

As a general rule slight calcination produces the quickest-setting cements, and prolonged calcination those which have the greatest strength.

Hydraulic Limestones.—When a stone yielding hydraulic lime is subjected to too high a temperature, the effect will be to partly fuse the particles, which prevents them from absorbing water and slaking at once. They thus form either a totally inert substance, or one which slakes after a lapse of some considerable time.

Hydraulic limestones should therefore be burnt at a moderate temperature.

Cement Stones containing a small amount of Clay.—With a stone containing the smallest quantity of clay required to form a cement, a *slight calcination* will not carry the combination far enough to form a strong cement, and the result will probably be either a hydraulic lime which slakes on the addition of water and sets afterwards, or a mixture of quicklime and quick-setting cement, the latter of which sets first, and is then broken up by the slaking of the lime.

A high degree of calcination produces a cement of great strength ; the best Portland cement is therefore produced by burning at a high temperature.

If, however, the calcination be carried too far, the extreme heat will vitrify the cement, and make it almost entirely inert.

Cement Stones containing a large proportion of Clay.—Stones containing much clay give, on the other hand, the best result with a *slight calcination*, many indeed at a point short of the expulsion of the carbonic acid. A higher degree of heat, sufficient to make the whole of the lime caustic, sometimes gives a mixture of lime and cement like that produced by under-burning a slow-setting cement, or it may give a slow-setting cement.

The point of vitrification is reached much sooner in such a stone, especially if the clay contain much soda, potash, or iron.

Roman cement and others of the same class are produced from stones containing a large proportion of clay and of iron, and are therefore burnt at a low temperature.

The foregoing is only a general sketch of the results of the burning process, to which there are many exceptions caused by peculiarities of composition.

Some stones yield—1, a cement ; 2, an intermediate lime ; 3, a cement ; 4, an

---

[1] General Scott ; *R.E. Corps Papers*, vol. xi.

inert substance : 5, a cement ; and 6, an inert substance again, in the order given, and at progressive increasing degrees of calcination.

SLAKING.—This action also is influenced by the proportion of clay contained in the lime.

If the burnt stone contains so small a proportion of clay that the silicates and aluminates cannot combine with all the lime, a certain proportion of quicklime is left in an uncombined or free condition. This causes a modified slaking action, more or less marked in proportion to the amount of free lime that is present.

In proportion as the amount of clay increases, and therefore in proportion as the free quicklime diminishes, *e.g.* in hydraulic and eminently hydraulic limes, this slaking action is more and more suppressed.

Finally, in the case of cements, where the quantity of uncombined lime is reduced to a minimum, the slaking action entirely disappears, and the setting process begins immediately upon the addition of water.

SETTING.—We see, then, that after the processes of burning and slaking hydraulic limes containing clay, there is left within them a mixture of pure lime and silicates, or of pure lime, silicates, and aluminates, ready to combine, if a proper communication is provided to bring them into contact.

When the lime is placed under water, this is effected, for the water at once commences to disseminate and in some degree to dissolve the particles of pure lime ; these mingle with the silicates (many of which are also partially soluble), and combine with them, one by one, by slow degrees, to form a new set of hydrated silicates.

It is evident, then, that the water acts by enabling the particles of alumina and silica to get at the particles of lime, and thus to attack them ; whereas if the particles remained in a dry state, they would lie within a short distance of one another, without ever combining.

A certain proportion of the water also plays another part, by itself combining with the silicates and aluminates to form hydrated compounds, which set by crystallising, and pass into the solid state.

This explains why mortars of hydraulic limes should not be allowed to dry too quickly. The dissemination and dissolution of the particles is thereby stopped, and the setting process impeded.

A properly burnt lime, containing sufficient clay, when saturated with moisture after calcination, or when quite immersed, is therefore in a favourable condition for forming the hard and insoluble compound above mentioned ; in fact its composition adapts it for setting under water.

Even in the case of fat limes the presence of moisture for a certain time is useful, for it enables them more readily to absorb carbonic acid from the air. In hot countries it is necessary that work in which they are used should be kept moist for some time, otherwise the mortar will be in a granular crumbly state, it will not readily absorb carbonic acid, and the lime will not enter the crystalline condition which is essential for proper setting.

Proportion of Clay.—To give a perfect result there must be sufficient clay to combine with *all* the lime in the mixture ; otherwise some of the lime, having nothing to combine with, remains pure and soluble, and reduces the average setting property of the whole.

This occurs in those hydraulic limes which contain from 8 to 15 per cent of silica When all this silica has entered into combination, there is still some quicklime left (more or less in inverse proportion to the amount of clay). This remaining lime develops the slaking action as before explained, and impedes the action of setting.

On the other hand, there must not be too much clay, or, after the lime is turned into silicate, there will be a surface of free clay left—having in itself no hardening property, and which will decrease the strength of the resulting cement.

The proportions required to produce a cement vary, however, within tolerably wide limits (22 to 36 per cent in the raw material) ; as a general rule, the quicker-setting cements are produced from stones containing most clay.

Composition of Clay.—Those clays which contain a large proportion of iron and alumina cause the lime in which they occur to set with greater rapidity than do ordinary clays. "Clays deficient in iron and alumina, and in which the silica is present in the shape of finely divided quartz, are apt to form insoluble silicates of lime at a high temperature owing to the want of suitable bases to combine with the silica, which also renders them unfit for making Portland cement. Such clays also when calcined at too low a temperature yield hardly any soluble silicates, form therefore no protection to, and seem in no way to prevent the hydration of the lime, and produce a material devoid of hydraulic properties." [1]

The following TABLE sums up approximately, and in a concise form, the mutual

Proportion of Clay before burning.	Composition of Clay.	Degree of Calcination.	Setting Properties.	Examples of the Class.
0 to 8 p. c.	...	Very low.	Absorb carbonic acid from air.	Fat limes.
8 to 18 p. c.	Various. Those with most iron and alumina set most quickly	Moderate.	Moderately quick setting. No great strength.	Lias and other hydraulic limes
20 to 30 p. c.	Iron and alumina. Silicic acid.	Very high.	Sets slowly. Very strong.	Portland cement.
Do.	Do.	Extreme.	Become inert.	Do. over-burnt.
28 to 55 p. c.	Iron and alumina. Silicic acid.	Low.	Sets very quickly. No great strength.	Roman cement and others of that class (see p. 163).
Do.	Do.	High.	Become inert.	Do. over-burnt.

relations of the proportion of clay, composition of clay, degree of calcination, and setting properties in different classes of limes and cements.

We have considered the effect of clay in conferring hydraulic properties upon limes, which it does by presenting silica in a state fit for combination, but clay is not the only substance which has this effect.

POZZUOLANA, etc.—As before noticed, the presence of several other forms of soluble silica and pozzuolana will also answer the purpose in a greater or less degree.

The general nature of the reactions that take place in the setting of limes containing these substances are much the same, and produce effects similar to those already described.

It has been recommended that mortar made with substances of this kind should be allowed to remain in paste for some time before use. The reason for this is that in consequence of the clay and lime not having been burnt together, none of the silicates have been formed, as they are in ordinary hydraulic limes burnt in the kiln. Every facility should therefore be given to the silica to attack the lime through the intervention of the water (see p. 239), and thus to form silicates, before the mortar is used.

---

[1] Scott and Redgrave, *M.I.C.E.*, vol. lxii., p. 80.

CARBONATE OF MAGNESIA.—Carbonate of magnesia is a substance very similar to carbonate of lime ; it loses its carbonic acid in burning, combines with silica, etc., and behaves generally in the same way, with one important exception, viz. that the calcined magnesia will not slake on the addition of water, but combines with it gradually and quietly, and sets to some extent in doing so. When silica is present it combines with the magnesia, and with the lime, forming a double silicate of lime and magnesia, which is of greater strength than either silicate of lime or silicate of magnesia separately.

Besides this, the magnesia and lime, even without the intervention of the silica, will combine and harden under water.

The hydraulic mortar that is produced from magnesian limestones and dolomites (see p. 59) owes its properties to the different combinations above mentioned.

Several failures that have recently occurred in Portland cement after use have been attributed to an excess of magnesia in the cement.

As this has caused considerable mistrust of the material, the following remarks on the subject by Mr. Dent will be valuable :—

" When the lime is associated with magnesia, the magnesia should be regarded as to some extent taking the place of the lime, and the quantity of the lime should be proportionately diminished.

" A well prepared Portland cement, such as is made on the Thames or the Medway, should not contain any appreciable quantity of magnesia, say about 1 per cent. Although any large proportion of magnesia in Portland cement cannot be considered desirable, yet it must not be forgotten that magnesia is capable of forming hydrates of great permanence and hardness, and that some very good hydraulic cements contain as much as 8 per cent of magnesia, such, for example, as the well known Rosendale cement of the United States of America.

" There can be little doubt but that the assertions that have been frequently made as regards the tendency of cements containing magnesia to disintegrate, may sometimes have arisen from overlooking the fact that the results observed might be due to excess of basic constituents in the cement. In a recent statement put forward as to the injurious action of magnesia, the cement referred to contained 72 per cent of lime and magnesia, and it could scarcely be regarded as extraordinary that such a cement should prove a complete failure, since it is well known that such a proportion as 72 per cent of lime would render Portland cement so unsafe as to cause it to be condemned." [1]

It should be mentioned, moreover, that in the cases of failure which have occurred the magnesia has been found by analysis after the cement has been for some time under sea-water, and that it may not have been in the cement when originally deposited, but introduced by the chemical action of the sea-water upon the uncombined lime existing in the cement. If this is so, the best safeguard against the evil would be extreme care to avoid overlimed cements, and to cool and aerate the cements thoroughly before use (see p. 181).

With regard to the action of sea-water upon cement, Mr. Dent says :—

" From a recent report of Professor Brazier on the cause of the failure of some cement used in the construction of a graving dock in Aberdeen harbour, it would appear that the reaction which takes place between the magnesium chloride contained in sea-water and lime may, under certain conditions, be sufficient to cause the disintegration of some descriptions of Portland cement, the lime in the cement being dissolved.

SULPHATES.—Lime can also be made to unite with water by the presence of a small quantity of any sulphates, and the employment of this property by a suitable process will considerably increase its setting power.

The setting of lime thus treated is essentially distinct from that produced by combination with silica, inasmuch as it depends on combination with water only (which becomes solid) and the resulting substance (which is simply hydrate of lime) is entirely soluble in water, though with more difficulty than the ordinary slake lime, owing to its superior density.

Sulphate of magnesia [2] (commonly known as Epsom salts) is very soluble in water.

---

[1] Dent's *Cantor Lectures*, 1887. [2] *Magnesium sulphate.*

EFFLORESCENCE ON WALLS.

The surfaces of walls are often covered with an efflorescence of an unsightly character.

This efflorescence is formed by a process known as *saltpetreing*. It shows itself chiefly in the case of newly built walls, but also in those parts of older walls which are exposed to damp. It varies somewhat in appearance and also in chemical composition, and is most apparent in dry weather.

Appearance.—It is generally white in colour and crystalline in structure ; the crystals presenting the appearance of very fine fibres or needles, or looking like a thin coating of snow or white sugar.

Composition.—Chemical analysis has shown that these crystals vary considerably in composition. They often consist of sulphate of magnesia, also of sulphate of lime ; of carbonate, sulphate, or nitrate of soda ; of chlorides of soda and potash, and carbonate of potash.

Causes.—Efflorescence is attributable sometimes to the bricks or stones of a wall, sometimes to the mortar. Dampness is favourable to its formation. Cold as low as the freezing point stops it.

In bricks burnt with coal fires, or made from clay containing iron pyrites (bisulphide of iron), the sulphur from the fuel converts the lime or magnesia in the clay into sulphates. When the bricks are wet these dissolve ; when dry, they evaporate, leaving crystals on the surface. The sulphate of magnesia is generally found in much greater quantity than the sulphate of lime, as it is far more soluble in water.

Many limestones contain magnesia (see p. 154) ; these are acted upon during calcination by the sulphur in the fuel ; sulphates are formed, which find their way into the mortar and produce effects similar to those above mentioned.

Again, the sulphur acids evolved from ordinary house fires attack the magnesia and lime in the mortar joints of the chimney ; these dissolve and evaporate on the surface.

The formation of chlorides is nearly sure to take place if sea sand or sea water be used, or in bricks made from clay which has been covered by salt water.

In some situations the formation of the nitrates has been attributed to the absorption of ammonia from the air.

The potassium and sodium salts are supposed in many cases to be derived partly from the limestone used for the mortar, and partly from the fuel employed in burning the lime.

Disadvantages.—Not only does the efflorescence present a disagreeable appearance, but it causes damp patches on the surface of the wall, it will eat through any coat of paint that has been applied after the efflorescence has once commenced, and will even detach small fragments of the materials composing the wall.

Remedies.—Prevention in this case is better than any attempt at cure.

The best plan is to avoid all the materials above mentioned as likely to give rise to efflorescence.

In the case of bricks, clay containing pyrites or much magnesia should not be used ; special bricks may be burnt with coke or wood.

As regards mortar, the use of limestones containing magnesia to any great extent may generally be avoided.

If, however, it does occur in spite of all precautions, the following remedies may be tried :—

In the case of ashlar-work :—1. The surface may be covered with a wash of powdered stone, sand, and water, which is afterwards cleaned off. This fills up the pores of the stone, and temporarily stops the efflorescence. When the wash is removed the saltpetreing will recommence, but in a weaker degree than before.[1]

2. Painting the surface is sometimes efficacious if it is done before the efflorescence commences.

The mortar before use may be treated to prevent it from causing efflorescence—

1. By mixing with it any animal fatty matter. General Gillmore recommends 8 to 12 lbs. of fatty matter, 100 lbs. quicklime, and 300 cement powder.

2. Potash salts may be rendered harmless by adding hydrofluosilicic acid.[2]

ANALYSIS OF LIMES AND CEMENTS.

The strictly chemical view of this subject is beyond the scope of these Notes, and the reader desirous of analysing a lime or cement, and possessing the necessary chemical knowledge is referred for directions to *Notes on the Chemistry of Building Materials,* by Captain Abney, R.E., F.R.S.

The following Table gives analyses of a few cements (made by a friend of the writer's) taken from actual specimens of fair quality met with in practice :—

	PORTLAND CEMENT.			ROMAN CEMENT.		MEDINA CEMENT.
	Heavy slow-setting.	Light quick-setting.[3]	Average good cement.	Specimen 1.	Specimen 2.	
Clay unacted upon . .	·3	4·4	Traces.	9·7	7·9	5·3
Soluble Silica . . .	22·3	20·6	22·0	16·0	17·2	19·0
Oxide of Iron[4] . . .	3·2	3·5	3·5	} 22·2	} 21·5	} 16·6
Soluble Alumina . .	7·2	10·9	8·0			
Sulphuric Acid . .	1·4	2·4	1·0			
Lime . . .	63·0	50·0	62·0	41·2	. 46·1	49·8
Magnesia	·6	·2	1·0	1·7	1·6	
Alkalies	1·3	1·5	1·5	} 9·2	} 5·7	} 9·3
Carbonic Acid . . .	...	5·0	Traces.			
Moisture and loss . .	·7	1·5	1·0			

---

[1] Burnell *On Limes and Cements.*   [2] Gillmore *On Limes and Cements.*
[3] This cement evidently contains too little lime.   [4] Ferric oxide.

PLASTERS, ETC.

Materials used by Plasterers.—A great variety of compositions are used by plasterers, some of which will be described.

Among the most important of these are cements of various
kinds. Many of these are used also for building purposes, and
have already been considered. Others are very deficient in strength
and weathering properties, and are suitable only for covering the
surfaces of internal walls. These will now be described.

In addition to these there are several mixtures made up of
lime, sand, and other materials, distinguished by various names,
and also used for covering surfaces of walls.

The description of these was, to a slight extent, necessarily
anticipated in Part II., but will here be repeated.

Materials used by the plasterer in common with other trades,
such as size, laths, etc., will be described in Chapter IX.

Cements, etc., used as Plasters.—GYPSUM.—The basis of most plasters
is a native hydrated sulphate of lime occurring as a soft stone, usually of a
more or less crystalline texture, and varying in colour from white through
shades of brown and grey to black. White and light shades are the commonest in England, where it is found in Derbyshire, Nottinghamshire,
Cheshire, and Westmoreland. It is also found in great abundance in the
neighbourhood of Paris.

The very fine-grained pure white varieties are termed " alabaster," or, when
transparent, " selenites."

The raw stone is prepared either by simple calcination, or by calcination
and combination with various salts of the alkalies.

PLASTER OF PARIS is produced by the gentle calcination of gypsum to a
point short of the expulsion of the whole of the moisture. The raw stone is
sometimes ground in the first instance and calcined in iron vessels.

Paste made from it sets in a few minutes, and attains its full strength in
an hour or two.

At the time of setting it expands in volume, which makes it valuable for
filling up holes and other defects in ordinary work.

It is also added to various compositions in order to make them harden
more rapidly.

Plaster of Paris is used for making ornaments for ceilings, etc., which are
cast by forcing it, in a pasty state, into wax or gutta-percha moulds.

Where it is plentiful, as in the neighbourhood of Paris, it is used in all
parts of house-construction where it will be free from exposure to the
weather, for which exposure it is unfit, as it is very soluble in water.

There are three qualities of plaster of Paris in the market—the "*superfine*,"
"*fine*," and "*coarse*;" the two former being whiter and smoother in grain
than the last. The superfine is sold in casks, and the other qualities in casks
or sacks. Both casks and sacks contain 2 cwt.

PORTLAND CEMENT is much used by plasterers for external rendering (see
Chap. ix., Part II.).

As before mentioned, the lighter varieties of Portland cement, weighing from 95 to 105 lbs. per bushel, are those best adapted for this purpose. They set more quickly, and thus save expense not only in their first cost, but also in the labour that is bestowed upon them by the plasterer.

ROMAN CEMENT, and others of the same class, described at page 157, are used for external rendering, as mentioned at page 209, Part II.

KEENE'S CEMENT is a plaster produced by recalcining plaster of Paris, after soaking it in a saturated solution of alum, or a strong solution of borax and cream of tartar.[1]

One pound of alum is dissolved in a gallon of water, and in this solution are soaked 84 lbs. calcined plaster of Paris in small lumps ; these lumps are exposed eight days to the air, and then recalcined at a dull red heat.

The addition of half-a-pound of copperas gives the cement a cream colour, and is said to make it better capable of resisting the action of the weather.

This cement is harder than the other varieties made from plaster of Paris, and is consequently used for floors, skirtings, columns, pilasters, etc.; it is also frequently painted to imitate marble.

Keene's cement is made in two qualities, the coarse and the superfine : the former is white, and capable of receiving a high polish ; the latter is not so white, or able to take so good a polish, but sets hard. The superfine quality is sold in casks containing $3\frac{1}{2}$ bushels, and the coarse in casks of the same size, and in sacks containing 3 bushels.

PARIAN CEMENT, sometimes called *Keating's Cement*, is said to be produced by mixing calcined and powdered gypsum with a strong solution of borax, then recalcining, grinding, and mixing with a solution of alum.

There are two qualities of Parian cement in the market—the " *superfine* " and the " *coarse.*" They are sold in casks and sacks of the same sizes as those used for Keene's cement.

" Parian is said to work freer than either Keene's or Martin's cement, and is therefore preferable for large surfaces, which have to be hand-floated before trowelling ; but the two latter cements are *fatter*, and produce sharper arrises and mouldings." [2]

As Keene's and Parian cement are not used for mortar, their tensile strength is of no practical importance. When allowed to set in air their strength was found by Mr. Grant to be as follows per sectional area of $2\frac{1}{4}$ inches :—

	Keene.	Parian.
	lbs.	lbs.
Seven days . .	546·0	642·3
Fourteen days .	585·8	671·2
Three months .	720·5	853·7

MARTIN'S CEMENT is made in a similar way to Parian—carbonate of potash (pearl-ash) being used instead of borax, and hydrochloric acid being sometimes added.

It is made in three different qualities—*coarse, fine,* and *superfine*—the coarser kinds being of a reddish-white colour, and the finer pure white. It is said to cover more surface in proportion to its bulk than any other similar material.[3]

---

[1] Redgrave.　　　　[2] Seddon.　　　　[3] Papworth.

ROBINSON'S CEMENT is made from alabaster (sulphate of lime) found in the Inglewood Forest near Carlisle. It has somewhat similar properties to Keene's and Parian cements, and can be used for similar purposes—for decorations, plastering, etc.

METALLIC CEMENT " has a metallic lustre, is suitable for outside work, and is intended to dispense with colouring or painting, but is not much used."[1]

One variety is made by mixing ground slag from copper-smelting works with ordinary cement stone.

Portland Cement Stucco is a mixture of Portland cement and chalk. It is of a good colour and close texture ; weaker than Portland cement, but not so liable to crack.

Lias Cement is produced from Lias shales containing a large proportion of soluble silica. It resembles Lias lime in appearance ; sets in eight or ten minutes, and is used for lining water-tanks, or other purposes for which a light quick-setting cement is required.

JOHN'S STUCCO CEMENT is used as a wash or paint, and when mixed with three parts of sand as a stucco. It is said to adhere well, to be hard when set and impervious to wet, and to be fit for mouldings or castings.[2]

Uses.—The Keene's, Parian, and similar " cements " or plasters are largely used for the best class of internal plastering, and, as they set very quickly, they can be painted within a few hours, which is a great advantage.

They are capable of receiving a very high polish, to obtain which the surface is rubbed down with gritstones of various degrees of coarseness ; afterwards stopped or paid over with semi-liquid neat cement which fills up the pores ; rubbed again with snake-stone, and finished with putty powder.

The plasters should not be used in situations much exposed to the weather, on account of their solubility. This consideration, combined with their cost, and the moderate strength they attain even under favourable circumstances, makes them unsuitable for most engineering works.

MASTICS are a species of cements consisting of brick, burnt clay, or limestone powdered, mixed with oil and litharge, or some other drier.

In former years they were much used for covering external mouldings, etc. They were applied in a thin coat with great care, and looked well, but required painting periodically to compensate for the evaporation of the oil.

Several varieties were used on the Continent, but that best known in England was called *Hamelin's Mastic*.

This material, however, was expensive, and has been superseded by Portland cement.

The Materials used in Ordinary Plastering are laid on in successive coats, which differ from one another in composition.

In all of them the lime used should be most thoroughly slaked,

[1] Seddon. [2] Papworth.

or it will throw out blisters after being spread. For this reason the "stuff" is generally made long before it is required, and left for weeks to cool.

Pure or fat limes are generally used for the sake of economy, and for safety. Hydraulic limes would require special attention to prevent them from blowing. Moreover, the surface of plaster made with fat lime is more absorbent, and less liable to encourage condensation, than that of plaster made with hydraulic lime.

Salt water and sea-sand should not be used, as the salts they contain would cause permanent dampness and efflorescence.

HAIR.—The hair used by the plasterer in order to make his "coarse stuff" hang together is obtained from the tanner's yard.

It should be long, sound, free from grease and dirt, thoroughly separated, beaten up, or switched with a lath, so as to separate the hairs, and dried.

It is classed according to quality as Nos. 1, 2, and 3, the last being the best. A bushel weighs from 14 to 15 lbs.

White hair is selected for some work, but as it should all be thoroughly covered by the coats subsequent to that in which it occurs, its colour is not of importance.

COARSE STUFF is a rough mortar containing 1 or 1½ part of sand to 1 of slaked lime by measure.

This is thoroughly mixed with long sound ox hair (free from grease or dirt, and well switched, or immersed in water to separate the hairs) in the proportion of 1 lb. hair to 2 cubic feet of the stuff for the best work, and 1 to 3 for ordinary work.

The sand is generally heaped round in a circular dish form; the lime, previously mixed with water to a creamy consistence, is poured into the middle. The hair is then added, and well worked in throughout the mass with a rake, and the mixture is left for several weeks to "cool," *i.e.* to become thoroughly slaked.

"If mixed in a mill the hair should only be put in at the last moment, or it will get broken and torn into short pieces.

"If there is sufficient hair in coarse stuff for ceilings, it should, when taken up on a slate or trowel, hang down from the edges without dropping off.

"For walls the hair may be rather less than in top stuff for ceilings." [1]

FINE STUFF is pure lime slaked to paste with a small quantity of water, and afterwards diluted with water till it is of the con-

[1] Seddon.

sistence of cream.   It is then allowed to settle; the water rising
to the top is allowed to run off, and that in the mass to evaporate until the whole has become thick enough for use.   For some
purposes a small quantity of hair is added.

PLASTERER'S PUTTY is pure lime dissolved in water, and then
run through a fine sieve.   It is very similar to fine stuff, but prepared in a more careful manner, and is always used without hair.

GAUGED STUFF, also called "*Putty and Plaster*," contains from
$\frac{3}{4}$ to $\frac{4}{5}$ plasterer's putty, the remainder being plaster of Paris.

The last-named ingredient causes the mixture to set very
rapidly, and it must be mixed in small quantities, not more being
prepared at a time than can be used in half an hour.

The proportion of plaster used depends upon the nature and
position of the work, the time available for setting, the state of the
weather, etc., more being required in proportion as the weather is
damp.   An excess of plaster causes the coating to crack.

It is used for finishing walls and for cornices.   In the latter
the putty and plaster should be in equal proportions.

Selenitic Plaster is made with selenitised lime, otherwise known as
selenitic cement.

This material has been described at page 184.

The method of mixing the material for the first coat of plastering on
brickwork is exactly similar to the process as carried out for mixing mortar.

This process has been described at pages 211, 212 ; and also in Chap. ix.,
Part II., and need not therefore be repeated.

For plastering on lath work and other coats the following directions of the
patentees should be rigidly followed.

They have already been given in Part II., but are here repeated to make
these Notes more complete in themselves.

"*For Plastering on Lath Work.*—To the same quantities of water and prepared
lime, as given, add only 6 or 8 bushels of clean sharp sand and 2 hods of
well-haired lime putty ; the hair being previously well hooked into the lime
putty.   When the mill is used, the haired putty should only be ground sufficiently to ensure mixing.   Longer grinding destroys the hair.

" Lime putty should be run a short time before being used, to guard against
blisters, which will sometimes occur.

" *N.B.*—This mixture will be found to answer equally well for ceilings as for
partitions.   If the sand is very sharp, use only 6 bushels of sand for covering
the lath, and when sufficiently set, follow with 8 bushels of sand for floating
(or straightening).

" *Setting Coat and Trowelled Stucco.*—For common setting (or finishing coat
of plastering), the ordinary practice of using chalk lime putty and washed
sand is recommended.   But if a hard selenitic face is required, care must be
taken that the prepared selenitic lime be first passed through a 24 by 24
mesh sieve, to avoid the possibility of blistering, and used in the following
proportions :—4 pails of water ; 2 bushels of prepared selenitic lime (pre-

viously sifted through a 24 by 24 mesh sieve) ; 2 hods of chalk lime putty ; 3 bushels of fine washed sand.

" This should be treated as trowelled stucco ; first well hand-floating the surface, and then well trowelling. A very hard surface is then produced.

" *Selenitic Clay Finish.*—5 pails of water ; 1 bushel of prepared selenitic lime ; 3 bushels of prepared selenitic clay ; 2 bushels of fine washed sand ; 1 hod of chalk lime putty.

"This mixture, well hand-floated to a fair face, and then well trowelled, will produce a finished surface equal to Parian or Keene's cement, and will be found suitable for hospital walls, public schools, etc. Being non-absorbent, it is readily washed.

" The use of ground selenitic clay improves the mortar, and renders it more hydraulic.

" When the selenitic clay is used, 2 bushels may be added to 1 bushel of prepared selenitic lime, the proportion of sand, ballast, etc., being the same as for prepared selenitic lime. The use of selenitic clay effects a considerable saving, as it is much cheaper than lime.

"*For Outside Plastering* use 6 or 8 bushels only of clean sand, and for finishing *rough stucco face* use 4 or 5 bushels only of fine washed sand, to the proportions of lime and water given."

Rough Cast is composed of washed gravel mixed with hot hydraulic lime and water. It is applied in a semi-fluid state, as described in Chap. ix., Part II.

Stucco.—This term is very loosely applied to various substances which differ considerably from one another. These may be classed as follows :— 1. Compounds of hydraulic lime, formerly much used for external covering to walls. 2. Mixtures of lime, plaster, and other materials for forming smooth surfaces on internal walls, chiefly those intended to be painted. 3. All sorts of calcareous cements and plasters used for covering walls.

These latter have been described under their several heads.

COMMON STUCCO consists of three parts clean sharp sand to one part of hydraulic lime.

It was much used at one time as an external covering for outside walls, but has to a great extent been superseded by cements of recent introduction.

The method of applying this and the other compositions mentioned below is described in Chap. ix., Part II.

TROWELLED STUCCO is used for surfaces intended to be painted, and is composed of two-thirds fine stuff (without hair) and one-third very fine clean sand.

BASTARD STUCCO is of the same composition as trowelled stucco, with the addition of a little hair.

ROUGH STUCCO contains a larger proportion of sand, which should, moreover, be of a coarser grit. The surface is roughened as described in Chap. ix., Part II., to give it an appearance like that of stone.

Artificial Marbles may be produced by skilful workmen by working colours in with almost any of the white cements or rather plasters mentioned at pages 244, 245.

Certain processes for imitating marbles are, however, known by distinctive names, and one or two of the more important of these will now be briefly noticed.

SCAGLIOLA is a coating applied to walls, columns, etc., to imitate marble. It is made of plaster of Paris, mixed with various colouring matters dissolved in glue or isinglass; also with fragments of alabaster or coloured cement interspersed through the body of the plaster.

The method of applying and finishing this material is described in Chap. ix., Part II.

MAREZZO MARBLE is also a kind of plaster made to imitate marble.

A sheet of plate-glass is first procured, upon which are placed threads of floss silk, which have been dipped into the veining colours previously mixed to a semi-fluid state with plaster of Paris. Upon the experience and skill of the workman in placing this coloured silk the success of the material produced depends. When the various tints and shades required have been put on the glass, the body colour of the marble to be imitated is put on by hand. At this stage the silk is withdrawn, and leaves behind sufficient of the colouring matter with which it was saturated to form the veinings and markings of the marble. Dry plaster of Paris is now sprinkled over to take up the excess of moisture, and to give the plaster the proper consistence. A canvas backing is applied to strengthen the thin coat of plaster, which "is followed by cement to any desired thickness; the slab is then removed from the glass and polished.

"Imitation marble of this description is employed for pilasters and other ornamental work, and is now used by Mr. George Jennings in the manufacture of a variety of articles."

"The basis of Marezzo marble, as well as of Scagliola, being plaster of Paris, neither of them is capable of bearing exposure to the weather." [1]

"The *Artificial Marble* now manufactured in London is made on the same principle as the Marezzo, but differs from it in the character of the cement used. A less expensive table is also substituted for the plate glass, and the canvas backing is altogether omitted." [1]

Other artificial marbles are described at page 76.

Enrichments.—The plasterer requires a great variety of mouldings, ornaments, pateras, flowers, and other enrichments for the decoration of his work.

These may be made either in plaster of Paris composition or in papier-maché.

PLASTER ORNAMENTS are cast either in wax or in plaster, the latter process being used chiefly for large ornaments which have an undercut pattern.

The ornament is in either case first modelled in clay and well oiled.

In making wax moulds, the wax is melted, mixed with rosin, and poured in upon the model, arrangement having been made to prevent its escape; the whole is then steeped in water, and the wax becomes detached in one mass.

When plaster is used as the material for the mould, it is laid on to the

---

[1] Dent

model in plastic pieces fitted together, and then the whole, when dry, is immersed in boiled linseed oil.

In casting, the plaster in a semi-fluid state is dabbed with a brush into the mould.

COMPOSITION ORNAMENTS are made with a mixture of whiting, glue, water, oil, and resin.

The oil and resin are melted together and added to the glue, which has been dissolved in water separately. This mixture is then poured upon pounded whiting, well mixed, and kneaded up with it to the consistency of dough.

When used the material is warmed to make it soft, and is forced into box-wood moulds carved to the patterns required.

PAPIER-MACHÉ is a much lighter material for ornaments than either composition or plaster, and it is much used for the purpose.

Cuttings of paper are boiled down and beaten into a paste, mixed with size, placed in a mould of metal or sulphur, and pressed by a counter-mould at the back, so as to be reduced to a thickness of about $\frac{1}{4}$ inch, the inner surface being parallel to the outer surface, and roughly formed to the same pattern.

Papier-mache is sometimes made of sheets of paper glued together, and forced into a metal mould to give the pattern required.

In some cases a composition of pulp of paper and rosin is first placed in the mould. This adheres to the paper ornaments moulded as above described, and takes the lines and arrises of the mould more sharply than the paper alone would do.

CARTON PIERRE is a species of papier mache made with pulp of paper, whiting, and size, pressed into plaster moulds.

FIBROUS PLASTER consists of a thin coating of plaster of Paris on a coarse canvas backing stretched on a light framework, and formed into slabs.

This material has great advantages. Large surfaces can be quickly covered without much preparation for fixing, as it is less than $\frac{1}{4}$ the weight of plaster, and it can, if required, be painted at once.

Dennett's Fireproof Material.—The material used for Dennett's patent fireproof construction is a concrete of broken stone or brick imbedded in a matrix of plaster produced by calcining gypsum at a strong red heat (see Chap. viii., Part II.). Being fireproof, it is much used for theatres. It sets at about the same rate as ordinary Portland cement, and attains a strength nearly equal to that of the original gypsum.

ASPHALTES.

Asphaltes are combinations of bitumen and calcareous matter sometimes found in nature, sometimes artificially formed.

Natural asphaltes are superior to artificial imitations, probably because in them the bitumen is more thoroughly incorporated with the limestone or other calcareous matter.

The natural asphalte is generally ground, mixed with sand and a further proportion of bitumen, and run into moulds. When thus mixed it is known as *mastic.*

In the preparation of mastic mineral pitch (bitumen) must be used, not coal-tar pitch; the latter is brittle, easily softened, and weak.

Uses, Advantages and Disadvantages. — Patent asphalte (or mastic) is waterproof, fireproof, easily applied, and to some extent elastic, it can therefore be used with advantage for many purposes.

It is an admirable material for the damp-proof courses of walls (see Chap. iv., Part I.), also as a waterproof layer over arches or flat roofs, or for lining tanks. It is useful for floors that require a very smooth surface, as in racket courts; also for those that have to resist water, as in wash-houses, and for skirtings of such floors. When spread and brought to a smooth surface it wears well in footpaths, makes substantial and almost noiseless carriage-ways, but is very slippery in damp weather.

It is also used for the joints of pavements of stone and other materials, and prevents the penetration of wet, but makes such pavements more noisy.

Characteristics.—Good *mastic* should be proof against frost and damp, tough not brittle, and uninflammable. It should withstand a temperature of from 140° to 160° Fahr. without softening to any appreciable extent, and should not become so fluid as to run down below a temperature of 260° Fahr.[1]

Laying.—Any details regarding the laying of asphalte would be out of place in these Notes, which relate to the characteristics of materials, not to the manner of using them.

The following remarks are necessary, however, in order to understand the peculiarities of the different kinds of asphalte described below.

[1] Dant.

There are two principal methods by which asphaltes may be applied to a surface : (1) by being melted, spread, and rubbed to a smooth surface.

(2) By being ground to powder, spread, and consolidated by ramming.

Of these methods the first is the more convenient in many positions, but asphaltes laid as compressed powder appear to be the most durable under considerable wear, as in carriage-ways.[1]

In all cases asphalte should be laid on a good base of concrete or other solid material.

When the surface is at a slope exceeding about $\frac{1}{10}$, the asphalte is apt to run if exposed to the sun, unless a good key can be obtained.

For steep inclinations and for vertical work (such as the linings of tanks) the face must be roughed, the joints well raked out and filled with asphalte, the whole surface free from moisture and warmed ; the asphalte is then applied in successive thin coatings. Where the moisture cannot be got rid of, it is necessary to build the face of the wall with asphalte joints, to which the covering asphalte adheres. Plates of asphalte are sometimes used.

" Minute holes are noticeable in compressed asphaltes shortly after they are laid, which seem after a time to close up or disappear, while others open. The cause of these has not been satisfactorily explained." [2]

Varieties in the Market.—There are several different asphaltes in the market. A few of them will now be described.

Seyssel Asphalte, known also as *Claridge's Patent Asphalte,* is made from a bituminous rock found at Pyrimont Seyssel, in the Jura mountains.

It is a limestone saturated with bitumen, and contains about 90 to 92 per cent carbonate of lime and 10 to 8 per cent of bitumen.

This material is ground, mixed with grit and with heated mineral tar until the mass has thoroughly amalgamated and become reduced to a mastic. It is then run into moulds to form blocks.

These blocks are 18 inches square, 6 inches deep, and weigh about 125 lbs. each ; countersunk on two sides with the words PYRIMONT AND SEYSSEL as the trade mark.

The asphalte is imported in this form by the Pyrimont Seyssel Asphalte Company, from whose circular most of the following information is obtained :—

QUALITIES.—There are three qualities in the market—

1. *Fine,* without grit, used for magazine floors and as a cement for very close joints in brickwork.

2. *Fine-gritted,* for covering roofs and arches, lining tanks, as a cement for brickwork, and for running the joints of stones.

3. *Coarse-gritted,* containing more and larger grit ; used for pavements and floorings where great strength is required, as gun-shed floors, tun-room floors, margins of stall floors, etc. In gateways for heavy carriage traffic small pieces of granite chippings, etc., are introduced.

---

[1] Report of Engineer, City of London, 1871.   [2] Clark on Roads.

MIXING.—The blocks of asphalte are broken up into pieces of not more than 1 lb. weight each, and melted in iron caldrons heated by wood or peat.

Coal is objectionable on account of the smoke it creates ; coke injures the material and destroys the caldron.

The following directions are from the circular of the company :—

" The fire having been lighted in the caldron, put into the boiler 2 lbs. of mineral tar, to which add 56 lbs. of asphalte, broken into pieces of not more than 1 lb. each. Mix the asphalte and tar together with the stirrer, till the former becomes soft, and then place the lid on the caldron, keeping up a good fire. In a quarter of an hour repeat the stirring, and add 56 lbs. more asphalte, in similar sized pieces, distributed over the surface of that in the caldron. Again cover the caldron for ten minutes, after which keep the contents constantly stirred, adding by degrees asphalte in the proportion of 112 lbs. to 1 lb. of tar, until the caldron is full and the whole is thoroughly melted.[1] When fit for use the asphalte will emit jets of light smoke and freely drop from the stirrer."

The asphalte is removed from the cauldron in ladles, poured over the concrete foundation, or other place where it is to be applied, brought to a smooth surface with wooden rubbers, and finished, either with a mixture of slate-dust and silver sand in equal parts, or roughened by grit stamped in while the asphalte is soft.

Val de Travers Asphalte is from a rock found at Neuchatel in Switzerland.

It is said to be richer in bitumen than the asphalte from Seyssel, containing from 11 to 12 per cent, and sometimes as much as 20 per cent.

The material is laid in two different ways—either in powder, compressed, by ramming, into a solid condition, or by melting and spreading, as in the case of Seyssel asphalte.

Hot Compressed Process.—The natural rock having been ground to powder, is subjected to great heat in a revolving boiler. The boiler may be on the spot, or the powder may be brought in a hot state in closed iron carts.

A foundation of Portland cement concrete having been formed, its surface is spread over with the powder, which is then compressed by means of hot iron rammers into one homogeneous layer without joints, and impervious to moisture.

Carriage-ways are generally laid by this method.

Liquid Process.—The material used is composed of Val de Travers rock, mixed with a large quantity of clean grit about the size of a split pea.

The asphalte is melted in boilers as above described, a small quantity of bitumen being gradually added.

---

[1] Practice, however, best regulates the quantity of tar to properly flux the asphalte. In exposed situations, particularly on the coast during cold and other unfavourable weather, a strong fire is necessary to be kept up, and at such times the asphalte work is longer in execution. On this account the tar is more quickly consumed, and a small quantity will have to be added. A somewhat larger proportion of tar is also necessary in the application of asphalte to brickwork, and also in running the joints of stones. In warm climates an excess of tar must be avoided. From the first lighting of a caldron about 3½ hours will be occupied before the entire mass with which it is to be filled will become melted. The subsequent operation will occupy about half an hour less time.

It is carried in ladles from the caldrons to the concrete foundation prepared for it, and spread in a liquid state over the surface and allowed to cool.

About 18 parts asphalte and 2 parts grit are used for roofs, linings, tanks, etc.

About 16 parts asphalte and 2 parts grit for flooring, footways, stalls, etc. etc.

Rather more bitumen is added in the roofs than the floors, but the amount depends, of course, upon circumstances.

Limner Asphalte is obtained from Limner, near Hanover.

The asphalte is broken up and mixed with clean grit, together with a small quantity of bitumen.

The mixture is melted in caldrons, and laid in two thicknesses, the lower stratum having coarser grit in it than the other.

Brunswick Rock Asphalte is obtained from mines at Vorwohle, in Brunswick, Germany.

Montrotier Asphalte is a French production, and is laid in compressed powder. **Mastic Asphalte** comes from Spain, and is laid in small blocks.

There are several other so-called asphaltes in which the natural substance is mixed with various ingredients.

Among these may be mentioned the following :—

Barnett's Liquid Asphalte is made from natural or artificial asphaltes, mixed with powdered oxide of iron and a small proportion of mineral tar.

The materials are melted and laid, as before described, on a concrete foundation.

Trinidad Asphalte is a mixture of Trinidad pitch, broken stone, chalk, and other ingredients, and is laid hot, in the form of powder.

Patent British Asphalte is a mixture of quicklime, pitch, sawdust, and ground iron slag, heated and laid in a semi-liquid state.

Inferior Asphaltes are also made with coal-tar pitch boiled with chalk and sand.

Pitch plays an important part in asphaltes, and it will be well to distinguish between the different varieties.

Mineral Pitch, or bitumen, is the constituent that makes asphalte so valuable.

In fact, strictly speaking, solid bitumen is asphalte ; the rock asphalte, generally known by engineers as asphalte, is merely stone saturated with asphalte.

It used to be found in large quantities on the Dead Sea (*Lacus asphaltites*), and thus obtained the name of *Bitumen of Judea.*

Natural bitumen is found also in the island of Trinidad.

Bitumen contains an oil which in coal tar is very volatile, and escapes, leaving the tar brittle.

Coal tar is very brittle at the freezing point, and softens at 115° (Fahr.), whereas true bitumen is tough at 20°, and will not soften at 170° (Fahr.)

Coal Tar Pitch is the residue obtained by distilling coal tar.

This material is sometimes used instead of bitumen for mixing with asphalte.

It is, however, brittle, softens more under heat, is easily crushed, and is altogether inferior.

WHITENING AND COLOURING.

WHITEWASH is made from pure white lime mixed with water.

It is used for common walls and ceilings, "especially where, for sanitary reasons, a frequent fresh application is considered preferable to any coating which would last better. It readily comes off when rubbed, will not stand rain, nor adhere well to very smooth or non-porous surfaces. It is cheap, and where used for sanitary reasons should be made up of hot lime and applied at once, under which conditions it also adheres better." [1]

Whitewash is improved by adding 1 lb. of pure tallow (free from salt) to every bushel of lime.

The process is generally described as *lime whiting.*

The following is a method recommended for making whitewash for outside work.

"Take a clean water-tight barrel, and put into it half a bushel of lime. Slake it by pouring water over it boiling hot, and in sufficient quantity to cover it 5 inches deep, and stir it briskly till thoroughly slaked. When the slaking has been effected, dissolve it in water, and add 2 lb. sulphate of zinc and 1 of common salt ; these will cause the wash to harden, and prevent its cracking." [2]

COMMON COLOURING is prepared by adding earthy pigments to the mixtures used for lime whiting.

The following proportions [2] may be used per bushel of lime ; more or less according to the tint required :—

Cream Colour.—4 to 6 lbs. of ochre.

Fawn Colour.—6 to 8 lbs. umber ; 2 lbs. Indian red ; 2 lbs. lampblack.

Buff or Stone Colour.—6 to 8 lbs. raw umber, and 3 or 4 lbs. lampblack.

WHITING is made by reducing pure white chalk to a fine powder.

It is mixed with water and size, and used for whitening ceilings and inside walls. It will not stand the weather.

"The best method of mixing it is in the proportion of 6 lbs. whiting to 1 quart of double size (see p. 447), the whiting to be first covered with cold water for six hours, then mixed with the size and left in a cold place till it becomes like jelly, in which condition it is ready to dilute with water, and use."

"It will take 1 lb. jelly to every 6 superficial yards." [1]

Whiting is made in three qualities—*common, town,* and *gilders.* It is sold by weight in casks containing from 2 to 10 cwts., in sacks containing 2 cwts., in firkins (very small casks), in bulk, and in small balls.

DISTEMPER is the name for all colouring mixed with water and size.

White Distemper is a mixture of whiting and size.

The best way of mixing is as follows :—Take 6 lbs. of the best whiting and soak it in soft water sufficient to cover it for several hours. Pour off the water, and stir the whiting into a smooth paste, strain the material, and add 1 quart of size in the state of weak jelly ; mix carefully, not breaking the lumps of jelly, then strain through muslin before using ; leave in a

---

[1] Seddon.   [2] Burn.

cold place, and the material will become a jelly, which is diluted with water when required for use.

Sometimes about half a tablespoonful of blue black is mixed in before the size is added.

It is sometimes directed that the size should be used hot, but in that case it does not work so smoothly as when used in the condition of cold jelly, but on the contrary drags and becomes crumpled, thus causing a rough surface.

When the white is required to be very bright and clean, potato starch is used instead of the size.

Coloured Distemper is tinted with the same pigments as are used for coloured paints (see page 420), whiting being used as a basis instead of white lead or zinc white.

In mixing the tints the whiting is first prepared, then the colouring pigment, the latter being introduced sparingly, size is then added, and the mixture is strained.

The colours are classed as " Common," " Superior," and " Delicate," in the same way as described at page 420.

Quantity of Materials used for Plastering, etc.—The quantity of materials required for plastering, rendering, etc., depends upon the nature of the materials used, the degree of roughness of the walls, and other circumstances. Information on this subject will be found in the Builders' Price-books, Hurst's *Surveyor's Pocket-Book*, etc. The following Table was carefully compiled from practical observation for Colonel Seddon's *Notes on the Building Trades*, etc.—

TABLE of the QUANTITY of MATERIALS used in PLASTERING, RENDERING, etc.

10 Yards Superficial.	Chalk Lime Slaked.	Hydraulic Lime.	Sand.	Hair.	Water.	Ptld. Cement.	Lath and Half-Laths.
	Feet cube.		Feet cube.	Lbs.	Gallons.	Bshls.	Bundles
Render float and trowel, 1 Portland cement, 2 sand	...	...	6	...	15	2½	
Render one coat, and set with fine stuff	5	...	5	2½	20		
Render float, and set with fine stuff	6¼	...	6¼	3	25		
Lath, plaster, and set with fine stuff	5½	...	5½	2¾	22	...	2¼
Lath, plaster, float, and set with fine stuff	6¾	...	6¾	3¼	27	...	2¼

WEIGHT OF LIMES, CEMENTS, ETC.

The weights of various limes and cements are given approximately below. The precise weight varies, of course, according to degree of freshness, size of lumps, fineness of grinding, etc.

Quicklime in Small Lumps (Fresh).	Weight per Foot Cube in Lbs.
White chalk lime	39
Grey ,, ,, (Halling) . . .	44
Portland stone lime	47
Blue lias ,, ,, (various) . . .	58 to 70

Quicklime, Ground (Fresh).	Weight per Foot Cube in Lbs.	Weight per Striked Bushel in Lbs.[1]
Blue Lias lime (various) . . .	49 to 68	63 to 87
Grey chalk lime	43	55
Arden ,, ,, ...	68	87

CEMENTS, ETC.	Weight per Foot Cube in Lbs.	Weight per Trade Bushel in Lbs.[1]	Weight per Striked Bushel in Lbs.
Portland . . .	74 to 101½	100	95 to 130
Roman	60 to 62½	70	77 to 80
Medina	61	68	78
Keene's	64	75	82
Parian	60	66	77
Plaster of Paris . .	50	Sold by weight.	64
Whiting	64	Do.	82

[1] See page 163.

Chapter IV.

METALS.

THE metals used by the engineer and builder are iron, copper, lead, zinc, tin, and some of their alloys.

Ores.—These metals are not found to any great extent in the pure metallic state, but chiefly in the form of oxides, carbonates, or sulphides, called " ores."

Dressing.—The ores are broken up, and separated from the earthy matters adhering to them, by stamping or crushing in mills, and by washing in a stream, which carries away the lighter impurities, leaving the ore, which is then said to be " dressed."

Calcination and Roasting.—The next step is, as a rule, to roast the ore in heaps or in kilns, in order to drive off the moisture and carbonic acid, and to fit it for smelting.

Smelting.—The ore is mixed with a substance called a " flux," selected in consequence of its tendency to combine with the particular impurities of the ore. The mixture is then thrown into a furnace and subjected to intense heat, upon which the metal sinks down in a fluid state, while the impurities combine with the flux, and run off in a light and fusible slag.

IRON.

Production.—ORES.—Iron ores are generally carbonates, hydrates, or oxides of the metal, the latter being the best.

British iron is obtained from ores found in several strata, but chiefly in those of the coal-bearing or carboniferous series, in which they are most conveniently interspersed with the fuel (coal) and the flux (limestone) necessary for their reduction.

The following are the principal British iron ores :—

Clay Ironstone is a carbonate of iron of clay-like appearance. This is a very impure ore, containing not only clay, but pyrites and sulphur, and producing in some cases as little as 20 per cent of iron. However, on account of the large quantities in which it is found, and in consequence of its being near coal and limestone, it is the most important iron ore worked in Great Britain.

It occurs chiefly in the coal measures of Derbyshire, Staffordshire, Shropshire, Yorkshire, Warwickshire, and South Wales ; also in the Lias formations of Yorkshire (Cleveland). The ores vary greatly in quality, having a yield of iron which ranges from 20 to 40 per cent.

Blackband is clay ironstone darkened by from 10 to 25 per cent of bituminous and carbonaceous matter, which makes it cheaper to smelt. It is found chiefly in Lanarkshire and Ayrshire, where it yields about 40 per cent of iron ; also in Staffordshire, Durham, North Wales, where the yield varies considerably, being generally less than in Scotland.

Red Hæmatite is an oxide of iron found in many forms, often in globular or kidney-shaped masses of red colour.

This is the richest British iron ore, the chief impurity being silica ; it yields from 50 to 60 per cent of iron.

This ore is found in the carboniferous limestone of Cumberland (Cleator Moor, Whitehaven), Lancashire (Ulverston), and in Glamorganshire.

Some of these ores are greatly in demand for making Bessemer steel.

Brown Hæmatite is also an oxide of iron (hydrated), and of a brown colour. It contains some 60 per cent of iron, and is found in Gloucestershire (Forest of Dean), Cumberland (Alston Moor), in Durham, Devonshire, Northamptonshire, and on the Continent.

Magnetic Iron Ore is seldom found in this country. A little occurs in Devonshire, but it exists in large quantities in Sweden and Norway.

Spathic Ore is a crystallised carbonate of iron, generally mixed with lime, found in Durham (Weardale), Devonshire (Exmoor), and Somersetshire (Brendon Hills). It yields about 37 per cent of iron.

Such of these ores as are rich in manganese are used for the manufacture of Spiegeleisen (see p. 304).

Foreign Ores which cannot be described in detail are much used in connection with those of home production, such as the Spanish ores for steel making.

Iron ores are not sufficiently valuable to pay for their being washed and dressed.

Those that occur in large masses, such as clay ironstone, are roasted to drive off the carbonic acid, and to render them more easy to break up.

SMELTING.—The extraction of the metal from the ore is effected in a large upright furnace lined with firebrick.

Into this furnace a strong blast of air is forced.

In former years the air for the blast was supplied at its ordinary temperature. This is still done in some few instances, the process being called the "*cold blast*," and the resulting material *cold-blast iron.*

The *hot-blast* process was patented by Neilson in 1828. In this the air is raised to a temperature of some 800° or 900' Fahr. (sometimes to 1200° or 1400°) before being forced into the furnace. By this a very great saving of fuel is effected, and a greater heat obtained. Moreover, calcining may sometimes be dispensed with, coal may be used instead of coke, and altogether the process is far more economical.

The object of smelting is to free the metal from its combinations, and to get (as far as possible) all impurities out of the ore in the form of a fusible slag.

To effect this a *flux* is added of a nature suited to combine with the impurities or "*gangue*" in the ore.

If the gangue is chiefly clay, as it often is in this country, limestone is added as a flux. If the gangue is chiefly quartz, an argillaceous iron ore and limestone are added. If the gangue itself is limestone, clay or clayey ores are added.

The furnace is filled to a certain height with fuel. When this is burning, ore mixed with flux is introduced from the top, and then layers of fuel and ore, with flux, alternately.

When the furnace is fully heated, the molten iron sinks to the bottom, being covered by the lighter and more fusible impurities in the form of "*slag.*"

A furnace once lighted is not allowed to go out until it requires thorough repair, but is continually replenished with fuel and ore at the top.

When a considerable quantity of molten iron has collected, the furnace is tapped, and the iron is run into a long channel formed in sand, having branches on each side, called the sow and her pigs—hence the bars produced are called "pig-iron."

Comparative Advantages of Hot and Cold Blast Iron.—The very high temperature produced by the *hot blast* enables many of the impurities in the ore to be reduced to a molten state, and run out with the metal.

If this is taken advantage of, the impurities are retained in the resulting metal, instead of being got rid of in the smelting process, and a very weak inferior iron is produced.

It is evident, then, that the hot blast *may* be used to produce a very inferior material, and this for some time brought it into disrepute.

It has, however, been shown by experience that the temperature of the blast has, in itself, but little effect upon the iron produced, and that, with the same care in the selection of materials and conduct of the process, iron may be produced by the hot blast of as good quality, and as reliable, as that from a cold-blast furnace.

After a great many experiments on the relative strength of hot-blast and cold-blast iron, Sir William Fairbairn came to the following conclusion :-

" From the evidence here brought forward it is rendered exceedingly probable that the introduction of a heated blast in the manufacture of cast irons has injured the softer irons, while it has frequently mollified and improved those of a harder nature ; and considering the small deterioration that the irons of quality No. 2 [1] have sustained, and the apparent benefit to those of No. 3, together with the great saving effected by the heated blast, there seems some reason for the process becoming general, as it has done." [2]

There are but few cold-blast furnaces now in the country. Among them may be mentioned those at the celebrated Lowmoor and Bowling works ; also some at Blænavon in South Wales. Cold-blast iron is, however, still in use for the production of high-class wrought iron, and of castings of great soundness and tenacity.

---

[1] See page 265.　　[2] *Iron Manufacture*, etc., by Sir William Fairbairn.

PIG-IRON.

Pig-iron is the name given to the rough bars of unpurified iron run from the blast furnace.

In this form it is sold to the founder or to the iron manufacturer. By them it is subjected to various processes, which will hereafter be described.

Different Materials produced from Pig-iron.—The result of these processes is the production of materials which, though originally from the same ore, and still of nearly the same chemical composition, differ very widely in their mechanical properties and characteristics.

These materials may be divided into three general classes :—

> Cast iron,
> Wrought iron,
> and Steel.

The different processes required for the production of these three classes of material, and those connected with the conversion of the metals generally into the forms suited for the market—such as pigs or ingots, plates and sheets, bars of different sections, etc.— ·will be very lightly touched upon, the details of these processes being of less value to the practical designer than the results of the mechanical or other tests which he specifies in order to prove the suitability of the finished material he desires to employ.

Foreign Substances in Pig-iron.—Pig-iron always contains foreign substances, among which are

Carbon, Silicon, Sulphur, Phosphorus, and Manganese, besides many others in smaller proportion.

Of these foreign bodies that which plays by far the most important part is *carbon.*

The great differences (which will presently be pointed out) that exist between

> Cast iron,
> Steel,
> Wrought iron,

depend chiefly upon the amount of carbon they respectively contain.

The other substances may generally be regarded as impurities.

Each, however, when present, plays an important part (see pp. 264, 265), and in some cases their presence is beneficial.

With regard to the influence of carbon, Dr. Percy makes the following remarks :—

" Of all the compounds of iron none are to be compared with those of carbon in practical importance. . . . When carbon is absent, or only present in very small quantity, we have *wrought iron*, which is comparatively soft, malleable, ductile, weldable, easily forgeable, and very tenacious, but not fusible except at temperatures rarely attainable in furnaces, and not susceptible of tempering like steel. When present in certain proportions, the limits of which cannot be exactly prescribed, we have the various kinds of *steel*, which are highly elastic, malleable, ductile, forgeable, weldable, and capable of receiving very different degrees of hardness by tempering, even so as to cut wrought iron with facility, and fusible in furnaces. And lastly, when present in greater proportion than in steel, we have *cast iron*, which is hard, comparatively brittle, and readily fusible, but not forgeable or weldable. The differences between these three well-known sorts of iron essentially depend upon differences in the proportion of carbon, though—as we shall learn hereafter—other elements may and do often concur in modifying, in a striking degree, the facilities of this wonderful metal." [1]

It is very important for the proper understanding of this subject that the student should, from the outset, bear in mind the fact that

Cast Iron contains a large percentage of carbon (about 2·0 to 6·0 per cent).

Steel contains a small percentage of carbon (from about ·10 per cent, or even less in extra soft steel, to 1·50 per cent in razor steel).

Wrought Iron, when perfectly pure, is quite free from carbon. Practically, however, it contains a small quantity—not exceeding 0·25 per cent.

Between these main classes there are several gradations, merging gradually one into the other, and to which no definite limits (as to percentage of carbon) can be assigned.

There are also several varieties of each class, varying according as the percentage of carbon varies within the limits of that class.

These minor distinctions will presently be referred to, but at present it will be only necessary to remember that the three great divisions—cast iron, steel, and wrought iron—differ chiefly according to the proportion of carbon they contain.

THE EFFECT OF CARBON UPON CAST IRON.—There are many varieties of pig-iron, which themselves also differ pretty much according to the proportion of carbon contained by them.

These differences depend upon the quantity of fuel used in the reduction of the ore, the heat at which the reduction was effected, and other particulars.

Before proceeding to consider the different varieties, it is necessary to understand that there are two distinct forms in which carbon occurs in cast iron.

---

[1] Percy's *Metallurgy*, p. 102.

1. *In the state of Mechanical Mixture.*—In this state the carbon is visible in the shape of little black specks interspersed throughout the mass, which give the iron containing them a dark-grey colour.

These little black specks are particles of free carbon, otherwise known as graphite or plumbago.

2. *In the state of Chemical Combination.*—The carbon in this state is not visible, and can be detected only by analysis.

The properties of cast iron depend not upon the absolute *amount* of carbon it contains, but upon the *condition* in which that carbon exists.

The varieties containing a large proportion of free carbon are of a dark-grey colour, are soft, and run freely into moulds.

When the carbon is all, or nearly all, in chemical combination with the iron, there are no black specks; the metal is white, very hard, brittle, and forms, when fused, a somewhat pasty mass, which will not freely fill a mould.

The former of these classes merges gradually into the latter, and between them there are several gradations. Some varieties contain both free and combined carbon.

White cast iron sometimes contains as much carbon as the grey varieties (about 4 per cent), but of this very nearly all is in a state of chemical combination, whereas in the grey iron a very large proportion of it is free, in the shape of distinct specks of plumbago, only about 1 per cent being in chemical combination with the iron.

Impurities in Pig-iron.—The impurities mentioned below are originally derived either from the ore or fuel, and unless eliminated in subsequent processes, they will injure the respective metals produced in the manner stated.

SILICON is, next to carbon, the most common constituent of pig-iron. It is derived from the ore and from the fuel. A good deal of it is got rid of in the slag produced by smelting, and also during the refining and puddling processes.

In many respects silicon resembles carbon, and it affects *cast iron* in nearly the same way.

Wrought Iron is rendered by it hard and brittle. To obtain good wrought iron the silicon must be removed as far as possible by repeatedly heating and working the iron.

Steel.—$\frac{1}{2000}$ part makes it cool and solidify without bubbling and agitation; more makes it brittle. $\frac{5}{10}$ per cent makes it unforgeable.

PHOSPHORUS is very readily taken up by the iron during the smelting process, and is one of the worst impurities it can contain.

Cast Iron is hardened by it, but is made more readily fusible. Its tenacity is reduced.

Wrought Iron is injured by it in proportion to the quantity present.

$\frac{1}{10}$ per cent does not reduce the strength of wrought iron; and improves its capacity for welding. $\frac{3}{10}$ per cent makes it harder, but not weaker. $\frac{5}{10}$ per cent makes it "cold short" (see p. 277). $\frac{8}{10}$ decidedly cold short. 1 per cent makes it very brittle, and unfit for any but special purposes.

Steel is injured by a very minute proportion.

$\frac{1}{200}$ per cent makes it unfit for the best cutlery. $\frac{1}{10}$ per cent makes it cold short, and useless for tool-making of any kind. $\frac{4}{100}$ to $\frac{6}{100}$ per cent are found in analyses of mild steel for structural purposes.

MANGANESE nearly always exists in *cast iron*. It tends to produce the white variety, in which a large proportion is generally to be found.

In *Wrought Iron and Steel* it counteracts red shortness, probably by encouraging the departure of the sulphur and silicon (see page 304).

Its presence is essential in the manufacture of Bessemer Steel, and in some other processes.

SULPHUR is derived from the pyrites in the ore and coal.

In *Cast Iron* it tends to produce the mottled and white varieties.

In *Wrought Iron* $\frac{3}{10}$ to $\frac{4}{10}$ per cent produces red shortness.

In *Steel* more than $\frac{2}{10}$ per cent unfits it for forging; but makes it more fluid, and better for casting. $\frac{1}{10}$ per cent produces red shortness. $\frac{2}{100}$ to $\frac{6}{100}$ per cent are found in structural mild steel.

Copper has the following effects :—

In *Cast Iron* $\frac{2}{10}$ per cent does no harm.

In *Wrought Iron* $\frac{3}{100}$ per cent reduces tenacity. $\frac{5}{10}$ per cent makes it red short

In *Steel* $\frac{5}{10}$ per cent makes it red short. 2 per cent makes it brittle.

Arsenic is not a very frequent impurity in iron.

In *Cast Iron* a smaller proportion is said to be good for chilled castings.

In *Wrought Iron* it causes red shortness.

Among the impurities met with more rarely, or in smaller quantities, are

Tin, which makes wrought iron cold short.

Tungsten, which imparts hardness and elasticity to cast steel, and renders it more capable of retaining magnetism.

Antimony, which makes wrought iron both hot and cold short.

Titanium, which tends to produce mottled cast iron. The so-called "titanic steel" contains no traces of titanium. The good qualities attributed to it must arise from some indirect action.[1]

Classification of Pig-Iron.—The different varieties of pig-iron are sometimes classed under three general heads.

Bessemer Pig.—A distinct variety of pig-iron made from hæmatite ores for conversion by the Bessemer process (see p. 302). It should be as free as possible from sulphur, phosphorus, or copper ; but a small percentage of manganese and of silicon improves it for the purpose.

Foundry Pig, including all pigs having a fracture of a grey colour, containing a considerable proportion of free carbon, and being therefore adapted for the use of the ironfounder.

This iron is produced when the furnace is at a high temperature and properly provided with fuel.

Forge Pig, consisting of those pigs which are almost free from uncombined or graphitic carbon, and are therefore unfit for superior castings, being useful only for conversion into wrought iron.

This description of iron occurs when the temperature is low, or the fuel insufficient, also when there is much sulphur in the ore or fuel.

Forge iron is generally run from the blast furnace into iron moulds (instead of sand), by which it is kept free from the impurities of the sand, and also chilled, and thus rendered brittle and easy to break up for further treatment.

The pig-iron of commerce is more carefully divided into six or sometimes eight varieties.

The exact classification varies at different works.

The following is condensed from one given in Wilkie's *Manufacture of Iron in Great Britain,* and quoted by Mr. Matheson in his *Works in Iron :*—

No. 1.—The fracture of this quality of pig is of a dark-grey colour, with high metallic lustre ; the crystals are large, many of them shining like particles of freshly cut lead.

This iron is of the best description, and the highest in price. The amount of carbon it contains is from 3 to 5 per cent, which makes it fusible and specially fitted for foundry work.

No. 2 is intermediate in quality between Nos. 1 and 3.

No. 3 contains much less carbon than No. 1. The crystals shown in a

---

[1] Bauerman's *Metallurgy.*

fracture of this iron are smaller and closer than in No. 1, but are larger and brighter in the centre than nearer the edges of the fracture.

The colour is a lighter grey than that of No. 1, with less lustre.

No. 4 or Bright.—This iron has a light-grey fracture, and but little lustre, with very minute crystals of even size over the whole fracture. It is not fusible enough for foundry purposes, but it is used in the manufacture of wrought iron.

It is the cheapest of the grey irons.

When inferior in quality, and nearly passing into the variety called mottled, there is usually a thin coat or "list" of white iron round the exterior edges of the fracture.

No. 5, Mottled is intermediate between No. 4 and white iron, the fracture being a dull dirty white, with pale greyish specks, and with a white "list" at the edges. It is fit only for the manufacture of wrought iron.

No. 6, White.—This is the worst, most crude, hard, and brittle of the pig-irons, the fracture being metallic white, with but little lustre, not granulated, but having a radiating crystalline appearance. This iron is largely used in the manufacture of inferior bar iron.

Cinder Iron is an inferior material obtained from the slag of the puddling furnace, technically called "cinder."

This cinder contains a large proportion of iron ; but also the phosphorus and sulphur which have been extracted in making the better iron.

Such iron can only be extracted by the hot blast, and has done a great deal to discredit the material produced by that process.

It is, however, very fusible, and therefore valuable to mix with other irons, and is useful in itself for castings which do not require much strength.

Mine Iron is a name given to iron smelted from the ore only, without admixture of slag.

When iron is specified as "hot-blast—all-mine," it means that no cinder-iron or slag has been used in its production.

CAST IRON.

Cast Iron is obtained by remelting the foundry pig-iron of commerce, and running it into moulds of the shape required as hereinafter described.

In some cases the metal is run into the moulds direct from the blast furnace, but in superior work it is generally specified that the cast iron is to be of the "second melting ; " that is, from pigs remelted in a cupola.

The cupola is somewhat similar to a small blast furnace, and acts in the same way. A little limestone is added as a flux, which combines with some of the impurities left in the pigs, and removes them in the form of slag.

There are several varieties of cast iron—made from the different qualities of pig-iron—and they are classified by engineers in a somewhat similar manner.

Grey Cast Iron is made from foundry pigs Nos. 1, 2, 3 of the classification at page 265, and is itself generally divided into three classes according to the nature of the pigs from which it is made.

No. 1 is of a dark-grey colour, caused by the profusion of specks of graphitic carbon throughout its mass; it melts into a very fluid state, which adapts it for very fine sharp delicate castings not requiring much strength.

It is, however, not so strong as the other varieties of cast iron, and is very soft, yielding readily to a chisel.

When broken it gives out a somewhat dull leaden sound, and shows a large, dark, bright grain.

No. 2 contains less free carbon than No. 1, is therefore lighter in colour, closer in the grain, and more difficult to melt; but being harder when cold is better for machinery, girders, castings to carry weight, or in any position where strength and durability are required.

No. 3 is of a lighter grey, with less lustre, and contains still less carbon than No. 2. It is therefore harder and more brittle, and is employed in heavy castings.

White Cast Iron is made from forge pigs; it contains very little free carbon; is of a silvery hue, extremely hard and brittle, and is unfit for castings, except those of the very commonest kind, such as sash weights.

White cast iron can be converted into the grey variety by melting and slowly cooling it, and grey cast iron can be converted into granular white cast iron by melting and suddenly cooling it.

Mottled Cast Iron contains both the grey and white varieties, which can easily be distinguished. The fractured surface is either chiefly white with grey specks, or grey with white spots and patches.

Grey cast iron may be distinguished from white cast iron by treating the surface of a fracture with nitric acid. On grey iron a black stain will be produced, on white iron a brown stain.

White and mottled cast iron are less subject to be destroyed by rusting than the grey kind.

They are less soluble in acids, are hard, brittle, and not so elastic as the softer kinds.

Chilled Iron.—It is sometimes advisable to produce a casting, some parts of which are required to have the hardness of white iron, while others are required to be of the toughest grey iron.

This effect may be produced by placing in the mould over those parts where a hard skin is required, pieces of cold iron of suitable shapes, thinly coated with loam. Where these are touched by the molten metal its surface is suddenly chilled and converted into white iron.

Thus the running surface of a cast-iron wheel is sometimes chilled, and covered with a hard skin of white iron, while the remainder of the wheel is of tough grey iron.

Malleable Cast Iron is made by extracting a portion of the carbon from ordinary cast iron in order to assimilate it to the composition of wrought iron, and thus increase its toughness. This is generally done, in the case of very small castings, by embedding them in powdered hæmatite ore, or in scales of oxide of iron, and raising to a bright red heat in an annealing oven.

Malleable castings " may be easily wrought cold, but become very brittle when heated, breaking to pieces under the hammer at an incipient white heat ; at a higher temperature the kernel of unaltered cast iron melts, so that articles that have been subjected to the process cannot be united by welding, but may be brazed without difficulty."

Mr. Kinnear Clark states that the tensile strength of annealed malleable cast iron is "guaranteed by manufacturers to 25 tons per square inch," and that it "is capable of supporting a tensile stress of 10 tons per square inch without distortion." [1]

Castings treated by this process, though they have not the peculiar fibrous structure characteristic of wrought iron, become to a certain extent malleable, and can be hammered or bent when cold without fracture.

They are specially suitable for intricate forms which could not be forged in wrought iron without much difficulty and expense.

The depth to which the casting is effected by this process depends upon the time during which it is exposed. Pieces about half an inch thick are rendered malleable throughout ; thicker pieces have merely a skin of wrought iron, the interior remaining unaltered.

This process is applied to the manufacture of buckles, gun-locks, snuffers, pokers, tongs, etc. ; and on a larger scale it has been used for toothed wheels of machinery, screw propellers, and other purposes where a certain amount of toughness is required combined with intricate forms.

Mr. Matheson recommends that malleable iron castings should be used for the shoes and connecting pieces in roof structures.[2]

Toughened Cast Iron is produced by adding to the cast iron and melting amongst it from $\frac{1}{4}$ to $\frac{1}{7}$ of its weight of wrought iron scrap.

DESCRIPTIONS OF PIG-IRON FOR CASTINGS.—Great experience is required in order to know exactly what descriptions of pig-iron to choose in order to make castings for any particular purpose.

Mixtures of pigs classed under different numbers, and even selections from different localities and makers, are recommended for large and important castings.

---

[1] Clark's *Tables.*   [2] Matheson.

Sir William Fairbairn recommends [1] the following mixture as being of "great value in castings, such as girders for bridges, beams for buildings, etc., where rigidity and strength are required :—

Low Moor, Yorkshire,	No. 3,	30	per cent.
Blaina, or Yorkshire,	No. 2,	25	,,
Shropshire or Derbyshire,	No. 3,	25	,,
And good old scrap,		20	..
		100	

Many other recommendations as to different mixtures were made before the Royal Commissioners who reported on the employment of iron in railways.

It is now, however, generally considered better by engineers to stipulate that the iron shall stand certain tests, leaving the mixture to be used to the judgment of the ironfounder.

Castings.—The description of the art of the ironfounder does not come within the range of these Notes.

The few remarks which follow are intended only to give such a general idea of the process of ironfounding as will enable the student to understand the points to be observed in examining and testing castings of different kinds.

CASTING IN SAND.—Castings, such as are used in building and engineering works, are generally made by pouring molten iron into sand, in which an impression of the article required has been formed by means of a wooden pattern.

The sand is of a fine loamy character, free from oxides, and is filled into iron frames or boxes, without tops or bottoms, called "*flasks,*" made in two similar parts, one of which fits over the other.

The "*pattern*" having been accurately formed in wood (a little larger than the required casting, so as to allow for contraction in cooling, see p. 353), is placed in the lower flask, and the space round it is tightly filled with damp sand, the surface of the pattern having been dusted with dry "parting sand."

The upper flask is then placed upon the lower one, and in its turn filled with damp sand rammed round the pattern.

The box is then opened, the pattern taken out, and the halves carefully put together again without disarranging the sand, an orifice being left for the fluid metal, which is poured through it, into the space, in the sand, previously occupied by the pattern.

In order to prevent the metal from being chilled (see page 268) by contact with the sand, the inside of the mould is painted with a blacking made of charred oak, which evolves gases under the

[1] *Application of Iron to Building Purposes*, p. 85.

action of the hot iron, and prevents too close a contact between the metal and sand.

The sand is also pierced with holes to allow of the escape of the air, and of gases evolved when the metal is poured in. If these are allowed to force their way through the metal, they will cause it to be unsound and full of flaws.

The passages through which the molten iron is poured into the mould should be so arranged that the metal runs together from different parts at the same time. If one portion gets partially cool before the adjacent metal flows against it, there will be a clear division when they meet, the iron will not be run into one mass, but will form what is called a *cold shut*.

The above is the simplest form of the process.

When a casting is to be hollow, a pattern of its inner surface, called a "*core*" is formed in sand, or other material, so that the metal may flow round it.

This leads to arrangements in the pattern which are somewhat complicated, and which cannot here be fully described.

The core for a pipe consists of a hollow metal tube, having its surface full of holes. This is wound round with straw bands, and the whole is covered with loam turned and smoothed to the form of the inside of the pipe.

The strength of a casting is increased if it be run with a *head* or superincumbent column of metal, which by its weight compresses the metal below, making it more compact and free from bubbles, scoriæ, etc. These rise into the head, which is afterwards cut off.

For the same reason pipes and columns are generally specified to be cast vertically, that is when the mould is standing on end. This position has another advantage, which is that the metal is more likely to be of uniform density and thickness all round, than if the pipe or column is run in a horizontal position.

In the latter case the core is very apt to be a little out of the centre, so as to cause the tube to be of unequal thickness.

In casting a large number of pipes of the same size iron patterns are used, as they are more durable than wooden ones, and draw cleaner from the sand. Socket pipes should be cast with their sockets downwards, the spigot end being made longer than required for the finished pipe, so that the scoriæ, bubbles, etc., rising into it may be cut off. Pipes of very small diameters are generally cast in an inclined position.[1]

CASTING IN LOAM.—Large pipes and cylinders are cast in a somewhat different way.

A hollow vertical core of somewhat less diameter than the interior of the proposed cylinder is formed either in metal or brickwork.

The outer surface of this is plastered with a thick coating of loam (which we may call A), smoothed and scraped to the exact internal diameter of the cylinder (by means of a rotating vertical template of wood), and covered with

---

[1] Humber.

"parting mixture." Over this is spread a layer of loam (B) thicker than the proposed casting, the outer surface of B is struck with the template to the form of the exterior of the proposed casting, and dusted with parting mixture.

This surface is then covered with a third thick covering of loam (C) backed up with brickwork, forming a *cope* built upon a ring resting on the floor, so that it can be removed.

The outer brick *cope*, with C attached to it, is then temporarily lifted away upon the ring. The coating (B) is cleared out, and the cope is replaced so that the distance between its inner surface and the outer surface of A is equal to the thickness of the casting.

The metal is then run in between C and A. When cool C and A can be broken up, and the casting extracted.

The core, etc., have to be well dried in ovens before the metal is run. B is often dispensed with, and the inner surface of C struck with the template.

FORM OF CASTINGS.—The shape given to castings should be very carefully considered.

All changes of form should be *gradual*. Sharp corners or angles are a source of weakness. This is attributed to the manner in which the crystals composing the iron arrange themselves in cooling. They place themselves at right angles to the surfaces forming the corner, so that between the two sets of crystals there is a diagonal line of weakness. All angles, therefore, both external and internal, should be rounded off.

There should be no great or abrupt differences in the bulk of the adjacent parts of the same casting, or the smaller portions will cool and contract more quickly than the larger parts.

When the different parts of the casting cool at different times, each acts upon the other. The parts which cool first resist the contraction of the others, while those which contract last compress the portions already cool.

Thus the casting is under stress before it is called upon to bear any load.

The amount of this stress cannot be calculated, and it is therefore a source of danger in using the casting.

In some cases it is so great as to fracture the casting before it is loaded at all.

Thus in cast-iron girders whose section has been improperly designed, as shown in Fig. 111, the web being very thin would cool and contract first. The subsequent contraction of the thick flanges would be resisted by the already cold and rigid web. The flanges would therefore be kept by the web in a state of tension, and the web would be kept in a state of compression, the amount of which is unknown; moreover, the sharp angles between the flanges and the web would also be a source of weakness.

When the girder is properly designed, as in Fig. 112, the change of thickness is gradual, and the unequal contraction does not occur. (See Part I. p. 245.)

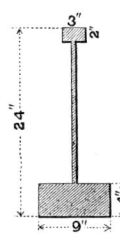

Fig. 111.

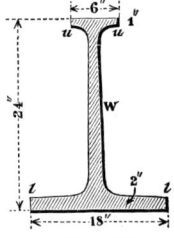

Fig. 112.

In a cast-iron girder of ornamental character such as that in Fig. 113, with an open web and moderately thin flanges, the flanges and verticals contract first, then the subsequent contraction of the diagonals brings them into tension, and they are very liable to break across, being resisted by the outer flanges.

On the other hand, if the diagonals contract first, they prevent the flanges from contracting, and cause a rupture in them by throwing them into tension.

Fig. 113.

The internal stress, produced by unequal cooling in the different parts of a casting, sometimes causes it to break up spontaneously several days after it has been run.

A case is mentioned by Mr. Anderson,[1] which actually occurred in practice.

The casting was of the form shown in Fig. 114. It was duly delivered by the maker without any apparent flaw, but after lying by for a day or two it suddenly split through the middle to within a few inches of the outer edges. On inquiry, it was found that the cooling of the mass had been hastened. The outer edges cooled first ; the thicker inner portion remained hot and prevented the outer edges from contracting, so they became stretched. When the interior became cooled it attempted to contract, but the outer edges being rigid cracked in the attempt.

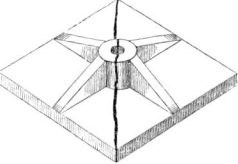

Fig. 114.

Castings should be covered up and allowed to cool as slowly as possible ; they should remain in the sand until cool. If they are removed from the moulds in a red-hot state, the metal is liable to injury from too rapid and irregular cooling.

The unequal cooling and consequent injury caused by great and sudden differences in the thickness of parts of a casting, are sometimes avoided by uncovering the thick parts so that they may cool more quickly, or by cooling them with water.

It is generally thought that molten cast iron expands slightly just at the moment when it becomes solid, which causes it to force itself tightly into all the corners of the mould, and take a sharp impression.

As molten iron cools down it shrinks about $\frac{1}{96}$ in all its dimensions ; the patterns must therefore be made proportionately larger.

The exact amount of contraction depends, however, upon the size and thickness of the casting, and upon the quality of the iron. The amount of contraction differs considerably in other metals, and the patterns should vary in size accordingly (see p. 353).

The patterns should also be slightly bevelled (about $\frac{1}{8}$ inch to the foot), so that they may be easily drawn out of the sand.

Superior castings should never be run direct from the furnace. The iron should be remelted in a cupola. This is called the *second melting*, and is generally prescribed in specifications. It greatly improves the iron, and gives an opportunity for mixing different descriptions which improve one another.

---

[1] *Proceedings Society of Arts.*

Castings required to be turned or bored, and found to be too hard, are softened by being heated for several hours in sand, or in a mixture of coal dust and bone ash, and then allowed to cool slowly.

EXAMINATION OF CASTINGS. — In examining castings, with a view to ascertaining their quality and soundness, several points should be attended to.

The edges should be struck with a light hammer. If the blow make a slight impression, the iron is probably of good quality, provided it be uniform throughout.

If fragments fly off and no sensible indentation be made, the iron is hard and brittle.

Air bubbles are a common and dangerous source of weakness. They should be searched for by tapping the surface of the casting all over with the hammer. Bubbles, or flaws, filled in with sand from the mould, or purposely stopped with loam, cause a dulness in the sound which leads to their detection.

The metal of a casting should be free from scoriæ, bubbles, core nails, or flaws of any kind.

The exterior surface should be smooth and clear. The edges of the casting should be sharp and perfect.

An uneven or wavy surface indicates unequal shrinkage, caused by want of uniformity in the texture of the iron.

The surface of a fracture examined before it has become rusty should present a fine-grained texture, of an uniform bluish-grey colour and high metallic lustre.

Cast-Iron Pipes should be straight, true in section, square on the ends and in the sockets, the metal of equal thickness throughout. They should be proved under a hydraulic pressure of four or five times the working head. The sockets of small pipes should be especially examined, to see if they are free from honeycomb. The core nails are sometimes left in and hammered up. They are, however, objectionable, as they render the pipe liable to break at the points where they occur.[1]

TESTS FOR CAST IRON.—For small girders and other castings intended to carry weight, it is usual to test a certain proportion of the number supplied by loading them till they break, and noting the weight under which they give way.

For large castings this system of testing would be too expensive. Small bars are therefore cast from the same metal and at the same time as the castings, and these are tested to fracture by a weight applied at the centre.

Some engineers require that the test bar should be cast with the main casting, and not broken from it until they have seen it.

The test bars are usually about 3 feet 6 inches long, 2 inches deep, and 1 inch wide, with a clear bearing of 3 feet.

---

[1] Humber.

The test weight varies, according to the opinion of the engineer, from 16 to 35 cwt.

It is important, however, to ascertain not only the weight that will break the test bar, but also the amount of deflection that will occur before fracture.

The reason for this is that a very hard iron will often bear a considerable cross strain when it is steadily applied, though it would be so brittle as to be unfit for any position in which it would be liable to slight vibration or shocks of any kind.

With regard to this point Mr. Matheson says :—

" A strength capable of enduring 25 cwt. on the test bar without fracture should be the minimum quality allowed even for short and heavy columns ; but for other purposes a load of from 28 to 30 cwt., and a deflection of $\frac{5}{16}$ inch, should be demanded.

" The deflection will vary from ·3 to ·5 inch.

" There is no difficulty in getting such iron, and higher qualities can be given if neces-sary, breaking strains of 30 to 35 cwt. being obtainable with judicious mixtures of the best kinds of iron ; and in testing such iron it will generally be found that some of the bars will endure as much as 38 cwt." [1]

Mr. Stoney points out " a singular fact that there is an excess of about 16 per cent in the weight that a 2-inch by 1-inch test bar will support when cast on edge and proved as cast, over that which it will support when proved with the underside as cast placed at the top as proved, and 8 per cent over the weight which the same test bar will sup-port if cast on its side or end, and proved on edge.

" Hence cast-iron girders should be cast with the tension flange downward in the sand." [2]

Dr. Pole has pointed out that small cast bars do not give a fair indication of the strength of larger castings run at the same time.

The cast-iron sleepers for the Great Indian Peninsula Railway were tested by a falling weight ; and test bars, of the ordinary form, cast at the same time, were broken by cross strain ; others, having a central section one inch square, were broken by tension.

WROUGHT IRON.

Wrought Iron is, or should be (as before mentioned), very nearly the pure metal, containing not more than about 0·15 per cent of carbon.

It may, by a peculiar process, be procured direct from the ore, but is generally obtained from the harder descriptions of pig-iron by a succession of processes, the object of which is to get rid of the carbon, and . of the phosphorus, silica, and other impurities, which injure the iron and make it brittle.

In order to expel these foreign substances the finest qualities of wrought iron are refined and then puddled : the inferior quali-ties are puddled only.

Forge pig is generally used for the manufacture of wrought iron, and can be converted at once by the puddling process.

Grey iron, however, contains graphite and silicon. The latter makes it difficult to puddle, and it is often removed by the pre-liminary process of refining described below.

[1] Matheson's *Works in Iron.*   [2] Stoney *On Strains,* p. 477.

REFINING consists in keeping the pig-iron in a state of fusion on an open hearth with coke, for about two hours, with a strong current of air directed upon it. It is at the same time well stirred, so that all parts of it are brought into contact with the air and oxidised.

The oxygen in the air deprives the cast iron of part of its carbon, and at the same time converts the silicon into silica, which combines with some of the oxide of iron to form a fusible slag, that runs off.

The liquid iron is then run into cast-iron moulds lined with loam, and kept cool with water circulating below them, so that it is chilled and easily broken up into what is technically known as "*plate metal.*"

The resulting fine metal greatly resembles white cast iron in its characteristics, but the percentage of impurities will be found to have been considerably reduced by the refining process.

PUDDLING consists in melting the pig-iron in a reverberatory furnace, by means of which the metal is subjected to the heat of the flame and a strong current of air, and kept quite clear of the fuel.

The molten metal is at the same time well mixed with oxidising substances, such as hæmatite ore, oxide of iron, forge scales, etc., and sometimes with limestone and common salt. The oxygen in these combines with the remnant of carbon left in the iron, and the silicon is also oxidised, passing off in slag.

As the carbon is removed the iron becomes less fusible, and clotty lumps of pure iron appear, which are collected by the puddler and pressed together with the tool until they are formed into *puddle-balls* weighing about $\frac{3}{4}$ cwt. or more.

In order to reduce the labour in puddling, rotatory furnaces and other ingenious inventions have been introduced. These, however, need not be further referred to.

SHINGLING.—The lumps or balls formed in the puddling furnace are at once placed under a helve or a tilt-hammer, the blows of which force out the cinder and consolidate and weld the particles of iron together, forming it into what is called a *bloom.*

Inferior descriptions of iron generally have the slag removed by a squeezer, a machine something like the jaws of an alligator, after which animal it is sometimes named.

On many works the steam-hammer is used for this purpose, and it can be made to do the work very effectually. It may, however, be used to produce very inferior iron, because it can be adjusted to give the mass such very light blows that the slag is not squeezed out, but left in the iron to its very great detriment.[1]

ROLLING.—Directly after this the red-hot slab of iron, or "bloom," is passed between grooved rollers, which convert it into *puddled bars* about 3 or 4 inches wide, $\frac{3}{4}$ to 1 inch in thickness, and 10 or 12 feet long.

The puddled bars thus formed are wrought iron, but of the lowest class. They possess hardly any of the characteristics of the higher qualities, and require to be greatly improved by subsequent processes of piling, reheating, and rolling.

Before referring to these processes and to the different qualities of iron produced by them, it will be well to glance at the effect of rolling upon the structure and strength of the iron.

---

[1] Pole.

EFFECT OF ROLLING IRON.—All wrought iron, after fusion, or after having been exposed to high temperatures sufficient to induce softening or pastiness, which is the case when iron is reheated to a white heat, consists of an aggregation of crystals of a cubical form.

In the act of rolling, these crystals are elongated into fibres, which form the mass of all good wrought iron.

Some authorities consider that when bar iron is subjected to continued vibration, constantly repeated loads, shocks, or blows, its structure becomes altered, and that it returns to a crystalline condition. On this point, however, there is considerable doubt (see p. 330).

The chemical constitution of the iron as well as its mechanical structure is altered during the process of rolling. When heated the surface is exposed to the oxidising influence of the atmosphere, the amount of carbon is considerably reduced, and a large proportion of other impurities may be got rid of.

Some experiments made at Woolwich on Bessemer wrought iron showed that this iron, when fused and run into a mould, had a tensile strength of 18·412 tons per square inch, but when the same iron was rolled its tensile strength became 32·4 tons per square inch, by which it appears that the operation of rolling has the effect of nearly doubling the strength of the iron.

The effect of rolling is illustrated also by the example given at page 331.

Iron, however, will not bear to be rolled too often, for it appears from Sir W. Fairbairn's experiments that it gains strength only up to the fifth reheating, and then its strength begins to fall off.

Professor Rankine says —" Good bar iron has in general attained its maximum strength, and the desired size and figure should be given to it with the least possible amount of reheating and working."

Different Qualities of Bar Iron.—The products of the rolling process are classified as follows :—

PUDDLED BARS, also known as No. 1 or *rough bars*.

The puddled bar obtained by the processes above described is of a very weak and inferior quality.

It has a coarse crystalline structure, and very small tensile strength (see Table, p. 316), but is of a harder texture than the better kinds of bar iron about to be mentioned.

In order to improve the quality of the material, the puddled bar is cut up into short lengths and subjected to the processes of piling, reheating, and rolling.

The effect of these processes is—1st, To drive out the slag ; 2d, To give uniformity of structure, weak parts of one bar being brought alongside the strong parts of other bars ; 3d, To produce a finished surface.

Some of the harder kinds of iron are, however, worked chiefly by the hammer, the bar being passed through the rolls only at the last when it is to receive its finished section.

MERCHANT BAR or *Common Iron,* known also as No. 2, is produced by piling up short lengths of puddled bars, raising them to a welding heat, and passing them through rollers. This amalgamates them into a single bar, and gives the iron a fibrous structure which greatly increases its strength.

This quality of bar is, however, still very inferior, being hard and brittle

It can be forged only with difficulty, and is useful only for the commonest purposes.

BEST BAR is produced by cutting up merchant bars, and repeating the processes of piling, reheating, and rolling.

In some cases the top and bottom of the pile are made with bars that have been twice rolled.

Best bar is far tougher and more easily worked than merchant bar, and is generally used for ordinary good work.

BEST BEST and *Best Best Best* iron bars are those which have respectively been submitted to three and four repetitions of the processes of piling, welding, and rolling.

SCRAP BARS are made from short pieces that are useful for no other purpose welded and rolled together into a single bar.

When the scraps used are old pieces not thoroughly cleaned, the resulting bar is of an inferior description.

If, however, the scraps used are of new and clean iron, such as the short ends cut off finished bars, an iron of capital quality is produced, which is known as *best scrap* or *best best scrap*.

MANUFACTURE OF T AND I IRON.—In manufacturing iron of T, I, or other sections, a pile of bars is formed, heated, and welded together under a steam hammer.

This is then rolled, in the roughing or cogging rolls, into a *bloom* of about half or two-thirds the sectional area of the original pile.

The bloom is reheated, and rolled down in grooved finishing rollers, each approaching more and more nearly to the section of the finished rail.

The bars are then cut to a length, straightened, and finished.

Wrought-iron girders can be rolled with ease up to a depth of from 12 to 15 inches. When they are required of greater depth than this, the upper and lower portions are sometimes rolled separately, and then united by inserting a piece of iron containing more carbon, and which is therefore more fusible. This piece is subjected to a fierce heat from blowpipes, and at the same time hammered on both sides, so as to weld the upper and lower portions of the girder together. Mild steel joists can now be obtained, as a result of the increased power of rolling mills, up to 20″ in depth, and from 7″ to 8″ in width of flange.

Contraction of Wrought Iron.—" When a bar of wrought iron is heated to redness and quenched in water it becomes permanently shorter than before. This fact is well known to practical men, who sometimes avail themselves of it when a wrought-iron crank, etc., has been accidentally bored out too large for its shaft ; by one or more heats it may be reduced so as to be a good fit." [1]

Cold Rolled Iron.—Wrought iron bars and plates, rolled cold under a great pressure, acquire a polished surface, and have their tensile strength increased, and their ductility reduced as shown in the Table, p. 317.

Defects in Wrought Iron.—COLD SHORT iron is very brittle when cold, and cracks if bent double, though it may be worked at a high temperature.

This defect generally appears in an iron produced from a poor ore, or is caused by an excess of phosphorus.

[1] Box *on Heat.*

RED SHORT or *Hot Short* iron cracks when bent or finished at a red heat, but is sufficiently tenacious when cold. The defect is generally caused by sulphur from the fuel. Red short iron, though useless for welding and for many other purposes, is tougher, when cold, than other iron.

Arsenic, copper, and several other impurities also produce red shortness.

TESTS FOR WROUGHT IRON.

General Remarks.—There are several ways in which the quality of a piece of wrought iron may be ascertained.

It may be broken by direct slow tension, or by a falling weight, the breaking stress, elongation, contraction of area, and other particulars being noted. In the absence of facilities for breaking it, it may be subjected to certain rough tests which will be presently described.

Where these tests cannot be applied, some idea may be formed of the quality of the iron by the appearance of the fractured surface.

Wrought iron is used in many structures in which it is liable to receive sudden and often-repeated shocks. This is the case, for example, in bridges, and to a certain extent in roofs. It must, therefore, be able not only to resist a great tensile stress, but also to withstand sudden concussion or continued vibration.

A very hard iron will withstand a very high tensile stress, but is brittle, and will snap under a sudden strain.

A good iron must, therefore, not only possess great tensile strength, but must be *ductile,* that is, able to stretch before it gives way. This ductility may be measured either by the proportion borne by the permanent elongation to the original length of the iron, or by the amount of contraction of area of section caused by the stretching.

A specimen of such iron when torn asunder by slow tension

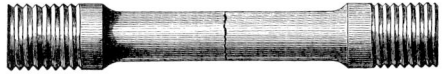

Fig. 115.

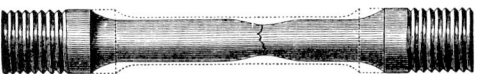

Fig. 116.

will not break off short as in Fig. 115,[1] but will draw out as in Fig. 116,[1] not only becoming longer, but also being reduced in diameter and sectional area at the centre. The dotted lines in Fig. 116 show the original size of the specimen.

In order that both strength and ductility may be secured, it is now usual for engineers to require that the iron for bridges and similar important work should fulfil at least two conditions :—

(1) That it shall not break with a tensile stress less than a certain specified amount.

(2) That before breaking it shall elongate not less than a named proportion of its original length ; or

That before breaking its sectional area shall be reduced (as a consequence of its stretching) by not less than a certain named proportion of its original area.

Of these two forms of the test for ductility, the measurement of the elongation is generally simpler and more easily managed than the measurement of the reduction of area.

With ordinary irons, as a rule, that specimen which has the greatest tensile strength is the hardest, and will contract least in sectional area, or lengthen the least before breaking.

Iron can, however, be made which will possess both qualities in a very high degree.

In addition to ascertaining the strength and ductility of the iron, it is desirable to know how the iron will behave when reheated and worked.

This is ascertained by bending or otherwise distorting the iron when hot, as described at page 282, under the head of Forge Tests.

Such tests are especially valuable when the iron is to be forged into different shapes before use in the structures for which it is intended.

Mr. Kirkaldy's Experiments.—At one time it was thought that the tensile stress required to break a piece of iron was all that was necessary to be known in order to ascertain its quality.

The investigations of Mr. Kirkaldy founded upon an elaborate series of experiments made by him on iron of every description and quality, led him, however, to the following conclusions,[1] among many others, some of which will be referred to presently :—

---

[1] From Kirkaldy's *Experiments on Iron and Steel.*

"1. The breaking strain does *not* indicate the quality as hitherto assumed.

" 2. A *high* breaking strain may be due to the iron being of superior quality, dense fine, and moderately soft, or simply to its being very hard and unyielding.

" 3. A low breaking strain may be due to looseness and coarseness in the texture, or to extreme softness, though very close and fine in quality.

" 4. The contraction of area at fracture, previously overlooked, forms an essential element in estimating the quality of specimens.

" 5. The respective merits of various specimens can be correctly ascertained by com· paring the breaking strain *jointly* with the contraction of area.

" 6. Inferior qualities show a much greater variation in the breaking strain than superior.

" 7. Greater differences exist between small and large bars in coarse than in fine varieties.

" 8. The prevailing opinion of a rough bar being stronger than a turned one is erroneous.

" 9. Rolled bars are slightly hardened by being forged down.

" 10. The breaking strain and contraction of area of iron plates are greater in the direction in which they are rolled than in a transverse direction." (In Mild Steel Plates the strength and extension are practically equal in both directions.)

UNIFORMITY.—In choosing iron for railway bridges and similar structures it is not only important that the iron should be strong and tough, but also that it should be *uniform* in quality.

Iron structures should be so proportioned that an equal stress shall come upon every square inch of the section of every part. It is of no advantage that the iron in one part should be so good as to enable it to take more than this working stress, when at the same time another part would give way if the stress were applied.

GENERAL METHODS OF TESTING.—Upon receiving a quantity of iron for any work, a certain proportional number of pieces are usually selected at random and tested to breaking in the manner before described, the percentage of elongation being at the same time observed, or of the contracted area. In addition to tensile tests, hot and cold forge tests are also applied, and in the case of steel boiler plates, tests for tensile strength and temper bending are frequently applied to each individual plate.

This is the usual method of testing—all the particulars required to be known with regard to the iron may be ascertained—and though some bad pieces may escape detection, yet the general average of the whole, and the degree of uniformity which exists, is pretty well arrived at.

In order that the iron may be uniform in ductility as well as in tensile strength, it has been recommended that a *maximum* percentage of elongation or contraction of area should be specified as well as a *minimum*. This, however, is not done, the minimum only being referred to in most specifications

Built-up girders, rolled joists, and trusses, such as roof principals, are frequently tested, either by hydraulic pressure or dead weight, and much valuable information, relative to the behaviour of the finished structure, is to be obtained by these means.

Testing Machines.—Numerous machines for the accurate testing of the strength of materials are now to be found both in public and private establishments. Many of these are remarkable, not only for the great range of stress which can be applied, but also for their sensitiveness and accuracy of measurement.

Among these may be mentioned the well-known machine of Mr. Kirkaldy of Southwark ; the 450 ton Emery Machine at Watertown Arsenal, U.S.A. ; the 600 ton machine of the Union Bridge Co., Pa., U.S.A. ; and many more designed on the principles of Werder, Wicksteed, Fairbanks, Riehle Bros., Greenwood and Batley, Buckton, and others. The methods of application of the load are usually by hydraulic pressure, a system of levers with weights, or a combination of both.

Autographic apparatus for recording the behaviour of material under stress is frequently applied and used in the delineation of stress-strain diagrams, the determination of the elastic limit and yield-point, etc.

The arrangements of these machines is such that specimens can be subjected to compressive, tensile, transverse, and torsional stresses, while in the larger machines full-sized members, such as columns of wood, iron, steel, or brickwork, and steel or iron tension members can be tested to destruction.

For a detailed description of the various types of machine and their principles of design, the student is referred to Professor Unwin's *Testing of Materials of Construction.*

Tensile Tests for Wrought Iron.—The Tables at page 316 give the tensile strength, the contraction of area, and other particulars with regard to several different descriptions of iron.

These particulars differ in nearly every case. It is not usual to make shades of difference in the tests applied, so that they do not vary with each minute difference in the description of iron that is to be used.

The following Tables, showing the tests that are applied to the various classes of iron by the different Government departments, will therefore be useful.

India Office.—The following Table is extracted from one prepared for the India Office by Mr. Kirkaldy : [1]—

SCALE OF TENSILE TESTS FOR IRON OF VARIOUS QUALITIES.

	CLASS C.		CLASS D.		CLASS E.		CLASS F.		CLASS G.	
DESCRIPTION.	Ultimate stress per square inch.	Contraction of area at fracture.	Ultimate stress per square inch.	Contraction of area at fracture.	Ultimate stress per square inch.	Contraction of area at fracture.	Ultimate stress per square inch.	Contraction of area at fracture.	Ultimate stress per square inch.	Contraction of area at fracture.
	Tons.	Per cnt.	Tons.	Per cnt.	Tons.	Per cnt.	Tons.	Per cnt.	Tons.	Per cnt.
Bars, round or square . .	27	45	26	35	25	30	24	25	23	20
Bars, flat . .	26	40	25	30	24	25	23	20	22	16
Angle and Tee or T	25	30	24	22	23	18	22	15	21	12
Plates, grain lengthways	24 ⎱ 23 ⎰	20 ⎱ 16 ⎰	23 ⎱ 21½ ⎰	15 ⎱ 12 ⎰	22 ⎱ 20½ ⎰	12 ⎱ 9½ ⎰	21 ⎱ 19½ ⎰	10 ⎱ 7½ ⎰	20 ⎱ 18½ ⎰	8 ⎱ 5½ ⎰
Plates, grain crossways	22	12	20	9	19	7	18	5	17	3

SWEDISH BARS.

Ultimate Stress ⎱ 22 tons.   Contraction of ⎱ 60 per cent.
per square inch. ⎰   area at fracture. ⎰

This Table may be compared with the Tables of strength given at page 316, of various descriptions of malleable iron.

The best Yorkshire Iron might be expected to stand the tests under Class C, as regards the contraction of area.

---

[1] Wray's *Theory of Construction.*

The Best Best irons of the market should stand the test under Class E. The ordinary Best iron of the market should stand those of Class G.

A and B are reserved for special qualities of iron which might be required at any future time, and the Classes D to F would be for qualities intermediate between the others. Recent India Office specifications are summarised as in the following Table, which shows the ultimate tensile stress per square inch, and the contraction of area and percentage of elongation for each description of iron.

	B.B. Staffordshire.		B.B.B. Staffordshire.		Yorkshire.		Miscellaneous.		For Iron roofing.		
	Ultimate tensile stress per sq. inch.	Contraction of area per cent.	Ultimate tensile stress per sq. inch.	Contraction of area per cent.	Ultimate tensile stress per sq. inch.	Contraction of area per cent.	Ultimate tensile stress per sq. inch.	Contraction of area per cent.	Ultimate tensile stress per sq. inch.	Contraction of area per cent.	
	Tons.		Tons.		Tons.		Tons.		Tons.		
Bars round and square .	23	30	24	40	23	50	24	40	24	20	Flat bars over 5″ wide.
Do. flat	22	25	23	35	22	45	...	...	22	15	Flat bars under 5″ wide.
Angle iron . . .	22	25	23	35	22	45	22	20	22	15	
T or H ,, . . .	22	25	23	35	22	45	22	20	20	10	
The percentages for Plates and Sheets are for elongation in 8″.											
Plate { grain lengthways	20	10	22	12	21	20	21	10	18	5	
{ grain crossways .	17	5	18	7	19	12	18	5			
Sheet { grain lengthways	20	10	22	12	21	20					
{ grain crossways .	19	5	18	7	19	12					

Admiralty.—The Admiralty Tests for iron may be tabulated as follows :— (See Tests for Mild Steel, p. 308.)

Tensile Strain per square inch.

BB or 1st class plate iron and sheet iron ¼ inch thick and above . . { (grain lengthways) . . 22 tons. { (grain crossways) . . 18 ,,

Do., boiler plate, do. do. . { (grain lengthways) . . 21 ,, { (grain crossways) . . 18 ,,

B or second class plate and sheet iron { (grain lengthways) . . 20 ,, { (grain crossways) . . 17 ,,

Angle, Bulb, T, Angle-bulb, Tee-bulb or other iron of ordinary form } (grain lengthways) . . 22 ,,

All the above are in addition to the forge tests enumerated at page 281.

BB bar iron, moulding, sash bar, half round, and segmental iron. Fire bar iron } Do. . . . . 22 ,, and such forge tests, hot and cold, as may be deemed expedient.

Cable Iron, sizes under 2¼″ diameter . . . . . . 23 tons.
Do. Do. Do. above 2½″ ,, ... 22 ,,
With an elongation of 20 per cent in 8 inches for all sizes.

Rough Tests for Wrought Iron.—There are several very useful tests which may be applied to iron of different forms in addition to the tensile tests.

Forge Tests.—Plate iron may be bent either hot or cold, with or across the grain. The bending is done upon a cast-iron rectangular slab, having the corner slightly rounded off. The angle *through which the plate should bend* without cracking depends upon the quality and thickness of the iron, and is shown in the following tables, which, together with the tests following for angle irons, etc., have been extracted from the Admiralty directions for testing iron.

PLATE IRON.					
	HOT.	COLD.			
	THICKNESS.				
	1 inch thick and under.	1 inch.	¾ inch.	½ inch.	¼ inch.
B B, grain lengthways . . .	125°	15°	25°	35°	70°
„ „ crossways . . .	90°	5°	10°	15°	30°
B, grain lengthways . . .	90°	10°	20°	30°	65°
„ „ crossways	60°	...	5°	10°	20°

SHEET IRON.		
	HOT.	COLD.
B B, grain lengthways . . .	125°	90°
„ „ crossways . . .	90°	40°
B, grain lengthways . . .	90°	75°
„ „ crossways	60°	30°

N.B.—It should be noticed that the angle mentioned above in each case is the angle through which the plate is bent, commencing at the horizontal, not the angle between the two sides of the plate after it is bent.

Different descriptions of iron may be tested as follows :—[1]

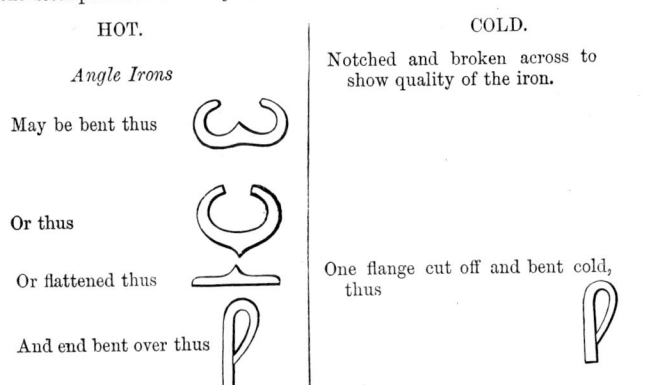

HOT.

Angle Irons

May be bent thus

Or thus

Or flattened thus

And end bent over thus

COLD.

Notched and broken across to show quality of the iron.

One flange cut off and bent cold, thus

---

[1] These are tests used by the Admiralty.

Tee Irons

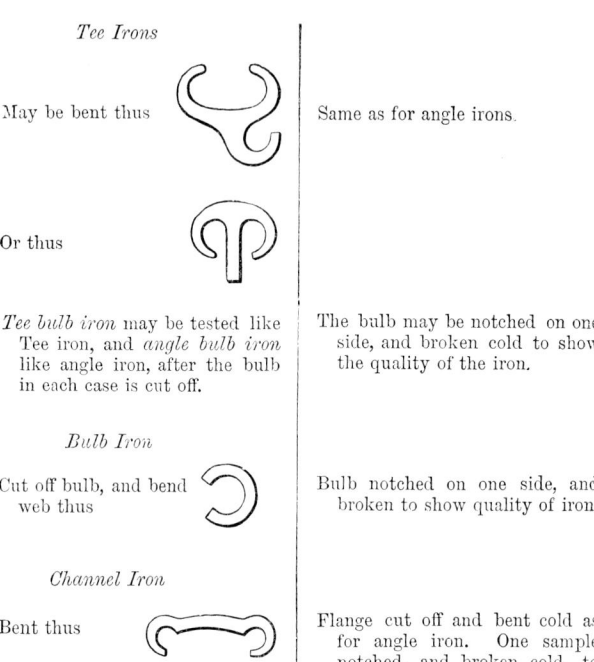

May be bent thus   Same as for angle irons.

Or thus

Tee bulb iron may be tested like Tee iron, and *angle bulb iron* like angle iron, after the bulb in each case is cut off.	The bulb may be notched on one side, and broken cold to show the quality of the iron.

Bulb Iron

Cut off bulb, and bend web thus	Bulb notched on one side, and broken to show quality of iron.

Channel Iron

Bent thus	Flange cut off and bent cold as for angle iron. One sample notched, and broken cold, to show quality of iron.

Rivets of good quality should double when cold without showing any signs of fracture. The heads " when hot should stand being hammered down to less than $\frac{1}{8}$-inch thickness without cracking at the edge. Rivets should also stand having a punch of nearly their own diameter driven right through the shank of the rivet when hot without cracking the iron round the hole." [1]

Appearance of the Fractured Surface of Wrought Iron.—At one time it was thought that a fibrous fracture was a sign of good tough wrought iron, but that a crystalline fracture showed that the iron was bad, hard, and brittle.

Mr. Kirkaldy's experiments led him, however, to the following conclusions.[2]

" 1. Whenever wrought iron breaks *suddenly*, a *crystalline* appearance is the invariable result ; when *gradually*, invariably a *fibrous* appearance.

" 2. Whether, on the one hand, it is finely or coarsely crystalline, or on the other, the fibre be fine and close, or coarse and open, depends upon the quality of the iron.

" 3. When there is a combination in the same bar or plate of two kinds—the one harder or less ductile than the other—the appearance will be partly crystalline and partly fibrous, the latter produced by the gradual drawing asunder action previous to and at the time of rupture ; whilst in the former the iron breaks suddenly, without elongating at time of rupture.

" 4. When the proportion of the harder is considerably less than the softer, the former snaps suddenly, whilst the latter continues stretching ; but when nearly equal, or the less ductile predominates, both portions break together, or almost at the same moment ;

---

[1] Graham Smith in *Proceedings Liverpool Engineering Society*, from *"Engineer."*
[2] Kirkaldy's *Experiments on Wrought Iron and Steel.*

the one part, gradually arriving at its limit of endurance, breaks with a fibrous appearance, whilst a greatly increased strain consequently coming on the remaining portion, it suddenly gives way producing a crystalline appearance.

" 5. The relative qualities of various irons may be pretty accurately judged of by comparing their fractures, provided they have all been treated in precisely the same way, and all broken under the same sort of strains similarly applied.

" 6. By varying either the shape, the treatment, the kind of strain, or its application, pieces cut off the same bar will be made to present vastly different appearances in some kinds of iron, whereas in others little or no difference will result."

It will be seen then that the appearance of the fractured surface of wrought iron is to a certain extent an indication of its quality, provided it be known how the stress was applied which produced the fracture.

Good iron may be either crystalline or fibrous, according as the stress which caused fracture was sudden or gradual, but it should be remarked that bad iron is never fibrous.

Small uniform crystals of a uniform size and colour, or fine close silky fibres, indicate a good iron.

Coarse crystals, blotches of colour caused by scoriæ or other impurities, loose and open fibres, are signs of bad iron, and flaws in the fracture surface are signs that the piling, welding, and rolling processes have been imperfectly carried out.

Fractures examined should be those of bars at least half-an-inch thick, or they will become distorted and will not exhibit the characteristic peculiarities to be seen in larger bars.

The fibres of wrought iron are readily exposed by immersing the specimen for a few days in very weak hydrochloric or nitric acid, which eats away the material between the fibres, leaving the latter exposed.

Test by means of Falling Weight, or IMPACT TEST.—In testing iron for very important situations, where it will be subject to sudden shocks, it is well to subject it to the tension produced by a weight falling from a height, so as to imitate as nearly as possible the action of the force to which it will be subjected.

This is done in the case of bolts for fastening the thick iron plates of armour-plated forts. These are tested by means of a ton weight falling through a distance of 30 feet. The testing apparatus is so arranged that the blow acts in the direction of the length of the bolt. This, it is found, will pull asunder a 3-inch bolt in six or seven blows. The fracture is required to be " silky fibrous, not crystalline in any degree," and the contraction of area 40 per cent.

Steel rails are also frequently tested by a falling weight.

DESCRIPTIONS AND MARKET FORMS OF MILD STEEL AND WROUGHT IRON, AND THEIR RELATIVE VALUE

DESCRIPTIONS OF WROUGHT IRON.

The following are the different kinds of wrought iron most generally known in this country.

Swedish Iron is made from pure magnetic iron ore—chiefly from Dannemora—smelted with charcoal.

It excels any iron made in this country with regard to tenacity and tough ness, but its great cost precludes it from use in engineering and building structures.

Best Yorkshire Iron—produced by certain well-known firms of long standing (see p. 297). This iron can be more thoroughly relied upon for strength, toughness, and uniformity than any other made in this country. It is generally specified for important work intended to withstand an unusual stress, or to resist sudden shocks, or changes of temperature.

Other Yorkshire manufactures — Staffordshire Iron, Scotch Iron, Cleveland Iron, Newcastle Iron, Middlesborough Iron, Welsh Iron, etc. etc., are other descriptions of bar and plate iron in the market. They vary considerably in quality ; some of them possess considerable strength and toughness, are only half the cost of best Yorkshire iron, and are more generally used for ordinary purposes.

These different varieties are generally distinguished by marks or brands, which are described at p. 294. The qualities of the different kinds of iron are further referred to in connection with the brands under which they are sold.

MARKET FORMS OF MILD STEEL AND WROUGHT IRON.

Mild Steel and Wrought Iron are prepared for the market in several convenient forms.

Ordinary Dimensions are those generally made and kept in stock. Everything required of different dimensions from these must be paid for at a higher price.[1]

Dead Lengths and Exact Dimensions are also charged for extra, that is when a bar must be the exact length specified within $\frac{1}{2}$ inch, or when a plate must be a special unusual length by a special unusual width in disproportion.

Iron of irregular or unusual figure or dimensions, or cut according to sketches, is also charged extra.

Bar Iron includes simple sections—round, square, or flat.

■ *square.* Ordinary dimensions are generally from $\frac{1}{2}$ inch to 3 inches
● *round.* diameter, or sides, increasing by $\frac{1}{16}$ of an inch each size.

If under $\frac{1}{2}$ an inch diameter, they are classed as *rods ;* or if under $\frac{3}{16}$ inch diameter, as *wire.*

▬▬ *Flat Bars.*—The ordinary dimensions are generally from 1 foot by $\frac{1}{4}$ of an inch to 6 inches by 1 inch, the width increasing $\frac{1}{8}$ of an inch and the thickness increasing $\frac{1}{16}$ of an inch (at the same time) in the various sizes.

Bars of these sections may be readily obtained of to 22 feet in length without extra charge.

◣ *Half round,* ◗ *Oval,* ◂ *Convex,* ◀ *Half-oval,* ⬢ *Hexagon,*

● *Octagon,* and ▬◖ *Tyre iron,* are other sections which are useful for different purposes, but which need not be more fully described.

Bar iron is classified as to quality in the manner described at p. 276.

USES.—*Best Yorkshire Bar Iron* is used for locomotives, and in superior shipbuilding, also for bolts and fastenings of very important structures.

Best Best Bar Iron of other descriptions is used for all very important work where the expense of Lowmoor, Bowling, or other best Yorkshire irons excludes their use.

---

[1] For remarks on Standard Sections, see p. 292.

Best Staffordshire Bar Iron is used for ordinary work.

Common Bar Iron is used where the iron requires hardly any forging, and is not expected to offer much resistance. It suffices for hurdles, standards, and work of similar class.

Angle and T Irons.—Iron of these sections is most useful in a great many building and engineering structures, such as roofs, girders, bridges, etc. etc.

The sections are made of a great variety of dimensions. Iron merchants generally publish lists showing those that they keep in stock.

Figs. 117.   118.   119.   120.   121.   122.   123.

The sides *a b*, *b c* are sometimes equal, as in Fig. 117 ; sometimes unequal, as in Fig. 118.

These forms of iron are obtainable in lengths up to about 40 feet.

It will be seen that extras are charged for the smaller sized angle and T irons, also for sections exceeding 8 *united inches*, that is sections in which the sum of the length of the sides, or of the length of both and stem, is more than 8 inches.

Extras are also charged for sections having an obtuse or an acute angle, as in Figs. 121, 122 ; or " round-backed," as in Fig. 123.

Angle irons cannot well be rolled of a thickness less than ⅛ of the width of one side. On the other hand, if they are very thick, there is a considerable percentage of loss of metal in the rivet holes.

They should have sides or flanges of equal thickness—holding up the full thickness to the ends of the flanges, not feather-edged.

Channel Iron, known also as half H iron, is a form frequently used in lattice girder bridges and similar structures.

The *united inches* in channel iron consist of the width added to twice the height.

Fig. 124.

Rolled Girder Iron, known also as *Rolled Joist Iron, Beam Iron, I Iron,* or *H Iron.*

This is one of the most useful sections of iron for fireproof and other floors, parts of bridges, roofs, etc., and is rolled in depths of from 3 to 20 inches.

An endless variety of sections is kept by different makers who generally publish full-size sections of their iron joists, showing the weight per foot run of each joist, and the distributed load that it will support.

Fig. 125.

Miscellaneous Sections.—Besides the above-mentioned there are a great many forms of wrought iron, some of which are in common use, others required only for special purposes.

A few of these will now be briefly mentioned. Their sections are shown below.

Figs. 126.   127.   128.   129.   130.   131.

Bulb Beam (Fig. 126) is chiefly used for shipbuilding.

Bulb Tee Beam or *Deck Beam* (Fig. 127) is also used for ships, and sometimes for roofs.

Bulb Angle (Figs. 128, 129) also for ships.

Square Root (Fig. 130), used for riveted structures in which the rounded edge would cause an empty space.

Double Angle or Z Sections (Fig. 131), used for riveted structures instead of a flat and two angle irons.

Rail Bars.—These are made of various sections, principally in steel ; some of them may be illustrated, but need not be described.

Figs. 132.   133.   134.   135.   135*a*.

Double-headed Rail (Fig. 132).—The prevailing section, on English lines, usually "bull-headed," *i.e.* with greater area in the upper flange.

Vignoles or *flat-bottomed Rail* (Fig. 133), used especially on Continental lines, in India, and the Colonies.

Bridge Rail (Fig. 134), used formerly by the Great Western Railway Company for the old broad-gauge traffic.   Figs. 133 and 134 are useful sections for Traveller Roads.

Tram Rail (Figs. 135 and 135*a*), are only two of the many forms of sections used for tramways in streets.

Sash and Fancy Iron.—This class includes a great variety of forms of iron

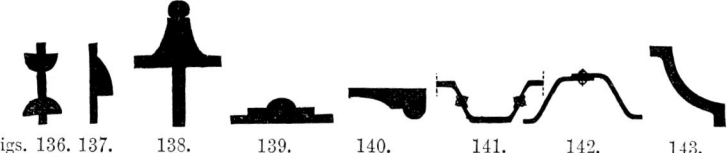

Figs. 136. 137.   138.   139.   140.   141.   142.   143.

sash bars, such as those in Figs. 136 to 138.   Trough sections (Figs. 141, 142), for flooring or bridge decking ; *Beading Iron*, for ornamental work, as in Figs. 139, 140 ; *Quadrant Iron* (Fig. 143), for building up piles or columns : and other sections useful for different purposes.

MARKET SECTIONS.—Before ordering L, T, ⊔ iron, joists, or iron of the various other sections, it is well to ascertain the dimensions of the usual sections kept in stock by the merchants, or for which the works have rolls.   Iron merchants give their customers printed lists of all such sections, which include almost every variety that can possibly be required ; for example,

Angle irons are shown from $\frac{1}{2}$ inch × $\frac{1}{2}$ inch × $\frac{1}{8}$ inch thick up to 8 inches × 8 inch × $\frac{3}{4}$ inch thick.

T irons from $\frac{7}{16}$ inch table × $\frac{3}{4}$ inch stem × $\frac{1}{8}$ inch thick, up to 12 inch table × $3\frac{1}{2}$ inch stem × $\frac{7}{8}$ inch thick.

⊔ irons from $\frac{5}{8}$ inch wide × $\frac{3}{8}$ high × $\frac{3}{16}$ thick, to 12 inches wide × 4 inches high × $\frac{1}{2}$ inch thick.

I iron from $2\frac{3}{4}$ inches high × 1 inch wide × $\frac{1}{4}$ inch thick, to 20 inches high × 8 inches wide × $1\frac{1}{4}$ inch thick.   The disadvantages arising from an undue multiplicity of sections has, however, given rise to the Lists of British Standard Sections issued by the Engineering Standards Committee, to which further reference is made on p. 292.

Rivet Iron, Chain Iron, Horse-shoe Iron, Nail Iron, are special qualities manufactured for the purposes indicated by their respective names.

Plate Iron is made in thicknesses between $\frac{1}{8}$ inch and 1 inch.

The different thicknesses vary $\frac{1}{16}$th inch each in succession.

When beyond $\frac{3}{4}$ inch thick the plates are generally of ordinary quality, unless specified " best " or " best best."

Extras.—Large or heavy plates are more expensive because they require more care and labour in manufacture.

The extras charged upon plates vary slightly in the different districts, as will be seen by the list of extras.

Thus, in Staffordshire, an extra is charged if the plate is more than 15 feet long or 4 feet wide, or if it contains more than 30 square feet surface, or weighs more than 4 cwt. In the North of England many of these extras are given up, and a proportion (say 10 per cent) of an order for plates, weighing as much even as 10 cwts. per plate, will be rolled without extra charge.

Plates less than $\frac{1}{4}$ inch thick are charged extra in the Cleveland district.

USES.—*Common plates* are used for shipbuilding, and called "ship plates." [1]

Best plates, also for shipbuilding where more tensile strength is required, and for girder work.

Best best plates, for the better class of shipbuilding, also for boilers of engines

Treble best plates are used in boilers of superior construction, and first-class work generally.

CHARCOAL PLATE is produced by a peculiar process of refining with charcoal instead of coke.

It is very tough and strong, and can be bent either way, with or against the grain, and is used chiefly for the manufacture of utensils which are stamped out of it.

TIN PLATES are coke or charcoal iron plates coated with tin. [2]

TERNE PLATE is the same plate coated with an amalgam of lead instead of tin, and wears a less brilliant look, but suffices for lining packing-cases, etc.

MALLET's BUCKLED PLATES.—These are plates of any shape in plan, arched from the edges towards the centre ; the arch has a very slight rise, and forms a dome or groined surface, according as the plate is round or square.

Such plates will bear a very great weight, and are applicable to fireproof floors, bridges, and several other purposes.

FLITCH PLATES are made in widths up to 18 inches for use in flitch girders (see Chap. vii., Part I.) They are generally of common iron, as they require no bending or smithing, nothing but a few holes punched.

Sheet Iron is so called when the material is of a thickness equal to or less than No. 4 B.G.—*i.e.* ·250 inch ; above that thickness the material is called plate iron.

It is generally of superior quality and higher price (as there are so many more sheets to the ton, and consequently extra labour in rolling, etc., than in plate iron) and its thickness is specified in terms of the Birmingham gauge (B.G.).

The following table shows the classification of sheet iron as to thickness :—

---

[1] Mild steel in the construction of the hulls of ships has practically superseded the use of wrought iron.   [2] For the use of mild steel in this connection see p. 346.

Name of Class.	B. Gauge.		Thickness.	
	From	To	From	To
Singles . . .	3/0	20	·500	·039
Doubles . . .	21	24	·035	·025
Lattens . . ,	25	27	·022	·017

Sheet iron is not much required for engineering purposes, but in many places it is used for roofing churches, houses, sheds, etc.

Corrugated Sheet Iron is made by passing sheets between grooved rollers, or dies, which force and bend them into a series of parallel waves or corrugations. These enormously increase the stiffness and strength of the sheets, and adapt them for several purposes for which the flat sheets would be too weak.

The sheets must be of good quality to stand the process, or they will crack.

The sheets are in sizes, generally about 6 feet by 3 feet 2 inches, or 8 feet by 3 feet 2 inches, before corrugation ; with corrugations 5 inches apart, which reduce the width from 3 feet 2 inches to 2 feet 6 inches.

The thicknesses and weights are as follows :—

Birmingham Wire Gauge. B.W.G.	Thickness in inches.	Weight per square in lbs.	Width of Flutes.	Uses.
No. 16 . . .	·065	380	5 in.	Where great strength is required.
,, 17 . . .	·056	320	,,	⎫
,, 18 . . .	·049	280	,,	⎬ For first-class work generally.
,, 19 . . .	·042	252	,,	⎭
,, 20 . . .	·035	224	3 in.	
,, 21 . . .	·032	205	,,	
,, 22 . . .	·020	185	,,	
,, 23 . . .	·025	165	,,	⎫
,, 24 . . .	·022	150	,,	⎬ Sent to Colonies.
,, 26 . . .	·020	112	,,	⎭

" The flutes or corrugations are made of various widths, those most usual in England being 3, 4, and 5 inch. Sheets with 5 inch flutes are commonly preferred by engineers. The depth D is generally ¼ of the width A, and the proportions can only be modified in the manufacture by making special new dies. Sheets with flutes wider than 5 inch are occasionally used when great strength is required, but in such cases the thinner gauges of iron should not be employed." [1]

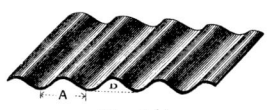

Fig. 144.

The ordinary form of corrugation is shown in Fig. 144, the sheets, when used for roof-covering, being laid with the corrugations parallel to the slope. A special form, shown in Fig. 145, is sometimes made for sheets intended to be laid with the corrugations parallel to the ridge.

Corrugated iron is generally galvanised (see next paragraph).

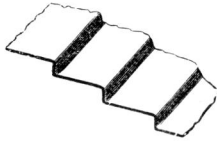

Fig. 145.

[1] Matheson's *Works in Iron.*

Galvanised Iron is iron covered with a coating of zinc by the process described at p. 333.

The quality of galvanised sheets depends upon the kind and thickness of the Iron, the purity of the zinc, and the care with which the process has been conducted.

Hoop Iron is made of from $\frac{1}{2}''$ to 6" wide, and from 12G to 23G.

It is not much use in building, except as an additional bond in brickwork, for which purpose it is generally about $1\frac{1}{2}$ inch wide and of No. 16 BG, tarred and sanded, and laid as described in Chap. iv., Part I.

Mitis Wrought Iron Castings are produced by Mr. Nordenfeldt's patent process. They are made from scrap iron containing a very small quantity of carbon. This is melted in crucibles, and has to be raised to a very great heat produced by the use of naptha as fuel. The molten metal is poured into the moulds from ladles in which the contents are kept hot by means of a surface blast. The best castings are made from the highest class of irons, such as Swedish or Best Yorkshire. Raw material containing as much as $\frac{1}{4}$ per cent of phosphorous makes the castings too brittle. These castings may be readily run into any form required, and the inventor claims for them that their strength is 20 per cent greater than that of the best forgings.

A specimen mentioned by Mr. Warren[1] in his paper on cast steel gave the following results :

Tensile test, 28 tons per square inch.
Elongation in 2 inches, 12·8 per cent.

RELATIVE VALUE OF DIFFERENT FORMS OF MILD STEEL AND WROUGHT IRON.

The price of iron of all kinds fluctuates continually according to the state of the market. These Notes do not profess to deal with the cost of materials ; and the following remarks are intended only as a general guide to the student as to the relative value of various sections in iron, and the limits of dimension or weight (differing somewhat in different localities) beyond which extras are usually charged over and above the normal market rate for what is called the " basis " size of the section under consideration.

Extras are usually charged when the section for :

Rounds is below $\frac{1}{2}$ in. diameter, or above 3 in. diameter.

Squares is below $\frac{1}{2}$ in. side, or above 3 in. side.

Half-rounds and Convex is below 1 in. wide, or less than $\frac{1}{2}$ in. thick.

Flats is below 1 in. wide, or above 6 in. wide.

Angles exceeds 8 to 9 united inches, or 30 ft. in length.

Tees exceeds 8 united inches, 25 to 30 ft. in length.

Plates exceeds 5 cwt. in weight, or 4 ft. in width, or 56 ft. in area.

[1] In paper read before the Institute of Naval Architects, 21st May 1886.

Angles are somewhat more expensive than bars, and tees than angles.

Cutting to exact lengths is usually an extra, as also the shaping of plates to forms other than rectangular, or having other than a regular taper.

Economy in the cost of structural iron or steel work will also usually be found to accompany the repetition of forms as far as possible, and the avoidance of small differences in sections (arising from over-refinements in calculation) or of an excess of riveting, the avoidance of closed cells or positions difficult of access, irregular shapes, and details involving expensive smith-work.

With respect to the selection of sections in the design of structural steelwork, the draughtsman will in future do well to confine himself to those included in the Lists of British Standard Sections issued by the Engineering Standards Committee. These Lists include Equal Angles, Unequal Angles, Bulb Angles, Bulb Tees, Bulb Plates, Z Bars, Channels, Beams and T Bars, based on a system of standard dimensions, thicknesses, and profiles.

These Lists are further accompanied by Tables of the Mechanical Elements of the Sections, including their moments of Inertia and Resistance, Radii of Gyration, etc., for purposes of calculation in design. To these Lists the attention of the student is directed.

BRANDS ON IRON.

Pig-Iron Brands.

The different pig-irons in the market are distinguished by brands, which indicate the locality from which the iron was procured.

The brands, which are in raised letters on the pig, serve as a guide to the quality and place of manufacture. The brands are sometimes the initials of the manufacturer, but more commonly refer to the place of production. Thus the brands Blaenavon, Gartsherrie, Weardale, indicate at once to the initiated that the first is a Welsh, the second a Scotch, and the third a North of England pig-iron.

G.M.B. are letters often quoted in price-lists, etc.; they do not refer to any particular locality, but stand for " Good Marketable Brands."

As already stated, the engineer has but little personal concern with the peculiarities and shades of difference between the brands of the same or different districts. It is sufficient for him to specify that the finished iron he requires must stand certain tests, and the selection of ores, pig-iron, and other raw material, required to effect this is better left to the manufacturers.

It may be useful, however, to mention a very few of the principal brands in each district, with their characteristics. The classification of pigs by numbers has been described at p. 265.

YORKSHIRE BRANDS.—There are several ordinary descriptions. The best works, such as *Lowmoor, Bowling, Farnley, etc.*, use the local ores mixed with other kinds.

SCOTCH BRANDS, such as *Calder, Clyde, Govan, Carnbroe, Carron, Gartsherrie, Langloan, Coltness.* This pig-iron, chiefly from clay ironstone, and used largely for foundry purposes ; sometimes mixed with N. England pig-iron to improve its strength.

NORTH OF ENGLAND BRANDS, such as *Acklam, Ridsdale, Southbank, Weardale*, etc., are chiefly from ores of the Carboniferous system, tougher and stronger than Scotch, and used chiefly for forge purposes. *Cleveland Iron* is from ore of the Lias formation.

WELSH BRANDS, such as *Blaenavon, Gadlys, Ystalyfera, Pentyrch*, are from good but lean ores, generally mixed with Spanish, Cumberland, or other hæmatite ores, and used in the district for rails.

Blaenavon is a cold-blast pig used for engine cylinders and other special purposes.

HÆMATITE PIG-IRON—*Askam, Barrow, Workington, Cleator, Carnforth, Harrington*—is made from the rich ores of Cumberland. Is used largely by steel, tin-plate, and sheet-iron manufacturers. A special quality—" *Bessemer Pig* " (see p. 265)—is made for the Bessemer steel workers.

NORTHAMPTONSHIRE PIG-IRON is made from poor ore, but useful to mix with others.

SHROPSHIRE BRANDS are *Lilleshall, Madeley Wood, Old Park*, etc.

STAFFORDSHIRE PIG-IRON differs very much in quality, and is mostly used in the district. Much of the Staffordshire iron is made from ores from other counties.

Wrought Iron Brands.

In order to understand the different qualities of British wrought iron in the market, the relative cost of the different forms, and to form some idea of the brands by which they are distinguished, it will be well to examine the Lists issued from time to time by manufacturers and merchants.

It will be seen from such lists that the prices of iron vary according to the shapes in which it is manufactured, according to the locality it comes from, and also with its quality, *i.e.* best, best best, or treble best.

Before proceeding further it should be mentioned that all good iron has some sort of mark upon it to indicate where it came from, although it does not follow that all marked iron is good ; an unmarked iron may be suspected to be bad ; the maker is probably ashamed of it.

In most cases the only sure way of ascertaining the quality of a piece of iron is to test it as described at p. 278 ; but as some engineers still specify certain brands in order to secure a good quality of iron, it will be as well to know what some of those brands mean.

It may be stated with regard to brands generally, that, though they differ in detail, they frequently consist of the maker's initials, name, or device stamped upon the bar or plate near its end, immediately after which is stamped either *best*, *best best*, or *best best best*, to indicate the quality ; the letters B, BB, or BBB are often used instead of the words. The crown which is introduced in many brands has no special signification ; indeed, several crowns have been known on very bad iron.

Some large exporters use their own marks for particular markets. Thus Messrs. Bolling and Lowe sell certain irons branded Bird with a horseshoe, crown, or similar device, or with stars so placed that they can always trace the works whence the iron came.

Staffordshire Brands.—It will be seen that these irons are divided into three classes, which are (putting them in order of price) *list brands*, *good marked iron*, and *common iron*.

LIST BRANDS are those used by some of the oldest established manufacturers, known as the *list makers*, who having years ago obtained a good character for their iron, and thereby secured a connection for their productions, are able, as a rule, to dispose of it and to fix their own prices. Of late years many of the old Staffordshire firms have died out, and other districts are coming to the front as making special classes of iron suitable for special purposes. Many engineers however, who have heard of the good qualities of the iron in former days, or are unwilling to take the responsibility of a change, still specify the brands to which they have been accustomed, though they pay a higher price for doing so.

As specimens of *list brands*, the following are given :—[1]

The iron from the Earl of Dudley's Round Oak Works is marked

and also best, double best, treble best, according to quality.

---

[1] It will be understood that a few of the brands of each sort are given as specimens ; it is not intended to assert that they are the best in the market.

is the brand used by Messrs. Bradley of Stourbridge, known by engineers as " SC Crown Iron."

BBH is the mark used by Messrs. W. Barrow and Sons, of South Staffordshire.

I B stands for John Bagnall and Sons (Limited), Staffordshire.

is the brand of the New British Iron Company.

 the brand of various makers and firms with or without their initials.

 is the brand of Messsrs. Philip Williams and Sons, Wednesbury Oak Works, known as " Mitre Iron."

 are brands used by Messrs. Brown and Freer (late Hunt and Brown), Leys Ironworks near Stourbridge.

is the brand on galvanised iron of Messrs. William Lee and Company, Gospel Oak Works.

GOOD MARKED IRON is that produced by many manufacturers of repute, who do not belong to the *list makers;* their brands are well known, but not so long established. Their iron can be purchased at 5s. to 15s. a ton cheaper than the *list brands,* though it may be of exactly the same quality when gauged by the result of tests. They frequently brand iron with their names, initials, or devices, and with the usual additions to denote quality—such as *best, best best, best best best.* A good deal of this iron is made in North Staffordshire.

Uses.—Of the qualities of iron above mentioned, BBB is used for rivets and chains. BB iron is used for bridges, roofs, and similar important structures.

B iron, or iron of equal quality, is generally specified for other ordinary structures in iron.

Under this head the following brands are given as specimens :—

SHELTON used by the Shelton Iron, Steel, and Coal Company, Stoke.

GRANVILLE   do.   do.

KIRKSTALL.   Kirkstall Forge Company, Kirkstall, Leeds.

 Kinnersley and Company, Clough Hall Iron Works, Kidsgrove.

T K

   by the Birchills Hall Iron Company, Walsall.

B.I.C.
SWAN
SPC   are other brands used by the same Company.
BILSTON

 Samuel Groucutt and Sons, Bilston.

SC&S

SPARROW XXX.   W. and J. S. Sparrow and Sons, Bilston.

COMMON IRON is made by manufacturers who, being slovenly and careless in manufacture, or having but a small range of sizes, or being known to buy common pig-iron and raw material, or having been found unreliable or wanting in uniformity of quality, etc., have to submit to lower rates.   Common iron is, or should be, used only for unimportant work which requires but little heating or forging to bring it into the shape required.

Midland and other Districts.—The remarks upon the iron manufactured by the less known makers in Staffordshire apply also to the qualities of iron produced in these districts.

[crown brand image] is the brand used by the Midland Iron Company (Limited), Rotherham.
MIC

PARKGATE ♛ ♛ is used by the Parkgate Iron Company, Rotherham.

♛ is the brand of the Pearson and Knowles Coal and Iron Company
B F (Limited), Warrington, for bars and plates.

♛ is the brand of the same company for sheets and hoops.
WIW

    North of England.—The principal make in this locality is in rails, ship plates, angles, and T irons for shipbuilding, also common boiler plates. The better varieties of this iron are marked with a crown.

    Well-known brands are—

 used by Bolckow, Vaughan, and Company (Limited), Middlesborough, Cleveland, and Wilton Park Works.

 by Jno. Abbot and Company (Limited), Gateshead-on-Tyne.

PALMERS. ♛ Palmer's Shipbuilding and Iron Company (Limited), Jarrow.

WEARDALE. Weardale Iron and Coal Company, Spennymoor.

    " CONSETT," by the Consett Iron Company, Newcastle-on-Tyne.

    Yorkshire.—The term " best Yorkshire qualities " is generally understood to refer to iron made by the following works or manufacturers :—Lowmoor Works and Bowling Works, near Bradford, Farnley Works, Taylor Brothers, S. C. Cooper's and Monkbridge Works—the last four in or near Leeds. These makers owe their reputation to the good ore and coal used by them, and, above all, to the great care with which their iron is made.

    Every process is carefully watched. The different qualities and textures of iron produced at the various stages of manufacture are carefully selected, sorted, and blended together, so that the resulting iron as turned out is thoroughly homogeneous and uniform in quality, and may be relied upon as equally good throughout.

    Iron of this class is generally branded with the name of the place where it is produced in full : thus—LOWMOOR ; but it has no marks relating to quality, because it is all of the very best description.

    The above-mentioned works make a good deal of bar iron, but their chief manufacture is in plates for boilers, and railway carriage and waggon tyres, locomotive axles, armour bolts, etc.

    Their iron costs about double the price of ordinary brands, principally because it bears such a high character that it is generally specified for parts of structures which have to be subjected to great heat, changes of temperature, or sudden shocks, and which demand a material of great ductility and trustworthiness.

    Wales.—The manufacture of bar iron and plates in this locality had of late years nearly ceased, the chief business carried on being in making rails. The best known brands are WC (Crawshays), GL (Dowlais Company), JBS and LIC (Llynvi Iron Company, Limited), TW and Company, BJ and Company, Clydach.

Scotland.—Manufactured iron brands of ordinary quality are "Coats" (short for "Coatbridge.") Somewhat better brands are Glasgow and Monkland.

Swedish Iron is marked, and the brands are noted and classified in an official book. Examples of them will be seen in Percy's *Metallurgy*.

STEEL.

Steel has been defined by Dr. Percy as "iron containing a small percentage of carbon, the alloy having the property of taking a temper; and this definition is substantially equivalent to those found in the works of Karsten, Wedding, Grüner, and Tunner."[1]

On this point, however, there are many opinions, some of which will presently be briefly referred to.

The amount of carbon in steel, as used in engineering, varies from about ·10 for very soft to 1·5 per cent for very hard steels.

It contains, therefore, less carbon than cast iron, but more than wrought iron.

Practically, steel often contains other substances besides iron and carbon.

These substances are generally got rid of, as far as possible, in the process of manufacture. When, however, they remain in the steel, they influence its characteristics in the manner described at p. 264.

In consequence of the practical existence of these impurities, and for other reasons, it is difficult to give an exact definition of steel.

Several definitions have been proposed, some depending upon the chemical composition, and some upon the physical characteristics of the material. None, however, has at present been universally accepted.[2]

M. Adolphe Grenier, of Seraing, has classified the irons and steels, according to the proportion of carbon they contain, in the following manner :—[2]

Percentage of carbon.	0 to 0·15.	0·15 to 0·45.	0·45 to 0·55.	0·55 to 1·50, or more.
Series of the irons.	Ordinary irons.	Granular irons.	Steely irons or puddled steels.	Cemented steels. Styrian steel.
Series of the steels.	Extra soft steel.	Soft steel.	Half soft steels.	Hard steels.

Sir Joseph Whitworth has pointed out that a definition based upon chemical composition is unsatisfactory. He proposes to do away with all distinctive names such as

---

[1] Dr. Siemens' Address, Iron and Steel Institute, 1877.

[2] *Journal Iron and Steel Institute*, 1873.

blister steel, shear steel, cast steel, etc., and "to express what is wanted to be known by two numbers which should represent tensile strength and ductility. . . . He would suggest that the limit of tensile strength be taken at about 18 tons per square inch, so that the metal exceeding this strength should be called 'steel,' while any description of iron falling below this limit of tensile strength should be known as 'wrought iron.'" [1]

An International Committee sitting at Philadelphia in 1876 recommended a somewhat elaborate nomenclature for different descriptions of iron and steel. Dr. Siemens, alluding to this, says—

"Difficulties . . . have hitherto prevented the adoption of any of the proposed nomenclatures, and have decided engineers and manufacturers in the meantime to include under the general denomination of cast steel *all compounds consisting chiefly of iron which have been produced through fusion and are malleable.* Such a general definition does not exclude from the denomination of steel materials that may not have been produced by fusion, and which may be capable of tempering, such as shear steel, blister steel, and puddled steel ; nor does it interfere with distinctions between cast steels produced by different methods, such as pot steel, Bessemer steel, or steel by fusion on the open hearth." [2]

In a paper on "Steel for Structures" Mr. Matheson said, "Steel for the purposes of the present paper is any variety of iron or alloy of iron which is cast while in the liquid state into a malleable ingot, and to go further, which will when rolled in a plate or bar, endure from 26 to 40 tons before fracture." [3]

To the engineer some practical definition which would enable him to know exactly what material he would receive under a certain specification would be of great value.

In whatever way steel may be defined, it is of the utmost importance that the characteristic differences between it and iron, both cast and wrought, should be clearly understood. Some of these will now be pointed out.

Characteristics of Steel.— HARDENING. — The characteristic difference between steel and pure wrought iron is as follows :—

When steel is raised to a red heat and then suddenly cooled, it becomes hard and brittle. This process, which is known as *hardening,* has no effect upon pure wrought iron.

TEMPERING is a characteristic of steel which distinguishes it from cast iron. If steel has been hardened by being heated and suddenly cooled, as above described, it may be softened again by applying a lower degree of heat and again cooling. This is known as *tempering.*

Cast iron, on the contrary, though it is hardened by the first process, cannot be softened by the second.

When a bar of steel is struck it gives out a sharp *metallic ring,* quite different from the sound produced by striking wrought iron.

Other characteristics of steel are its great *elasticity* and its *retention of magnetism.*

Amount of Carbon in Steel.—It has already been stated that the peculiarities of cast iron, wrought iron, and steel are caused by the difference in the amounts of carbon which they respectively contain.

Pure wrought iron contains no carbon. The wrought iron of commerce contains a minute quantity, steel contains more, while the largest percentage is found in the softer kinds of grey cast iron.

---

[1] *Proceedings Mechanical Engineers,* 1875.
[2] Dr. Siemens' Address, Iron and Steel Institute, 1877.
[3] *M.I.C.E.,* vol. lxix. p. 1.

The transition from one class to the other is so gradual and insensible that it is difficult to say where one ends and the other begins, but the following remarks bear with them the high authority of Dr. Percy.

" When the carbon reaches ·5 per cent and other foreign matters are present in small quantity, iron is capable of being hardened sufficiently to give sparks with flint, and may then be regarded as steel. But in the case of iron perfectly free from foreign matters, not less than ·65 of carbon is required to induce this property.

" Iron containing from 1·0 to 1·5 per cent is steel, which, after hardening, acquires the maximum hardness combined with the maximum tenacity.

" When the carbon exceeds the highest of these limits still greater hardness may be obtained, but only at the expense of tenacity and weldability.

" When the carbon rises to 1·9 per cent or more, the metal ceases to be malleable while hot ; and 2 per cent of carbon appears to be the limit between steel and cast iron, when the metal in the softened state can no longer be drawn out without cracking and breaking to pieces under the hammer." [1]

As a general rule it may be said that the varieties of steel containing the larger proportions of carbon are harder, stronger, more brittle, and more easily melted. Those containing less carbon are tougher, more easily welded and forged, but are weaker as regards tenacity. [2]

VARIETIES OF STEEL.

Methods of Making Steel.—Steel may be produced either by adding carbon to wrought iron, or by partially refining pig-iron, thus removing a portion of its carbon until the proper amount only remains.

There are several ways in which these processes may be carried out, the result being that there are several descriptions of steel in the market. Of these, however, only a few of the most important can here be described.

Blister Steel is produced by placing bars of the purest wrought iron in a furnace between layers of charcoal powder, and subjecting them to a high temperature for a period varying from five to fourteen days, according to the quality of steel required.

This process is called *cementation*.

Swedish iron is generally used for the purpose, that marked Ⓛ , from the Dannemora mines, being the best.

The steel differs greatly from the bar iron from which it was produced.

Its distinctive name is derived from the appearance of its surface, which is covered with blisters due to the evolution of carbonic oxide.

The bars are now brittle, the fracture is of a reddish or yellowish tinge, with but little lustre.

The structure is no longer fibrous but crystalline ; " the finer the grain and

---

[1] Percy's *Metallurgy.*   [2] Rankine.

the darker the colour, the more highly carbonised, or harder, will be the steel produced."

" When the blisters are small and tolerably regularly distributed the steel is of good quality ; but when large, and only occurring along particular lines, they may be considered as indicative of defective composition, or want of homogeneity in the iron employed." [1]

Uses.—Blister steel is full of fissures and cavities, which render it unfit for forging except for a few rough purposes. It is used for welding to iron for certain parts of machines, for facing hammers and steeling masons' points, etc., but not for edge tools. Most of the blister steel made is used for conversion into other descriptions of steel.

Spring Steel is blister steel heated to an orange red colour, and rolled or hammered.

Shear Steel, sometimes called *Tilted steel.*—By the process of cementation just described, the exterior only of the bars is carbonised. To produce steel of uniform quality throughout its mass, bars of blister steel are cut into short lengths ; these are piled into bundles or faggots, sprinkled with sand and borax, placed at a welding heat under a tilt hammer, which by rapid blows removes the blisters, closes the seams, beats and amalgamates the faggots into a bar of *single shear steel.*

In order still further to improve the quality of the metal, this bar is doubled or faggoted, and again subjected to the processes of hammering and rolling, the result being a bar of *double shear steel.*

The oftener the processes of faggoting and hammering are repeated, the more uniform is the resulting steel, but at the same time it loses carbon during these operations, and therefore becomes softer.

Characteristics.—The processes to which the steel has been subjected restore its fibrous character. It is still weldable, is more malleable, and tougher, is close grained, and capable of receiving a finer edge and higher polish than blister steel.

Uses.—Shear steel can be forged into such tools as are required to be tough without extreme hardness, such as large knives, scythes, plane irons, shears, etc., and it is useful for such instruments as are composed of iron and steel.

Cast Steel.—There are several ways of producing cast steel, some of which will now be mentioned.

The ingots produced by any of these processes generally contain cavities. In order to get rid of these, they are reheated at a low temperature and hammered into bars, being increased in length and reduced in section, by which they are made compact, solid, and homogeneous.

" The appearance of the fractured surface of ingots of cast steel varies with their hardness or relative proportion of carbon. The softer kinds are bright and finely granular. The harder qualities often show crystalline plates of a certain size, arranged in parallel stripes or columns at right angles to the surface of the mould, so that in a square ingot the columns intersect, forming a cross." [2]

CRUCIBLE CAST STEEL may be made by melting fragments of blister steel in covered fireclay crucibles, and running the metal into iron moulds. This process was originally introduced by Huntsman of Sheffield.

Most crucible steel, however, is now made direct from bars of the best wrought iron (often Swedish iron produced from pure magnetic ores). The bars are broken

---

[1] Bauermann's *Metallurgy.*   [2] Bauermann.

into lengths and placed in the crucibles together with a small quantity of char coal, the amount varying according to the temper of the steel to be produced. Spiegeleisen (see below), or oxide of manganese is subsequently added.

Characteristics.—Cast steel is the strongest and most uniform steel that is made. It is much denser and harder than shear steel, but requires more skill in forging.

Cast steel made in this way should never be raised beyond a red heat, or it will become brittle, so that it cannot easily be forged. It is unweldable, for it will fly to pieces when struck by the hammer.

In making tools, after forging, the cutting edge should be well hammered down, so as to close the pores or grain of the metal.

The fracture of cast steel should have a slaty-grey tint almost without lustre, the crystals being so fine that they are hardly distinguishable.

Uses.—It is used for the finest cutlery, for cutting tools composed of steel only, especially those in which great hardness is required.

Heath's Process is an improvement on the method just described, and consists in adding to the molten metal a small quantity of carburet of manganese.

"After this addition the cast steel possesses much more tenacity at a high temperature, and can be welded either to itself or to wrought iron, so that it may be employed for the fabrication of many implements which were formerly obliged to be made of shear steel. Thus the blades of table knives can be made of cast steel, welded on to an iron *tang*, as that part of the knife is called which is fixed into the handle." [1]

Heaton's Process consists in adding nitrate of soda to molten pig-iron, thus removing most of the carbon and silicon.

Mushet's Process.—In this malleable iron is melted in crucibles with oxide of manganese and charcoal.

With the preceding methods of manufacture the designer of constructional steelwork is not so much concerned (except so far as regards the production of tools and implements) as with those processes, next to be described, by which the great bulk of Mild Steel or "ingot iron" is produced, now used so largely in the construction of bridgework, railway material, shipbuilding, and builders' ironwork, and which has practically displaced wrought iron in these and other branches of construction.

BESSEMER PROCESS.—By this process steel is made from pig-iron. The whole of the carbon is first removed so as to leave pure wrought iron, and to this is added the precise quantity of carbon required for the steel.

The pig-iron used should be dark grey, containing a large proportion of free carbon, and a small percentage of silicon and manganese. It should be almost free from sulphur and phosphorus.

The pigs are melted in a cupola,[2] and run into a *converter*, which is a large pear-shaped iron vessel hung on hollow trunnions, and lined with firebrick, fireclay, or "*ganister.*" [3]

A blast of air is then blown through the metal in the converter for about twenty minutes.

---

[1] Bloxam's *Metals.*   [2] In some works the melted metal is carried direct from the blast furnace to the converter.

[3] A sandstone from the coal measures much used in a powdered state for this and similar purposes.

This removes all the carbon, after which from 5 to 10 per cent of spiegel-eisen [1] (a variety of cast iron rich in carbon and manganese) is added.

The blowing may then be resumed for a short time, in order to thoroughly incorporate the two metals, the steel is run off into a ladle, and thence into moulds.

The colour of the flame issuing from the mouth of the converter indicates the moment at which all the carbon has been removed, or this may be accurately ascertained by examining the flame with a spectroscope.

The ingots produced contain air-holes, and are not sufficiently dense. They are therefore kept hot and rendered compact by the blows of a steam hammer, after which they may be rolled or worked as required for the purpose for which they are intended.

Uses.—Bessemer steel is used largely for rails, axles, and tyres for the wheels of railway carriages, also for common cutlery and tools, such as hatchets, hammers, etc.

It is sometimes used for the members of roofs and trussed bridges, also for the expansion rollers of such structures, and for boiler plates.

THE BASIC PROCESS, by Messrs. Thomas and Gilchrist, resembles the Bessemer process, but that the converters into which the fluid pig-iron is run are lined with basic material, generally magnesian limestone or some refractory substance as free as possible from silica. By this process the less pure ores of the Cleveland district, containing a large proportion of phosphorous, can be converted into steel. Lime having been added, the blowing in the converter commences, the silicon passes off first, then the carbon, and then the phosphorous. When the operation is nearly completed, a small sample ingot is cast, cooled, and broken, and by the fracture the amount of phosphorous still remaining is estimated.

SIEMENS' PROCESS.—In this process pig-iron and ore are the ingredients employed to produce steel by fusion upon the open hearth of a regenerative gas furnace. The pig metal is first melted upon the hearth of the furnace, and after having been raised to a steel-melting temperature, rich and pure ore (such as Mokta ore [2]) and limestone are added gradually, whereby a reaction is established between the oxygen of the ferrous oxide and the carbon and silicon contained in the metal. The silicon is thus converted into silicic acid, which with the lime forms a fusible slag, whereas the carbon in combining with oxygen escapes as carbonic acid, causing a powerful ebullition in the bath.

Modification of Siemens' Process.—According to another modification of the process, the iron ore is treated in a separate rotatory furnace with carbonaceous material, and converted into balls of malleable iron, which are transferred in the heated condition from the rotatory to the bath of the steel-melting furnace. This latter process is suitable for the production of steel of very high quality, because the impurities, such as sulphur and phosphorous, in the ore are separated from the metal in the rotatory furnace.[3]

SIEMENS-MARTIN PROCESS.—In another important modification of the same process, which is known generally as the *Siemens-Martin Process*, a bath of highly-heated pig metal is prepared in the furnace, and three or four times

---

[1] Mirror-iron ; so-called from its shining appearance.   [2] From Algeria.

[3] See paper by Dr. C. William Siemens, "Some further Remarks regarding Production of Iron and Steel by direct process," read at Newcastle meeting of Iron and Steel Institute, September 1877.

its weight of scrap-iron or steel is gradually added (preferably in a highly heated condition) and dissolved in the fluid bath.

Towards the end of these various operations, samples are taken from the bath in order to ascertain the percentage of carbon still remaining in the metal, and ore is added in small quantities to reduce the carbon to about $\frac{1}{10}$th per cent. At this stage of the process the furnace contains from 6 to 12 tons of fluid malleable iron, to which siliceous iron, spiegeleisen, or ferromanganese is added in such proportions as to produce steel of the required degree of hardness. The metal is thereupon discharged, either by tapping into a ladle, or more generally directly into ingot moulds by the ascensional process.

Uses.—This material is now very extensively used for shipbuilding, rails, tyres, boilers, forgings for engines, tin-plates, etc., as well as rolled joists, girders, trough decking for floors, and constructional work generally.

WHITWORTH'S COMPRESSED STEEL.—It has already been stated that ordinary steel, as first cast, is porous, full of small cavities, which have to be removed by hammering before a sound metal is produced.

In order to remedy this evil, Sir Joseph Whitworth subjects the molten steel to a pressure of some six tons per square inch, by which all cavities are closed up, the gases contained in them driven out, the metal is compressed to about $\frac{7}{8}$ of its bulk, its density and strength being greatly increased.

Sir J. Whitworth gives the steel a maximum ductility of about 30 per cent. He considers that more is unnecessary, "for cylinders of such metal do not fly into pieces when hurt, but simply open out or tear like paper, and a metal of greater ductility could not be required for structural purposes."[1] The strength and ductility of the different varieties is given at p. 321.

Puddled Steel is produced by stopping the puddling process used in the manufacture of wrought iron before all the carbon has been removed.

The small amount of carbon that is left, *i.e.* from ·3 to 1·0 per cent, is sufficient to form an inferior steel.

It is used chiefly for making inferior boiler plates and plates for shipbuilding.

A similar product resulting from imperfect refining is known as *Natural Steel* or *German Steel*.

Mild Steel or "Ingot Iron" contains from ·10 per cent, or even less, to say ·40 per cent of carbon. When more carbon is present it may be considered as *Hard Steel* relatively to the softer grades used for constructional purposes.

Mild steel is stronger and more uniform in texture than hard wrought iron, and superior to it in nearly every way.

The practical applications of mild steel, as prepared by the Bessemer and Siemens-Martin processes, have already been alluded to.

Tungsten, Manganese, and Chromium (or chrome) Steels are made by adding a small percentage of the metals named to crucible steel ; the result in each case being a steel of great hardness and tenacity, suitable for drills and other special tools.[2]

---

[1] *Proceedings Inst. Mech. Engineers,* 1875.

[2] Steel has lately been made containing 13·75 per cent of Manganese, and having a tensile strength of 60 tons per square inch, combined with 50 per cent elongation. It bids fair to become an important material.

Homogeneous Metal is a name that was formerly given to a variety of cast steel containing about ·25 per cent of carbon.

This material "welds with facility, and, with proper precautions, may be joined to iron or steel at a very high welding heat." [1]

It is used for rifleproof shutters, guns, etc.

Hardening Steel.—It has already been mentioned that steel plunged into cold water when it is itself at a red heat becomes excessively hard. The more suddenly the heat is extracted the harder it will be.

This process of *hardening*, however, makes the steel very brittle, and in order to make it tough enough for most purposes it has to be *tempered*.

Tempering.—The process of tempering depends upon another characteristic of steel, which is that if (after hardening) the steel be reheated, as the heat increases, the hardness diminishes.

In order then to produce steel of a certain degree of toughness (without the extreme hardness which causes brittleness), it is gradually reheated, and then cooled when it arrives at that temperature which experience has shown will produce the limited degree of hardness required.

Heated steel becomes covered with a thin film of oxidation, which becomes thicker and changes in colour as the temperature rises. The colour of this film is therefore an indication of the temperature of the steel upon which it appears.

Advantage is taken of this change of colour in the process of tempering, which for ordinary masons' tools is conducted as follows :—

Tempering Masons' Tools.—The workman places the point or cutting-end of the tool in the fire till it is of a bright red heat, then hardens it by dipping the end of the tool suddenly into cold water. He then immediately withdraws the tool and cleans off the scale from the point by rubbing it on the stone hearth. He watches it while the heat in the body of the tool returns, by conduction, to the point. The point thus becomes gradually reheated, and at last he sees that colour appear which he knows by experience to be an indication that the steel has arrived at the temperature at which it should again be dipped. He then plunges the tool suddenly and entirely into cold water, and moves it about till the heat has all been extracted by the water.

It is important that considerable motion should be given to the surface of the water while the tool is plunged in, after tempering, otherwise there will be a sharp straight line of demarcation between the hardened part and the remainder of the tool, and the metal will be liable to snap at this point.

Tempering very small Tools.—In very small tools there is not sufficient bulk to retain the heat necessary for conduction to the point after it has been dipped. Such tools, therefore, are heated, quenched, rubbed bright, and laid upon a hot plate to bring them to the required temperature and colour before being finally quenched.

In some cases the articles so heated are allowed to cool slowly in the air, or still more gradually in sand, ashes, or powdered charcoal. The effect of cooling slowly is to produce a softer degree of temper.

Table of Temperatures and Colours.—The following Table shows the temperature at which the steel should be suddenly cooled in order to produce the hardness required for different descriptions of tools. It also shows the colours which indicate that the required temperature has been reached.

---

[1] Mushet on *Iron and Steel.*

Colour of Film.	Temp. Deg. Fahr.	Nature of Tool.
Very pale straw yellow .	430	Lancets and tools for metal.
A shade of darker yellow .	440	Razors and do.
Darker straw colour .	470	Penknives.
Still darker straw yellow .	490	Cold chisels for cutting iron, tools for wood.
Brownish yellow . .	500	Hatchets, plane irons, pocket knives, chipping chisels, saws, etc.
Yellow tinged with purple	520	
		Do. do. and tools for working granite.
Light purple . . .	530	Swords, watch springs, tools for cutting sand- stone.
Dark purple . . .	550	
Dark blue . . .	570	Small saws.
Pale blue . . .	600	Large saws, pit and hand saws.
Paler blue with tinge of green	630	Too soft for steel instruments.

The tempering colour is sometimes allowed to remain, as in watch springs, but is generally removed by the subsequent processes of grinding and polishing.

A blue colour is sometimes produced on the surface of steel articles by exposing them to the air on hot sand. By this operation a thin film of oxide of iron is formed over the surface, which gives the colour required.

Steel articles are often varnished in such a way as to give them an appearance of having retained the tempering colours.

The exact tempering heat required to produce the same degree of hardness varies with different kinds of steel, and is arrived at by experience. It would be impossible to go very fully into the subject in these Notes. The above remarks will give some idea of the process, and the effects produced upon the strength and ductility of steel by tempering in different ways is shown in the Table, page 323.

Different Methods of Heating.—There are several ways of heating steel articles both for hardening and tempering.

They may be heated in a hollow or in an open fire, exposed upon a hot plate, or in a dish with charcoal in an oven, or upon a gas stove.

Small articles may be heated by being placed within a nick in a red-hot bar.

If there is a large number of articles, and a uniform heat of high degree is required, they may be plunged into molten metal alloys, or oil raised to the temperature required.

Degree of Heat for Hardening.—In hardening steel care must be taken not to overheat the metal before dipping. In case of doubt it is better to heat it at too low than too high a temperature.

" The best kinds require only a low red heat. If cast steel be overheated it becomes brittle, and can never be restored to its original quality." [1]

If, however, the steel has not been thoroughly hardened it cannot be tempered. The hardness of the steel can be tested with a file.

The process of hardening often causes the steel to crack. The expansion of the inner particles caused by the heat is suddenly arrested by the crust formed in consequence of the cooling of the outer particles, and there is a tendency to burst the outer skin thus formed.

Cooling.—When the whole bulk of any article has to be tempered, it may either be dipped or allowed to cool in the air. " It matters not which way they become cold, providing the heat has not been too suddenly applied, for when the articles are removed from the heat they cannot become more heated, consequently the temper cannot become more reduced." But those tools in which a portion only is tempered, and in which the heat for tempering is sup-

---

[1] Edes, 80.

plied by conduction from other parts of the tool (as described at p. 305) "must be cooled in the water directly the cutting part attains the desired colour, otherwise the body of the tool will continue to supply heat and the cutting part will become too soft." [1]

HARDENING AND TEMPERING IN OIL.—When toughness and elasticity are required rather than extreme hardness, oil is used instead of water both for hardening and tempering, and the latter process is sometimes called *toughening*.

The steel plunged into the oil does not cool nearly so rapidly as it would in water. The oil takes up the heat less rapidly. The heated particles of oil cling more to the steel, and there is not so much decrease of temperature caused by vaporisation as there is in using water.

Sometimes the oil for tempering is raised to the heat suited to the degree of hardness required.

When a large number of articles have to be raised to the same temperature they are treated in this way.

Blazing.—Saws are hardened in oil, or in a mixture of oil with suet, wax, etc. They are then heated over a fire till the grease inflames. This is called being *blazed*. After blazing the saw is flattened while warm, and then ground.

Springs are treated in somewhat the same manner, and small tools after being hardened in water are coated with tallow, heated till the tallow begins to smoke, and then quenched in cold tallow. [2]

Annealing or *Softening Steel* is effected by raising hardened steel to a red heat and allowing it to cool gradually, the result of which is that it regains its original softness.

Case-Hardening is a process by which the surface of wrought iron is turned into steel, so that a hard exterior, to resist wear, is combined with the toughness of the iron in the exterior.

This is effected by placing the article to be case-hardened in an iron box full of bone dust or some other animal matter, and subjecting it to a red heat for a period varying from half an hour to eight hours, according to the depth of steel required.

The iron at the surface combines with a proportion of carbon, and is turned into steel to the depth of from $\frac{1}{16}$ to $\frac{3}{8}$ of an inch.

If the surface of the article is to be hardened all over, it is quenched in cold water upon removal from the furnace. If parts are to remain malleable it is allowed to cool down, the steeled surface of those parts removed, and the whole is then reheated and quenched, by which the portions on which the steel remains are hardened.

Gun-locks, keys, and other articles which require a hard surface, combined with toughness, are generally case-hardened.

A more rapid method of case-hardening is conducted as follows :—The article to be case-hardened is polished, raised to a red heat, sprinkled with finely powdered prussiate of potash. When this has become decomposed and disappeared, the metal is plunged into cold water and quenched.

The case-hardening in this case may be made local by a partial application of the salt.

Malleable castings (see p. 268) are sometimes case-hardened in order that they may take a polish.

To DISTINGUISH STEEL FROM IRON.—Steel may be distinguished from wrought iron by placing a drop of dilute nitric acid (about 1 acid to 4 water) upon the surface. If the metal be steel a dark grey stain will be produced, owing to the separation of the carbon. [3]

Tests for Steel.—Steel to be used in important work should be tested as to its strength, ductility, and other qualities. The

[1] Edes, 86.   [2] Holtzapffell.
[3] Bloxam's *Chemistry.*

methods of testing are similar to those adopted for wrought iron and described at p. 278.

Fractured Surface.—Many people think that they can judge of steel by the appearance of the fracture.

Mr. Kirkaldy found that "the conclusions respecting wrought iron are equally appropriate to steel, viz.—Whenever rupture occurs *slowly*, a silky fibrous, and when *suddenly*, a granular appearance, is invariably the result, both kinds varying in fineness according to quality.

"The surface in the latter case is even, and always at right angles with the length ; in the former, angular and irregular in outline.

"The colour is a light pearl gray, slightly varying in shade with the quality ; the granular fractures are almost entirely free of lustre, and, consequently, totally unlike the brilliant crystalline appearance of wrought iron.[1]

The appearance of the fracture is, however, at the best but a vague and uncertain guide, and, without great experience on the part of the observer, almost useless.

Trial.—A better test in the hands of a practical man is to heat the steel, and try it with regard to its tenacity, welding powers, and resistance to crushing when struck with a hammer upon a hard surface.

Tensile Tests.—The only certain test, however, for tensile strength and ductility is by direct experiment.

The tensile strength of steel may be tested in the same way as that of wrought iron.

The varieties of steel are, however, even more numerous than those of wrought iron, and their strength differs accordingly. Moreover, it is greatly influenced by the treatment to which the steel has been subjected.

ADMIRALTY TESTS FOR STEEL.—Plates for shipbuilding, bars, angles, angle-bulbs, tees. tee-bulbs or tee-bars, made by the open-hearth process, either acid or basic.

Tensile Test.—Strips cut lengthways (or for plates either lengthways or crossways, or in round bars a piece from the bar), "to have an ultimate tensile strength of not less than 26, and not exceeding 30 tons per square inch of section, with an elongation of 20 per cent in a length of 8 inches."

Forge Test.—"Such forge tests, both hot and cold, as may be sufficient in the opinion of the receiving officer to prove soundness of material and fitness for the service."

Tempering Test.—Strips cut lengthwise (or for plates either lengthwise or crosswise) "1½ inches wide, or in round bars a piece from the bar, heated uniformly to a low cherry red and cooled in water of 82° Fahr., must stand bending in a press to a curve of which the inner radius is one and a half times the thickness of the steel tested.

"The strips are all to be cut in a planing machine, and to have the sharp edges taken off.

"The ductility of every bar is to be ascertained by the application of one or both of these tests to the shearings, or by bending them cold by the hammer.

"The pieces cut for testing are to be of parallel width from end to end, or for at least 8 inches in length."

Percussive Test for Round Bars.—A specimen bar of 2″ diameter is taken, when required by the overseer, from every charge, or from every 50 bars or portions of 50, and subjected to a percussive test. The test for a bar of 2″ diameter should be the fall of 15 cwt. through 30 feet, or 20 cwts. through 22½ feet, whichever may be most convenient. Sample must stand at least one blow without injury, and the following facts must be noted.

   a. The number of blows to break the bar.
   b. The character of the fracture.
   c. The reduction in diameter after each blow.
   d. The reduction in sectional area at point of fracture.
   e. The elongation in 8 inches and in the inch containing the fracture.

Welding Tests for all Bars.—Sample pieces will be taken for testing the welding qualities of the steel, by welding two pieces together and bending it in the way of the weld when cold.

---

[1] Kirkaldy's *Experiments on Wrought Iron and Steel.*

LLOYD'S TESTS.—The steel plates used in ships to be classed in the register of Lloyd's Insurance Corporation have to stand a tensile stress of 28 to 32 tons per square inch, with 16 per cent elongation in 8 inches ; the angles and beams 28 to 33 tons, with 16 per cent elongation, and the same tempering tests as required by the Admiralty.

Board of Trade Tests for Boiler Steel.—Furnace plates, and plates exposed to flame 26 to 30 tons per square inch, with 20 per cent elongation in 10 inches. Shell plates and plates not exposed to flame 27 to 32 tons per square inch, with 20 per cent elongation in 10 inches. From every plate exposed to flame a shearing is taken, heated to a low cherry red, and cooled in water of 82° Fahr., to stand bending to a curve the inner radius of which equals $1\frac{1}{2}$ times the plate thickness. From plates not exposed to flame the test piece is bent cold to a curve of similar radius.

Bureau Veritas.—Ship plates, angles, etc., 27 to 32 tons per square inch, with 20 per cent elongation in 8 inches. Boiler plates, 26 to 30 tons per square inch. Elongation at 26 tons, 24 per cent, and at 30 tons, 20 per cent in 8 inches. Tempering tests as for Admiralty

British Corporation.—Ship plates and angles, 28 to 32 tons per square inch, with an elongation of at least 16 per cent in 8 inches. Boiler plates, 27 to 32 tons per square inch, with an elongation of 20 per cent in 8 inches, for shell plates and straps. Flanged or welded plates 25 to 30 tons, with 20 per cent elongation. Temper-bending tests as for Admiralty.

Test by Dead and Falling Loads. Steel rails are sometimes tested by dead loads, and frequently by a falling weight. The following extract is from a recent specification for a steel rail weighing 79 lbs. per yard, requiring both tests :—

" A length of 6 feet will be cut off from each sample rail and tested as follows :—

" *1st.* A piece will be placed in the position it would assume for traffic, on solid supports 3' 6" apart in the clear, and equidistant from the ends, and a weight of one ton will be allowed to fall freely upon the centre of the rail from a height of 12 feet 6 inches. The rail must bear two such blows without showing the least sign of fracture, and the permanent set caused by the first blow must not exceed 2 inches.

" *2d.* A piece of rail is to be placed on supports as before, and a weight of 18 tons is to be applied at the centre, when the deflection must not exceed $\frac{3}{16}$ of an inch."

The War Office percussive test for bolts is given at p. 285.

Steel for Bridges, Roofs, Girders, and builders' steelwork, when combined with sufficient ductility, may have a high elastic limit which will enable it to endure a high working stress. Good steel for such purposes can be obtained having an ultimate tensile strength of from 26 to 32 tons per square inch, a limit of elasticity of from 13 to 18 tons, and with 20 per cent elongation on the test specimen of 8 inches length. Such a steel would endure a working stress of from $6\frac{1}{2}$ to 8 tons to the inch in tension,[1] and may be expected to contain from about ·15 to ·25 per cent of carbon, when the other chemical constituents are contained in their usual proportions.

Recent specifications from the India Office for large steel bridges contain the following requirements :—

Steel Bars and Plates must weld perfectly, and not crack or crumble at all when hammered at a welding heat. The strips, 1 inch wide and 8 inches long, to have a tensile strength not less than 28 tons or more than 31 tons per square inch, an elongation of not less than 20 per cent, and a limit of elasticity of 15 tons per square inch. The same tempering tests as the Admiralty require, except that the radius of the curve to which the steel is bent is three inches instead of $1\frac{1}{2}$ inches.

Buckle Plates of Roadway to bear a concentrated load of 12 tons at centre without permanent set, and of 24 tons at centre without fracture.

Rivets.—Tensile strength 26 to 28 tons per square inch, in test pieces of 10 diameters, elongation not less than 25 per cent. A piece of bar heated to cherry red,

---

[1] In this country Board of Trade rules restrict the working stress on steel in bridges to $6\frac{1}{2}$ tons per square inch.

quenched in water of 82° Fahr., to bear being doubled quite close without injury. A piece heated to full red or orange, dropped into a hole in a cast iron block, so that 1½ to 2 diameters project, to bear having the end hammered out to a thin edge all round without showing signs of cracking.

A comparison of the specification for steel used in Bridge Construction by several of the leading railroads in the United States, shows the following outside limits within which the stipulated requirements are found to range for the several classes of steel enumerated.

	Ultimate Strength. Tons per sq. in.	Elastic Limit. Tons per sq. in.	Elongation in 8 inches per cent.	Reduction of Area per cent.
Tension Members . .	23 to 31	12 to 17	20 to 26	30 to 50
Rolled Beams . . .	25 to 29	13 to 15	20	30
Rivets	22 to 28	13 to 16	25 to 28	44 to 50
Compression Members .	24 to 39	15 to 19	18 to 25	30 to 50

As regards the process of manufacture some companies require the use of the open hearth process, while others do not restrict manufacturers to this method. A somewhat larger percentage of phosphorus is allowed in the acid process than in the basic, being about 0·08 per cent in the former, and from 0·04 to 0·06 in the latter. Sulphur from 0·04 to 0·06 per cent. The reaming of punched holes is a general but not universal requirement, some companies restricting this process to plates exceeding ⅝″ to ¾″ in thickness.

The Bending Tests range from a bend of 180° flat to a bend of 180°, in which D ranges from 1 to 3 times t, where D = diameter of curve of bend, and t = thickness of piece, the higher values of D applying to the harder qualities of steel. The drifting of holes ranges from 1¼ to 2 times the original diameter of the hole.[1]

MARKET FORMS, RELATIVE VALUE OF DIFFERENT KINDS, AND BRANDS ON STEEL.

MARKET FORMS.—Steel may be obtained in most of the forms adopted for wrought iron, and described at page 286. Angle and T of all sizes up to 4 inches × 4 inches are easily obtained, but many sections in the market beyond that size are made. [2]

Steel bars (Bessemer or Siemens-Martin process) can be made in the following sizes :—
Rounds.   ¼ in. to 1½ in., advancing by 1/16 in.   1½ in. to 4 in., advancing by ⅛ in.
    4 in. to 6 in., by ¼ in.
Squares.   ¼ in. to 1½ in., advancing by 1/16 in., and 1½ in. to 4 in. by ⅛ in.   4 in. to 6 in. by ¼ in.
Flat bars.   1 in. to 2 in., advancing by ⅛ in.   Thickness, ⅛ in. to 1 in.
    2¼, 2⅜, 2½, 2⅝, 2¾, 3, 3¼, 3½, 3¾ in.   Thickness, 3/16 in. to 1 in.
    4, 4½ in.   Thickness, 3/16 in. to 1¼ in.
    5, 5½, 6, 6½, 7, 8, 9, 10, 11, 12 in.   Thickness, ¼ in. to 1½ in.
Steel hoops.   ⅜ in. to 6 in. wide.   Thickness, usual gauges.
Maximum dimensions to which plates and sheets are rolled.—
Steel plates. 1/16 in. thick can be rolled 30 super. ft. Max. length, 14 ft. Max. width, 4 ft.

3/32	,,	,,	45	,,	,,	18 ,,	,,	4½ ,,
⅛	,,	,,	60	,,	,,	22 ,,	,,	5 ,,
5/32	,,	,,	75	,,	,,	25 ,,	,,	5¼ ,,
3/16	,,	,,	90	,,	,,	30 ,,	,,	5½ ,,
¼	,,	,,	100	,,	,,	34 ,,	,,	6 ,,
5/16	,,	,,	120	,,	,,	38 ,,	,,	6¾ ,,

---

[1] H. J. Lewis, *Engineering News.*
[2] For remarks on Standard Sections, see p. 292.

Maximum dimensions to which plates and sheets are rolled—Continued.

Steel plates.			super. ft.			Max. length			Max. width
$\frac{3}{8}$ in. thick can be rolled			135 super. ft.	Max. length	42 ft.	Max. width	$7\frac{1}{2}$ ft.
$\frac{7}{16}$	,,	,,	150	,,		,,	45 ,,		8 ,,
$\frac{1}{2}$	,,	,,	165	,,		,,	48 ,,		$8\frac{1}{2}$,,
$\frac{5}{8}$	,,	,,	200	,,		,,	50 ,,		$9\frac{1}{2}$,,
$\frac{3}{4}$	,,	,,	225	,,		,,	52 ,,		10 ,,
$\frac{7}{8}$	,,	,,	250	,,		,,	56 ,,		10 ,,
1	,,	,,	230	,,		,,	60 ,,		10 ,,
$1\frac{1}{8}$	,,	,,	210	,,		,,	60 ,,		10 ,,
$1\frac{1}{4}$	,,	,,	200	,,		,,	60 ,,		10 ,,

Other sizes can be made by special arrangement.

No plate is rolled to contain both the maximum length and width. The thickness and width being given the length equals $\frac{\text{area}}{\text{width}}$, or the length being given, width equals $\frac{\text{area}}{\text{length}}$.

Extras on Steel Plates.—The following list gives the extras upon steel plates, but steel is generally sold to specification, as so much depends upon the *proportion* of different sizes :—

Limit.	*Extra.*
Weight, 30 cwts.	5s. per 5 cwt. or part.
Length, 25 feet.	5s. per 5 feet or part.
Width, 6 feet.	5s. for every 3 inches or part.
Thickness, $\frac{3}{8}''$ to $1\frac{3}{4}''$	Under $\frac{3}{8}''$ 5s. per $\frac{1}{16}$ in. or part.
Sketch Plates.	10 per cent allowed. Above 10 per cent 25s. per ton.
	(9″ taper allowed before counting as sketch.)

A margin of 5 per cent, *i.e.* $2\frac{1}{2}$ per cent over or under calculated weight, to be allowed for rolling.   In boiler plates 5 per cent over weight and nothing under is allowed. Sometimes 5 per cent under for plates $\frac{1}{4}''$ thick and upwards, and 10 per cent for plates under $\frac{1}{4}''$ is allowed.

BRANDS ON STEEL.—There are no *list brands* for steel (see p. 296).   Each maker has his own trade mark, generally the name of his firm, with or without name of his work.   Thus—(Atlas), Jno. Brown and Company, Limited ; (Cyclops), Cammell and Company, Limited ; (Globe), Ibbotson and Company ; (Norfolk), Messrs. Thomas Firth and Sons, Limited ; (Vickers), Messrs. Vickers, Sons, and Company, Limited ; (Hallside), Steel Company of Scotland ; (E. V.), Ebbwvale Steel, Iron, and Coal Company ; Dalzell Steel and Iron Works, Motherwell ; Glasgow Iron and Steel Company, Wishaw, and others.

The following are some of the marks used :—

 Thomas Jowitt's double shear steel for the trade generally.

"BRADES."   William Hunt and Son's "Brades Company."

 Mushet's borer steel.   Titanic.

L Blister steel, known as Hoop L Swedish brand, that iron being used in the manufacture.   Used by many firms.

 Turton and Sons.

Osborn.

The Sheffield merchant steels have usually a paper wrapping with the maker's name and address in full.

CRUCIBLE CAST STEEL.—The ingots have each a paper label attached, on which is marked the purpose for which the steel is best adapted, as follows :— *borer, welding, tool, rivet.*

SHEAR AND DOUBLE SHEAR steel bars are marked with the words *shear* and *double shear* in indented letters on each bar.

BESSEMER STEEL has no marks. Rails of this material are generally stamped with the maker's name, and the word *steel.*

STRENGTH OF CAST IRON, WROUGHT IRON, AND STEEL.

It is beyond the province of these Notes to enter upon the general subject of the physical properties of materials. The meanings of a few of the terms used in connection with those properties are given at pp. 465, 468, and the subject will be further entered upon in Part IV.

The value of iron and steel to the engineer is, however, so entirely dependent upon their strength, ductility, etc., that a few observations on these points will be necessary in order to clear the way for an intelligent selection and testing of these materials.

In considering the strength of materials care must be taken to distinguish between the *ultimate strength*—that is, the stress per square inch of section which will cause rupture—and the *working strength*, or the stress per square inch which the material can safely bear in practice.

In the following pages the ultimate strength, as found by experiment, will first be given for various descriptions of iron and steel.

The effect upon this strength, produced by various circumstances, will be briefly mentioned.

The working stresses that may be permitted in practice will then be stated.

Finally, one or two points regarding the effect of vibration, cold, etc., will be merely glanced at.

ULTIMATE STRENGTH AND DUCTILITY.

The tests which are applied in practice to cast iron, wrought iron (of different classes), and steel, have been described in previous pages.

In order to apply these tests intelligently, it is necessary to know something of the peculiarities of the different descriptions of iron ordinarily met with, to see what their actual ultimate strength or resistance to rupture has been found to be by experiment, and to understand how that ultimate strength is modified by slight differences in their composition, form, treatment in working, and other surrounding circumstances.

The strength of iron and steel will be considered only with reference to their resistance to tension, compression, shearing, bearing, and transverse stress.

Their resistance to torsion, though of importance in machines, does not come into play in buildings of any kind, and will, therefore, not be considered.

The breaking stresses, found by experiment and given in the following tables, were, in all cases, produced by a *dead load*, gradually applied. Very much smaller *live loads, i.e.* stresses suddenly applied, would cause rupture (see p. 330).

Strength of Cast Iron.

The average ultimate strength of the ordinary varieties of cast iron found in the market may be taken as follows :—

Tons per Square Inch.

Tension	9
Compression	48
Transverse	$13\frac{1}{2}$ B.
Shearing	$8\frac{1}{2}$ S.[1]

The above figures are intended to give a low average.

The following extracts from the most important experiments on the strength of cast iron show the wide differences that occur in different specimens.

The Table below is condensed from the records of Mr. Eaton Hodgkinson's experiments made for the Commission on the use of iron in railway structures.[2]

The experiments were made by crushing cylinders $\frac{3}{4}$ in. diameter, some $\frac{3}{4}$ in. and some $1\frac{1}{2}$ in. high. The figures given below show the resistance of the cylinders $1\frac{1}{2}$ inch high; the shorter cylinders offered a greater resistance.

TABLE GIVING CRUSHING AND TENSILE STRENGTH OF DIFFERENT DESCRIPTIONS OF CAST IRON.

DESCRIPTION OF IRON.	Crushing Strength.	Tensile Strength.
	In tons per square inch.	
Lowmoor Iron, No. 1	25·2	5·7
No. 2	41·2	6·9
Clyde, No. 1	39·6	7·2
No. 2	45·5	7·9
No. 3	46·8	6·5
Blenavon, No. 1	35·9	6·2
No. 2	30·6	6·3
Calder, No. 1	33·9	6·1
Coltness, No. 3	45·4	6·8
Brymbo, No. 1	33·8	6·4
No. 3	34·3	6·9
Bowling, No. 2	33·0	6·0
Ystalyfera No. 2 (Anthracite)	42·7	6·5
Ynis-cedwyn No. 1 do.	35·1	6·2
No. 2 do.	33·6	5·9
Mean of irons tested by Mr. Hodgkinson in his experimental researches	49·5	7·38
Morris Stirling's iron tested by Mr. Hodgkinson—mean	55·6	11·0

[1] S. Stoney.   B. Barlow.
[2] *Report of Commissioners appointed to inquire into the application of Iron to Railway Structures*, 1849.

The mean of experiments made by the Ordnance authorities, as analysed by Professor Pole, give

	Breaking weight in tons per square inch.		
	Max.	Min.	Mean.
Tension . . .	15·3	4·2	10·4
Compression . .	62·5	19·8	40·6
Transverse[1] . .	20·0	4·6	12·6

The specimens tried were generally *samples* received from the makers, of the second or third melting. The iron subsequently supplied in larger quantities was often inferior in strength to the samples.

Influence of various circumstances upon the Strength of Cast Iron.—
Size of Section.—The iron close to the surface of a casting has been found to be harder and stronger than that within. In a small bar the amount of this hard skin is greater in proportion to the section than in large castings, and hence the average strength is greater.

Again, the interior of large castings is more spongy and open than that of small castings. Mr. Eaton Hodgkinson found the relative tensile strength—per square inch—of bars 1 in. 2 in. and 3 in. square to be 100, 80, 77.

Repeated Remeltings.—Repeated remelting of cast iron increases its strength, probably in consequence of the carbon being burnt out of it, thus tending to assimilate it in composition to wrought iron. Sir William Fairbairn, experimenting upon Scotch iron, obtained the following results :—

Its resistance to cross breaking increased up to the twelfth remelting, and then fell off; at the twelfth remelting its strength was $\frac{7}{8}$ of what it originally possessed. Its resistance to crushing was a maximum at the fourteenth remelting, *i.e.* nearly 2¼ times its original strength. Its resistance then fell off, until at the eighteenth remelting it possessed only twice its original strength.

In America the iron is kept in a state of fusion for two or three hours at each remelting. Major Wade found the result to be as follows :—

Strength of pigs . . . 5 to 6½ tons per square inch.
First melting . . . . 9·3 ,, ,,
Fourth melting . . . . 12·4 ,, ,,

The effect of remelting varied considerably, being greatest in No. 1 soft grey pig-iron. This question can rarely be of any great importance to the engineer, though it might possibly have to be considered in using old iron.

Effect of Temperature.—Sir William Fairbairn's experiments led him to the following conclusion :—"Cast iron of average quality loses strength when heated beyond a mean temperature of 120°, and it becomes insecure at the freezing point, or under 32° Fahrenheit." [2]

At a red heat its original strength is diminished by $\frac{1}{3}$. A mass of cast iron raised to a red heat will crumble to pieces when struck. This property may be taken advantage of in breaking up large pieces of old cast iron, such as guns.

---

[1] This is the value of the co-efficient C in the formula, $W = C\frac{bd^2}{l}$

Where W = breaking weight in tons.
b = breadth ⎫
d = depth ⎬ of beam in inches.
l = span ⎭
This subject will be explained in Part IV.

Application of Iron to Building Purposes, by Sir William Fairbairn, p. 73.

An increase or decrease of temperature amounting to 27° Fahr. causes such expansion or contraction that it would bring a stress of 1 ton per square inch upon the metal, if it was rigidly secured at the ends before the change of temperature took place.

The *Effect of mixing Different Brands,* when judiciously done, is doubtless to increase the strength of the iron beyond that of any single brand. The exact increase depends, of course, upon the mixtures used. As before mentioned, this is a question better left alone by the engineer.

Strength and Ductility of Wrought Iron.

The strength and ductility of wrought iron depend upon the quality of the material and the care with which it is manufactured.

A very small proportion of carbon is practically always present ; if this is increased, the strength of the iron is considerably augmented, and its power of welding diminished,—in fact it approximates more to steel in its characteristics.

The presence of other impurities occasions the defects mentioned at pp. 264, 265.

The strength of different descriptions differs so greatly that an average is somewhat likely to be misleading in any particular case ; but the following may be taken as a low average for the ultimate strength of wrought iron under different stresses.

		Tons per sq. inch.				Tons per sq. inch.
Tension	Bars	25		Compression	. .	16 to 20
	Plates { lengthways	21		Shearing		20
	{ crossways	20				

Tensile Strength.—The following Table shows the tensile strength, contraction of area and elongation after fracture, ascertained by experiments upon some of the more important descriptions of iron found in the market.

AVERAGE TENSILE STRENGTH and DUCTILITY of IRON PLATES and BARS made by several noted Manufacturers.[1]

MANUFACTURERS AND DESCRIPTION OF IRON.	Tensile strength per square inch of original section.	Contraction of area fractured.	Ultimate elongation.
	Tons.	Per cent.	Per cent in 10 inches.
Round Oak Iron Works (see p. 294)—	{ 24·94 to	48·2 to	28·8
L.W.R.O. bars	{ 26·57	44	27·5
Best bars	24·67	45·3	25·4
Best best bars	23·35	45·2	29·7
Best best best bars	23·60	46·9	30·7
Best rivet iron bars	24·75	45·7	26·6
Best best best rivet iron bars . .	24·26	47·2	27·4
Shelton Iron & Steel Co., Stoke-on-Trent—			in 12 inches.
Best boiler plates, ½-in. thick, lengthways	22·3	10·3	7·3
„ „ „ crossways	18·7	4·6	4·2
Best best boiler plates, ⁷⁄₁₆-in. thick,			
„ „ lengthways	23·6	16·2	8·8
„ „ „ crossways	20·6	10·4	5·2
Rivet iron	25·0	40·0	27·0
Angle iron	26·5	34·1	27·0
N. Hingley and Sons, Dudley—	{ 22·6 to	45·0 to	30·0
Netherton crown best bar iron . .	{ 23·8	35·0	24·0
„ „ „ rivet iron .	23·5	50·0	20·0

[1] Extracted from Tables in Hutton's *Practical Engineer's Handbook.*

TABLE giving the TENSILE STRENGTH and DUCTILITY of various Descriptions of MALLEABLE IRON.   From Mr. Kirkaldy's Experiments.[1]

[1] Kirkaldy's *Experiments on Wrought Iron and Steel*, Tables G, J, K.   In these and all the other selections from Mr. Kirkaldy's experiments the stresses given by him in lbs. have been reduced to tons.

District.	Names of Makers or Works and Brands.	Description.	Tearing weight per sq. inch of original section, in tons.	Contraction of area fractured.	Ultimate elongation or tensile set after fracture.
				Per cent	Per cent
Rolled Bars.					
Yorkshire . .	Lowmoor . . .	Rolled Bars, round, 1″ diameter	27·59	53·1	26·5
Do. . .	Bowling . . .	Do.	27·86	45·3	24·4
Do. . .	Farnley . . .	Do.	28·07	50·6	25·6
Staffordshire .	J. Bradley & Co., L circle (charcoal) . .	Do.	25·54	60·9	30·2
Do. .	Do. B.B. scrap	Do.	26·5	52·0	26·6
Do. .	Do. S.C. 👑 .	Do. ⅞″ dia.	27·78	36·2	22·2
Do. .	J. Bagnall, J.B. . .	Do. 1¼″ do.	24·55	27·0	17·3
Scotland . .	Govan, Ex. B. Best .	Do. 1″ do.	26·39	40·0	22·3
Do. . .	Do. B. Best . .	Do.	28·05	28·9	19·1
Do. . .	Do. ✳ . . .	Do.	26·53	25·1	16·4
Do. . .	Glasgow, B. Best . .	Do.	26·29	39·6	23·2
Wales . .	Ystalyfera (puddled) .	Flat strips.	17·20	2·4	2·0
Rivet Iron.		Diameter.			
Yorkshire . .	Lowmoor . . .	Round. 11⁄16″	26·82	52·2	20·5
Do. . .	Bradley & Co., 👑 S.C.	Do. ¾″	25·32	49·5	22·5
Lancashire .	Ulverstone, Rivet Best .	Do. ¾″	24·00	48·6	21·6
Staffordshire .	Thorneycroft & Co., TNS	Do. 13⁄16″	26·46	40·4	22·4
Do. .	Lord Ward, L 👑 W. . R. O.	Do. 11⁄16″	26·69	37·6	18·6
Scotland . .	Glasgow, Best Rivet .	Do. ⅞″	25·49	40·7	23·7
Iron Plates.		Thickness.			
Yorkshire . .	Lowmoor . . .	L. 5⁄16″	23·21	19·7	13·2
		C. 5⁄16″	22·55	12·1	9·3
Do. . .	Farnley . . .	L. ⅜″	25·00	17·8	14·1
		C. ⅜″	20·63	13·2	7·6
Do. . .	Bowling . . .	L. ⅜″	23·32	15·3	11·6
		C. ⅜″	20·73	6·9	5·9
Staffordshire .	Bradley & Co., 👑 S.C.	L. ½″	24·92	17·2	12·5
		C. ½″	22·52	9·0	5·5
Do. .	Thorneycroft, Best Best	L. 15⁄16″	24·48	12·5	11·2
		C. 15⁄16″	20·35	4·6	4·6
Do. .	Lloyds, Foster, & Co., Best	L. 6⁄16″ to 7⁄16″	20·07	8·7	5·3
		C. Do.	19·92	6·9	4·6
North of England	Consett, Best Best .	L. ¾″	22·88	13·1	8·9
		C. ¾″	20·85	10·2	6·4
Scotland . .	Glasgow, Best Best .	L. ¼″ to ¾″	23·84	10·6	9·0
		C. Do.	18·65	3·7	2·6
Angle Iron.					
Yorkshire . .	Farnley . . .	Thickness 9⁄16″	27·34	41·4	20·9
Staffordshire .	Albion 👑 Best . .	Do. ⅝″	25·07	19·1	14·0
Do. .	Do. Best . .	Do. ⅝″	23·28	22·8	14·1
Do. .	Eagle	Do. 21⁄32″	22·34	15·3	8·8
	Do. *Best Best* . .		24·42	23·4	13·7
Durham . .	Consett . . .	Do. 9⁄16″	22·68	11·7	5·8
Do. .	Do. Best Best .	Do. ½″	23·90	18·3	12·6
Scotland . .	Glasgow, Best scrap	Do. ⅝″	25·04	20·1	15·0
Do. .	Do. Best Best	Do. ⅝″	24·78	11·0	8·5

L. signifies lengthways—in the direction of the grain ; C. crossways—across the grain.

From the results above recorded, it will be seen that the average of ordinary qualities of bar iron is nearly 20 per cent stronger than that of the same qualities of plate iron, and its elongation under a given stress is $2\frac{1}{2}$ times as great ; also that plate iron has a greater strength in the direction of the fibre or grain than across the grain, the difference being on the average about 10 per cent.

The Elastic Limit of a few different classes of iron is shown in the following Table (see also p. 329) :—

DESCRIPTION OF IRON.	Elastic limit in tension, tons per square inch.	Breaking tensile stress per square inch.	Elongation in 10 inches per cent.
Bowling iron [1] with grain . . .	12·6	20·9	19·8
" " across grain . . .	11·1	18·4	7·5
Barrow B.B.H.[1]	14·5	25·4	22·3
Cleveland [2] ⅜-inch and ½-inch plates—			in 8 inches.
With grain . . . . ⎱	13·7	{ 21·0	7·5
Across grain . . . . ⎰		{ 18·2	3·0
Belgian joists [3]	{ 16·9 to { 15·4	22·4 to 20·9	... in 3 inches.
Wrought iron from crank shaft . .	13·3	14·18	4·0

The Crushing Strength of wrought iron varies in different specimens with the hardness of the iron.

"Ordinary wrought iron is completely crushed, i.e. bulged, with a pressure of from 16 to 20 tons per square inch." [4]

The best soft wrought iron begins to bulge sensibly with about 12 tons per square inch.[5]

The Shearing Strength of wrought iron has been proved by experiment to be practically equal to the tensile strength of the material. See Chap. vii., Part IV.

Effect of different Processes and Circumstances upon the Strength of Wrought Iron.—It has already been stated that the strength and elasticity of wrought iron depend not only upon its quality, but upon the treatment to which it has been subjected in working, and upon other surrounding circumstances.

The following Table shows concisely the effect produced by different modes of working, by changes of temperature, etc.

The conclusions given are founded upon a large number of experiments by Mr. Kirkaldy and others. Those by Mr. Kirkaldy are clearly classified in Mr. Kinnear Clarke's *Rules and Tables for Mechanical Engineers*.

·	Tensile Strength.	Ductility.
Reducing diameter by rolling, forging, or hammering . . .	Increased . . .	Reduced.
Turning or removing skin . .	No alteration . .	No alteration.
Annealing	Reduced . . .	Increased.
Welding	{ Reduced from between { 4·1 and 43·8 per cent	} Reduced.
Stress suddenly applied . . .	Reduced 18·5 per cent	{ Reduced in nearly all { cases.
Hardening in water or oil . .	Increased . . .	Reduced.
Cold rolling—*plates* . . .	Doubled . . .	Destroyed.
bars . · . .	Increased 50 per cent	Reduced 60 per cent.
Galvanising	No difference.	...
Effect of frost 23° F. . . .	Reduced 2·3 per cent	Reduced 8 per cent.
Effect of frost, stress suddenly applied	Reduced 3·6 per cent	{ Reduced between 0 { and 30 per cent.

Effect of Temperature.—Sir William Fairbairn found that the strength of wrought iron was practically the same at all temperatures between 0· and 400° Fahr.[6]

[1] Kennedy, *Iron*, 11th May 1883.
[2] *Institute Mechanical Engineers*, 4th August 1885.
[3] *Architect*, 18th February 1882.   [4] Stoney.   [5] Downing.
[6] *Useful Information for Engineers*, Series ii.

Strength and Ductility of Steel.

The strength and ductility of steel varies greatly in different descriptions. It depends not only upon the original composition of the metal, but also upon the treatment to which it has been subjected. The following Tables give some idea of the variety to be met with in different specimens.

Average Strength.—The great differences in strength caused by varieties in the amount of carbon and in temper make it useless to attempt to arrive at an average strength for all steels.

The following Tables, selected from different records of experiments, show the great variation that there is in the strength and ductility of different descriptions of steel :—

TENSILE STRENGTH, ELASTIC LIMIT, and DUCTILITY of CAST STEEL.

	Ultimate or breaking tensile stress per sq. inch.	Elastic limit in tension.	Contraction of area per cent.	Elongation per cent.
	Tons.	Tons.		
C. Bessemer steel (average of different qualities for tyres, axles, and rails)				
C. Hammered . .	33·9	22·2	46·9	12·0
C. Rolled . . .	32·0	19·0	36·1	18·0
C. Crucible steel (average of different qualities for tyres, axles, and rails)				
C. Hammered . .	38·2	21·9	22·8	7·0
Rolled (for axles) .	30·6	18·7	10·1	10·6
C. Bessemer steel tyres and axles	33·7			
C. Crucible cast-steel from Swedish bar-iron, chisel temper . . .	52·8	26·0	...	5·3
C_1 Crucible cast-steel, rolled	34·43	20·5	...	2·0
C_1 ,, ,, hammered	37·05	25·0	...	13·5
C_1 Cast-steel, piston rods .	33·7	26·75	...	0·9

C. *Experiments on Steel by a Committee of Civil Engineers*, 1868. The bars experimented upon were turned down from 2-inch square bars to a diameter of $1·382 = 1\frac{1}{2}$ square inch.

C_1 Further experiments of the same Committee results, bound up with the report just quoted.

WHITWORTH'S COMPRESSED STEEL.[1]

Purposes for which the Steel is available.	Tensile Strength in tons per sq. inch.	Ductility or percentage of Elongation.
Axles, boilers, connecting rods, rivets, railway tyres, gun furniture and barrels, and gun carriages	40	32
Cylinder linings, parts of large machines, and hoops and trunnions for ordinance	48	24
Large, planing and lathe tools, large shears, smiths' punches and dies and sets, small swages, cold chisels, screw tools, corn mill rollers, armour-piercing shells	58	17
Boring tools, finishing tools for planing and turning . .	68	10
Alloyed with tungsten for particular purposes . .	72	14

[1] *Proceedings Institute of Mechanical Engineers.*

TENSILE STRENGTH and DUCTILITY of STEEL of different descriptions.
Selected from Sir W. Fairbairn's Experiments.[1]

Manufacturers and Description of Steel.	Breaking tensile stress per square inch of section.	Corresponding ultimate elongation.	Contraction or set due to compression under 100·7 tons per square inch.
	Tons.	Per cent.	Per cent.
Messrs. J. Brown and Company.			
Best cast steel from Russian and Swedish iron for turning tools .	30·53	·56	25·3
Do. milder . .	40·85	1·50	26·3
English tilted steel made from English and foreign pigs	26·57	7·6	55·3
Messrs. C. Cammell and Company.			
Specimen of cast steel, termed "Diamond Steel"	49·13	1·77	23·3
Specimen of cast steel termed "Tool Steel"	48·69	2·06	26·3
Specimen of cast steel termed "Chisel Steel"	53·75	2·81	31·3
Specimen of cast steel termed "Double Shear Steel"	43·15	2·50	30·3
Messrs. Naylor and Vickers.			
Cast steel called "Axle Steel" . .	39·58	6·25	42·3
Do. do. "Tyre Steel" . .	40·85	4·75	38·8
Do. do. "Vickers' Cast Steel, special" . .	59·87	1·00	15·3
Do. do. "Naylor and Vickers' Cast Steel" .	52·70	2·87	18·3
Messrs. S. Osborne and Company.			
Specimen of best tool cast steel .	44·17	1·56	20·3
Specimen of best double shear steel .	39·25	2·43	32·3
Extra best tool cast steel . . .	38·26	0·37	19·3
Cast steel for boiler plates . . .	49·85	10·62	33·3
H. Bessemer and Company.			
Specimen of hard Bessemer steel .	46·02	1·87	22·3
Do. milder do. . .	39·36	10·93	44·3
Do. soft do. . .	35·09	9·81	47·3
Messrs. T. Turton and Sons.			
Specimen of double shear steel . .	32·70	0·87	29·3

[1] *Iron Manufacture,* 1869.   *British Association Report,* 1867.

TENSILE STRENGTH and DUCTILITY of STEEL PLATES With and Against
the Grain.   From Mr. Kirkaldy's Experiments.[1]

L. signifies lengthways of the grain ; C. across the grain.

Names of Makers or Works.	Thickness.	Description.		Tearing weight per square inch of original area.	Ultimate elongation or tensile set after fracture.	Contraction of area at fracture.
	Inch.			Tons.	Per cent.	Per cent.
Turton and Sons, cast steel .	$\frac{1}{4}$	Samples.	L	42·09	5·71	5·6
			C	42·99	9·64	13·4
Moss and Gambles, cast steel .	$\frac{3}{16}$ to $\frac{4}{16}$		L	33·74	19·82	28·2
			C	30·84	19·64	38·6
Shortridge and Co., do.	$\frac{3}{16}$		L	42·98	8·61	15·6
			C	43·37	8.93	14·8
Shortridge and Co., puddled steel	$\frac{1}{4}$		L	32·32	5·93	11·5
			C	32·84	3·21	5·7
Mersey Company, puddled steel (ship plates), . . .	$\frac{2}{16}$ to $\frac{3}{16}$		L	45·29	2·79	6·4
			C	37·93	1·25	4·4
Mersey Company, puddled steel, "Hard"	$\frac{1}{4}$	Samples.	L	45·80	4·86	4·5
			C	38·11	3·30	4·7
Mersey Company, mild steel	$\frac{1}{4}$		L	34·39	6·16	12·5
			C	30·22	5·72	8·5
Mersey Company, mild steel (ship plates) . . .	$\frac{9}{32}$		L	31·93	3·57	7·5

TENSILE STRENGTH and DUCTILITY of STEEL BARS.   Selected from Mr.
Kirkaldy's Experiments.[2]

Names of Makers or Works.	Description.	Average breaking weight per square inch of original area.	Ultimate elongation or set after fracture.	Contraction of area at fracture.
		Tons.	Per cent.	Per cent.
Turton's cast steel for tools . .	Forged .	59·0	5·4	4·7
Jowitt's double shear steel . .	Do.	53·0	13·5	19·6
Bessemer's patent steel for tools .	Do.	49·7	5·5	22·3
Naylors, Vickers, and Co., cast steel for rivets	Rolled . .	47·59	8·7	32·8
Wilkinson's blister steel bars . .	Forged .	46·5	9·7	21·4
Jowitt's cast steel for taps . .	Do.	45·1	10·8	28·8
Krupp's cast steel for bolts . .	Rolled . .	41·8	15·3	34·0
Shortridge and Co.'s homogeneous metal	Do.	40·5	13·7	36·6
Jowitt's spring steel . . .	Forged .	32·3	18·0	24·0
Mersey Co., puddled steel . .	Do.	31·91	19·1	35·3
Blochairn puddled steel . .	Rolled . .	31·32	11·3	19·4
Do. do. . . .	Forged .	29·13	12·0	19·0

[1] Kirkaldy's *Experiments on Wrought Iron and Steel,* Table H.   [2] *Ibid.* Table F.

AVERAGE TENSILE STRENGTH and DUCTILITY of MILD STEEL PLATES and BARS.

Manufacturers and Descriptions of Steel.	Tensile Strength per square inch of original area. Tons.	Limit of Elasticity.	Contraction of Area per cent.	Ultimate Elongation per cent.
W. Beardmore and Company [1]—				In 8 inches.
Bridge plates	29	...	...	23
Angles and bars . . .	31	...	...	17½
Rivet steel	27½	...	...	33
Bolton Iron and Steel Company [1]—				
Bridge plates	29½	...	40	20
Angles, tees, bulb, beams for bridge and shipbuilding	29¾	...	40	20
Rivet steel	27	...	50	30
Landore Steel [2]—				
Plates and angles (mean of 101 samples)	28·16	...	...	24·25
Plates lengthways,[3] unannealed	31·1	14·5	41·1	23·4
,, crossways . .	31·2	14·4	40·5	23·5
Dalzell Steel Works [4]—				In 10 inches.
Plates lengthways, ⅜″ thick .	28·1	14·1	50·2	29·9
,, ,, ¾″ ,, .	27·1	13·2	56·8	32·6
,, ,, 1″ ,, .	26·2	13·1	57·7	33·4
Plates crossways, ⅜″ thick .	28·8	15·0	43·4	26·3
,, ,, ¾″ ,, .	27·5	14·1	49·6	31·9
,, ,, 1″ ,, .	26·2	12·7	51·2	33·7
Rivet steel	28·9	16·5	61·6	...
Weardale Iron and Coal Coy. [4]—				
Plates lengthways, ⅜″ thick .	27·1	16·3	54·1	26·8
,, ,, ¾″ ,, .	29·3	16·8	52·5	28·5
,, ,, 1″ ,, .	26·9	14·3	43·6	29·7
Plates crossways, ⅜″ thick .	27·1	16·1	46·5	22·7
,, ,, ¾″ ,, .	29·4	16·0	51·8	26·7
,, ,, 1″ ,, .	26·7	14·3	49·2	30·6
Rivet steel	30·0	17·9	54·4	...
Steel Company of Scotland [4]—				
Plates lengthways, ¼″ thick .	31·1	19·6	46·4	22·6
,, ,, ½″ ,, .	29·0	15·8	51·9	29·7
,, ,, 1″ ,, .	27·9	14·8	48·8	30·3
Plates crossways, ¼″ thick .	32·2	19·9	39·9	19·8
,, ,, ½″ ,, .	28·8	15·8	39·8	28·3
,, ,, 1″ ,, .	28·0	14·8	40·9	25·2
Rivet steel	31·1	18·5	62·9	...

These experiments show that the difference in the strength of the steel, when tested lengthwise and crosswise of the grain is almost imperceptible.

---

[1] Hutton, *Practical Engineer's Handbook.*   [2] Riley.
[3] Kirkaldy, *Proceedings Institute of Naval Architects.*
[4] Traill, *Experiments on Mild Steel.*

The **Elastic Limit** of steel plates having a tensile strength of from 26 to 32 tons per square inch either way of the grain, may be taken at from 50 to 65 per cent of the ultimate tensile strength.[1] The elastic limit of other forms of steel are given at pages 318, 329.

Steel Wire is sometimes made for special purposes, e.g. for pianos and for wire rope, with a tensile strength of 120 or even as much as 150 tons per square inch, and with an elongation of about 33 per cent.[2]

The **Crushing Strength** of steel varies greatly, according to the quality of the steel and the hardness to which it has been tempered.

Some cylinders of cast steel (of a height $= 2\frac{1}{2}$ diameters) cut from the same bar[3] were crushed under the weights given below.

	Crushing weight per inch of section.
Not hardened	89 tons.
Hardened—low temper, suitable for chipping chisels . .	158 ,.
Hardened—high temper, suitable for tests for turning hard steel	166 ,,

In the experiments of the Committee of Civil Engineers (see p. 318), steel cylinders of 1 inch area and 1 diameter in height bulged but did not crack under 89 tons, and cylinders of the same area, but with height of 4 diameters, crushed with weights averaging 20 tons.

It must be remembered that the steel begins to fail when its elastic limit is passed. This was found by Mr. Berkley to be about 17 tons for Bessemer steel. In the experiments of the Committee of Civil Engineers it ranged between 27 and 15 tons, the average as deduced by Mr. Stoney being 21 tons[4] (see also p. 319).

Shearing Strength of Steel.—Mr. Kirkaldy's experiments led him to the following conclusion :—" The shearing strain of steel rivets is found to be about a fourth less than the tensile strain." [5]

The steel he experimented upon broke under a tensile stress of $38\frac{1}{2}$ tons per square inch of area, and the mean strain required to shear the rivets was $28\frac{1}{2}$ tons per square inch.

"**The tests** on torsion and transverse strain, tension and compression, show that the relations which subsist between the resistances to these strains in steel correspond very nearly with those found by previous experiments in wrought iron ; that is to say, a bar of steel which has 50 per cent more tensile strength than a similar bar of wrought iron will also have approximately 50 per cent more strength in resisting compression, torsion, and transverse strain." [6]

Effect of different Processes and Circumstances upon the Strength of Steel. —*Effect of Tempering.*—After a series of experiments Mr. Kirkaldy came to the following conclusions as to the influence upon steel caused by its treatment in different ways.

" 35th. Steel is reduced in strength by being hardened in water, while the strength is vastly increased by being hardened in oil.

" 36th. The higher steel is heated (without of course running the risk of being burned) the greater is the increase of strength by being plunged into oil.

" 37th. In a highly converted or hard steel the increase in strength and in hardness is greater than in a less converted or soft steel.

" 38th. Heated steel, by being plunged into oil instead of water, is not only considerably *hardened*, but *toughened* by the treatment."

The following are extracts from the results of the experiments which led to these conclusions :—

---

[1] See preceding Table, p. 321.   [2] Percy, *Iron and Steel Institute.*
[3] By Major Wade, U.S. Army.   [4] Stoney *On Strains.*
[5] Kirkaldy's *Experiments in Iron and Steel.*
[6] *Report by Committee appointed by Board of Trade,* etc. etc.

CAST STEEL FOR CHISELS.

	Breaking weight per sq. inch, in tons.	Contraction of area per cent.	Elongation per cent.
Highly heated, and cooled in oil . . .	96	3·5	3·3
Do. do. in water . . .	40	0·0	0·0
Do. do. ,, yellow temper	45	0·0	0·0
Do. do. ,, spring . .	47	0·0	0·7
Do. do. ,, blue . .	50	0·0	0·7
Do. do. in ashes, slowly . .	54	12·9	7·0
Medium heat, and cooled in oil	82	3·5	2·7
Do. do. in tallow . . .	79	3·4	2·7
Do. do. in coal tar . . .	75	6·4	6·0
Do. do. slowly . . .	53	9·4	7·7
Low heat, and cooled in oil	73	6·8	5·0
Do. do. in tallow	64	6·6	7·0
Do. do. in coal tar . . .	63	13·1	8·7
Do. do. slowly	56	16·5	10·0

Effect of Annealing Steel Plates.—Hard steel plates are greatly improved in ductility by being annealed. With soft steel, however, the increase of ductility is not necessary, and the tensile strength is lessened.

Influence of Carbon upon Strength of Steel.—The following Table [1] contains the relation between the specific gravity and tensile strength of Bessemer steel of various degrees of carbonisation, made at Sandriken, in Sweden :—

Percentage of Carbon.	Specific Gravity.		Tensile Strength.
	Soft.	Hardened.	Tons per sq. inch.
1·5	7·785	7·736	34-39
1·2	7·832	7·771	37-40
0·9	7·874	7·808	56-59
0·6	7·879	7·807	37-41
0·4	7·893	7·839	30-34

The absolute strength appears to be greatest when the steel contains from 1 to 1¼ per cent of carbon.

---

[1] From Bauermann's *Metallurgy.*

SAFE OR WORKING STRESSES FOR CAST IRON, WROUGHT IRON, AND STEEL.

The limiting or working stresses that can be safely applied in practice to cast iron, wrought iron, and steel respectively, depend not only upon the quality and characteristics of the material, but upon the nature of the load which causes the stresses, and in many cases also upon the form of the member or structure under stress.

These points, and many others which bear upon the question, cannot here be entered upon without anticipating the information to be given in Part IV., where the subject is more fully discussed.

Factors of Safety.—It will be sufficient at present to call attention to the following Table, which shows the " factors of safety "[1] recommended by eminent engineers for application in various cases that arise in practice.

Autho-rity.	Nature of Structure.	Nature of Load.	Factor of Safety.
	Cast Iron.		
B.	Girders	Dead.	3-6
S.	Do.	Do.	6
S.	Pillars	Do.	6
S.	Water tanks	Do.	4
S.	Crane posts or machinery .	Live.	8
S.	Pillars subject to vibration .	Do.	8
S.	Do. do. transverse shock	Do.	10
	Wrought Iron.		
S. R.	Girders	Dead.	3
S.	Do.	Live.	6
B.	Bridges	Mixed.	4 in tension.
U.	Roofs	Do.	4
S.	Compression bars subject to shocks	Live.	6
S.	Compression bars not subject to shocks	Dead.	4
	Steel.		
C.	Bridges	Mixed.	4

B. Board of Trade.   S. Stoney.
U. Unwin.   C. Commissioners.

[1] See page 465.

The working stresses are obtained by dividing the known breaking strength of the particular class of material to be used, by the factor of safety applicable to the structure and load for which it is to be used.

The breaking strength is found by experiment, or taken from tables giving the results of experiments on iron or steel of a similar class (see pp. 312 to 323).

The factor of safety is varied according to judgment and experience, or, in the absence of these, may be taken from the Table above.

It is necessary here to state that the working stress should in no case exceed the elastic limit of the material. The reasons for this are given at p. 327.

It will be seen, however, that the elastic limit is generally about $\frac{1}{2}$ of the ultimate strength, whereas the working stress is seldom more than $\frac{1}{4}$ of the same, so that if the factors of safety are carefully applied there is no danger of passing the elastic limit of any ordinary material.

In the absence of experimental knowledge with regard to the particular material about to be used, the engineer takes care to calculate for a low working stress, so that he may be sure not to overtax the strength of the material.

Working Stresses.—The following working stresses may be used in practice :—

CAST IRON.—For girders, etc., to carry a dead load—

Compression	. . .	8 tons per square inch.
Tension	. . . .	$1\frac{1}{2}$,,
Shearing	. . . .	2 ,,

An allowance of 30 per cent should be made to cover defects, such as air-holes, etc., in the castings.

Cast iron is not well adapted for structures intended to carry a live load, but if used for such, the working stresses would be reduced in the proportion shown by the factors of safety for the different cases given in the Table, p. 324.

WROUGHT IRON.—The working stresses practically applied to wrought iron are as follows :—

Built-up Plate-Iron Girders and similar structures—

[1] Tension	. .	5 tons	per square inch.	
[1] Compression	. .	4 ,,	,,	
[2] Shearing	. .	4 to $4\frac{1}{2}$ tons	,,	
Bearing	. .	8 ,,	,,	in riveted joints.

These working stresses are in practice applied to girders with dead loads, and also to those carrying moderate live loads. This, of course, is not theoretically correct. When the load is all dead the working stresses may safely be higher—equal to $\frac{1}{4}$ the breaking stress of the material ; and when the live load becomes large in proportion to the weight of the girder (not a common case in girders connected with buildings), the working stresses must be reduced by a method which will be explained in Part IV.

For *rolled girders* the stresses may be taken slightly higher, *i.e.* at about 6 tons in tension and 5 tons in compression or shearing.

Where part of the load is *live* it is converted into an equivalent amount of dead load as described at page 466.

When *bar iron* is used, as in roofs and braced girders, the working stresses in tension may be considerably higher, because bar iron is, as a rule, stronger than plate iron (see p. 315).

---

[1] In calculating the area of the sections to which these stresses are applied, the rivet holes are deducted in the tension flange, but are not generally deducted in the compression flange. Some engineers deduct them in both flanges.

[2] The shearing stress might be taken as high as the tensile stress, but that the former generally acts upon a group of rivets, some of which often get a larger share of the stress than the others, so that a lower limit is taken in order to be on the safe side.

Thus, with good bar iron (such as $\frac{W}{SC}$, see Table, p. 316) a factor of safety of 4 for a dead load would give a working stress of $\frac{27.78}{4}$, or nearly 7 tons per square inch of section.

However, taking into consideration the sudden shocks caused by the wind, a working stress of 6 tons is high enough ; and where the iron is of an unknown quality, it is better to allow only 5 tons per square inch.

Board of Trade Rule.—Though the construction of bridges is a subject entirely beyond the limits of any part of these Notes, it may be as well to mention here the Board of Trade rule as to the working stress for bridges, because this rule has governed the practice with regard to bridges, and has to a great extent influenced it in other structures.

"In a wrought-iron bridge the greatest load which can be brought upon it, added to the weight of the superstructure, should not produce a greater strain on any part of the material than 5 tons per square inch."

Practically this rule is modified by taking the working stresses, as given above, all of them except the tensile stress being lower than the limits laid down by the rule.

Bearing Strength.—The resistance of wrought iron to indentation by bolts or rivets varies, of course, according to the quality of the iron.

For most ordinary work the safe statical pressure per square inch of bearing surface may be taken at 5 tons,[1] but in chain-riveted joints it may be taken at 8 tons.

Mr. Stoney takes it at $1\frac{1}{2}$ times the safe tensile stress, or $7\frac{1}{2}$ tons for all structures.

STEEL.—The factor of safety applied to steel structures should depend (*cæteris paribus*) on the nature of the steel and its temper.

Thus a very hard steel, with high tensile strength and slight ductility, should be worked at a smaller proportion of its breaking stress than a mild and soft steel.

Working Tensile Stress.—Mr. Stoney recommends a working stress of 8 tons per square inch for mild steel plates, being about $\frac{1}{4}$ of their ultimate tensile resistance (see Table, p. 320).

Opinion of Committee appointed by the Board of Trade.[2]—The following extracts from their report will give the conclusions at which this Committee arrived :—

" As regards the ordinary steel of commerce, there appears to be no difficulty in obtaining the usual amount of tensile strength, varying from 29 to 35 tons per inch. A point requiring equal attention is the toughness or malleability." . . .

" We assume that with steel, as with iron, the engineer will take care that, as well as the required strength, he secures a proper amount of ductility. . . .

The steel employed should be cast steel, or steel made by some process of fusion,

---

[1] Latham *On Wrought Iron Bridges.*

[2] Sir John Hawkshaw, C.E., F.R.S. ; Colonel W. Yolland, R.E., F.R.S. ; W. H Barlow, Esq., C.E., F.R.S.

subsequently rolled or hammered, and that it should be of a quality possessing considerable toughness and ductility." . . .

"The greatest load which can be brought upon the bridge or structure, added to the weight of the superstructure, should not produce a greater strain in any part than $6\frac{1}{2}$ tons per square inch."

From other parts of their report it appears that they consider that the working stress upon steel should bear the same proportion to its ultimate strength that the working stress upon iron does to its ultimate strength.

Thus, taking the ultimate strength of iron at 20 tons per inch, and the working stress allowed by the Board of Trade for bridges at $\frac{20}{4} = 5$ tons, they infer that the ultimate strength of steel may very safely be taken at 26 tons per inch, and the working stress applied to it at $\frac{26}{4} = 6\frac{1}{2}$ tons.

Working Stress in Compression.—With regard to the working stress in compression Mr. Stoney says :—

"The crushing strength of tool steel is so high that 12 or even 15 tons per square inch is perhaps a safe working load for pillars of tool steel, when they are so short that they will not fail from flexure." [1]

As regards the softer grades of mild steel for structural purposes, experiments on struts and columns appear to show a tendency to equality of resistance as between iron and steel as the column increases in length. For further information on the safe working stress in steel columns the student is referred to Part IV.

Shearing Stress.—The shearing strength of mild steel rivets in bars of $\frac{5}{8}''$ to $\frac{3}{4}''$ diameter, possessing a tensile strength of 25 to 29 tons per square inch, has been found to be from 70 to 82 per cent of the ultimate tensile resistance in single shear, and from 66 to 76 per cent in double shear. [2]

Limit of Elasticity.—In investigating the properties of a specimen of iron or steel a very important point to be ascertained is its *limit of elasticity.*

The meaning of this term has been defined in several different ways.

Mr. Stoney's definition is the one perhaps best suited to the engineer. He says—" The limit of elasticity may be defined to be the greatest stress that does not produce an appreciable set."

A short explanation will perhaps make the meaning of the term more clear than the definition alone would do.

If a small weight be suspended from a bar so as to cause a tensile stress in the direction of its length, the bar will at once begin to elongate.

It will stretch a certain proportion of its own length. This proportion will vary according to the description and quality of the material, and to the amount of weight applied.

If a weight of 1 ton be hung from the end of a wrought iron bar of average quality, having a sectional area of 1 square inch, the bar will stretch about $\frac{1}{12000}$ part of its original length.

If the weight be removed, the bar will soon recover itself—that is, it will return to its original length. [3] If measured by any ordinary means of measurement, it will be found to be of the same length that it was before the weight was imposed upon it.

This recovery of the bar occurs, however, only up to a certain point. If the load be increased until it amounts to a considerable proportion of the breaking weight, the result produced is very different.

For example, if, instead of 1 ton, a weight of 12 tons be applied to the bar just men-

---

[1] Stoney *On Stresses.*
[2] Hunt "On Method of Testing Structure Steels," *Proc. Am. Soc. C.E.*, vol. xxx.
[3] See page 315.

tioned, the iron will stretch about $\frac{1}{1000}$ of its length.   Upon removal of the weight, how-ever, it will not entirely recover itself, but will be found, upon measurement, to be a little longer than it originally was.

This slight increase upon the original length of the bar is called the *permanent set*.

The greatest stress that can be applied to the bar without causing an appreciable per-manent set is called the *limit of elasticity*, or the *elastic limit*.

It is evident then, that there is a very important line to be drawn.   On one side of it are weights, the application of which will produce no appreciable permanent set ; on the other side are the weights which produce an appreciable permanent set.

This line of demarcation is called the *Limit of Elasticity*, or the *Elastic Limit*.

It is, as before said, a certain proportion of the breaking load for the material, and its value is generally stated in lbs. or tons per square inch.

The proportion which the limit of elasticity bears to the breaking load varies very con-siderably in cast iron, wrought iron, and steel, and even in different specimens of the same classes.

The above remarks have been made with regard to a tensile stress, but the same thing occurs with a bar under compression.   Weights placed upon the end of the bar produce no permanent contraction or *set* up to a certain point.   Weights greater than this per-manently shorten the bar.   This point is called, as before, the elastic limit, or limit of elasticity.

The exact point at which the permanent set commences varies according to the quality and characteristics of the material.   A hard brittle iron has a high limit of elasticity, it will not stretch much before breaking ; on the other hand, a soft ductile iron soon takes a slight permanent set, but stretches considerably before breaking.   Practically, for ordi-nary good wrought iron, the limit of elasticity may generally be taken at about $\frac{1}{2}$ the breaking stress.

So long as ductility is not sacrificed it is important to have material with a high limit of elasticity for nearly all structures, but especially for those which are subjected to loads constantly repeated, as in the case of railway bridges.   The reasons for this are given below.

Fatigue of Iron.—Many careful experiments made by Sir W. Fairbairn and others have led to the conclusion that a load may be applied to a wrought iron bar, removed and reimposed thousands of times without the slightest injury to the bar, so long as the stress per square inch does not exceed the elastic limit of the material.

Directly this limit is exceeded, the first application of the load produces a permanent set ; each repeated application increases that set, until at last rupture takes place.

The failure of iron under repeated loads or blows of this kind is known as the *fatigue of iron*.

It will be useful to notice one or two other points connected with the elastic limit.

In wrought iron, steel, and indeed in most other building materials, the temporary elongations produced before the limit of elasticity is reached are proportional to the loads which produce those elongations.

Thus, in the bar above referred to, if a load of 1 ton produce an elongation of $\frac{1}{12000}$ in the length, 2 tons will produce $\frac{2}{12000}$, 3 tons $\frac{3}{12000}$, and so on, until 12 tons pro-duce an elongation of $\frac{12}{12000} = \frac{1}{1000}$ the length.

At this point, however, the permanent set occurs, and beyond it the elongations are not in proportion to the load, but increase more rapidly than the loads increase.   Thus 13 tons will produce more than $\frac{13}{12000}$ elongation, and so on.

In cast iron, however, the temporary elongations caused, even by small loads, are from the first irregular, not in proportion to those loads, and an appreciable set is noticed at a very early stage.

False Permanent Set.—In some cases, after imposing upon a bar a load far within the elastic limit, a permanent set seems at first to have been caused, but upon leaving the bar unloaded for a short time this set disappears, and the bar slowly returns to its original length.

Set caused by Continued Load.—It has been found that a load within the elastic limit, which will not cause a permanent set if imposed and quickly taken off, will nevertheless cause a set if it be allowed to remain for a considerable time.

To put it in another way, the elastic limit is lower for a continued stress than for a temporary one.

Elastic Limit raised by different Processes.—It has been shown that the processes of hammering, rolling, and drawing iron or steel, when cold, into bars or wire, increase the tenacity and the elasticity of the material.

Elastic Limit raised by stretching.—Again, it has been shown that when a bar of iron has been subjected to a load less than the elastic limit, and continued for several hours, so that a permanent set ensues, the elastic limit of the bar thus altered is considerably raised. For example, General Uchatius tested a bar of soft steel, and found the following results :—[1]

	Limiting Stress. Tons per square inch.		Ultimate Elongation per cent.
	Absolute.	Elastic.	
Bar of soft steel	24·38	13·81	25·3
Same loaded for 24 hours so as to elongate 3·3 per cent	24·38	17·77	21·5
Same oil-hardened	48·13	17·77	10·6

Other Definitions of the Limit of Elasticity.—It should here be mentioned that Mr. Eaton Hodgkinson's experiments led him to the conclusion that the very smallest load produces a permanent set. His conclusions have been questioned by more recent investigators, but even supposing they are correct, they do not affect the engineer. The permanent sets, if any, produced by loads less than the limit of elasticity are so small that they cannot be measured by an ordinary instrument—in fact, they are inappreciable.

When such loads are constantly repeated, though they may produce an inappreciable set as regards the original length of the bar, yet it is not an *increasing set*, does not lead to rupture, and may therefore practically be ignored.

When, however, the load is greater than the limit of elasticity, an *increasing set* takes place upon each application, which eventually leads to rupture.

Elastic Limit of Cast Iron, Wrought Iron, and Steel.—*Cast Iron* is very imperfectly elastic, that is, even a very small load will produce in it an appreciable permanent set. There is no clearly-defined elastic limit. The permanent sets are, however, very small at first, and may be practically ignored until the load applied is about ⅓ of that required to produce rupture.[2] The sets then become partially appreciable.

Wrought Iron.—The elastic limits for different descriptions of wrought iron vary according to the nature of the iron.

As an average, however, it may be said that the elastic limit, both in compression and tension, is as follows :—

Bars . . . ·5 of ultimate strength.
Plates . . . ·6 „ „

being about 13 tons per square inch for "Best Yorkshire" iron, and about 11½ tons for Staffordshire crown iron[3] (see p. 317).

Steel.—The elastic limit of different kinds of steel varies considerably, according to the nature of the material and the degree of temper to which it has been subjected.

It ranges from about 12¾ tons in annealed Landore mild steel plates (see p. 321) to 26¾ tons in very hard cast steel (see p. 318), the proportion of the elastic limit to the ultimate strength varying from ·45 to ·8.

Live and Moving Loads.—To consider the effect of moving and live loads upon the strength of iron and steel would open up an interesting subject, which, however, is outside the scope of these Notes.

---

[1] *Proceedings Institute of Civil Engineers,* vol. xlix.
[2] Pole.   [3] From *Experiments of Committee of Civil Engineers.*

Live Loads.—Such loads are seldom met with in buildings, except perhaps in the effect of wind upon roofs ; but they are of frequent occurrence in railway bridges and other engineering structures.

With regard to the effect of live loads, it will be sufficient to say that such loads have a greater effect than if they were gradually applied as dead loads.

In practice, the effect of a live load is generally taken as equal to twice that of the same load considered as dead.

Live and moving loads frequently produce stresses (upon any member of a structure) which vary considerably in intensity from time to time—*e.g.* a bar in a bridge may be subject to a stress of 3 tons per inch of section when a light train is passing, and 5 tons per inch when a heavy train is passing.

Again, moving loads sometimes cause the stresses upon a particular bar to differ in kind. Thus, trains passing over a bridge may cause a bar to be in compression and tension alternately.

It has been shown by Wohler that in either case the intensity of stress that the bar can bear is much lower than what it can bear when the stress is of the same kind (either tension or compression) throughout, and also of the same intensity.

To put it in another way, the stresses produced are much more trying to the bar than a stress which is unvarying in kind (being either compression constantly or tension constantly), and which is also unvarying in amount.

Repeated Loads.—It has already been pointed out that repeated loads do not tend to cause rupture so long as they are kept below the limit of elasticity of the material.

Vibration.—The effect of such loads, or of vibration, has been commonly supposed to be dangerous, and eventually to cause fracture by changing the internal structure of the iron from a fibrous to a crystalline structure. There is still considerable difference of opinion on the subject.

Dr. Percy, who has carefully considered the cases bearing upon this question, says :—
" The question will naturally suggest itself whether gentle vibration—the result of very frequently repeated light blows, or of vibration without impact, caused by jarring grinding action—as in an axle working in badly lubricated bearings, or of straining and torsion in shafts, etc., very much less intense than would be produced by heavy hammering—would tend to incline permanent disaggregation of the crystals of iron, and consequent tenderness. . . .

" Opinions are divided upon it, and I am not acquainted with any precise experimental data to justify any very positive conclusion on the subject. . . .

" Another point remains to be considered, namely, whether vibration, caused by impact or otherwise, may induce a crystalline arrangement which did not previously exist, or was only imperfectly developed. I have not met with any evidence to justify an answer in the affirmative." [1]

Extreme Cold.—The effect of extreme cold upon the strength of iron and steel is another open question.

It has already been pointed out (see p. 272) that in some castings, the bulkier parts, being the last to cool, are left in a state of tension.

Now, if such castings are exposed to cold, the parts already in a state of tension may endeavour to contract still farther, and rupture may ensue.

With regard to the effect of cold upon wrought iron and steel many experiments have been made, but they afford up to the present time very conflicting data.

The discrepancies between the results obtained seem to have been caused in some measure by differences in the composition of the materials experimented upon, the presence of phosphorus especially having a marked influence.

---

[1] Percy's *Metallurgy.*

Iron tyres, chairs, and other parts of a railway which are made of iron or steel, break more frequently during frosty weather than at other times. This, however, has been accounted for by pointing out that the hardness and rigidity of the ground during such weather causes the shocks to have much greater effect upon the permanent way.

As a rule practical men incline to the opinion that frost and extreme cold have a weakening effect upon iron and steel, and render them specially liable to be broken by a sudden shock or concussion.

Thus it is the custom to pass the chains used for lifting heavy weights through the fire on frosty days ; and there is no doubt that while the question is unsettled it is safe to take some precaution of this kind.

Forging.—Forging metal consists in raising it to a high temperature and hammering it into any form that may be required.

It is not proposed to describe the process, but merely to mention one or two points, the neglect of which will seriously impair the strength of the material.

FORGING IRON.—Good wrought iron may be seriously injured by want of care or skill in forging it to different shapes.

Repeated heating and reworking increases the strength of the iron up to a certain point ; but overheating may ruin it (see below) ; the iron should therefore be brought to the required shape as quickly as possible.

The form given to forgings is also important ; there should be no sudden change in the dimensions—angles should be avoided—the larger and thicker parts of a forging should gradually merge by curves into the smaller parts. Experiments have shown that the " continuity of the fibres near the surface should be as little interrupted as possible ; in other words, that the fibres near the surface should lie in layers parallel to the surface." [1]

Overheating.—If wrought iron be "burnt," *i.e.* raised to too high a temperature, its tensile strength and ductility are both seriously reduced. These qualities may, however, be to a great extent restored by carefully reheating and rerolling the iron.

This is well illustrated by the experiments made upon a specimen of bolt iron now before the writer—of which the results are shown below in a tabular form.

	Tensile strength per square inch.	Elongation.	Remarks.
	Tons.	Per cent.	
Original specimen as tested, 1⅛ inch diameter	26·5	68·0	Fine fibrous fracture.
Overheated and fractured by slow tension . .	14·0	20·0	Burnt leaden-looking fracture.
Reheated, rolled down to ¼ inch diameter, and fractured by slow tension	26·8	18·0	Fine grey fibre.

FORGING STEEL requires still more care in order to avoid overheating.
Each variety of steel differs as to the heat to which it can safely be raised.
Shear Steel will stand a white heat.
Blister Steel will stand a moderate heat.
Cast Steel will stand a bright red heat.

Welding is the process by which two pieces of metal are joined together with the aid of heat.

---

[1] Rankine, *Civil Engineering.*

There are several different forms of *weld*.

It is not proposed here to describe the shape of the joint, or the process by which it is made, but merely to give an indication of the principles upon which the welding of metals depends. These are laid down in Dr. Percy's valuable work on *Metallurgy*, from which the information here given is extracted.

It will be sufficient to say that in welding generally the surfaces of the pieces to be joined having been shaped as required for the particular form of weld, are raised to a high temperature, and covered with a flux to prevent oxidation. They are then brought into intimate contact and well hammered, by which they are reduced to their original dimensions, the scale and flux are driven out, and the strength of the iron improved.

WELDING WROUGHT IRON.—The property of welding possessed by wrought iron is due to its continuing soft and more or less pasty through a considerable range of temperature below its melting point.

When at a white heat it is so pasty that if two pieces at this temperature be firmly pressed together and freed from oxide or other impurity they unite intimately and firmly. The flux used to remove the oxide is generally sand, sometimes salt.

WELDING STEEL.—" The facility with which steel may be welded to steel diminishes as the metal approximates to cast iron with respect to the proportion of carbon ; or, what amounts to the same thing, it increases as the metal approximates to wrought iron with respect to absence of carbon.

" Hence in welding together two pieces of steel—*cæteris paribus*—the more nearly their melting points coincide—and these are determined by the amount of carbon they contain—the less should be the difficulty."

Puddled steel welds very indifferently, and so does cast steel containing a large percentage of carbon. The mild cast steels, also shear and blister steel, can be welded with less difficulty.[2]

In forging and welding and tempering steel tools, more than the requisite heat is detrimental, as it opens the grain of the steel and makes it coarse. The heat should be applied regularly, irregular heat causes fracture and irregular grain.

In welding cast steel borax or sal-ammoniac, or mixtures of them, are used as fluxes.[1]

WELDING STEEL TO WROUGHT IRON.—If the melting points of two metals "sensibly differ, then the welding point of the one may be near the melting point of the other, and the difference in the degree of plasticity, so to speak, between the two pieces may be so considerable that when they are brought under the hammer at the welding point of the least fusible, the blow will produce a greater effect upon the latter, and produce an inequality of fibre."

" This constitutes the difficulty in welding steel to wrought iron.

" A difference in the rate of expansion of the two pieces to be welded produces unequal contraction, which is a manifest disadvantage."[3]

Hard cast steel and wrought iron differ so much in their melting points that they can hardly be welded together.

Blister and shear steel, or any of the milder steels, can, however, be welded to wrought iron with ease, care being taken to raise the iron to a higher temperature than the steel, as the welding point of the latter is lower in consequence of its greater fusibility.

WELDING OTHER METALS.—It is not certain that other metals do not become pasty before fusion, but the range of temperature through which it occurs is so small that it would be scarcely possible to hit upon it with any certainty in practice.

---

[1] 16 parts borax, 1 part sal-ammoniac, boiled over a slow fire, and when cold ground to powder, may be used.

[2] Mr. James Christie, M.A.S.C.E., in his remarks on the treatment of mild steel (*Transactions, Am. Soc. C.E.*, vol. xxx.) says, " No welding should be allowed on any steel that enters into structures."

[3] Percy's *Metallurgy.*

CORROSION AND PRESERVATION OF CAST IRON WROUGHT IRON, AND STEEL.

Corrosion.—The different varieties of iron and steel will not oxidise in dry air, or when wholly immersed in fresh water free from air, but they all rust when exposed to the action of water or moisture and air alternately.

" Very *thin* iron oxidises more rapidly than thick iron, owing to the scales of rust on the former being thrown off as soon as formed in consequence of the expansion and contraction from alterations of temperature.

" Iron plates are more durable when united in masses than when isolated. The oxidation of iron is to a great extent arrested by vibration.[1]

" The comparative liability to oxidation of iron and steel in moist air, according to Mr. Mallet, is—[2]

Cast iron	.	.	.	.	100
Wrought iron	.	.	.	.	129
Steel	.	.	.	.	133."

Cast Iron does not rust rapidly in air. When immersed in salt water, however, it is gradually softened, made porous, and converted into a sort of *plumbago*.[3]

Mr. Mallet found that the rate of corrosion decreased with the thickness of the casting, being from $\frac{1}{10}$ to $\frac{4}{10}$ inch during a century in depth for castings 1 inch thick. Mr. D. Stevenson found the decay to be more rapid than this.

Wrought Iron oxidises in moist air more rapidly than cast iron.

The evidence as to its rate of corrosion in salt water is rather contradictory.

Mr. Rennie found that it corroded less quickly than cast iron, but Mr. Mallet's experiments showed that it corroded more quickly.

Steel rusts very rapidly in moist air, more quickly but more uniformly than wrought iron, and far more quickly than cast iron. Low shear steel corrodes more quickly than hard cast steel.[4]

Recent experiments show that steel immersed in salt water is at first corroded more quickly than wrought iron, but that its subsequent corrosion is slower, and the total corrosion after a long period of immersion is less than that of wrought iron.

Preservation.—*Galvanising* consists in covering the iron with a thin coating of zinc.

The iron is cleaned by being steeped for some eight hours in water containing about 1 per cent of sulphuric acid, then scoured with sand, washed, and placed in clean water.

After this the iron is heated, immersed in chloride of zinc to act as a flux, and then plunged into molten zinc, the surface of which is protected by a layer of sal ammoniac.

The process differs slightly according to the size and shape of the article. It is a simple one, and may be applied to small articles in any workshop.

Mr. Kirkaldy found that galvanising does not injure iron in any way.

The zinc protects the iron from oxidation so long as the coating is entire ; but if the sheet iron be bad, or cracked, or if the zinc coating be so damaged that the iron is exposed, a certain action is set up in moist air which ends in the destruction of the sheet.

"The sheets are generally galvanised before they are corrugated ; but as in process of corrugation the sheets, especially the thicker ones, sometimes crack slightly on the surface (unless the iron is of the very highest quality), it is an advantage with all sheets thicker

---

[1] *Proceedings Inst. Civ. Eng.*, vol. xxvii.   [2] Hurst.
[3] A form of carbon known as *graphite* or *blacklead*.
[4] Mr. Mallet in *Proceedings Inst. Civ. Eng.*, vol. ii.

than 20 gauge (see p. 353) to galvanise after corrugation, so as to fill up with zinc any cracks that may have occurred. As, moreover, a larger quantity of zinc adheres to the corrugated than to the flat sheets, they have, when so coated, a distinctly higher value." [1]

PAINTING is an effectual method of preserving iron from oxidation, if the paint is good and properly applied, and the iron in a proper condition to receive it. In order that the protection by painting may continue, the surface should be carefully examined from time to time, so that all rust may be removed. The paint may be renewed directly it is necessary (see Chap. x., Part II.).

The following hints on the subject are condensed chiefly from the eminently practical book entitled *Works in Iron*, by Mr. Matheson.

Cast Iron should be painted soon after it leaves the mould, before it has time to rust. The object of this is to preserve intact the hard skin which is formed upon the surface of the metal by the fusing of the sand in which it is cast. After this a second coat should be applied, and this should be renewed from time to time as required. In any case, all rust upon the surface of castings should be carefully removed before the paint is applied. Small castings are often *japanned* (see p. 433).

Wrought Iron.—Before painting wrought iron care must be taken to remove the hard skin of oxide formed upon the surface of the iron during the process of rolling, and which, by the formation of an almost imperceptible rust, becomes partly loose and detached from the iron itself. An attempt to prevent this rusting is sometimes made by dipping the iron, while still hot, in oil. This plan, however, is expensive, and not very successful. The scale is sometimes got rid of by "pickling," the iron being first dipped in dilute acid, to remove the scale, and then washed in pure water.[2] "If the trouble and expense were not a bar to its general adoption, this is the proper process for preparing wrought iron for paint, and it is exacted occasionally in very strict specifications." "But somewhat the same result may be obtained by allowing the iron work to rust, and then scraping off the scale preparatory to painting. If some rust remains upon the iron the paint should not be applied lightly to it, but by means of a hard brush should be mixed with the rust." Ordinary lead paints, especially red lead, are often used for protecting iron work, but they are objected to on the ground that galvanic action is set up between the lead and the iron.

Mr. Matheson recommends oxide of iron paints for iron work generally, and bituminous paints for the inside of pipes or for ironwork fixed under water. The precautions to be taken in using these paints, and the objections to ordinary lead paints, are given in chapter VI. The ironwork for roofs, bridges, and similar structures, generally receives one coat of paint before it leaves the shops, and two or three more after it is fixed.

Dr. Angus Smith's process is an admirable means for preventing corrosion in cast-iron pipes. The pipes having been thoroughly cleaned from mould, sand, and rust, are heated to about 300° Fahr. They are then dipped vertically into a mixture consisting of coal-tar, pitch, about 5 or 6 per cent of linseed oil, and sometimes a little resin, heated to about 300° Fahr.[3] After remaining in the mixture, the pipes are gradually withdrawn and allowed to cool in a vertical position. Perfect cohesion should take place between the coating and the pipe, and the former should be free from blisters of any kind.

In practice the heating of the pipes before immersion is found to be very expensive, and is usually omitted. However, some engineers consider it essential for really good work.[4]

The Bower-Barff processes[5] protect the surfaces of iron and steel by covering them with a coating of black magnetic oxide. In the original process, invented by Professor *Barff*, this was effected by subjecting the articles to be coated in a heated muffle to the action of superheated steam. The heated metal decomposes the steam and combines with some of its oxygen to form the coating of magnetic oxide. A similar effect is produced by Mr. *Bower's* patent, under which the gas from a

---

[1] Matheson.   [2] This process is also adopted for mild steel plates.
[3] *Specification of Robert Angus Smith, Application and Preparation of Coal Tar, A.D. 1848, No. 12,291.*
[4] For further remarks on the alternative processes of dipping pipes when hot, and dipping them when cold, see Humber, *Water Supply of Cities and Towns;* Burton, *Water Supply of Towns;* Fanning, *Water Supply Engineering;* Wegmann, *Water Supply of the City of New York; Report of the American Waterworks Association, Philadelphia, 1891; Report of Committee on Specification for Cast Iron Water Pipes.*
[5] *Proceedings Soc. Engineers,* 1884, p. 59, Mr. Bower's paper.

producer is burnt with a slight excess of air, and taken into a brick chamber, in which the articles to be coated are placed, a red coating of sesquioxide is produced soon after the articles are red, but after about 40 minutes the air is shut off, and the producer gases only admitted, when, in 20 minutes more, the sesquioxide is converted into magnetic oxide.

"This alternate treatment goes on for different periods, depending upon the nature of the articles and the purpose for which they are required.

"For indoor work 4 hours are sufficient, but the time varies from 4 to 8 hours, or about half that necessary for coating by the aid of steam."[1]

Both processes are now worked by Mr. Bower, Prof. Barff's process being better for wrought iron, and that of Mr. Bower, which is much cheaper, for cast iron.

These processes are said not to impair the strength or other qualities of the iron, and to protect it thoroughly against oxidation or corrosion from damp earth, salt-water, or other causes.

BRIGHT IRONWORK.—The portions of ironwork that have been turned or fitted, and all tooled surfaces, should be protected by a coating of tallow, mixed with white lead to prevent it from easily melting and running off the metal.

"Dr. Percy recommends for the same purpose common rosin melted with a little Gallipoli oil and spirits of turpentine. The proportions, which may easily be found by trial, should be such as will make it adhere firmly and not chip off, and yet admit of being easily detached by cautious scraping."[2]

Bronzing is done with bronze powder, paint, or varnish, but does not stand the weather well.

Gilding has to be done with special care, or the gold will be destroyed by rust. The surface of the iron, having been very carefully cleaned, is painted with two coats of iron oxide paint, then with two coats of lead paint of light colour as a basis for the "oil gold size" upon which the gold leaf is placed. When properly done the gilding will last fifteen or twenty years.[3]

CHARACTERISTICS AND USES OF IRON AND STEEL.

The student will have perceived that the products of the iron manufacturer may be divided into three classes—cast iron, wrought iron, and steel, the differences in which are caused partly by the amount of carbon they respectively contain, and also by the processes they have undergone.

The following Table, from Bauermann's *Metallurgy*, gives the proportion of carbon in different varieties of iron and steel according to Karsten :—

NAME.	PERCENTAGE OF CARBON.	PROPERTIES.
1. Malleable iron . . .	0·25	Is not sensibly hardened by sudden cooling.
2. Steely iron . . .	0·35	Can be slightly hardened by quenching.
3. Steel	0·50	Gives sparks with a flint when hardened.
4. Do.	1·00 to 1·50	Limits for steel of maximum hardness and tenacity.
5. Do.	1·75	Superior limit of welding steel.
6. Do.	1·80	Very hard cast steel, forging with great difficulty.
7. Do.	1·90	Not malleable hot.
8. Cast iron	2·00	Lower limits of cast-iron cannot be hammered.
9. Do.	6·00	Highest carburetted compound obtainable.

---

[1] *P.I.C.E.* 1884, p. 59.   [2] Pole.   [3] Matheson.

The great differences in the characteristics of cast iron, and wrought iron and steel, are briefly recapitulated below, and these determine the uses to which they are respectively applied.

Cast Iron has little tensile strength, but affords great resistance to compression.

It is hard, brittle, wanting in toughness and elasticity, and gives way without warning, especially under sudden shocks or changes of temperature. It is easily melted and run into various shapes.

The castings thus produced are liable to air-holes and other flaws, which reduce their strength. Small castings are stronger in proportion to their size than large ones.

Cast iron can be cut or turned with edge tools, but is not malleable either when cold or hot, nor is it weldable.

It is not so easily oxidised in moist air as wrought iron. In salt water, however, it is gradually softened and converted into plumbago.

Cast iron is peculiarly adapted for columns, bedding plates, struts, chairs, shoes, heads, and all parts of a structure which have to bear none but steady compressive strains ; also for gutters, water pipes, railings, grate fronts, and ornamental work of nearly every description.

It has been much employed for girders, but is an untrustworthy material for those of large size, or in important positions. It is liable to crack and give way without warning under sudden shocks, and also under extreme changes of temperature, such as occur in the case of buildings on fire, where the girders may become highly heated, and then suddenly cooled by water being poured on them.

MALLEABLE CAST IRON possesses originally the fusibility of cast iron, and eventually acquires some of the strength and toughness of wrought iron. It may be used for heads, shoes, and other joints in roofs, and for all articles in which intricacy of form has to be combined with a certain amount of toughness.

Wrought Iron has many most valuable qualities, though these differ considerably as to degree in different varieties of the material.

Its tensile strength is three or four times as great as that of cast iron, but it offers not half the resistance to compression.

It is, however, very tough and ductile, and therefore gives way gradually instead of suddenly snapping.

Its elastic limit is equal to about half its ultimate strength, and it will bear repeated loads below that limit without injury.

Wrought iron is practically infusible, is malleable hot and cold, is weldable at high temperatures, and can be forged into various shapes.

It is subject to " hot and cold shortness " produced by impurities, and to other defects. Large sections are more likely to contain flaws than small ones. Bars are, as a rule, stronger than plates, and plates are stronger with the grain than across it.

Malleable iron rusts quickly in moist air, but under certain conditions stands salt water better than cast iron.

The great tensile strength of wrought iron leads to its employment for tie-rods, bolts, straps, and all members of any structure which are exposed to tensile stress. Prior to the introduction of Mild Steel it was employed in compression members such as columns and struts, for all important beams and girders, especially those exposed to sudden shock, and was largely used in roofs, braced girders, and other iron structures. Its use is still maintained in the manufacture of corrugated sheets for roof coverings, and in other ways.

Steel differs even more than wrought iron in the characteristics of its several varieties.

It has a high tensile strength, much greater than that of wrought iron. Its resistance to compression is also much greater. Moreover, it has a harder surface, and is better able to resist wear and tear.

Hard steels, containing a large proportion of carbon, are fusible, easily tempered, have a high tenacity and elastic limit. Their resistance to compression is enormous, especially when they are tempered, but they cannot be easily welded or forged, are brittle, and very uncertain in quality.

Soft mild steels have a tenacity and resistance to compression, and an elastic limit somewhat higher than wrought iron. They can be hardened and tempered, but not easily. They are weldable and easily forged, and afford a very reliable and ductile material adapted for structures subject to sudden shocks.

Steel is more easily oxidised than wrought iron, and far more easily than cast iron.

The importance of Mild Steel or " Ingot Iron " in various branches of construction has already been alluded to. Examples of its successful application are to be found in many of the largest engineering structures in existence, while in building construction its use in the form of rolled joists, columns, and other structural details has superseded that of wrought iron.

COPPER.

Uses.—Copper is used by the builder chiefly for slate nails and bell wires, sometimes for rain-water pipes and gutters, for covering roofs, for lightning-conductors, and for dowels ; also for bolts and fastenings in positions where iron would be corroded or oxidised. Moreover, it forms most useful alloys with other metals.

Copper wire cord is sometimes used for sash lines, and also for lightning-conductors.

Ores.—It is frequently found in the metallic slate, and is also obtained from copper pyrites, grey and red copper ores, from copper glance, and other ores, by roasting, calcining, refining, and melting them with certain fluxes and oxidising agents.

The presence of sulphur and antimony decreases the malleability and ductility of copper. Small quantities of arsenic and phosphorus increase its toughness, but large quantities injure it.

Properties.—The red colour of copper is familiar to all. The metal is peculiarly malleable, and can be hammered or rolled into very thin sheets.

In tenacity it is inferior to wrought iron, but is superior to most other metals. The tensile strength of rolled copper annealed is about 13 to 14 tons per square inch with an elongation of 30 per cent in 4 inches, that of cast copper being $8\frac{1}{2}$ tons. It can be drawn into moderately fine wire with a tenacity when hard of 26 tons per square inch, or of 20 tons per square inch when annealed.[1] It can be worked either cold or hot—in the latter case it is easily oxidised—but it cannot be welded.

OXIDATION AND CORROSION.—Copper oxidises very slowly in air, being covered with a film of carbonate, commonly called *verdigris*.[2] The appearance of this film is well known to all ; it forms a protective coating which preserves the surface of the copper from further oxidation.

Copper is corroded by salt water if at the same time air has access to it ; the presence of a small proportion of phosphorus is said to retard the corrosion.

Market Forms.—*Sheet Copper.*—The most useful form for the builder in which copper is sold, is in sheets measuring about 4 feet by 2 feet (in Scotland 4 feet by 3 feet 6 inches), and described according to their thickness (by the Birmingham Wire Gauge), and their weight per foot superficial, or their weight per sheet.

The gauges of the sheets vary from No. 1 to 30 W.G. The weights of a few of the most useful thicknesses are given in the Table below :—

TABLE of WEIGHT of SHEET COPPER.

Birmingham Wire Gauge.	Weight per foot superficial in ounces.	Weight per sheet, 4 feet by 2 feet, in lbs.	Birmingham Wire Gauge.	Weight per foot superficial in ounces.	Weight per sheet, 4 feet by 2 feet, in lbs.
20	26	13	26	13	$6\frac{1}{2}$
22	20	10	28	10	5
24	16	8	30	8	4

Sheet copper, weighing from 12 to 20 oz. per square foot, is used for roofs, flats, and gutters. Copper wire from 17 to 19 B.W. gauge for bell-hanging. When used for roofing, copper is laid in a way somewhat similar to zinc (see p. 209, Part I.).

Copper Wire Cord.—The following are the working loads [3] for the different sizes :—

Circumference in inches . . $1\frac{1}{4}$, $1\frac{1}{8}$, 1, $\frac{3}{4}$, $\frac{5}{8}$, $\frac{1}{2}$, $\frac{3}{8}$, $\frac{1}{4}$.
Working load in lbs. . . 448, 336, 224, 168, 112, 75, 50, 34.

Copper Wire-covered Steel Ribbon Sash Line is also made under Hookham's patent in three sizes, having a breaking strain in cwt. as follows :—

No. 1   2   3
3 cwt. $4\frac{1}{2}$ cwt.   7 cwt.

---

[1] **Wertheim.**   [2] *Verdigris*, properly so called, is a basic acetate of copper.
[3] Sheffield Standard List.

LEAD.

Uses.—Lead is much used by the builder for cisterns, pipes, flat roofs, etc., and from it is prepared white lead, the basis of most ordinary paint. The engineer requires it as a bedding for the ends of girders, and for other minor purposes.

Ores.—Lead is not found in the metallic state, but is reduced chiefly from the ore called *galena* (the sulphide) by roasting or smelting in a reverberatory furnace, furnished with long flues to catch the particles of lead, which would otherwise be carried away in the smoke.

Properties.—Lead is extremely soft and plastic, very malleable, fusible, heavy, and very wanting in tenacity and elasticity.

Market Forms.—Lead may be purchased in cast pigs, sheets, or pipes.

Sheets are either " cast," or " milled," and are described according to their weight per foot superficial.

Cast lead is made in sheets from 16 to 18 feet long, and 6 feet wide ; it is thicker and heavier than milled lead, and has a harder surface.

It is, however, liable to flaws and sand holes, and is irregular in thickness, on account of which it should not be used of a lighter substance than 6 lbs. per square foot.

Cast lead is often made by the plumber himself out of the old lead waste pieces and clippings that accumulate in the course of his work.

Milled lead is rolled out thinner than the other, is more uniform in thickness, bends easily, and makes neater work, but cracks if much exposed to the sun. The sheets are from 25 to 35 feet long, and from 6 feet to $7\frac{1}{2}$ wide, Sheet lead is always described according to its weight in lbs. per foot superficial.

The following Table shows the thickness of sheet lead for different weights per square foot.

Table giving Weight and Thickness of Sheet Lead.

Weight in lbs. per superficial foot.	Thickness in inches.	Nearest simple fraction.	Weight in lbs. per superficial foot.	Thickness in inches.	Nearest simple fraction.
1	0·017	$\frac{1}{60}$	7	0·118	$\frac{7}{64}$
2	0·034	$\frac{1}{32}$	8	0·135	$\frac{1}{8}$
3	0·051	$\frac{1}{20}$	9	0·152	$\frac{9}{64}$
4	0·068	$\frac{1}{16}$	10	0·169	$\frac{5}{32}$
5	0·085	$\frac{5}{64}$	11	0·186	$\frac{11}{64}$
6	0·101	$\frac{3}{32}$	12	0·203	$\frac{3}{16}$
			15	0·255	$\frac{1}{4}$

The weights of sheet lead generally used are as follows (see Chap. xiv., Part I.) :—For aprons, 5 lb. lead ; for roofs, flats, and gutters, 7 or 8 lbs.; for hips and ridges, 6 or 7 lbs. ; thicker if much exposed.

Laminated Lead is a very thin description of sheet used, made for covering damp walls.

Action of Water upon Lead.—Soft water, especially when full of air, or when containing organic matter, acts upon lead in such a way that some of it is taken up in solution, and the water is poisoned.

This makes lead a dangerous material to use in many cases for cisterns and pipes connected with the supply of water for drinking purposes, or for roofs and flats whence that supply may be drawn.

Vitiated or impure air acts upon lead in a somewhat similar manner.

There has been a good deal of discussion with regard to the action of different kinds of water upon lead, as the subject is an important one, the following remarks are inserted. They are chiefly founded upon the valuable standard work on hygiene by the late Professor Parkes.

Pure water, not containing air, does not act upon pure lead.

When the water contains much oxygen, the lead is oxidised ; and oxide of lead, a highly poisonous substance, is to some extent soluble in water.

If there is much carbonic acid present it converts some of the oxide into carbonate of lead, which is almost insoluble and therefore comparatively harmless.

The waters which act most upon lead are the purest and most highly oxygenated, also those containing organic matter—nitrites, nitrates, and chlorides.

The waters which act least upon lead are those containing carbonate of lime and phosphate of lime, in a less degree sulphate of lime. Some of these form a coating on the inside of the pipe which protects it from further action.

Some vegetable substances contained in water, peaty matter for example, also protect the pipe by forming an internal coating upon it.

It appears therefore that hard waters, containing (as they generally do) carbonate of lime, do not readily affect lead.

Soft waters, such as rain water, and water obtained by distillation—water polluted with sewage—water in tanks having a muddy deposit—may all become poisoned when in contact with lead.

" The mud of several rivers, even the Thames, will corrode lead, probably from the organic matter it contains, but it does not necessarily follow that any lead has been dissolved in the water. Bits of mortar will also corrode lead." [1]

Vegetables and fatty acids arising from fruit and vegetables, cider, sour milk, etc., also act upon lead.

The poisonous effects of lead show themselves in other materials connected with building.

For example, white lead, the basis of most paints, is a highly poisonous substance, and leads to serious diseases among the workmen who manufacture the white lead, and among the painters who use it (see p. 404).

Lead Pipes are much used in connection with water supply, etc.

Pipes of large diameter are generally made by the plumber out of sheet lead.

Smaller pipes used to be cast in short lengths of considerable thickness, and then *drawn* out to the proper dimensions.

Now, however, they are generally formed by forcing the molten metal, by hydraulic pressure, through a die of the section required.

Soil pipes should always be " drawn," and are thus made of from $3\frac{1}{2}$ to 5 inches diameter, and of thicknesses equal to those of sheet lead varying in weight from 5 to 10 lbs. per square foot.

Water pipes.—The thickness and consequently the weight of lead pipes used for water supply should be regulated by the pressure of water they are intended to bear.

---

[1] Parkes' *Hygiene.*

The following Table shows the sizes and weights per yard run of pipes usually made and the heads of water to which they can be safely subjected in practice :—

Lengths in which made Feet.	Internal diameter of pipe in inches.	Overflows.		Heads about 50 feet.		Heads about 300 feet.		Heads about 500 feet.	
		Weight in lbs. per yard run.							
15 (or in coils of 60 ft.)	¼	...	...	...	...	...	...	...	3·9
,,	⅜	...	...	...	...	...	3·	3·9	4·8
,,	½	2·7	3·	3·6	3·9	4·5	4·8	5·7	6·0
,,	⅝	...	..	...	...	...	3·6	4·5	6·0
,,	¾	4·5	4·8	5·1	5·7	6·3	7·2	8·4	9·0
,,	1	...	6	7·2	5·1	9·6	11·1	12·0	12·9
12 (or in coils of 50 ft. or from 40 to 50 ft.)	1¼	...	...	...	9·0	10·5	12·0	12·9	15·0
,,	1½	...	...	9·0	12·0	14·1	18·0	21·0	24·0
,,	1¾	...	...	...	...	...	18·0	21·0	24·0
,,	2	...	9·0	14·1	18·0	21·0	24·0	27·9	30·0
10	2½	...	...	10·8	21·0	25·2	28·8	33·6	36·0
,,	3	12·6	18·0	24·0	30·0	33·6	36·0	39·0	42·0
,,	3½	16·8	27·0	33·6	36·0	39·0	45·0	48·0	54·0
,,	4	16·8	21·0	24·0	33·6	42·0	48·0	51·0	60·0
,,	4½	...	19·8	25·2	33·6	42·0	51·0	60·0	66·0
,,	5	...	...	...	51·0	60·0	70·2	76·2	84·0
Number of Column		1	2	3	4	5	6	7	8

The above are reduced from the price list of Messrs. John Bolding and Sons, manufacturers.

COATING LEAD PIPES TO PREVENT POISONING.—Several methods have been proposed for coating and lining the insides of lead pipes to prevent the water conveyed by them from being poisoned.

All of these are condemned by Professor Parkes as being objectionable, except the following :—

M'Dougal's Patent consists in applying an internal bituminous coating, which is said to have been successful.

Schwartz's Patent.—The pipe is boiled in sulphide of soda for fifteen minutes, by which the interior is coated with sulphide of lead (a substance insoluble in water).

Lead Encased Pipes.—Tin pipes, and copper pipes, lined with tin, have been proposed as substitutes for lead pipes, but they are too expensive.

The lead encased pipe, made under *Haines's patent*, has, however, been found to be perfectly successful.

This consists of an inner pipe of block tin, encased in a lead pipe as shown in section, Fig. 146. The two metals are so united that no joint between them is perceivable, and they cannot be separated by any amount of bending or twisting.

In consequence of the tin melting at a lower temperature than the lead it is somewhat difficult to make a soldered joint in these pipes. However, it may be done with care, or *Heap's* mechanical joints may be used, in which the union is effected by means of screwed couplings.

Fig. 146.

Weight of Lead-encased Pipe.—As the lead-encased pipe is stronger than ordinary lead pipe, it may be of less weight per yard for water supply under any given pressure. To meet the case in which water companies require pipes to be of a certain regulated weight according to the head, a special lead-encased pipe is made with a smaller proportion of tin. The weights of pipes of this class are shown in cols. 9, 10, 11 of the Table below. They are heavier and cheaper than the pipes with full proportion of tin, whose weights are given in cols. 6, 7, 8.

TABLE of WEIGHTS of LEAD-ENCASED PIPES IN LBS. PER YARD RUN.[1]

Internal diameter in inches.	Extra light Weights.				Weights suitable for supply of water under the heads stated.			Extra heavy weights with less tin for supply of water under heads stated.		
					50 feet head and under.	51 to 250 feet head.	251 to 500 feet head.	50 feet head and under.	51 to 250 feet head.	251 to 500 feet head.
$\frac{3}{8}$	1	$1\frac{1}{2}$	2	...	$2\frac{1}{2}$	3	$3\frac{1}{2}$	4	$4\frac{1}{2}$	5
$\frac{1}{2}$	2	$2\frac{1}{2}$	3	...	$3\frac{1}{2}$	4	$4\frac{1}{2}$	5	6	7
$\frac{5}{8}$	3	$3\frac{1}{2}$	4	...	$4\frac{1}{2}$	$5\frac{1}{4}$	6	7	8	9
$\frac{3}{4}$	$3\frac{1}{2}$	4	$4\frac{1}{2}$	5	$5\frac{1}{2}$	6	7	8	9	10
1	$4\frac{1}{2}$	5	$5\frac{1}{2}$	6	$7\frac{1}{4}$	8	9	10	11	12
$1\frac{1}{4}$	$6\frac{1}{2}$	7	8	...	9	10	12	$12\frac{1}{2}$	14	16
$1\frac{1}{2}$	8	9	10	...	11	$12\frac{1}{2}$	14	16	18	21
2	11	13	...	...	16	$18\frac{1}{2}$	21	23	26	30
No. of column.	2	3	4	5	6	7	8	9	10	11

Strength of Lead Pipes and Lead-Encased Pipes.—Mr. Kirkaldy found the strength of lead pipes and of lead-encased pipes to be respectively as follows :— [1]

Internal Diamr.	Lead Pipe.				Lead-encased Pipe.		
	Thickness.	Weight per foot.	Bursting pressure per sq. in. in lbs.		Thickness.	Weight per foot.	Bursting pressure per sq. in. in lbs.
$\frac{1}{2}$	·2	2·3	1579		·14	1·3	1859
$\frac{5}{8}$	·2	2·6	1349		·13	1·4	1454
$\frac{3}{4}$	·22	3·8	1191		·15	1·9	1416
1	·2	4·1	911		·14	2·4	1265
$1\frac{1}{4}$	·21	5·3	683		·13	2·7	835
$1\frac{1}{2}$	·24	7·1	734		·15	3·8	849
2	·21	9·2	498		·17	5·4	642

The tearing strength of lead pipe was 2159 lbs. per square inch, of lead-encased pipe 3759 lbs. per square inch.

Glass lined pipes may be mentioned here, though they are iron pipes lined with glass tubes. They are stated to be safe against lead poisoning, to require no soldering, to be rat proof, to have but little internal friction, and to be not liable to choke from corrosion like iron pipes. They are made from $\frac{1}{2}$ to $2\frac{1}{2}$ inches in diameter, in various lengths up to 6 feet, with ends screwed into sockets, and with asbestos washers.

Fret Lead, for glazing, is made (as described in Chap. x., Part II.) in *cames, i.e.* long strips, of H section, the width of the groove (*i.e.* the length of the cross bar of the H) : the width of the face (*i.e.* the side of the H), and the

---

[1] *Lead Poisoning of Water and its Prevention,* by A. M'Callum Gordon.

shape of the face differ, the latter is made flat or round. There are three classes of fret lead, known as *ordinary, narrow,* and *broad.*

The following Table, from Seddon's *Builder's Work,* gives a general idea of the sizes as obtained in the market :—

	Width of Grooves in inches.	Shape of Face.	Width of Face in inches.	Remarks.
Ordinary . .	$\frac{3}{32}$	Flat.	$\frac{7}{16}$	{ Used for ordinary lead lights up to 21 oz. sheet glass.
Narrow . .	$\frac{1}{8}$	Flat.	5 widths from $\frac{3}{16}$ to $\frac{1}{2}$	} Used for cathedral and thick antique glass, according to its thickness.
Narrow .	$\frac{3}{16}$	Round.	2 ,, $\frac{3}{16}$ to $\frac{1}{4}$	
Broad . .	$\frac{1}{4}$	Flat.	5 ,, $\frac{3}{16}$ to $\frac{1}{2}$	
Broad . .	$\frac{1}{4}$	Round.	2 ,, $\frac{3}{16}$ to $\frac{1}{4}$	

ZINC.

Uses.—Zinc is much used for roofs, for light gutters and pipes, for cisterns, chimney pots, ornaments, ventilators, etc.; for slating nails, for tubing, and for covering iron to protect it from oxidation. It also forms a component part of several useful alloys, and the oxide of the metal is used as a basis for zinc paint.

Ores.—The metal is produced from the ores known as " calamine " (the carbonate), " blende " or " black jack " (the sulphide), and red zinc ore (the oxide).

The ore is roasted, mixed with charcoal, and heated in peculiar retorts. The zinc is converted into vapour, condensed, and then fused. Most of the zinc used in this country comes from Belgium.

Properties.—Zinc is easily fusible. Cast zinc is brittle when cold. If pure it becomes malleable at about 220° F., and can be rolled into sheets, which retain their malleability. At very high temperatures, such as 400° F., it becomes very brittle again.

The presence of lead makes zinc too brittle to roll at any temperature.

Zinc should be cast at a low temperature, or the metal will become very hard, and some of it will pass off in vapour.

Zinc is easily acted upon by moist air ; a film of oxide is soon formed, which, however, protects the metal from further action.

If, however, the air contains acid, as it does near the sea and in large towns, the zinc is destroyed.

Soot is very destructive to zinc, forming with it a galvanic couple, which is brought into action by the moisture and acid in the air.[1]

Good sheet zinc is of an uniform colour, tough, and easily bent backwards and forwards without cracking.

Inferior zinc is of a darker colour than the pure metal, and of a blotchy

---

[1] *Proceedings Inst. Civ. Eng.* vol. **xxvii.**

appearance, caused by the presence of other metals, which set up a galvanic action and soon destroy the zinc.

There is no practical engineer's test for the quality of zinc. Good zinc should, however, be as free from iron as possible. The following is an analysis of Vieille Montagne zinc, which shows that it is practically pure :—

Zinc	. .	0·995
Iron	. .	0·004
Lead, etc.	.	0·001
		1·000

Zinc containing more than about 1 per cent of lead should be rejected.

Market Forms.[1]—Zinc is sold in sheets 7 feet by 2 feet 8 inches, 7 feet by 3 feet, or 8 feet by 3 feet, described by their thickness and weight in ounces per foot superficial (according to a special gauge which varies with different manufacturers).

ZINC GAUGE.—The following Table shows the weight of zinc per square foot for the various numbers of the *Zinc Gauge*, properly so called. This gauge originated in Belgium, and is sometimes called the *Belgian Zinc Gauge*, but it is known in the trade as the Zinc Gauge, and is used by Messrs. F. Braby and Company, the English agents of the Vieille Montagne Zinc Company, whose zinc, obtained from mines in Belgium, Sweden, and Spain, is of excellent quality and extensively used in this country.

The thickness of the sheets is also given in the Table ; those from Nos. 10 to 21 (except 18) have been accurately measured and kindly furnished by Messrs. Braby ; the others are calculated.

Gauge.	Approximate Weight per square foot.			Approximate Thickness.	Gauge.	Approximate Weight per square foot.			Approximate Thickness.
	Lbs.	Oz.	Dr.			Lbs.	Oz.	Dr.	
				Inch.					Inch.
1	0	1	2	·0018	14	1	2	12	·0326
2	0	2	4	·0036	15	1	5	12	·0364
3	0	3	7	·0055	16	1	8	12	·0400
4	0	4	9	·0073	17	1	11	11	·0437
5	0	5	11	·0091	18	1	14	11	·0478
6	0	6	14	·0110	19	2	1	11	·0509
7	0	8	0	·0128	20	2	4	10	·0581
8	0	9	2	·0146	21	2	8	2	·0728
9	0	10	5	·0165	22	2	12	14	·0764
10	0	11	7	·0180	23	3	1	1	·0800
11	0	13	5	·0217	24	3	5	3	·0896
12	0	15	2	·0254	25	3	9	5	·0992
13	1	0	15	·0290	26	3	13	7	·1088

Of the above sheets, Nos. 1 to 5 are rolled only to order and of special dimensions. The remaining gauges are made in sheets of all the three sizes mentioned above.

There are several other zinc gauges given in various Price Books, etc., but they are generally based upon the above, the range of numbers being smaller and the weight not so accurately given.

---

[1] The commercial name for zinc before it is converted into sheet and other useful forms is *Spelter.*

The thicknesses of zinc recommended for roofing purposes are given in Chap. xii., Part I.

The expansion and contraction of this metal with changes of temperature are greater than of any other, and should be carefully guarded against by laying sheets on roofs without rigid fastenings as described in Part I.

Zinc should not be allowed to be in contact with iron, copper, or lead. In either case voltaic action is set up, which destroys the zinc. This occurs especially, and more rapidly, when moisture is present.

Zinc should also be kept clear of lime or calcareous water, and of any wood, such as oak, which contains acid.

Zinc laid on flats or roofs where cats can gain access is also soon corroded.

An objection to zinc for roofs is that it catches fire at a red heat and blazes furiously.[1]

TIN.

USES.—Tin is used in building for lining lead pipes, occasionally as a protective covering for iron plates, and for small gas tubing.

ORES.—The metal is obtained from an ore called " tin-stone "—the binoxide, and also from tin pyrites. The ore is stamped ; roasted to expel sulphur and arsenic ; washed, mixed with flux, and smelted in a reverberatory furnace, whence the liquid metal is run into a basin, and thence into moulds. The ingots thus produced are refined and boiled.

Properties.—Tin is very soft, more easily fused than any other metal, very malleable, and very slowly oxidised, but its tensile strength and ductility are very low.

Tin may be distinguished from other metals by its crackling when bent. Its purity is tested by its extreme brittleness at high temperatures.

Tin Tubing is made of diameters varying from $\frac{1}{4}$ to 1 inch, and of light section, for the conveyance of gas, but is now not much used, having been superseded by the composition tubing described below. The cost of the metal makes it too expensive, if made strong enough, for water supply.

Composition Tubing is made from a mixture of tin, lead, and antimony. It is extensively used for the smaller branches of gas supply, being much less expensive than tin tubing, is easily bent to suit any position, and can be attached to connections by soldering

---

[1] Bloxam.

WEIGHT in Ounces per YARD RUN of TIN and of COMPOSITION TUBING.

Internal Diameter.	Weight per Yard in ounces.		Internal Diameter.	Weight per Yard in ounces.	
	Tin.	Composition.		Tin.	Composition.
$\frac{1}{8}$	...	5	$\frac{1}{2}$	17	29 to 34
$\frac{3}{16}$	...	8	$\frac{9}{16}$	...	36
$\frac{1}{4}$	7 to 8	11 to 13	$\frac{5}{8}$	23	44 to 52
$\frac{5}{16}$	9 to $9\frac{1}{2}$	14 to 16	$\frac{3}{4}$	30	52 to 68
$\frac{3}{8}$	11	18 to 21	$\frac{7}{8}$	38	64 to 76
$\frac{7}{16}$	14	23 to 26	1	47 to 48	80 to 88

Tin Plate is Siemens or Bessemer steel plate covered with a coating of tin by a process similar to that of galvanising, described at page 333, molten tin being used instead of zinc.

The terms *charcoal plate* and *coke plate* are still used to signify various qualities of tin plate, when made in steel, although having originally reference to the process of manufacture of iron, when that material was more largely used in the tin plate trade than it now is.

A good grade of steel for plates from which stamped articles are to be manufactured, is stated to have a chemical composition as follows : carbon 0·08 to 0·10 per cent ; sulphur below 0·05 ; phosphorous below 0·02 ; manganese 0·25 to 0·35 ; silicon a trace.

Terne Plate is described at page 289.
Block Tin or *Doubles* consists of tin plate with a much thicker coating of tin upon it. It is used for the best tin ware.
Crystallised Tin Plate is made by heating the surface of ordinary tin plate with hydrochloric and nitric acids, which gives it a variegated appearance. This is sometimes known as Moiré metallique.
Tinned Copper is often used for kitchen utensils. The surface of the copper is cleaned before tinning with sal-ammoniac.

ALLOYS.

Alloys are mixtures formed by melting two or more metals together.

They are not, however, mere mechanical mixtures, for they often exhibit properties different from those possessed by the metals in the mixture.

For example, copper and tin are both very malleable metals.

Two parts of copper with one of tin form a white alloy (*speculum metal*) so hard that it cannot be cut with steel tools, and as brittle as glass.

The tensile strength of this alloy is only $\frac{1}{5}$ that of tin and $\frac{1}{50}$ that of copper.

Nine parts of copper to one of tin make a tough, rigid *gun metal*, harder and more fusible than copper, but which cannot be rolled or drawn.

By adding tin (a softer metal than copper) to gun metal its hardness is increased !

In preparing alloys the most infusible metal should be melted first, and the others subsequently added.

If the metals are of different specific gravities they must be continually stirred while fluid, or the heavier will sink to the bottom and the alloy will not be homogeneous.

The specific gravity of an alloy is seldom equal to the mean of the specific gravities of the metals in the mixture. It is sometimes more and sometimes less dense.

The tensile strength of an alloy is generally much greater than that of the metals composing it.

Brass is an alloy composed of copper and zinc, the proportions of which vary according to the purpose for which the metal is required.

The zinc is melted first, and the copper added in small quantities. A little old brass in the crucible will facilitate the union of the metals. The crucible must be covered with charcoal powder and a close lid, or the zinc will pass away in vapour.

Colour.—The colour depends upon the proportions.

Common *yellow brass* contains 2 parts of copper to 1 of zinc. If the copper be in greater proportion than 4 to 1, the alloy is reddish ; if less than 3 to 1, it becomes of somewhat the colour of zinc.

Properties.—Brass is tough, as a rule, but is rendered brittle by continued vibration. The presence of iron injures its tensile strength and malleability.

It is more malleable than copper when cold, but cannot be forged at a red heat, because the zinc melts at a low temperature.

The fusibility of brass increases in proportion to the quantity of zinc it contains. The addition of a little phosphorus makes it very liquid and easily run into fine castings.

The proportions of the constituents for the different kinds of brass, and the uses to which these are applied, are shown in the Table on p. 348.

The name *Brass* is frequently given to all alloys of copper. Those containing tin should properly be called *Bronze*.

Muntz metal or *sheathing* is cheaper than common brass and more easily rolled. It is much used for sheathing ships, as it keeps cleaner than copper, and is sometimes employed as a covering for small roofs.

Muntz metal made, as it usually is, of 60 parts copper and 40 zinc, has been found to be attacked by salt water and to lose its zinc. An alloy is therefore used instead of 68 parts copper and 32 zinc.

DELTA METAL, sometimes called Dick's metal, is an improved brass, which can be made tough and hard ; it can be forged or rolled hot, or worked and drawn into wire when cold. It makes sound fine castings, is of the colour of gold alloyed with silver, and when exposed to the atmosphere tarnishes less than brass.

KIRKALDY'S TESTS.[1]

	Stress per square inch in tons.		Contraction of area at fracture per cent.	Extension in 10 inches per cent.
	Ultimate.	Elastic limit.		
Bar 1 as drawn .	33·6	22·1	15·0	8·8
„ 2 annealed .	27·2	8·8	19·9	17·5
Cast in sand . .	21·6	...	...	...

Bronze is a mixture of copper and tin, the proportions being varied for different purposes, as shown in the Table below.

The different specific gravities of the metals make it difficult to melt them together. The tin is first melted into twice its weight of copper to

[1] From Patentee's Circular.

make *hard metal*, which is then added to the proper proportion of copper separately melted.

Large castings in bronze are often not homogeneous throughout their mass, in consequence of the difference in fusibility of the copper and tin.

GUN METAL also differs in the proportions of its constituents according to the purpose for which it is intended.

At one time it was much employed for casting ordnance, from which it derives its name.

It is harder, more fusible, and stronger than copper, and is used for pump valves and parts of machines.

BELL METAL consists of copper and tin, in the proportion of from 3 to 1 to 5 to 1. Small house bells contain 5 copper to 1 tin. Large bells 4 copper to 1 tin. Large church bells $3\frac{1}{2}$ copper to 1 tin. The metal, after being cast, is heated to redness and quenched, then again heated and allowed to cool slowly.

ALUMINIUM BRONZE contains from 90 to 95 per cent copper and 10 to 5 per cent aluminium.

It may be cast or turned in a lathe, also forged cold or hot, but it cannot be welded.

TABLE giving the COMPOSITION of various ALLOYS.

ALLOYS.	PARTS BY WEIGHT.					
	Copper.	Zinc.	Tin.	Lead.	Antimony.	Iron.
Brass, ordinary . . .	2	1				
„ for locks and door handles	3	1				
„ „ turning and fitting .	3	1	...	$\frac{1}{2}$ [a]		
„ „ engraving . .	3	1	A little			
„ „ bushes and sockets .	18	1	...	...	1	
„ to bear soldering well .	$2\frac{2}{3}$	1				
„ pot metal [b]. . .	$2\frac{1}{2}$	...	...	1		
Bronze, hard, for bearings for machinery	8	...	1			
„ for stop cocks and valves	88	10	2			
„ „ wheel metal for small toothed wheels	10	...	1			
„ „ bearings for very heavy weights	32	1	5			Manganese. 2
Manganese bronze . .	88	...	10	...	...	...
Gun metal for ordnance .	$90\frac{1}{2}$	...	$9\frac{1}{2}$			
„ of maximum hardness for turning	5	...	1			
„ soft . . .	16	...	1			
Bell metal	4	...	1	Less than $\frac{1}{100}$		
Muntz metal [c] . . .	3	2	...			
„ nails for . .	87	4	9			
Gedge's metal . . .	60	38·2	...	...	...	1·8
Sterro-metal . . .	55 to 60	34 to 44	1 to 2	2 to 4		
Babbit's metal . . .	4	8	96			
White brass . . .	3	...	90	...	7	Bismuth.
„ . . .	1	...	...	7	7	muth.
Metal to expand in cooling .	...	...	...	9	2	... 1

[a] The lead prevents the filings from sticking to the tool, but renders the brass unfit for hammering.

[b] An inferior alloy, used for very common taps, etc., and called also *cock metal*.

[c] Composition varies between 50 copper and 50 zinc, and 63 copper and 37 zinc.

It is light, very malleable, ductile, and not easily tarnished, but its expense prevents it from being used for anything but instruments.

PHOSPHOR BRONZE is any bronze or brass alloy, together with a small proportion of phosphorus. Its qualities may be made to vary by altering the proportions of its constituents. It wears longer than gun metal in bearings, is very tough, and is useful in positions where it is subject to shocks.

The phosphorus preserves the metal from the effects of the atmosphere.

MANGANESE BRONZE is an alloy (usually white) of pure copper with from 2 to 30 per cent of manganese. It is made in different qualities for casting and for rolling. The latter has a tensile strength of some 30 tons per square inch, with an elongation of from 25 to 45 per cent ; it combines the strength and toughness of steel with resistance to oxidation.

STERRO-METAL varies in composition as shown in the Table p. 348.

This alloy has great tensile strength, and may be used instead of wrought iron.

BABBIT'S METAL is used for bearings of machinery. It is very soft, wears smooth, and reduces friction. If the journal becomes heated, the alloy melts.

WHITE BRASS is a name given to various alloys used for bearings, and intended to work smooth. These are made of various composition besides those given in the Table.

Pewter should consist of 4 or 5 parts of tin and one of lead.

It is used for drinking cups and other purposes, also sometimes for covering counters where liquor is sold.

It should be remembered, however, that cheap pewter generally contains an excess of lead, and in that case is apt to poison any liquid in contact with it.

Pewter consisting of 4 tin and 1 lead "has the specific gravity 7·8, so that specimens having a higher specific gravity than this will be known to contain more lead." [1]

Solder is the name given to several different alloys used for the purposes of making joints between pieces of metal.

The effect is not merely mechanical, for the solder itself combines with the metal to be united, and forms a fresh alloy.

The composition of the solder used in connection with the different metals varies immensely, and the proportions in which each different kind of solder is mixed also varies according to circumstances.

Every solder must be more fusible than the metals it is intended to unite.

Hard Solders are those which fuse only at a red heat, and which can be therefore used only to metals which will endure that temperature.

Soft Solders melt at very low degrees of heat, and may be used for nearly all the metals.

The more nearly the solder agrees with the metal in hardness and malleability, the stronger will be the joint.

Thus brass or copper united with soft solder could not be hammered without breaking the joint, whereas a joint in lead or tin, made with soft solder, can be safely hammered.

SOLDERING.—It is not proposed here to describe the operations connected with soldering of different kinds, but one or two points may be noticed with advantage.[2]

The surfaces to be united must be perfectly clean, and freed from oxide, which would prevent adhesion and the formation of an alloy between the solder and the metal.

As the surfaces when heated are very easily oxidised they must be protected at the time— this is done by means of a *flux* which covers the surface and protects them from the air.

The materials used for fluxes are mentioned at p. 351.

---

[1] Bloxam's *Metals.*

[2] Every particular connected with soldering of all kinds is fully described in Holtzapffel's *Mechanical Manipulation,* whence much of the information here given has been taken.

HARD SOLDERS—Of these there are two kinds in common use.

Spelter Solder made of copper and zinc in proportions which differ according to circumstances (see Table, below).

It is generally granulated by pouring it when melted through a bundle of twigs into water.

This solder is used for making joints in iron, copper, brass, and gun-metal ; the process is known as *brazing*.

Silver Solder is a mixture of silver with copper, or brass in varying proportions (see Table).

It is used for making fine and neat joints in iron, steel, brass, and gun-metal—to prepare the surface of metals for welding, also for joints in silver and other light-coloured metals.

BRAZING.—The process of brazing is conducted as follows :—

Granulated spelter and borax, ground together in water, are spread over the carefully cleaned surfaces of the joint, and exposed gradually to the heat of a clean open fire ; the borax fuses first, and then the solder.

With silver solder the joint is covered with borax and water, or dry powdered borax, and the solder, cut into little square plates, is laid along the joint.

SOFT SOLDERS are mixtures of tin and lead. The proportions vary as shown in the table.

Tin makes the solder fusible, but as it is more expensive than lead, only so much tin should be included in a solder as will make it fit for the purpose for which it is intended.

The addition of a little bismuth makes the solder still more fusible.

The more fusible solders are known in the trade as *fine*, and those containing less tin as *coarse solders*.

" Any zinc getting into plumber's solder will ruin it, by making it too brittle to work.

" Solder may be purified of any foreign matter, such as zinc, if only present in small quantity, by burning it out on the fire, letting the pot get red hot till it goes off in vapour and scum, which can be skimmed off the top." [1]

" In making solder the proportions of the metals can be judged of from the appearance of the alloy. When it contains a little more than one-third of its weight of tin, its surface on cooling exhibits circular spots due to a partial separation of the metals ; but these disappear when the alloy contains two-thirds its weight of tin." [2]

" It is never advisable to buy ready-made solder, as you cannot depend upon the alloy ; too much lead and too little tin, which is the dearer of the two, is almost sure to be put into plumber's solder ; besides which there is always plenty of scrap lead about, which can be used for the purpose.

" When a good deal of soldering is to be done, the plumber will often start with a little excess of tin in his solder, as by degrees it will pick up lead from the lead work on which it is used, which will decrease its fusibility." [1]

SOFT SOLDERING.—Soft solder is applied in several different ways.

For joints in lead the surfaces to be soldered are carefully cleaned and covered with tallow—the space around is smeared with a mixture of size and lampblack, called *soil*, to prevent the solder from adhering—melted solder is then poured on and the excess wiped off with a cloth or in other ways.

In joining thin sheets of tinned iron, zinc, copper, and other metals, the edges are cleaned and sprinkled with powdered rosin ; a tinned copper bolt or soldering iron is made hot and applied so as to heat the edges of the plates ; the stick of solder is at the same time forced against the bolt, and the solder as it melts is dropped into the joint.

The copper bolt is also used to supply the heat in soldering light work in lead, such as lattices. The soldering iron cannot be used for thick pieces of metal, as it will not impart sufficient heat to their edges.

When joints are to be made between thicker pieces, the latter must have their surfaces first tinned separately and then the solder run in between them.

A blowpipe flame is sometimes used as the source of heat in soldering the metals.

—————

[1] Seddon.   [2] Bloxam.

Solder for use with the copper bolt is cast in strips called "*strap-solder*," or in thin cakes for gasfitter's work.[1]

TABLE showing the PROPORTIONS of INGREDIENTS of different SOLDERS—
MELTING POINTS—PURPOSES for which used.

Description of solder.	Constituents and their melting points.							Melting point of solder, Fahr.	Uses.
	Tin. 446°	Lead. 617°	Zinc. 773°	Copper. 2000°	Bismuth. 507°	Silver. 1873°	Antimony. 1150°		
HARD SOLDERS.									
Brazing—									
Very fine .	2	...	...	...	...	...	1		
Fine . . .	1	...	3	4					
Spelter—soft .	...	...	1	1	...	...	...	...	For ordinary brass work.
Do. hard .	...	...	2	3 to 6	...	...	...	...	For copper, iron, and steel.
Silver solders—									
Hardest . .	...	...	...	1	...	4	...	...	For silver, copper, and brass.
Hard *a* . .	...	...	...	1 *a*	...	3			
Soft . . .	...	...	...	1	...	2			
SOFT SOLDERS.									
Plumber's—									
Fine . . .	1	1-1½	...	...	...	...	...	385	
Ordinary *b* .	1	2	...	...	...	...	...	441°	For lead work, cisterns, jointing pipes.
Coarse solder	1	3	...	...	...	...	...	482°	Used with copper soldering bolt.
Tinman's—									
Ordinary solder	2	1	...	...	...	...	...	340°	
Very fusible do.	4	1	...	...	...	...	...	320°	
Pewterer's—									
Fine . . .	2	1	...	...	1	...	...	201°	Used by pewterers.

a The brass is put into the melted silver, or the zinc would evaporate. One brass wire instead of one copper with 2 silver is used for soldering silver.

b Also called "pot metal;" is assayed by the Plumbers' Company, stamped as genuine, and sold in ingots, hence called "Plumber's sealed solder."

FLUXES.—The fluxes used are as follows :—

For hard soldering—Borax.

For soft soldering—(with solders of about 2 tin, 1 lead)—the flux is varied according to the metals to be united, as shown below :—[1]

Metals.	Fluxes.
Cast-iron, malleable iron, steel .	Borax or Sal ammoniac.[2]
Copper, brass, gun-metal .	Sal ammoniac, chloride of zinc, or rosin.
Tinned iron	Chloride of zinc or rosin.
Zinc	Chloride of zinc.
Pewter	Gallipoli oil.
Lead with coarse solder .	Tallow.
„ fine solder .	Rosin.

Soldering fluid is a concentrated solution of chloride of zinc.

---

[1] Seddon   [2] *Ammonium chloride.*

TABLES.

Tables showing the properties of metals, and giving the weights of plates, wires, tubes, angles, tees, and sections of various kinds, are to be found in Molesworth's, Hurst's, and other engineering pocket-books, and would be too voluminous for these Notes.

Only one or two tables are therefore inserted, giving the most necessary information in its simplest form :—

Properties of Metals.—TABLE showing some of the PROPERTIES of useful METALS.

METAL.	Specific gravity.	Weight of a cubic foot in lbs.	Weight as compared with Wrought Iron.	Resistance in tons per sq. inch. Tension.	Resistance in tons per sq. inch. Compression.	Modulus of elasticity.	Melting point, degrees Fahr.	Expansion between 32° and 212° Fahr.
Aluminium	2·67	166	0·34	12		lbs. per sq. inch.		
Bell metal .	8·0	502	1·04	1·4				
Bismuth .	9·8	614	1·28	1·45	...	...	507°	·0014
Brass, ordinary—2 copper, 1 zinc	8·3	519	1·08	13·0	5	9,170,000	1840°	·0019
Copper, cast .	8·6	537	1·12	10·0				
,, sheet	8·8	550	1·14	13·0	...	...	...	·0017
,, wrought	8·9	556	1·16	15·0	...	15,000,000	1990°	·00179
Gun-metal, 9 copper to 1 tin . .	8·5	528	1·10	14·0	50	9,900,000	1900°	·00181
Wrought iron Bar .	7·7	480	1·00	25	} 16 {	29,000,000	} 3280° {	·0012
Plate .	7·8	487	1·01	20		24,000,000		·0012
Cast .	7·2	450	·94	7	38	17,000,000	2700°	·0011
Lead, Cast .	11·35	709	1·47	8	3·1	...	...	·0028
Sheet .	11·4	713	1·48	1·5	...	720,000	612°	·0028
Phosphor bronze .	...	...	...	26	...	14,000,000		
Platinum .	21·5	1344	2·8	...	...	...	3280°	·0008
Steel, Mild .	7·88	489	1·02	26 to 32	...	30,000,000	3300°	·0012
Tin, cast .	7·3	456	·95	2·0	...	...	442°	·0023
Zinc, cast .	6·9	428	·89	3·0	...	...	770°	·0029
Muntz metal	8·2	511	1·06	22				

The above Table is compiled from the works of Rankine, Pole, Anderson, Unwin, Molesworth, and others, who have extended the results of the best experiments up to the present time. The figures given are merely approximate averages—liable to be materially altered by slight alteration in the composition of the metal and other circumstances.

It will be understood that there is great variation in the strength of different descriptions of the same metal. Particulars regarding these are given for the more important metals, such as iron and steel, in the Tables, pp. 315-323.

Contraction of Metals in Cooling.—

TABLE showing the CONTRACTION of different METALS in CASTING.

METAL.	CONTRACTION. In fractions of linear dimensions.	In parts of an unit per inch of linear dimensions.
Cast iron .	$\frac{1}{96}$	$\frac{1}{8}$
Copper .	$\frac{1}{60}$	$\frac{1}{5}$
Zinc . .	$\frac{1}{48}$	$\frac{1}{4}$
Gun metal .	$\frac{1}{72}$	$\frac{1}{6}$
Yellow brass .	$\frac{1}{64}$	$\frac{3}{16}$
Lead . .	$\frac{5}{192}$	$\frac{5}{16}$

Melting Points of Alloys of Lead and Tin.[1]

Degrees Fahr.

Temp.	Lead.	Tin.	Temp.	Lead.	Tin.
400	11	8	490	14	4
410	25	16	500	33	8
420	7	4	510	19	4
430	15	8	520	25	4
440	8	4	530	30	4
450	17	8	540	38	4
460	9	4	550	48	4
470	10	4	558	25	1
480	23	8	630	1	0

Gauges.—IMPERIAL STANDARD WIRE GAUGE.—The following Table gives the thicknesses of the *Standard Wire* and *Sheet Metal Gauges*—sometimes described as the SWG—approved by Her Majesty's order in Council 1st March 1883 to be Board of Trade standards from 1st March 1884 :—

NEW IMPERIAL STANDARD WIRE GAUGE.—Denominations of Standards.

Descriptive Number.	Equivalents in parts of an inch.	Descriptive Number.	Equivalents in parts of an inch.	Descriptive Number.	Equivalents in parts of an inch.	Descriptive Number.	Equivalents in parts of an inch.
7/0	0·500	9	0·144	24	0·022	39	0·0052
6/0	464	10	128	25	20	40	48
5/0	432	11	116	26	18	41	44
4/0	400	12	104	27	0·0164	42	40
3/0	372	13	0·092	28	148	43	36
2/0	348	14	80	29	136	44	32
0	324	15	72	30	124	45	28
1	300	16	64	31	116	46	24
2	276	17	56	32	108	47	20
3	252	18	48	33	100	48	16
4	232	19	40	34	0·0092	49	12
5	212	20	36	35	84	50	0·0010
6	192	21	32	36	76		
7	176	22	28	37	68		
8	160	23	24	38	60		

THE BIRMINGHAM WIRE GAUGE, known also as the *Birmingham Iron Wire Gauge*, the *Sheet Iron Gauge*, and the *Wire Gauge*, was at one time used for sheet iron, steel, hoop iron, tubes, and wire, but is now reserved for the three latter, and is generally expressed by the initials BWG or WG. The following Table gives the thicknesses as carefully measured by Mr. Holtzapffel, and given in his *Mechanical Manipulation*. The mark 00000 is not shown in his list, but is frequently added :—

[1] Extracted from Box *On Heat.*

Mark or No. of Guage.	Thickness in inches.	Mark or No. of Gauge.	Thickness in inches.	Mark or No, of Gauge.	Thickness in inches.	Mark or No. of Gauge.	Thickness in inches.
00000	0·500	7	0·180	18	0·049	29	0·013
0000	0·454	8	0·165	19	0·042	30	0·012
000	0·425	9	0·148	20	0·035	31	0·010
00	0·380	10	0·134	21	0·032	32	0·009
0	0·340	11	0·120	22	0·028	33	0·008
1	0·300	12	0·109	23	0·025	34	0·007
2	0·284	13	0·095	24	0·022	35	0·005
3	0·259	14	0·083	25	0·020	36	0·004
4	0·238	15	0·072	26	0·018		
5	0·220	16	0·065	27	0·016		
6	0·203	1 7	0·058	28	0·014		

" Although this gage seems only to possess 40 terms, in reality not less than 60 sizes of wire are made, as intermediate sizes are in many cases added, and occasionally, though the sizes are retained, their numbers are variously altered."[1]

WHITWORTH'S STANDARD WIRE GAUGE is given below. It will be seen that the number or mark of the gauge is the number of thousandths of an inch in the thickness :—

No. or Mark.	Thickness. Inch.	No. or Mark.	Thickness. Inch.	No. or Mark.	Thickness. Inch.	No. or Mark.	Thickness. Inch.	No. or Mark.	Thickness. Inch.
1	·001	14	·014	34	·034	90	·090	280	·280
2	·002	15	·015	36	·036	95	·095	300	·300
3	·003	16	·016	38	·038	100	·100	325	·325
4	·004	17	·017	40	·040	110	·110	350	·350
5	·005	18	·018	45	·045	120	·120	375	·375
6	·006	19	·019	50	·050	135	·135	400	·400
7	·007	20	·020	55	·055	150	·150	425	·425
8	·008	22	·022	60	·060	165	·165	450	·450
9	·009	24	·024	65	·065	180	·180	475	·475
10	·010	26	·026	70	·070	200	·200	500	·500
11	·011	28	·028	75	·075	220	·220		
12	·012	30	·030	80	·080	240	·240		
13	·013	32	·032	85	·085	260	·260		

The BIRMINGHAM METAL GAUGE, also called the Metal Gauge or the Plate Gauge, is intended for sheet metals—except sheet iron and steel—such as copper, brass, gold, silver, etc.

Copper is, however, frequently sold by the Birmingham Wire Gauge given above, and by the special gauges of manufacturers.

---

[1] Holtzapffel.

BIRMINGHAM PLATE GAUGE.

Mark or No.	Thickness in inches.	Mark or No.	Thickness in inches.	Mark or No.	Thickness in inches.	Mark or No.	Thickness in inches.
1	·004	10	·024	19	·064	28	·120
2	·005	11	·029	20	·067	29	·124
3	·008	12	·034	21	·072	30	·126
4	·010	13	·036	22	·074	31	·133
5	·012	14	·041	23	·077	32	·143
6	·013	15	·047	24	·082	33	·145
7	·015	16	·051	25	·095	34	·148
8	·016	17	·057	26	·103	35	·158
9	·019	18	·061	27	·113	36	·167

The *Sheet and Hoop Iron Gauge*, BG, was issued by the South Staffordshire Iron Masters' Association for the use of sheet and hoop iron makers, 1st March 1884, and is adopted by the trade. It is important that in all transactions in sheet and hoop iron the initial letters BG should appear, to distinguish the Sheet and Hoop Iron Gauge from the Imperial Standard Wire Gauge.[1]

Number of Gauge.	Thickness. Inch.	Number of Gauge.	Thickness. Inch.	Number of Gauge.	Thickness. Inch.
7/0	·6666	10	·1250	22	·0312
1/0	·3964	16	·0625	24	·0247
1	·3532	18	·0495	26	·0196
4	·2500	20	·0392	28	·0156

The more useful part of the Table is given above. The numbers continue to gauge No. 50, which has a thickness of ·0010 inch.

Weight of Metals.—WEIGHT in LBS. of a SQUARE FOOT of DIFFERENT METALS, in Thicknesses varying by $\frac{1}{16}$th of an Inch.

Thickness. Inches.	Wrought Iron.	Cast Iron.	Steel.	Copper.	Zinc.	Brass.	Gun Metal.	Tin.	Lead.
$\frac{1}{16}$	2·3	2·3	2·5	2·9	2·3	2·6	2·7	2·4	3·7
$\frac{1}{8}$	5·0	4·7	5·1	5·8	4·7	5·3	5·5	4·8	7·4
$\frac{3}{16}$	7·5	7·0	7·6	8·7	7·0	8·2	8·2	7·2	11·2
$\frac{1}{4}$	10·0	9·4	10·2	11·6	9·4	11·0	10·9	9·6	14·9
$\frac{5}{16}$	12·5	11·7	12·8	14·5	11·7	13·7	13·7	12·0	18·6
$\frac{3}{8}$	15·0	14·1	15·3	17·2	14·0	16·4	16·4	14·4	22·3
$\frac{7}{16}$	17·5	16·4	17·9	20·0	16·4	19·2	19·1	16·8	26·0
$\frac{1}{2}$	20·0	18·7	20·4	22·9	18·7	21·9	21·9	19·3	29·7
$\frac{9}{16}$	22·5	21·1	25·0	25·7	21·1	24·6	24·6	21·7	33·4
$\frac{5}{8}$	25·0	23·5	25·5	28·6	23·4	27·4	27·3	24·1	37·1
$\frac{11}{16}$	27·5	25·8	28·1	31·4	25·7	30·1	30·0	26·5	40·9
$\frac{3}{4}$	30·0	28·1	30·6	34·3	28·1	32·9	32·8	28·9	44·6
$\frac{13}{16}$	32·5	30·5	33·2	37·2	30·4	35·6	35·0	31·3	48·3
$\frac{7}{8}$	35·0	32·8	35·7	40·0	32·8	38·3	38·2	33·7	52·0
$\frac{15}{16}$	37·5	35·2	38·3	42·9	35·1	41·2	41·0	36·1	55·7
1	40·0	37·5	40·8	45·8	37·5	43·9	43·7	38·5	59·4

The weight per square foot to any gauge can easily be obtained from the above Table by multiplying the weight of a square foot of the metal 1 inch thick by the thickness of the gauge in inches or parts of an inch.

---

[1] Hutton's *Works Manager's Handbook*, where very useful Tables of the weight of iron according to this and the Imperial Standard Gauge are given.

CHAPTER V.

TIMBER.

A THOROUGH knowledge of the nature and properties of different kinds of timber is very important to the engineer or architect.

Before entering upon a description of the different varieties of timber under the forms in which they generally come into the market, it will be advisable to make a few remarks on the growth of trees. A very slight knowledge of this branch of the subject is necessary in order that other points more intimately connected with the practical use of timber may be clearly understood.

Growth of Trees.—The timber used in building and engineering work is obtained from trees of the class known by botanists as "*Exogens*," or outward growers.

In trees of this class the stem grows by the deposit of successive layers of wood on the outside under the bark, while at the same time the bark becomes thicker by the deposit of layers on its under side.

Upon examining the cross section of such trees (see Fig. 147) we find that the wood is made up of several concentric layers or rings, each ring consisting in general of two parts—the outer part being generally darker in colour, denser, and more solid than the inner part, the difference between the parts varying in different kinds of trees.

These layers are called "*annual rings*," because one of them is, as a rule, deposited every year, in a manner which will be presently explained.

In the centre of the wood is a column of pith, p, from which planes, seen in section as thin lines, $m\ m$ (in many woods not discernible), radiate toward the bark, and in some cases similar lines, $m\ m$, converge from the bark toward the centre, but do not reach the pith.

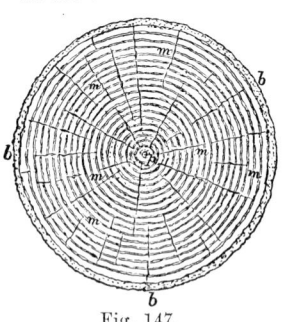

Fig. 147.

These radiating lines are known as "*medullary rays*" or "*transverse septa*." When they are of large size and strongly marked, as in some kinds of oak, they present, if cut obliquely, a beautiful figured appearance, called "*silver grain*" or "*felt*."

The wood is composed of bundles of cellular tubes, which serve to convey the required nourishment from the earth to the leaves.

The process of growth in a temperate climate is as follows :—

In the spring the root absorbs juices from the soil, which are converted into sap, and ascend through the cellular tubes to form the leaves.

At the upper surface of the leaves the sap gives off moisture, absorbs carbon from the air, and becomes denser ; after the leaves are full grown, vegetation is suspended until the autumn, when the sap in its altered state descends by the under side of the leaves, chiefly between the wood and the bark, where it deposits a layer of new wood (the annual ring for that year), a portion at the same time being absorbed by the bark. During this time the leaves drop off, the flow of sap then almost stops, and vegetation is at a standstill for the winter.

With the next spring the operation recommences, so that year after year a distinct layer of wood is added to the tree.

The above description refers to temperate climates, in which the circulation of sap stops during winter ; in tropical climates it stops during the dry season.

Thus, as a rule, the age of the tree can be ascertained from the number of annual rings, but this is not always the case. Sometimes a recurrence of exceptionally warm or moist weather will produce a second ring in the same year.

As the tree increases in age, the inner layers are filled up and hardened, becoming what is called " *duramen*," or " *heartwood.*" [1] The remainder is called " *alburnum*," or " *sapwood.*" The sapwood is softer and lighter in colour than the heartwood, and can generally be easily distinguished from it.

In addition to the strengthening of the wood caused by the drying up of the sap, and consequent hardening of the rings, there is another means by which it is strengthened—that is, by the compressive action of the bark. Each layer, as it solidifies, expands, exerting a force upon the bark, which eventually yields, but in the meantime offers a slight resistance, compressing the tree throughout its bulk.[2]

The sapwood is generally distinctly bounded by one of the annual rings, and can thus be sometimes distinguished from stains of a similar colour which are caused by dirty water soaking into the timber while it is lying in the ponds (see p. 388). These stains do not generally stop abruptly upon a ring, but penetrate to different depths, colouring portions of the various rings.

The heartwood is stronger and more lasting than the sapwood, and should alone be used in good work.

The annual rings are generally thicker on the side of the tree that has had most sun and air, and the heart is therefore seldom in the centre.

Felling.—While the tree is growing the heartwood is the strongest, but after the growth has stopped the heart is the first part to decay. It is important, therefore, that the tree should be felled at the right age.

The proper age varies with different trees, and even in the same tree under different circumstances. The induration of the sapwood should have reached its extreme limits before the tree is felled, but the period required for this varies with the soil and climate.

Trees cut too soon are full of sapwood, and the heartwood is not fully hardened.

---

[1] Sometimes called the " *Spine.*"   [2] Laslett.

The ages at which the under-mentioned trees should be felled are stated by Tredgold to be as follows :—

Oak	. .	60 to 200 years ; 100 years the best.
Ash	. .	⎫
Larch	. .	⎬ From 50 to 100 years.
Elm	. .	⎭
Spruce		⎫ From 70 to 100 years.
Scotch Fir	.	⎭

Oak bark, which is very valuable, is sometimes stripped in the spring, when it is loosened by the rising sap. The tree is felled in the winter, at which time the sapwood is found to be hardened like the heart. This practice is said by Tredgold to improve the timber.

Mr. Laslett says that " to select a healthy tree for felling we must seek for one with an abundance of young shoots, and the topmost branches of which look strong, pointed, and vigorous, this being the most certain evidence that it has not yet passed maturity."

The best season for felling timber is at midsummer or midwinter in temperate, or during the dry season in tropical climates, when the sap is at rest.

Squaring.—Directly the tree is felled it should be squared, or cut into scantling, in order that the air may have free access to the interior.

Characteristics of Good Timber. — The quality of timber depends greatly upon the treatment the tree has received, the time of felling, and, above all, on the nature of the soil in which it has grown.

These branches of the subject do not fall within the province of the engineer or builder, and will not here be entered upon ; it will be sufficient to point out some of the characteristics by which good timber may be known.

Good timber should be from the heart of a sound tree—the sap being entirely removed, the wood uniform in substance, straight in fibre, free from large or dead knots, flaws, shakes, or blemishes of any kind.

If freshly cut it should smell sweet ; " the surface should not be woolly, or clog the teeth of the saw," but should be firm and bright, with a silky lustre when planed ; a disagreeable smell betokens decay, and " a dull chalky appearance is a sign of bad timber."

The annual rings should be regular in form ; sudden swells are caused by rind-galls ; closeness and narrowness of the layers indicate slowness of growth, and are generally signs of strength. When the rings are porous and open, the wood is weak, and often decayed.

The colour of good timber should be uniform throughout ; when it is blotchy, or varies much in colour from the heart outwards or becomes pale suddenly towards the limit of the sapwood, it is probably diseased.

Among coloured timbers darkness of colour is said by Rankine to be in general a sign of strength and durability.

Good timber is sonorous when struck. A dull heavy sound betokens decay within (see p. 391). Among specimens of the same timber, the heavier are generally the stronger.

Timber intended for use in important work should of course be free from the defects mentioned in page 386. The knots should not be large or numerous, and on no account should they be loose.

The worst position for large knots is when they are near the centre of the balk required, and more especially when they are so situated as to form a ring round the balk at one or more points.

The sap should be entirely removed. According to Mr. Laslett, however, the heart of trees having the most sapwood is generally stronger and better in quality than the heart of trees of the same species that have but little sapwood.

The strongest part of the tree is generally that which contains the last-formed rings of heartwood, so that the strongest scantlings are obtained by removing no more rings than those containing the sap.

Timber that is thoroughly dry weighs less than when it was green (see p. 386); it is also harder, and consequently more difficult to work.

Defects in Timber.—There are several defects in timber caused by the nature of the soil upon which the tree was grown, and by the vicissitudes to which it has been subjected while growing.

Heartshakes are splits or clefts occurring in the centre of the tree. They are common in nearly every kind of timber. The splits are in some cases hardly visible ; in others they extend almost across the tree, dividing it into segments.

When there is one cleft right across the tree it does not occasion much waste, as it divides the squared trunk into two substantial balks. Two clefts crossing one another at right angles, as in Fig. 148, make it impossible to obtain scantlings larger than one-fourth the area of the tree.

The worst form of heartshake, however, is one in which the splits twist in the length of the tree, thus

Fig. 148.

making it impossible to convert the tree into small scantlings or planks.

Starshakes are those in which several splits radiate from the centre of the timber, as in Fig. 149.

Cupshakes are curved splits separating the whole or part of one annual

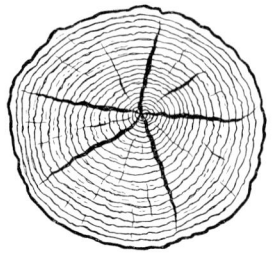 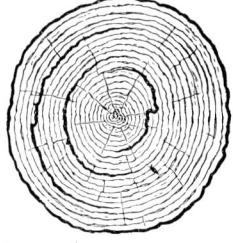

Fig. 149.   Fig. 150.

ring from another (see Fig. 150). When they occupy only a small portion of a ring they do no great harm.

Rind-Galls are peculiar curved swellings, caused generally by the growth of layers over the wound remaining after a branch has been imperfectly lopped off.

Upsets are portions of the timber in which the fibres have been injured by crushing.

Foxiness is a yellow or red tinge caused by incipient decay.

Doatiness is a speckled stain found in beech, American oak, and other timbers.

Twisted Fibres are caused by the action of a prevalent wind, turning the tree constantly in one direction. Timber thus injured is not fit for squaring, as so many of the fibres would be cut through.

CLASSIFICATION OF TIMBER.

The following classification of timber is a modification by Professor Rankine and Mr. Hurst of that originally proposed by Tredgold :—

CLASS I.—PINE WOOD (natural order *Coniferæ*).

Characteristics.	Examples.
Annual rings very distinct ; pores filled with resinous matter ; one part of each ring hard and dark, the other soft and light coloured.	Pine, Fir, Larch, Cowrie, Cedar, Cypress, Yew, and Juniper. Of these the first six only are in ordinary use, and will be described.

CLASS II.—HARD WOOD or LEAF WOOD (non-resinous and non-coniferous).

		Characteristics.	Examples.
Div. I.	With distinct large medullary rays.	*Subdiv. I.* Annual rings distinct ; one side porous, the other compact.	Oak.
		Subdiv. II. Annual rings not distinct ; texture nearly uniform.	Beech, Alder, Plane, Sycamore.
Div. II.	No distinct large medullary rays.	*Subdiv. I.* Annual rings distinct ; one side porous, the other compact.	Chestnut, Ash, Elm.
		Subdiv. II. Annual rings not distinct ; texture nearly uniform.	Mahogany, Walnut, Poplar, Teak, Greenheart.

With regard to the above Table Professor Rankine remarks :— " The chief practical bearings of this classification are as follows. " Fir wood, or coniferous timber, in most cases contains turpentine. It is distinguished by straightness in the fibre and regularity in the figure of the trees ; qualities favourable to its use in carpentry, especially where long pieces are required to bear either a direct pull or a transverse load, or for purposes of planking. At the same time the lateral adhesion of the fibres is small, so that it is much more easily shorn and split along the grain than hardwood, and is therefore less fitted to resist thrust or shearing stress, or any kind of stress that does not act along the fibres. Even the toughest kinds of firwood are easily wrought.

" In hard wood, or non-coniferous timber, there is no turpentine. The degree of distinctness with which the structure is seen, whether as regards medullary rays or annual rings, depends upon the degree of difference of texture of different parts of the wood. Such difference tends to produce unequal shrinking in drying, and consequently those kinds of timber in which the medullary rays and the annual rings are distinctly marked are more liable to warp than those in which the texture is more uniform. At the same time, the former kinds of timber are, on the whole, the more flexible, and in many cases are very tough and strong, which qualities make them suitable for structures that have to bear shocks." [1]

The classification shown above is that made by botanists and given by most writers on timber.

For many practical purposes, however, the timber used upon engineering and building works may be divided into two classes :—

Soft Wood, including firs, pines, spruce, larch, and all conebearing trees.

Hard Wood, including oak, beech, ash, elm, mahogany, etc.

Classification of Fir Timber.—The different trees included under the general head of " Fir Timber " are divided by botanists into the pines and firs, which produce timber of very different quality, and are distinguished in the growing tree by the leaves, the shape of the cones, and by other peculiarities.

The Pine (Pinus) has slender green needle-shaped leaves, growing in clusters of from two to six (according to the species) from the same stalk. It has one straight tap root, the trunk does not taper much, the wood is close grained, fibrous, very durable, full of resinous matter, and of a high bright colour. The cones have thick woody scales that do not fall away from the axis.

The Fir or Spruce (Abies) has straight short leaves, which come off singly from the stalks. The roots are ramified, the trunk tapers more than that of the pine, the shape of the tree is more pyramidal, the wood is of a much lighter colour, and not nearly so dur-

---

[1] Rankine, Civil Engineering, p. 440.

able. The cones are long and pendulous, with thin woody scales that do not fall away from the axis.

No attention is, however, paid to these botanical distinctions in the classification adopted on building or engineering works.

The carpenter generally gives the name *fir* to all red and yellow timber from the Baltic, somewhat similar timber from America he calls *pine*, whereas all white wood from either place is known as *spruce*.

Market Forms of Timber.—Before proceeding further, it will be well to describe the different forms to which timber is converted for the market.

A Log is a trunk of a tree with the branches lopped off.

A Balk is obtained by roughly squaring the log.

FIR timber is imported in the forms and under the designations mentioned below.

Hand Masts are the longest, soundest, and straightest trees after being topped and barked.

The term is technically applied to those of a circumference between 24 and 72 inches. " They are measured by the hand of 4 inches, there being also a fixed proportion between the number of hands in the length of the mast and those contained in the circumference taken at $\frac{1}{3}$ the length from the butt end." [1]

Spars or *Poles* have a circumference of less than 24 inches at the base.

Inch Masts are those having a circumference of more than 72 inches, and are generally dressed to a square or octagonal form.

Balk Timber consists of the trunk, hewn square, generally with the axe, (sometimes with the saw), and is also known as *square timber*.

Planks are parallel-sided pieces from 2 inches to 6 inches thick, 11 inches broad, and from 8 to 21 feet long.

Deals are similar pieces 9 in. broad and not exceeding 4 in. in thickness.

Whole Deals is the name sometimes given to deals 2 in. or more in thickness.

Cut Deals are less than 2 in. thick.

Battens are similar to deals, but only 7 in. broad.

Ends are pieces of plank, deal, or batten less than 8 feet long.

Scaffold and *Ladder Poles* are from young trees of larch or spruce. They average about 33 feet in length, and are classed according to the diameter of their butts.

Rickers are about 22 feet long, and under $2\frac{1}{2}$ in. diameter at the top end. The smaller sizes are called *Spars*.[2]

OAK is supplied as follows in her Majesty's dockyards.[3]

Rough Timber, consisting of the trunk and main branches roughly hewn to an octagonal section.

Sided Timber, being the trunk split down and roughly formed to a polygonal section.

Thick Stuff.—Not less than 24 feet long, and of an average length of at least 28 feet, from 11 to 18 inches wide between the sap in the middle of its length, and from $4\frac{1}{2}$ inches to $8\frac{1}{2}$ inches thick.

Planks.—Not less than 20 feet long, and of an average length of at least 28 feet, the thickness from 2 to 4 inches, and the width (clear of sap) required

---

[1] Laslett.   [2] Seddon's *Notes.*   [3] Laslett.

at the middle of the length varying according to the thickness, *i.e.* between 9 and 15 inches for 3, 3½, and 4 inch planks, between 8 and 15 inches for 2 and 2½ inch planks.

WANEY TIMBER is a term used for logs which are not perfectly square. The balk cut being too large for the size of the tree, the square corners of the balk are wanting, and their place is taken by flattened or rounded angles, often showing the bark, and called *wanes*.

COMPASS TIMBER consists of bent pieces, the height of the bend from a straight line joining the two ends being at least 5 inches in a length of 12 feet.

DESCRIPTIONS OF DIFFERENT KINDS OF TIMBER.[1]

PINE WOOD OR SOFT WOOD.

Northern Pine (*Pinus sylvestris*).—This timber, frequently known as " red or yellow fir," is from the " Scotch fir " tree.

The term *Northern Pine* has been introduced by Mr. Hurst for the reasons given in the following remarks, extracted from his *Handbook :*—

" Much confusion has arisen among architects and builders owing to the absurd practice of naming this timber after the ports of shipment, and also from confounding the pines (*Pinus*) with the firs (*Abies*), although they belong to distinct genera. . . . The *P. sylvestris* is essentially a wood of northern climates, and will thrive at greater elevations and in higher latitudes than even the fir ; hence the term ' northern pine ' given to it by the author in his edition of Tredgold's carpentry, and also adopted throughout this work."

This tree grows in Scotland, and also in the Baltic and Russia, whence most of the timber used in this country is imported, both in balks, and also in planks, deals, and battens.

Tredgold gives the following description of the appearance of this timber :—

" The colour of the wood of different varieties of Scotch fir differs considerably. It is generally of a reddish yellow, or a honey yellow of various degrees of brightness.

" It consists in the section of alternate hard and soft circles ; the one part of each annual ring being soft and light coloured, the other harder and dark coloured. It has no larger transverse septa, and has a strong resinous odour and taste. It works easily when it does not abound in resin ; and the foreign wood shrinks about ₃₀th part of its width in seasoning from the log.

" In the best timber the annual rings are thin, not exceeding ₁₀ inch in thickness. The dark parts of the rings are of a bright and reddish colour,

---

[1] Taken chiefly from the works of Tredgold, Hurst, Newland, Laslett, and Rankine. These works contain a great deal of information regarding various foreign timbers not used in this country, and also as to the less common varieties of home growth, which it is unnecessary here to enter upon.

the wood hard and dry to the feel, neither leaving a woolly surface after the saw nor filling its teeth with resin.   *   *   *

" The inferior kinds have thick annual rings—in some the dark parts of the rings are of a honey yellow, the wood heavy, and filled with soft resinous matter, feels clammy, and chokes the saw.

" Timber of this kind is not durable nor fit for bearing strains. Mar Forest timber is often of this kind. In other inferior kinds the wood is spongy, contains less resinous matter, and presents a woolly surface after the saw.

" Swedish timber is often of this kind, and is then inferior in strength and stiffness."

Mr. Fincham, quoted by Mr. Hurst, says further—

" If the timber is good, its parts, on being separated, appear stringy and oppose a strong adhesion, and the shavings from the plane will bear to be twisted two or three times round the fingers ; whereas if the stick is of bad quality, or in a state of decay and has lost its resinous substances, the chips and shavings come off short and brittle, and with much greater ease."

VARIETIES IN GENERAL USE.

Balk Timber.—The best balks of northern pine are imported from Dantzic, Memel, Riga.

DANTZIC TIMBER is grown chiefly in Prussia, and takes its name from the port where it is shipped.

Appearance.—Its general appearance answers to the description given above, though in colour it is rather whiter than other varieties.

Characteristics.—This timber is strong, tough, elastic, easily worked, and durable if well seasoned.

It contains, especially in small trees, a large proportion of sapwood, which in fresh timber can hardly be distinguished from the heartwood, and it frequently contains large and dead knots. The heart is often loose and " cuppy."

Market forms.—Dantzic balks are from 18 to 45 feet long, and generally 14 to 16 inches square.

The deals vary from 2 to 5 inches in thickness, and in length from 18 to 50 feet.

The classification of this timber, and of that from Memel, as to qualities, etc., is given at pp. 382, 383.

MEMEL TIMBER is very similar to that from Dantzic, but is considered hardly so strong. The scantlings of the balks are rather smaller, being from 13 to 14 inches square.

RIGA TIMBER is like the other varieties just described, but the annual rings are closer.

It is slightly inferior to Dantzic in strength, is remarkable for its straight growth, for the small proportion of sap it contains, and for its freedom from knots. It is, however, frequently a little shaky at the centre, and is therefore not so fit for conversion into deals as other varieties.

This timber is only once sorted for masts before it is exported, and is placed in the market without the brands described on p. 382.

NORWAY TIMBER is of small size, tough, and durable, but it generally contains a good deal of sapwood.

The balks are only about 8 or 9 inches square.

SWEDISH TIMBER somewhat resembles that from the Prussian ports, but the balks are generally tapering in form, of small size, and not of good quality.

Appearance.—The wood is of a yellowish-white colour, soft, clean, and straight in grain, with small knots and very little sap, but the balks are generally shaky at the heart, and therefore unfit for conversion into deals.

Mr. Laslett says—" There is little to recommend the Swedish fir to favourable notice beyond the fact of its being cheap and suitable for the coarser purposes in carpentry."

It is used chiefly for scaffolding.

Market forms.—The balks are generally from 20 to 35 feet long, and from 10 to 12 inches square.

The classification of Baltic timber is given at pp. 382, 383, in connection with the description of the marks upon it.

Planks, Deals, and Battens.—Planks, deals, and battens from the Baltic, when cut from the northern pine (*Pinus sylvestris*) are known as *yellow deal* or *red deal.* When cut from the spruce (*Abies*) (see pp. 361 and 369), they are called *white deals.*

It would be very difficult to give a list of all the different varieties of planks, deals, and battens of northern pine to be found in the market, with a detailed description of each.

The minute distinctions which exist in appearance and quality could not be described on paper, and any attempt to point out these differences would not be of any practical value.

Mr. Laslett says that taking deals, battens, etc., "in a general way, the order of quality would stand first or best with Prussia ; then with Russia, Sweden, and Finland ; and lastly with Norway."

Yellow Deals.—The following list mentions only a few of the principal ports from which manufactured timber is imported, and the salient or most marked characteristic, if any, which is peculiar to each kind :—

PRUSSIAN.—*Memel, Dantzic, and Stettin.*—The deals imported are very durable and adapted for external work, but they are chiefly used for shipbuilding.

The export of deal from the Prussian ports of Dantzic, Memel, Stettin, etc., is almost entirely confined to yellow planks and deck deals, called also red deals, 2 to 4 inches thick, used for shipbuilding."

" The reason for this is that the timber from the southern ports being coarse and wide in the grain, could not compete in the converted form, as deals, etc., with the closer-grained and cleaner exports from the more northern ports." [1]

RUSSIAN.—*Petersburg, Onega, Archangel, Narva.*—These are the best deals imported for building purposes. They are very free from sap, knots, shakes, or other imperfections ; of a clean grain, and hard well-wearing surface, which makes them well adapted for flooring, joinery, etc.

The lower qualities are, however, of course subject to defects.

---

[1] Seddon.

Petersburg deals are apt to be shaky, having a great many centres in the planks and deals, but the best qualities are very clean and free from knots." [1]

These deals are very subject to dry rot.

All the Russian deals are said [1] to be unfit for work exposed to damp. In those from Archangel and Onega " the knots are often surrounded by dead bark, and drop out when the timber is worked." [2]

Wyborg deals are sometimes of very good quality, but often full of sap.

FINLAND and NYLAND are stated by Newland to be 14 feet long, very durable, but fit only for the carpenter.

NORWEGIAN.—*Christiania, Dram.*—Yellow deals (as well as white, see p. 367) and battens are imported from Christiania, together with battens from Dram. They used to bear a high character, being clean and carefully converted, but are now very scarce.

A good deal of the Norwegian timber is imported in the shape of prepared flooring and matched boarding.

Dram battens are often found to be suffering from dry rot, especially when they are badly stacked.

SWEDISH.—*Gefle, Stockholm, Holmsund, Soderham, Gottenburg, Hernosand, Sundswall.*—" The greater portion of the Swedish timber is coarse and bad, but some of the very best Baltic deal, both yellow and white, comes from Gefle and Soderham."

" The best Swedish deals run more sound and even in quality than the Russian shipments, from the different way in which the timber is converted.

" A balk of Russian timber is all cut into deals of one quality, hence the numerous hearts or centres seen amongst them, which are so liable to shake and split ; whereas in Swedish timber the inner and the outer wood are converted into different qualities of deals. Hence the value of first-class Swedish goods.

" 4-inch deals should never be used for cutting into boards as they are cut from the centres of the logs. 3-inch deals, the general thickness of Russian goods, are also open to the same objection. Swedish 2½ and 2-inch deals of good quality are to be preferred to 3-inch, since they are all cut from the sound outer wood ; although, being a novelty in the market, and their value not understood, they are cheaper." [1]

It will be seen from the above quotation that the first qualities of Swedish deals have a high character for freedom from sap, etc. The lower qualities have the usual defects, being sappy, and containing large coarse knots.

Mr. Newland considers Swedish deals fit for ordinary carcase work, and Mr. Hurst says that from their liability to warp they cannot be depended upon for joiners' work.

Uses.—Swedish deals are commonly used for all purposes connected with building, especially for floors.

American Pine.—There are three or four descriptions of this timber in the market, which will now be described.

As a rule American pine is in many respects inferior to that from the Baltic. It is generally weaker, and comparatively wanting in durability. On the other hand, it is clean, free from defects, and easily worked.

American Red Pine (*Pinus rubra*), also *Pinus resinosa*, takes its name from the red colour of its bark, and is known generally as *Canada Red Pine.*

---

[1] Seddon.   [2] Newland.

Where found.—Canada.

Appearance.—Reddish white, clean fine grain. Very like Memel, but with larger knots.

Characteristics.—Small timber, very solid in centre, not much sap or pith, tough, elastic, does not warp or split, moderately strong, few large knots, very durable where well ventilated, adheres well to glue, not much loss in conversion.

Uses.—By cabinetmakers for veneering, sometimes for internal fittings of houses.

Market forms.—Logs 16 to 50 feet long, 10 to 18 inches square, and about 40 cubic feet in content ; classed as " large," " mixed," and " building " sizes.

American Yellow Pine (*Pinus strobus*) is produced from a straight and lofty tree found in North America ; used to be sometimes known as " Weymouth Pine," because it was first introduced into this country by Lord Weymouth. In America it is called *white pine* from the colour of its bark.

Its leaves grow in tufts of 5. The cones are very long, with loosely arranged scales.

Appearance.—The wood when freshly cut is of a white or pale straw colour, but becomes of a brownish yellow when seasoned. The annual rings are not very distinct, the grain is clean and straight ; the wood is very light and soft, when planed has a silky surface, and is easily recognised by the short detached dark thin streaks, like short hair lines, which always appears running in the direction of the grain.

Characteristics.—The timber is as a rule clean, free from knots, and easily worked, though the top ends of logs are sometimes coarse and knotty ; it is also subject to cup and heart shakes, and the older trees to sponginess in the centre. It adheres to glue, but does not hold nails well. This timber often arrives in this country in an incipient state of dry rot, and it is very subject to that disease.

It lasts well in a dry climate, such as that of America, but is not durable in England.

Uses.—Yellow pine is much used in America for carpenters' work of all kinds ; it is also used for the same purpose in Scotland, and in some large English towns, but in London and the neighbourhood it is considered inferior in strength to Baltic timber.

The great length of the logs and their freedom from defects causes this timber to be extensively used for masts and yards whose dimensions are so great that they cannot be procured from Baltic timber.

For joinery this wood is invaluable, being wrought easily and smoothly into mouldings and ornamental work of every description. It is particularly adapted for panels on account of the great width in which it may be procured, and it is also extensively used for making patterns for castings.

Market forms.—The best is imported as inch masts roughly hewn to an octagonal form.

Next come logs hewn square from 18 feet to 60 feet long, averaging about 16 inches square, and containing about 65 cubic feet in each log. A few pieces are only 14 inches square, and short logs may be had exceeding even 26 inches square. Some is imported as " waney timber " (see p. 363).

A few 3-inch deals are imported, varying in width from 9 to 24 inches, and even as wide as 32 inches.

Classification.[1]—American yellow deals are classed as follows :—

Brights   .   .   .   1st, 2d, and 3d quality,
Dry floated   .   .   ,,   ,,
Floated   .   .   .   ,,   ,,

their order of merit being first quality brights, first quality dry floated, first quality floated, then second quality brights, and so on.

Brights are sawn from picked logs and have not been discoloured by being floated down the rivers, and are therefore of a much cleaner and brighter yellow.

Floated deals, etc., have been floated or rafted down the rivers from the felling grounds.

Dry floated implies that the deals, etc., have been stacked and dried before shipment.

First quality yellow deals of each kind should be clean, straight grained, and quite free from shakes and knots.   Second quality are a little inferior in these respects, and third quality are inferior again.

Floating the deals damages them considerably, besides discolouring them. The soft and absorbent nature of the wood causes them to warp and shake very much in drying, so that floated deals should never be used for fine work.

The best ports are Quebec for yellow deals, and St. John for spruce deals. Goods from the more southern ports, such as Richibucto, Miramichi, Shedac, etc., are of an inferior quality.

Rafted or floated deals are shipped from all the Canadian ports except St. John, hence the superiority of St. John deals, which are always bright or unwatered.

Quebec Yellow Pine (*Pinus variabilis*) is imported chiefly from the place after which it is named.   It is used for masts and yards of large ships, but not much for other purposes.

Pitch Pine (*Pinus rigida*) has its leaves in threes, scales of cones rigid, sharp edges, rough bark.

The best of this timber comes from the southern states of North America, chiefly from the ports of Savannah, Darien, and Pensacola.

Appearance.—The wood has a reddish-white or brown colour ; the annual rings are wide, strongly marked, and form beautiful figures when the wood is wrought and varnished.

Characteristics.—The timber is very full of resinous matter, which makes it extremely durable, but sticky and difficult to plane.   It is hard, heavy, very strong, hard to work, free from knots, but containing a large proportion of sapwood.   It is subject to heart and cup shake, and soon rots in a moist atmosphere. The wood is brittle when dry, and its elasticity, strength, and durability are often reduced by the practice of " bleeding " or tapping the tree for the sake of the turpentine it contains.   It is too full of resin to take paint well.

Uses.—Pitch pine is used for the heaviest timber structures in

---

[1] From Seddon's *Builder's Work.*

engineering works, where great strength and lasting properties are required; also by shipbuilders for deep planks; by builders for floors (being very durable under wear), for window sills, and for ornamental joinery of all kinds. The heartwood is good for pumps.

Market forms.—Logs 11 to 18 inches square (averaging 16 inches square) and 20 to nearly 80 feet long; planks 3 to 5 inches thick, 10 to 15 inches wide, and 20 to 45 feet long. As it is subject to heart-shakes and cup-shakes, it is more economical to purchase it in the form of planks when it is required to be used in that form.[1]

White Fir or Spruce (*Abies excelsa*).—This timber is from trees found in Norway, in most of the mountainous parts in the north of Europe, in North America, and also in this country.

The peculiarities of the tree, leaves, etc., are given at page 361. The wood is generally known in this country as *white deal.*

Appearance.—The wood is of a yellowish-white, or sometimes of a brownish-red colour, becoming of a bluish tint when exposed to the weather. The annual rings are clearly defined, the surface has a silky lustre, and the timber contains a large number of very hard glossy knots, by which it may be easily recognised.

The sapwood is not distinguishable from the heart.

Characteristics.—This timber is tough, sometimes fine grained, light, and elastic, difficult to work, especially where the knots occur, shrinks but little, and takes a fine polish.

It, however, shrinks and twists, and warps very much, unless restrained when seasoning, and is wanting in durability.

It is moreover knotty; inferior in strength to the red and yellow pine, not so easily worked, and is apt to snap under a sudden shock.

Uses.—The deals are used for the coarser descriptions of joinery, cheap flooring boards, etc., for panels, also for packing-cases and other common work where cheapness is the first object.

" White deal " is a nice wood for tops of dressers, shelves, and common tables, but being liable to warp it should not be cut too thin, not under an inch if possible. For sticking mouldings and the finer kinds of joiners' work it is not fit, as the hard knots turn the plane iron."[1]

The trees being generally straight, strong, and elastic, are used for small spars for ships and boats, for ladders and scaffold poles.

Baltic Spruce comes chiefly from Norway, also Sweden, Russia, and Prussia.

---

[1] Seddon.

WHITE DEAL.—"Some of the best white deal comes from *Christiania;* but that from the other Norwegian ports is not to be relied upon, being apt to warp and split in drying." [1]

Both good and bad qualities are sent from *Dram.* The deals from the upland are more free from shakes than the other.

Spruce from *Frederikstadt* contains loose black knots.

"*Gottenburg* white deals are hard and stringy," only fit for packing-cases and temporary work. "The same remark applies in greater degree to those from *Hernosand* and *Sundsvall.*" [2]

The best Russian white deal is shipped from *Onega.* Very good deals come from *Narva, Petersburg;* white deals are fine and close in grain, but expand and contract with changes of weather.

Riga deals are coarser and more open-grained than the other Russian descriptions.

American Spruce.—There are at least four varieties of the tree from which this timber is produced :—The white spruce (*Abies alba*), which flourishes in the colder parts of North America ; the black spruce (*Abies nigra*), and the Wenlock spruce (*Abies Canadensis*), found chiefly in Lower Canada; and the red spruce (*Abies rubra*), imported from Nova Scotia.

The red spruce is sometimes known as "*Newfoundland red pine.*"

Appearance.—The timber greatly resembles the spruce from the Baltic, having the same characteristic glassy knots. The wood of the black and white varieties is the same in appearance—the difference of colour being only in the bark of the tree ; the black produces the largest and best timber.

Characteristics.—American spruce is inferior to that from Norway—it is not so resinous or so heavy—is tougher, warps and twists very much, and soon decays.

The Canadian spruce is better than that from New Brunswick.

Uses.—This timber is used for the same purposes as Baltic spruce.

The Larch (*Larix Europœa*) is found in various parts of Europe ; the finest varieties being in Russia.

Appearance.—The wood is honey yellow or brownish white in colour, the hard part of each ring being of a redder tinge, silky lustre.

There are two kinds in this country, one yellowish white, cross-grained, and knotty ; the other (grown generally on a poor soil or in elevated positions) reddish brown, harder, and of a straighter grain.

Characteristics.—"Decidedly the toughest and most lasting of all the coniferous tribe," [3] very strong and durable—shrinks very much—straight and even in grain, and free from large knots, very liable to warp, but stands well if thoroughly dry—is harder to work than Baltic fir—but surface is smoother, when worked. Bears nails driven into it better than any of the pines.

Uses.—Chiefly for posts and palings exposed to weather, railway sleepers, etc. ; also for flooring, stairs, and other positions where it will have to withstand wear.

AMERICAN LARCHES are the black variety (*Larix pendula*) known as *Hackmatack* or as *Tamarak;* and also the red variety (*Larix microcarpa*).

The timber from these trees resembles that from the European larch.

The Cedar (*Cedrus Libani*) properly so called, comes from Mount Lebanon, and Asia Minor, and is not much known in this country.

The wood generally known as cedar is from trees of the genus *Juniperus.*

These trees are found in Virginia, Bermuda, Florida, and also in India, Australia, etc.

---

[1] Seddon.   [2] Newland.   [3] Brown's *Forester*, p. 278.

Appearance.—The heartwood is a reddish brown, sapwood white, straight-grained, and porous.

Characteristics.—Very light and brittle, and wanting in strength ; Tredgold says it is about ¾ the strength of the best red pine ; is easily worked ; does not shrink much ; is very durable when well ventilated. Has a pungent odour which often unfits it for internal joinery, but protects it from being attacked by insects. A resinous substance exudes from the timber when freshly cut, and makes it difficult to work.

Uses.—For pencils, furniture, toys, carvings ; and in Bermuda for ship and boat building, for doors, window frames, sashes, and internal joinery. It is the best kind of wood to veneer upon.

Market forms.—Imported in logs from 6 to 10 inches square.

The Cypress (*Cupressus sempervirens*) furnishes a timber sometimes known as *cedar.*

It is found in Cyprus, Asia Minor, Persia, etc.

The wood is strong, very durable, has a strong odour, resists worms and insects, and is much used in Malta and Candia for building purposes.

The Oregon Pine or *Douglas Pine* (*Abies Douglasii*) is found in N.W. America.

It resembles Canadian red pine in appearance, but is slightly harder.

The Oregon pine from Puget Sound is stated to be obtainable in sizes so exceptional as 120 feet in length and 30 inches square.

The Kawrie, *Cowrie,* or *Cowdie Pine* (*Dammara Australis*) is found only in New Zealand.

Appearance.—The heartwood is yellowish white, fine and straight in grain, with a silky lustre on surface.

Characteristics.—Generally very free from defects ; may be obtained perfectly clean ; is very light, strong, and elastic ; has an agreeable odour when worked ; is less liable to shrink than most firs and pines, except when cut into narrow strips ; unites well with glue, and is very durable.

Uses.—Makes first-rate masts and spars ; is used for parts of military bridges ; is good for joinery.

HARD WOOD OR LEAF WOOD.

The varieties of timber of this class most in use for building purposes are oak, beech, ash, elm, mahogany, teak. These, with a few others, will now be described in more or less detail, according to their importance.

Oak.—Of this timber there are several varieties found, both in this country, and also in America, Holland, the Baltic, Austria, and Hungary.

British Oak.—The principal British varieties are—

The Stalk-fruited or *Old English Oak* (*Quercus robur* or *Quercus pedunculata*), in which the acorns have long stalks, and the leaves short stalks.

The Cluster-fruited or *Bay Oak* (*Quercus sessiliflora*), of which the acorns grow in close clusters with very short stalks, and the leaves have longer stalks, some nearly an inch long.

Durmast Oak (*Quercus pubescens*) has short stalks for the acorns and long stalks for the leaves, like the bay oak, but is distinguished by " the under side of the leaves being somewhat downy." [1]

Appearance.—Good oak is of a light brown or brownish-yellow colour, with a hard, firm, and glossy surface. A reddish tinge and dull surface are signs of decay. The annual rings are very narrow and regular, each having a compact and a porous layer, the pores in the latter being very small. Wide rings and large pores are signs of weakness. The medullary rays are hard and compact; where they are small and indistinct the wood is stronger.

When the timber is cut obliquely across, beautiful markings of *silver grain* appear, being caused by the cropping out of the large medullary rays.

Characteristics.—Sound heart of oak is very durable in earth or water. It has been known to last 1000 years when well ventilated.

The timber is very strong, hard, and tough, warps in seasoning. It is very elastic, easily bent to curves when steamed or heated. It is not easily splintered, but is rather liable to the attacks of insects.

It contains gallic acid, which makes it more durable, but corrodes iron fastenings.

Young oak is tougher, more cross-grained, and harder to work than old oak.

Uses.—Oak is used for all purposes where strength and durability are required in engineering structures.

The builder employs it for window and door sills, treads of steps, keys, wedges, trenails, etc., in common work, also for superior joinery of all kinds, for gateposts, etc.

Comparison of the Different Varieties.—It is generally considered that the timber from the stalk-fruited oak is superior to that from the Bay oak.

The respective characteristics of the two varieties, as given by Tredgold, Rankine, and other observers, are as follows :—

The wood of the stalk-fruited oak is lighter in colour than the other. It has a straight grain, is generally free from knots, has numerous and distinct medullary rays, and good silver grain ; it is easier to work and less liable to warp than the timber of the Bay oak, and is better suited for ornamental work, for joists, rafters, and wherever stiffness and accuracy of form are required ; it splits well and makes good laths.

The timber of the cluster-fruited oak is darker in colour, more flexible, tougher, heavier, and harder than that of the stalk-fruited oak ; it has but

---

[1] Laslett.

few large medullary rays, so that in old buildings it has been mistaken for chestnut ; it is liable to warp, and difficult to split ; it is not suited for laths or ornamental purposes, but is better than the other where flexibility or resistance to shocks are required.

Mr. Britton says that dry rot was introduced into ships by using the Bay oak.

Mr. Laslett says that the timber of the *sessiliflora* is a little less dense and compact than that of the *pedunculata*, but they so much resemble each other, that "few surveyors are able to speak positively as to the identity of either."

The Durmast oak is decidedly of inferior quality.

Felling.—Oak is sometimes felled in the spring for the sake of the bark (instead of being stripped in the spring and felled in the winter as described at p. 358). The tree being then full of sap, the timber it yields is not of a durable character.

American Oak.—There are many varieties of this timber, but that chiefly imported into this country is the *White Oak* (*Quercus alba*), so called from the white colour of its bark. It is this variety that is generally known in this country as *American Oak*, or *Pasture Oak*. It is found from Canada to Carolina ; the best comes from Maryland.

Appearance.—The wood has a pale reddish-brown colour, with a straighter and coarser grain than English oak.

Characteristics.—The timber is sound, hard, and tough, very elastic, shrinks very slightly, and is capable of being bent to any form when steamed. It is not so strong or durable as English oak, but is superior to any other foreign oak in those respects.

Uses.—This timber may be used for shipbuilding, and for many parts of buildings in which English oak is used.

Market forms.—It is imported in very large sided logs varying from 25 to 40 feet in length, and from 12 to 28 inches in thickness, also in 2 to 4 inch planks, and in thick stuff of $4\frac{1}{2}$ to 10 inches.

Other varieties of American oak are—

The Canadian or *Red Oak* (*Quercus rubra*) has wood of a brown colour, light and spongy in grain, moderately durable ; is used for furniture and cask staves, but is unfit for work requiring strength and durability.

The Live Oak (*Quercus virens*), with wood of a dark brown or yellow colour, fine grain, minute pores, distinct medullary rings, twisted grain. The logs are crooked, very strong and durable, suitable for ships. This wood makes good mallets and cogs for machinery. It is difficult to obtain in this country.

The Iron Oak (*Quercus obtusiloba*) is of great strength and durability, but of small size, and is chiefly used for posts and fencing.

The Baltimore Oak, with wood of a reddish-brown colour, is generally weak, and soon decays.

There are several other varieties of American oak, generally inferior to the above mentioned, and seldom met with in this country.

Dantzic Oak is grown chiefly in Poland, and shipped at the port after which it is named, also at Memel and Stettin.

Appearance.—It is of a dark brown colour, with a close, straight, and compact grain, bright medullary rays, free from knots, very elastic, easily bent when steamed, moderately durable.

Uses.—It is used for planking, shipbuilding, etc.

Market forms.—The timber is carefully classified as *crown* and *crown brack* qualities.

The planks are classed in the same way, the crown and crown brack marked respectively W and WW.

It is imported in logs from 18 feet to 30 feet long, 10 to 16 inches square, and in planks averaging 32 feet long, 9 to 15 inches wide, and 2 to 8 inches thick.[1]

French Oak is stated by Mr. Laslett to closely resemble British oak in colour, quality, texture, and general characteristics.

Riga Oak is grown in Russia, and is like that shipped from Dantzic, but with more numerous and more distinct medullary rays. It is valued for its silver grain, and is imported in logs of a nearly semicircular section.

Italian Oak—*Sardinian Oak.*—This timber is formed from several varieties of the oak tree. It is of a brown colour, hard, tough, strong, subject to splits and shakes in seasoning, difficult to work, but free from defects. It is extensively used for shipbuilding in her Majesty's dockyards.[1]

Austrian and Hungarian Oak is shipped from Fiume or Trieste in the Adriatic in large logs and wide planks of fine quality.

African Oak, known also as *African Teak* or *Mahogany*, is brought from Sierra Leone, and has many of the characteristics both of oak and teak.

It is of a dark red colour, hard, close grained, difficult to work, free from splits or defects.

It is much used for shipbuilding, but is too heavy for architectural purposes.

Wainscot is a species of oak, soft and easily worked, not liable to warp or split, and highly figured.

This last-mentioned characteristic is obtained by converting the timber so as to show the silver grain (see p. 356). It makes the wood very valuable for veneers, and for other ornamental work.

Wainscot is imported chiefly from Holland and Riga, in semicircular logs.

Clap Boarding is a description of oak imported from Norway, inferior to wainscot, and distinguished from it by being full of white-coloured streaks.

Beech (*Fagus sylvatica*) is known as black, brown, or white beech, all procured from the same species of tree, the difference in the wood being caused by variety in soil and situation.

This tree is found throughout England and Scotland, in the temperate parts of Europe, in America, and Australia.

Appearance.—Has remarkably distinct medullary rays; the annual rings are visible; each is a little darker on one side than the other, and is full of very minute pores. The colour is a whitish brown, darker or lighter according to the variety; the wood has considerable beauty, especially when the silver grain is exposed.

Characteristics.—The wood is of quick growth, light specific gravity, close texture; hard, compact, and smooth surface; is of fine grain, may be cut into thin plates, cleaves easily, is not difficult to work.

It is durable if quite dry or wholly submerged in water, but if subjected to alternate wet and dry becomes overspread with yellowish spots and soon decays. It rots quickly in damp places. It is very subject to the attacks of worms, and contains juices which corrode metal fastenings.

[1] Laslett.

The white variety is the hardest, but the black is tougher and more durable.

Uses.—This timber is not much used by the engineer except for piles under water, and wedges; also for mallets, carpenters' planes, and other tools, for cogs of machinery, cabinet work, and chairs.

Alder (*Alnus glutinosa*) is from a tree found in both Europe and Asia, generally near swamps or the low banks of rivers.

Appearance.—The wood is white when first cut, then becomes deep red on the surface, and eventually fades to reddish yellow of different shades. The roots and knots are beautifully veined.

Characteristics.—Very durable in water when wholly submerged, but when used above ground must be kept perfectly dry. Is soft, light, uniform in texture, with a smooth fine grain, and very easily worked. It is wanting in tenacity, and shrinks considerably.

Uses.—The wood is useful for piles, pumps, patterns, sides of stone carts, packing cases, etc.; also used for wooden bowls, turnery, and furniture. The roots and heart are used for cabinet work. The bark is valuable to tanners, and charcoal from the wood is used for making gunpowder.

Sycamore (*Acer pseudo-platanus*) is from a tree " generally called the *plane tree* in the north of England." It is very common in Great Britain, and is found in Germany.

Appearance.—The wood is white when young, but becomes yellow as the tree grows older, and sometimes brown near the heart.

The texture is uniform, and the annual rings not very distinct.

There are no large medullary rays, but the smaller rays are distinct.

Characteristics.—Compact, firm, not hard, durable when dry, does not warp, liable to be attacked by worms. In large trees the wood is generally tainted and brittle.

Uses.—For furniture, turnery, and wooden screws.

Chestnut (*Castanea vesca*).—This tree flourishes in sandy soils, and is found in most parts of England, in the south of Europe, in Africa, and North America.

Appearance.—The wood resembles that of oak in appearance, but can be distinguished from it, as chestnut has no distinct large medullary rays. The annual rings are very distinct, and the wood of a dark brown colour. The timber is of slow growth, and there is no sapwood.

Characteristics.—Is remarkably durable, easier to work than oak, does not shrink or swell so much; the young wood is hard and flexible, the old wood brittle.

Uses.—Formerly much used for roofs and other carpenters' work, and still valuable to coachmakers, wheelwrights, etc.; also for posts, hoops, etc.

Ash (*Fraxinus excelsior*).—This tree flourishes throughout Great Britain, in Asia, and America.

Appearance.—The colour of the wood is brownish white, with longitudinal yellow streaks; each annual layer is separated from the next by a ring full of pores.

Characteristics.—The most striking characteristic possessed by ash is that it has apparently no sapwood at all—that is to say, no difference between the rings can be detected until the tree is very old, when the heart becomes black.

The wood is remarkably tough, elastic, flexible, easily worked; very durable if felled in winter, well seasoned, and kept dry, but soon rots when exposed to alternate wet and dry. Is subject to the attacks of worms.

The timber is economical to convert, in consequence of the absence of sap. " Very great advantage will be found in reducing the ash logs soon after they are felled into plank or board for seasoning, since, if left for only a short time in the round state, deep shakes open from the surface, which involve a very heavy loss when brought on later for conversion." [1]

Uses.—This wood is too flexible for most building purposes, but is very useful for tool handles, shafts, felloes and spokes of wheels, wooden springs, and wherever it has to sustain sudden shocks.

CANADIAN AND AMERICAN ASH, of a reddish-white colour, is imported to this country chiefly for making oars. These varieties have somewhat the same characteristics as English ash. They are darker in colour. The Canadian variety is the better of the two.

Elm (*Ulmus*).—No less than five varieties of this tree are found in Great Britain, besides which it flourishes in many parts of Europe and in America.

The principal varieties of this timber are as follows :—

THE COMMON ENGLISH OR ROUGH-LEAVED ELM (*Ulmus campestris*), found in England, France, and Spain.

Appearance.—The colour of the heartwood is a reddish brown. The sapwood is of a yellowish or brownish white, with pores inclined to red. The medullary rays are not visible. The wood is porous and very twisted in grain.

Characteristics.—The wood is very strong across the grain; bears driving nails very well; is very fibrous, dense, and tough, and offers a great resistance to crushing. It has a peculiar odour, and is very durable if kept constantly under water or constantly dry, but will not bear alternations of wet and dry. Is subject to attacks of worms. None but fresh-cut logs should be used, for after exposure, they become covered with yellow doaty spots, and decay will be found to have set in. The wood warps very much on account of the irregularity of its fibre. For this reason it should

[1] Laslett.

be used in large scantling, or smaller pieces should be cut just before they are required; and for the same reason it is difficult to work. One peculiar characteristic of elm is that the sapwood withstands decay as well as the heart.

If elm timber is stored it should be kept under water to prevent decay.

The timber is very free from shakes, but frequently contains large hollow places caused by rough pruning and subsequent decay.

Uses.—Elm is used in many situations where it is subjected to continual wet—namely, for piles, parts of pumps, pulley blocks, keels and planks under water in ships, heavy naval gun carriages, coffins, naves and felloes of wheels, stable fittings, etc.; also for various purposes by carpenters, turners, and cabinetmakers.

THE WYCH ELM, of which there are two varieties, the broad-leaved (*Ulmus montana*), and the smooth-leaved wych elm (*Ulmus glabra*), is found chiefly in the north of England, Scotland, and Ireland.

The wood is of a somewhat lighter colour than the common elm. It is clean and straight in grain, tough and flexible, and used for the naves of wheels and for boatbuilding.

THE DUTCH ELM (*Ulmus major*) and the *Corkbarked Elm* (*Ulmus suberosa*) both furnish inferior timber.

The Canada Rock Elm (*Ulmus racemosa*) is grown in North America, and imported chiefly from Canada.

The wood is of a whitish-brown colour, with very close annual rings. It is very tough, flexible, free from knots and sap, with a fine smooth grain, durable under water, but liable to shrink and warp unless kept immersed, and to shakes if exposed to the sun and wind.

Uses.—Being flexible, it is used for boat building, also, on account of its clean appearance, for ladder steps, gratings, etc., on board ship.[1]

The sap is not durable like that of common elm, but subject to decay.

In selecting this wood only those logs should be taken which have an uniform whitish colour, any with dark annular layers full of moisture being left for inferior purposes.[1]

Common Acacia (*Robinia pseudo-acacia*) is found in America.

Appearance.—The wood is of a greenish-yellow colour, with reddish-brown veins. Its structure is alternately nearly compact and very porous, which marks distinctly the annual rings. It has no large medullary rays.

Characteristics.—It is very durable, heavy, hard, and tough, rivalling the best oak in these respects. The timber is generally of small size.

Uses.—It makes first-rate trenails, and very excellent durable posts for fencing, sills for doors, etc.

SABICU (*Acacia formosa*), or the true acacia, is found in the West Indies and Cuba.

Appearance.—It resembles mahogany, but is darker, and is generally well figured.

---

[1] Laslett.

Characteristics.—The wood is very heavy, weathers admirably ; is very free from sap and shakes.

Mr. Laslett says that the fibres are often broken during the early stages of the tree's existence, and that the defect is not discovered until the timber is converted, so that it is seldom used for weight-carrying beams.

This timber is much used in shipbuilding, and also by the cabinetmaker, but not in engineering works.

Poplar (*Populus*).—Of this tree there are several species common in England. The black and the common white poplar are the most esteemed. The Lombardy poplar is inferior.

Appearance.—The colour of the wood is a yellowish or brownish white. The annual rings are a little darker on one side than the other, and therefore distinct. They are of uniform texture, and without large medullary rays.

Characteristics.—The wood is light and soft, easily worked and carved, only indented, not splintered, by a blow.

It should be well seasoned for two years before use. When kept dry it is tolerably durable, and not liable to swell or shrink.

Uses.—The wood not being easily splintered is used for the sides of carts and barrows, for large light barn doors, for packing-cases, floors, etc.

Mahogany is imported from Honduras ("Bay-wood"), Cuba ("Spanish"), and St. Domingo ; formerly the chief sources of its supply ; also from Mexico, Yucatan, Tabasco, Panama, and Africa.

Honduras Mahogany is found in the country round the Bay of Honduras, the trees being of considerable size.

Appearance.—The wood is of a golden or red-brown colour, of various shades and degrees of brightness ; often very much veined and mottled. The grain is coarser than that of Spanish mahogany, and the inferior qualities often contain a large number of grey specks.

Characteristics.—This timber is very durable when kept dry, but does not stand the weather well. It is seldom attacked by dry rot ; contains a resinous oil which prevents the attacks of insects ; it is also untouched by worms. It is strong, tough, and flexible when fresh, but becomes brittle when dry. It contains a very small proportion of sap, and is very free from shakes and other defects. The wood requires great care in seasoning, does not shrink or warp much, but if the seasoning process is carried on too rapidly it is liable to split into deep shakes externally. It holds glue very well, has a soft silky grain, contains no acids injurious to metal fastenings, and is less combustible than most timbers.

It is generally of a plain straight grain and uniform colour, but is sometimes of wavy grain or figured.

Uses.—The builder uses this timber chiefly for handrails, to a small extent for joinery, and for cabinet work. It has sometimes been used for window sashes and sills, but is not fit for external work. " It has been largely used in shipbuilding, for beams, planking, and in many other ways as a substitute for oak, and found to answer exceedingly well." [1]

---

[1] Laslett.

Market forms.—Logs from 2 to 4 feet square, and 12 to 14 feet in length. Sometimes planks have been obtained 6 or 7 feet wide.

"Mahogany is known in the market as 'plain,' 'veiny,' 'watered,' 'mottled,' 'velvet-cowl,' 'bird's-eye,' and 'festooned,' according to the appearance of the vein-formations."[1]

Cuba or Spanish Mahogany, from the island of Cuba, is distinguished from Honduras mahogany by a white chalk-like substance which fills its pores. The wood is very sound, free from shakes, with a beautiful wavy grain or figure, and capable of receiving a high polish. It is used chiefly for furniture and ornamental purposes, handrails, etc., and also for shipbuilding.

Mexican Mahogany shows the characteristics of Honduras mahogany. Some varieties of it are figured. It may be obtained in very large sizes, but the wood is spongy in the centre, coarse in quality, and very liable to starshakes.

It was imported in balks 15 to 36 inches square, and 18 to 30 feet in length, but latterly of smaller sizes.

St. Domingo and *Nassau Mahogany* are hard, heavy varieties, of a deep red colour, generally well veined or figured, and used for cabinet works.

They are imported in very small logs from 3 to 10 feet long, and from 6 to 12 inches square.[2]

African Mahogany is of recent introduction. It is imported in logs 12 to 36 inches square, and up to 36 feet in length. It is shipped from Axim, Assinee, etc.

Jarrah, or *Australian Mahogany* (*Eucalyptus marginata*), comes from West Australia.

Appearance.—The wood is of a red colour, and close, wavy grain, with occasionally figure enough for ornamental purposes.

Characteristics.—Trees decay at centre ; wood is very brittle ; when sound contains a pungent acid repellent to the *teredo*, which is said never to penetrate beyond the sap. The Dutch Commission referred to at page 393 made, however, no exception in favour of this wood. It is also said to resist the white ant. It is full of defects like cupshakes, but filled with resin.

Uses.—It is admirably adapted for piers, jetties, dock gates, piles, and for shipbuilding, and has been extensively adopted for wood-block paving.

Market forms.—Imports to this country are increasing. The sound trees yield timber from 20 to 40 feet long and 11 to 24 inches square.[2]

Teak (*Tectona grandis*), sometimes called *Indian Oak,* is found in Southern India, Pegu, Java, Siam, and Burmah. The lightest, cleanest, and most flexible comes from Moulmein ; the heaviest and strongest from Johore ; and the most handsomely figured variety from the Vindhyan forests. The Malabar teak forests are nearly exhausted. The timber from these forests is darker and stronger than that from Moulmein, but very full of shakes.

Appearance.—The wood has a fine straight grain. It somewhat resembles English oak in appearance, but has no visible medullary rays. The annual rings are very narrow and regular. The colour varies from brownish yellow to dark brown. The texture is very uniform, though porous.

Characteristics.—The timber is stronger and stiffer than English oak, light, and easily worked, but splinters very readily, so

---

[1] Hurst.　　　　　　[2] Laslett.

that it must be worked with care. It contains a resinous aromatic oil, which makes it very durable, and enables it to resist the white ant and worms. It does not corrode, but rather preserves iron fastenings.

There are seldom shakes on the surface, but it is subject to heartshake, and is often worm-eaten.

The resinous oil which exists in the pores often oozes into and congeals in the shakes, and will then destroy the edge of any tool used in working the timber.

This oil is a preservative against rust, and teak is therefore used for backing armour plates and other iron structures.

The oil is sometimes extracted while the tree is growing by " girdling ; " that is, cutting away a ring of bark and sapwood. This practice makes the timber brittle and inelastic, and reduces its durability.[1]

Uses.—This timber is used extensively for shipbuilding, for armour-plated forts, and would be fit for many purposes for which oak is used in ordinary buildings, but that it is too expensive.

Market forms.—Teak is sorted in the markets according to size, not quality. The logs are from 23 to 40 feet long, and their width on the larger sides varies according to the class, as follows :—[1]

    Class A. 15 inches and upwards.
          B. 12 and under 15 inches.
          C. Under 12 inches.
          D. Are damaged logs.

Greenheart (*Nectandra rodiœi*) is found in British Guiana and in the N.E. portion of South America.

Appearance.—The section of this timber has a peculiar appearance, being of a fine grain, and very full of fine pores like the section of a cane. The annual rings are rarely distinguishable. The heartwood is of a dark-green or chestnut colour ; the centre portion a deep, brownish purple, often nearly black. The sapwood is dark green, and often not distinguishable from the heart

Characteristics.—Greenheart is the strongest timber in use. Its resistance to crushing is enormous, but when it gives way it does so suddenly. It is also apt to split and splinter, and therefore requires great care in working. The timber is clean and straight in grain, very hard and heavy. It contains an essential oil, and many authorities state that on account of this it is

[1] Laslett.

entirely free from the attacks of worms. The Dutch Commission that experimented some years ago on this subject reported that this is not the case,[1] and Mr. Laslett considers it doubtful. It appears, however, that in any case worms will only penetrate the sapwood. The presence of the oil above mentioned causes the wood to burn freely, so that it is known in Demerara as " torch-wood." [2]

Uses.—Greenheart is much used for shipbuilding, also for piles, jetties, piers, and other marine structures, and posts of dock gates.

Market forms.—The timber comes into the market roughly hewn, a great deal of bark being left upon the angles, and the ends of the butts are not cut off square The logs are from 12 to 24 inches square, and up to 50 feet in length.

Mora (*Mora excelsa*).—This timber comes from Guiana and Trinidad.

Appearance.—The wood is of a chestnut-brown colour, sometimes beautifully figured.

Characteristics.—The timber is very tough, hard, and heavy ; the grain is close, generally straight, but sometimes twisted so that the wood is difficult to split. An oil in the pores makes the wood very durable. It is free from dry rot, but subject to starshake.

Uses.—It is admirably adapted for shipbuilding.

Market forms.—Logs 18 to 35 feet long, and 12 to 20 inches square.

Hornbeam (*Carpinus betula*) is from a British tree.

Appearance.—The wood is white and close. The medullary rays are plainly marked, and there is no sap.

Characteristics.—The timber is hard, tough, and strong. When subjected to vertical pressure the fibres double up instead of snapping ; it stands exposure well. If cut from old or unseasoned trees the wood is worthless.

Uses.—This wood makes the best mallets. It is very good for turned articles, agricultural implements, cogs for wheels, etc. etc.

MARKS AND BRANDS UPON TIMBER.

There are several distinguishing marks used by the shippers and importers of timber. Some of them refer merely to the number of the balk and to its cubic content, others refer to the quality.

In general terms it may be said that Russian balk timber is marked with a scribe, *i.e.* letters or marks are cut upon it in thin scooped-out lines.

Russian deals are either unmarked or are stamped with small indented letters on their ends.

---

[1] Dent.　　　　　　[2] Haslett.

Swedish deals are marked with large red or black stencilled letters on their ends.

Inferior qualities are frequently without marks at all.

American deals are not generally branded, but are sometimes marked with one, two, or three red chalk marks, to indicate quality.

The letters used to indicate quality are liable to change year by year. A list of the principal marks in use is published annually in Laxton's *Price Book*, and other similar works.

Nearly all the information contained in the following remarks is taken from Colonel Seddon's *Builder's Work.*

"**Shippers' and Quality Marks.**—The different qualities of Memel and Dantzic timber are known as *crown, first* or *best middling, second* or *good middling, third* or *common middling ;* whilst inferior balks are classed as "short and irregular."

"Memel balks of first, second, and third qualities are almost always scribe-marked at one end of the balk ; but these marks must not be mixed up with the number of float or raft, which is also scribed at one end of each balk, and the distinguishing number of balk in the float, which, with the cubic content, is scribed about the centre of every balk floated in the docks, where timber of the same shipment and quality is roped together in separate floats or rafts, and an accurate registry kept of the cubic content, and what becomes of each piece.

"The scribe marks on Baltic timbers are often very numerous and perplexing, most of them being private marks put on by those through whose hands the timber has passed after being squared. On Dantzic they are much more numerous than on Memel or Riga timber ; but with these marks of ownership we have nothing to do ; all we care about are the bracker's or sorter's marks, distinguishing the different qualities from each other.

"The following are the recognised marks for the middling qualities. Very little crown timber is imported, being rarely used by builders, except perhaps for special Government purposes. Memel crown timber is marked as below, but with only a single stroke :—

QUALITY MARKS ON BALTIC TIMBER.

PORT OF SHIPMENT.	FIRST OR BEST MIDDLING.	SECOND OR GOOD MIDDLING.	THIRD OR COMMON MIDDLING.
Riga. (Scribed at centre.)	△	✳	⊗
Memel. (Scribed at end.)	‖	‖‖	‖‖‖
Dantzic. (Scribed at centre.)	✳	✳	✳
Stettin. (Scribed at end.)	✝	✢	✢

'Stettin timber is seldom marked unless to distinguish different qualities in the same cargo.

'Some Riga shippers always use the quality marks for best and good middling, and others only when different qualities are shipped in the same cargo. The common middling quality is rarely shipped from Riga.

" There is no absolute uniformity about these quality marks, as all shippers from the same port do not adopt them, many using private marks of their own, either alone or in addition to the ordinary marks, the latter being seldom omitted on Memel or Dantzic balks. The safest plan, in the case of large and important works, is to order the timber direct from the broker, selecting it out of shipments from houses who have earned a reputation, from the care with which their timber is *bracked* or sorted ; for there is a great difference in the same market quality of timber from different shippers ; one shipper's *good middling* being often nearly equal to another's *best middling*.

" If, amongst a lot of good middling logs, one or two marked as common middling or best middling, as the case may be, are found, it does not always follow that any deception has been practised, since the timber may have changed hands ; a balk here and there may have been considered by the last owner as too good for *common*, or too bad for *best* middling, and been shifted into a *good* middling float." [1]

The following private marks used by a well-known firm of shippers are given as an example :—

Crown	.	.	SK K	SKK R
Best Middling		.	SK	SK R
Good do.		.	SK |	SK | R
Common do.		.	SK ∥ SK	∥ R

As the letters are very roughly marked with the scribe, it will require some practice to recognise the marks. . . .

The addition of R to the SK marks indicates Russian timber shipped by the same firm (S. Kœhne).[1]

" **Baltic Planks, Deals, and Battens are**, speaking in general terms, classed in the market as *Crown, Crown Brack, First Quality, Second Quality*, etc., down to even *Fifth Quality*.

" Very few crown, or crown brack, goods come into market, there being little or no demand for them for building purposes. The different classes of deals, etc., will be found to vary very much in quality, one shipper's second quality being often equal to another's first quality. Hence some shippers have become well known for the greater care with which their goods are bracked or sorted, and their names or trade marks may be safely taken as a guarantee of a high standard in the different qualities into which they are classed."

Among the marks for Dantzic crown deck deals are — CSC. EH. EB.EB.EB.

MK.MK.MK. HP.HP.HP. 👑GCB 👑FGF. Some Dantzic crown brack deck deal marks are—FGF. MK. 👑RJ.

" **Russian and Finland Deals,** which are chiefly first and second quality, or according to the shippers *prima* and *secunda*, generally come unmarked into the market, or only *dry stamped* or marked at their ends with the blow of a branding hammer, such marks being also termed *hard brands*. Some good shipments from Uleaborg (Finland) are dry stamped U S for "mixed" (first and second quality unsorted) and U S in red paint for third quality goods. Onega and Archangel deals are dry stamped thus with the shipper's initials, or private mark, and often with a number in addition, which, however, does not denote the quality, but merely the number of the yard in which they were stored before shipping.

" In some cases, when the goods are not branded, the second quality have a red mark across the ends ; third being easily distinguished from first quality goods.

" The well-known Gromoff Petersburg deals are, however, marked with C. and Co., the initials of the shippers, Clarke and Company. Another good Petersburg brand is P B (Peter Belaieff) for best, and P B 2 for second quality.

" **Swedish Goods** are never hammer-marked, but invariably branded with letters or devices stencilled on the ends in red paint, which makes it difficult to judge of their quality by inspection, as they are stacked in the timber yards with their ends only showing. Some of the common fourth and fifth quality Swedish goods are left unmarked, but they may generally be distinguished from Russian shipments by the bluer colour of the sapwood.

" In the English market the first and second qualities, in Swedish deals, are classed

---

[1] Seddon.

District	Shippers	Mixed, or French 1st.	3ds, or French 2ds.	4ths, or French 3ds.	5ths, or French 4ths.	6ths, or French 5ths.	Unsorted, or French non-classes.
Sweden. Gefle	J. E. Francke . . . one end	E ✳ F / GEFLE	E A F / GEFLE	E B F / GEFLE	E C F / GEFLE	...	E (crown) B
Do.	Do. . . other end	K A B	N A S	GEFLE	GEFLE		B P
Do.	Kornäs Sågverks .	B (crown) P	B+P	B ♦ P	B P	...	B P
Do.	Strömsbergs Bruksegare .	Extra 1st / B (crown) P (crown)	1st / B (crown) P	2d / B + P	♦		♦ ♦ ♦
Do.	Do. . . floorings						
Gothenburg	A. Karberg and Co.	K (crown) K / J W W	S ♦ F / J W	T + F / W	− + −		
Do.	J. W. Wilson . English shipment						
Hernosand	J. E. Francke and others	BERGERE	BERGER	E L B	K + M	D (crown) B	B (crown) R
Stockholm	Lind, August, and Co.	F (crown) Co. or	F ♦ Co.	F ✚ Co. or	F o Co. or		
Sundswall	Axell and Co.	A (crown) L / AXELL	A ♦ L / A X L / D D D	A ✚ L / A X E / D	A o L / A X X / M F D		
Do. and Umea	J. Dickson and Co. English shipment	D B & Co.	J D & Co.	D D D	D D F		✳ W ✳
Do.	Do. . . flooring	D B & Co.					
Söderhamm	W. H. Kempe English shipment	W (crown) K	W K	K	K ✳ K		
Do.	J. Dickson and Co. do.	1st / D B & Co.	2d / J D & Co.	3d / D D D	4th / D D F	...	Unsorted.
Norway. Christiania[1]							
Drammen	Haneborg . flooring	1sts / A H O	2ds / A H A	3ds / D O L	4ths	5ths	
Do.	Haugan, O., and Co.	O + H	O H	H	O		
Do.	Do. . . flooring	HAU (crown) GAN	HAUGAN	H G N			
Russia. Archangel	Brandt, E. H., and Co	E H B & Co.[1]	E H B & Co.[2]	E H B & Co.[3]			
Do.	Do. . . flooring	E B & Co.	E B & Co.	E B & Co.			
Onega	Onega Wood Co.	1	0	+			Figures on flat in red chalk.
St. Petersburg .	Belaieff, successors and Co., Peter .	No mark	2	3			
Do.	Gromoff, F., and Co. .	P B S & Co.	P B S & Co.[2]	P B S & Co.[3]			

Several different marks.

[1] The marks on sawn wood shipped at Christiania are always red in colour; those on flooring mostly in black, but some in other colours.

together as 'mixed,' being scarcely ever sorted separately ; after which we get third, down to fifth quality goods.

" The French class the mixed as first, and our third as second quality, and so on."

" Except for temporary purposes, or for rough work such as slate boarding, no deals or a lower quality than *mixed* Swedish, or, as the timber merchants and contractors would call them, *best* Swedish, should be used on Government works."

The few brands on p. 384 are taken from the *Timber Trades Journal* List,[1] and are given merely as characteristic examples. As before mentioned, the marks are constantly changing, and any information regarding them should be renewed from year to year. A long list is published annually in Laxton's *Price Book*, and at intervals in the *Timber Trades Journal* Lists.

" To give an idea of the value of the different qualities, the *mixed* are worth from 15 to 20 per cent more than third, and third from 12 to 15 per cent more than fourth quality.

" It may be noticed in the above brands that three similar letters, when used, generally denote the shipper's third quality ; but a merchant would call these second quality goods, for it must be clearly understood that the term 'mixed' is confined to the shippers and brokers. Timber merchants always call the mixed 'best,' and the third quality second quality, and so on, or one class higher than that at which they were shipped.

" *The Norwegian marks* are very numerous, but, as the chief import is of cheap and very inferior battens (mostly $2\frac{1}{2} \times 6\frac{1}{2}$), they are not worth enumerating.

" From Christiania, however, come the very best white deals come, marked H M H for first quality, and H M M for second quality.

" Battens from Dram have several marks, among which are for 1st class HK JB, for 2d class HK and Co. I OO, for 3d class IW B, etc. etc.

" Norway also exports large quantities of cheap boards for flooring and other purposes, *match* or grooved and tongued boarding, mouldings, doors, window sashes, etc., all ready for fixing, which may often be used with advantage for inferior or temporary purposes.

" **American Goods** are not branded as a rule, though some houses use brands in imitation of the Baltic marks already described, though without following any definite rules. The qualities may, however, very often be known by red marks I II III upon the sides or ends, but the qualities of American yellow deals are easily told by inspection, the custom in the London Docks being to stack them on their sides, so as to expose their faces to view, and allow of free ventilation."

The following are marks upon some Quebec deals :—

	1st.	2d.	3d.	
Hamilton's bright dry floated deals	I	II	III	in red on flat
Gilmour and Co., pine deals, etc.	A	B	C	do. and on end.

Mahogany, Cedar, and other imported woods, are also marked with letters, a long list of which is given in Richardson's *Timber Importer's Guide.*

The following extract, from a valuable article in the *Building News*, shows the importance of the subject.

" From these remarks it will be seen that brands upon timber is a great and important subject. It is one in the hands of a small community of our traders, and is, consequently, a class of knowledge over which they are strict conservators. It is a subject new to authors, and that portion of our tradesmen whose office it is to buy and consume timber. This is somewhat strange, as the meaning of brands is well known on other goods that people engaged in trade are called upon to purchase. With architects, clerks of works, and builders generally, brands upon timbers are looked upon with perfect indifference. The current remarks are, ' I can tell a bit of good wood when I see it,' etc., and, in builders generally pursue the old-fashioned system of buying from inspection, the question carries but little importance.

" Were brands upon timber better known, architects would get better work and builders would obtain greater credit. The cheap builder would find his place, and what are termed 'old-fashioned builders' would again occupy the position they so richly merit."

Value of Timber, Deals, etc., and Method of Measuring.—The prices of different descriptions of timber, deals, etc., vary at the different ports. They are published weekly in the engineering and building journals, and also annually in the builders' price-books. The method in which timber is measured and the " standards " under which deals are sold, are described in Seddon's *Builder's Work*, Hurst's *Pocket Book*, and in works devoted to the subject of measuring and estimating.

---

[1] Published by W. Rider and Son, London, E.C.

SELECTION OF TIMBER.

In consequence of the great number of marks used in the timber trade, the difficulty of ascertaining what they mean, and the frequent changes that take place in them, the practical engineer or builder, as a rule, judges of the quality of the timber more by its appearance than by the way in which it is marked.

The characteristics of good timber and the defects to be avoided are given in general terms at p. 358, but a few remarks on selecting balks and deals may be useful. It should be remembered that most defects show better when the timber is wet.

Balk timber is generally specified to be free from sap, shakes, large or dead knots and other defects, and to be die-square.

In the best American yellow pine and crown timber from the Baltic there should be no visible imperfections of any kind.

In the lower qualities there is either a considerable amount of sap, or the knots are numerous, sometimes very large, or dead. The timber may also be shaken at heart or upon the surface.

The wood may be waterlogged, softened, or discoloured by being floated.

Wanes also are likely to be found which spoil the sharp angles of the timber, and reduce its value for many purposes.

The interior of the timber may be soft, spongy, or decayed, the surface destroyed by worm holes, or bruised.

The heart may be wandering—that is, at one part on one side of the balk, at another part on the other side. This interrupts the continuity of the fibre, and detracts from the strength of the balk. If on the same side of a balk sap is visible at one end and heart at the other, it shows that the heart is wandering; in good timber the " spine " or heartwood should be visible on 'all four sides. Again, the heart may be twisted throughout the length of the tree. In this case the annual rings which run parallel to two sides of the balk at one end run diagonally across the section at the other end. This is a great defect, as the wood is nearly sure to twist in seasoning.

Some of these defects appear to a certain degree in all except the very best quality of timber. The more numerous or aggravated they are, the lower is the quality of the timber.

Deals, planks, and battens should be carefully examined for freedom (more or less according to their quality) from sap, large or dead knots, and other defects, also to see that they have been carefully converted, of proper and even thickness, square at the angles, etc. As a rule, well-converted deals are from good timber, for it does not pay to put much labour upon inferior material.

The method in which the deals have been cut should be noticed, those from the centre of a log, containing the pith, should be avoided, as they are likely to decay (see p. 398).

SEASONING TIMBER.

The object of seasoning timber is either to expel or to dry up the sap remaining in it, which otherwise putrefies and causes decay.

One effect of seasoning is to reduce the weight of timber, and this reduction of weight is, to some extent, an indication of the success of the process.

Tredgold calls timber *seasoned* when it has lost $\frac{1}{5}$ of its weight, and says that it is then fit for carpenters' work and common purposes. He calls it *dry*, fit for joiners' work and framing, when it has lost ⅓ of its weight.

The exact loss of weight must depend, however, upon the nature of the timber and its state before seasoning.

Timber should be well seasoned before being cut into scantlings. The scantlings should then be further seasoned, and after conversion the wood should be left as long as possible to complete the process of seasoning before being painted or varnished.

Mr. Britton states that logs season better and more quickly if a hole is bored through their centre. This also prevents splitting.

There are several different methods of seasoning timber, the principal of which will now be briefly described.

Natural Seasoning is carried out by stacking the timber in such a way that the air can circulate freely round each piece, at the same time protecting it by a roof from the sun, rain, draughts, and high winds, and keeping it clear of the ground by bearers.

The great object is to ensure regular drying. Irregular drying causes the timber to split.

Timber should be stacked in a yard, paved if possible, or covered with ashes, and free from vegetation.

The bearers used should be damp-proof, and should keep the timber at least 12 inches off the ground. They should be laid perfectly level and out of winding, otherwise the timber will get a permanent twist.

If possible, the timber should be turned frequently so as to ensure equal drying all round the balks.

When a permanent shed is not available, temporary roofs should be made over the timber stacks.

Logs are stacked with the butts outwards, the inner ends being slightly raised so that the logs may be easily got out. Packing pieces are inserted between the tiers of logs, so that by removing them any particular log may be withdrawn.

Some authorities have stated that timber seasons better when stacked on end. This, however, seems doubtful, and the plan is practically difficult to carry out.

Boards may be stacked in the same way, laid flat and separated from one another by pieces of dry wood an inch or so in thickness and 3 or 4 inches wide. Any that are inclined to warp should be weighted or fixed down to prevent them from twisting.

Boards are, however, frequently stacked vertically, or inclined at a high angle.

Mr. Laslett recommends that they should be seasoned in "a dry cool shed, fitted with horizontal beams and vertical iron bars, to prevent the boards, which are placed on edge, from tilting over."

The time required for natural seasoning differs according to the size of the pieces, the nature of the timber, and its condition before seasoning.

Tredgold gives some algebraic formulæ for calculation of the time required, and a table deduced therefrom.

Mr. Laslett has, however, compiled a table from practical observation.

He says : " My experience of the approximate time required for seasoning timber under cover and protected from wind and weather is as follows :—

							Oak. Months.	Fir. Months.
Pieces 24 inches and upward square require about			.	.	26	13
,,	Under 24 inches to 20		,,		.	.	22	11
,,	,,	20	,,	16	,,	.	.	18	9
,,	,,	16	,,	12	,,	.	.	14	7
,,	,,	12	,,	8	,,	.	.	10	5
,,	,,	8	,,	4	,,	.	.	6	3

" Planks from ½ to ⅔ the above time according to the thickness."

Mr. Laslett further states that if the timber is kept longer than the periods above named the fine shakes which show upon the surface in seasoning " will open deeper and wider until they possibly render the logs unfit for conversion."

Tredgold says that the time required under cover is only $\frac{2}{5}$ of that required in the open.

Water Seasoning consists in totally immersing the timber, chaining it down under water, as soon as it is cut, for about a fortnight, by which a great part of the sap is washed out. It must then be carefully dried, with free access of air, and turned daily.

Timber thus seasoned is less liable to warp and crack, but is rendered brittle and unfit for purposes where strength and elasticity are required.

Care must be taken that the timber is entirely submerged. Partial immersion, such as is usual in timber ponds, injures the log along the water line.

Timber that has been saturated should be thoroughly dried before use ; when taken from a pond, cut up and used wet, dry rot soon sets in.

Salt water makes the wood harder, heavier, and more durable, but it should not be applied to timber for use in ordinary buildings, because it gives the wood a permanent tendency to attract moisture.

Boiling and Steaming.—Boiling water quickens the operation of seasoning, and causes the timber to shrink less,[1] but it is expensive to use, and reduces the strength and elasticity of the timber.

The time required varies with the size and density of the timber, and according to circumstances ; one rule is to allow an hour for every inch in thickness.

Steaming has very much the same effect upon timber as boiling, but the timber is said to dry sooner after the former process,[1] and it is by some considered that steaming prevents dry rot.

Mr. Britton says, however, " no doubt boiling and steaming partly remove the ferment spores, but *may not* destroy the vitality of those remaining."

Hot-air Seasoning, or *desiccation*, is effected by exposing the timber in an oven to a current of hot air, which dries up the sap.

This process takes only a few weeks, more or less, according to the size of the timber.

When the wood is green the heat should be applied gradually.

Great care must be taken to prevent the timber from splitting, the heat must not be too high, and the ends should be clamped.

Desiccation is useful only for small scantling ; the expense of applying it to larger timber is very great ; morever,." as wood is one of the worst conductors of caloric, if this plan be applied to large logs the interior fibres still retain their original bulk, while those near the surface have a tendency to shrink, the consequence of which would be cracks and splits of more or less depth." [2]

Desiccated wood should not be exposed to damp before use.

Mr. Laslett says that during this process ordinary woods lose their strength, and coloured woods become pale and wanting in lustre.

M'Neile's Process is one that has been some few years in operation.

It consists in exposing the wood to a moderate heat in a moist atmosphere charged with various gases produced by the combustion of fuel.

The wood is placed in a brick chamber, in which there is a large surface of water to produce vapour.

The timber should be stacked in the usual way, with free air-space round

---

[1] Tredgold.   [2] Britton.

every piece ; about $\frac{1}{3}$ of the whole content of the chamber should be air space.

Under the chamber is a fireplace.

The fire having been lighted, the products of combustion (among which is carbonic acid gas) circulate freely in a moist state around the pieces of timber to be seasoned.

The time required varies with the nature of the wood.

Oak, ash, mahogany, and other hard wood planks 3 inches thick, take about 3 weeks.

Oak wainscot planks 2 inches thick take from 5 to 6 weeks.

Deals 3 inches thick something less than a month.

Flooring boards and panelling about 10 days or a fortnight.

" The greener the wood when first put into the stove the better. As a rule, if too great heat be not applied, not a single piece of sound wood is ever split, or warped, or opened in any way. The wood is rendered harder, denser, and tougher, and dry rot is entirely prevented. The wood will not absorb by subsequent exposure to the atmosphere nearly so much moisture as does wood dried by exposure in the ordinary way, hence it is better for all purposes than air-dried wood." [1]

The process seemed to have no injurious effects upon the appearance or strength of the timber.

It has been adopted by some of the principal firms in London and elsewhere.

Smoke-drying.—It is said that if timber be smoke-dried over a bonfire of furze, straw, or shavings, it will be rendered harder, more durable, and proof against the attacks of worms. In order to prevent the timber from splitting and to ensure the moisture drying out from the interior, the heat should be applied gradually.

Second Seasoning.—Many woods require a second seasoning after they have been worked.

Floor boards should, if possible, be laid and merely tacked down for several months before they are cramped up and regularly nailed.

Doors, sashes, and other articles of joinery should be left as long as possible after being made, before they are wedged up and finished.

Very often a board that seems thoroughly seasoned will commence to warp again if merely a shaving is planed off the surface.

DECAY OF TIMBER.

To preserve timber from rot or decay it should be kept constantly dry and well ventilated. It should be clear of the influence of damp earth or damp walls, and free from contact with mortar, which hastens decomposition.

Wood kept constantly submerged is often weakened and rendered brittle, but some timbers are very durable in this state (see elm, beech, acacia, etc.)

Timber that is constantly dry is very durable. However, it also becomes brittle in time, though not for a great number of years.

[1] Patentee's Circular.

" When timber is exposed to alternate moisture and dryness it soon decays." [1]

The general causes of decay in timber are the presence of sap, exposure to alternate wet and dryness, or to moisture accompanied by heat and want of ventilation.

Rot in timber is decomposition or putrefaction, generally occasioned by damp, and which proceeds by the emission of gases, chiefly carbonic acid and hydrogen.

There are two kinds of rot generally known to practical men —*dry* rot and *wet* rot.

The chief difference between them seems to be that *wet rot* occurs where the gases evolved can escape. By it the tissues of the wood, especially the sappy portions, are decomposed. Dry rot, on the contrary, occurs in confined places, where the gases cannot get away, but enter into new combinations, forming fungi which feed upon and destroy the timber.

Tredgold says that wet rot may take place while the tree is standing, whereas dry rot takes place only when the wood is dead.

Dry Rot is generally caused by want of ventilation. Confined air, without much moisture, encourages the growth of the fungus, which eats into the timber, renders it brittle, and so reduces the cohesion of the fibres that they are reduced to powder. It generally commences in the sapwood.

An excess of moisture prevents the growth of the fungus, but moderate warmth, combined with damp and want of air, accelerates it.

" In the first stage of rottenness the timber swells and changes colour, is often covered with fungus or mouldiness, and emits a musty smell."

" When the fungus first appears on the sides and ends of timbers it covers the surface with a fine delicate vegetation called by shipwrights a mildew.

" These fine shoots afterwards collect together, and the appearance may then be compared to hoar-frost, and increases rapidly, assuming gradually a more compact form, like the external coat of a mushroom, but spreads alike over wood, brickwork, stone, and plastering in the form of leaves, being larger or smaller, most probably, in proportion to the nutriment the wood affords. The colours of the fungus are various, sometimes white, greyish white with violet, often yellowish brown, or a deep shade of fine rich brown." [2]

The positions in which dry rot occurs are, as already mentioned, those where the timber is exposed to warmth and damp stagnant air.

The principal parts of buildings in which it is found are—

In warm cellars, under unventilated wooden floors, or in basements, particularly in kitchens or rooms where there are constant fires. " All kinds of stoves are sure to increase the disease if moisture be present."

The ends of timbers built into walls are nearly sure to be affected by dry rot unless they are protected by iron shoes, lead, or zinc. The same result is produced by fixing joinery and other woodwork to walls before they are dry.

---

[1] Tredgold.   [2] Britton *On Dry Rot.*

Oilcloth, kamptulicon, and other impervious floorcloths, by preventing access of air and retaining dampness, cause decay in the boards they cover. Carpets do the same to a certain extent.

Painting or tarring cut or unseasoned timber has the same effect.

Sometimes the roots of large trees near a house penetrate below the floors and cause dry rot.

It is said that if two different kinds of wood—as, for example, oak and fir —are placed so as to touch end to end, the harder of the two will decay at the point of junction.

"There is this particular danger about the dry rot—viz., that the germs of the fungi producing it are carried easily, and in all directions, in a building where it once displays itself, without necessity for actual contact between the affected and the sound wood."

" Before dry rot has time to destroy the principal timbers in a building it penetrates behind the skirtings, dadoes, and wainscotings, drawing in the edges of the boards and splitting them both horizontally and vertically. When the fungus is taken off they exhibit an appearance similar, both in back and front, to wood that has been charred ; a slight pressure with the hand will break them asunder, even though affected with the rot but a short time, and in taking down the wainscot the fibrous and thin-coated fungus will generally be seen closely attached to the decayed wood. In timber of moderate length the fungus becomes larger and more distinctive in consequence of the matter congenial to its growth affording a more plentiful supply." [1]

Wet Rot occurs, as before mentioned, in the growing tree, and in other positions where the timber may become saturated with rain.

If the wood can be thoroughly dried by seasoning, and the access of further moisture can be prevented by painting or sheltering the timber, then wet rot can be prevented.

" The communication of the disease resulting from the putrefactive fermentation or the wet rot only takes place by actual contact," not by the dissemination of the germs of fungi as with dry rot.

Detection of Dry Rot.—In the absence of any outward fungus, or other visible sign, the best way is to bore into the timber with a gimlet or augur. A log apparently sound, as far as external appearances go, may be full of dry rot inside, which can be detected by the appearance of the dust extracted by the gimlet, or more especially by its smell.

If a piece of sound timber be lightly struck with a key or scratched at one end, the sound can be distinctly heard by a person placing his ear against the other end, even if the balk be 50 feet long ; but if the timber be decayed, the sound will be very faint, or altogether prevented from passing along the balk.

Imported timber, especially fir, is often found to be suffering from incipient dry rot upon arrival. This may have originated in the wood of the ship itself, or from the timber having been improperly stacked, or shipped in a wet state, or subjected to stagnant, moist, warm air during the voyage.

Sometimes the rot appears only in the form of reddish spots, which, upon being scratched, show that the fibres have been reduced to powder. After a long voyage, however, the timber will often be covered with white fibres of fungus.

Canadian yellow pine is very often found in this state.

The best way of checking the evil is to sweep the fungus off the timber, and restack it in such a way that the air can circulate freely round each piece. [1]

---

[1] Britton.

PRESERVATION OF TIMBER.

The best means for preserving timber from decay are to have it thoroughly seasoned and well ventilated.

Several processes have, however, been introduced at different times with a view of preventing decay in timber by excluding moisture, or by drying up or expelling the sap within it.

A few of these processes will now be described.

Painting preserves timber if the wood is thoroughly seasoned before the paint is applied. Otherwise the filling up of the outer pores only confines the moisture and causes rot. The same may be said with regard to *Tarring.*

Sometimes before the paint is dry it is sprinkled with sand, which is said to make it more durable.

Tredgold says—" For timber that is not exposed to the weather, the utility of paint is somewhat doubtful. . . . Wood used in outdoor work should have those parts painted only where moisture is likely to find a lodgment, and all shakes or cracks and joints should be filled up with white lead ground in oil, or oil putty, previous to being painted over."

Charring Timber.—The lower ends of posts put into the ground are generally charred with a view of preventing dry rot and the attacks of worms.

Care should be taken that the timber to which this process is applied is thoroughly seasoned, otherwise by confining the moisture it will induce decay and do more harm than good.

It may here be mentioned that posts should be put in upside down, with regard to the position in which they originally grew. The sap valves open upwards from the root, and when thus reversed they prevent the ascent of moisture in the wood.

Mr. Britton recommends that the charring process should be applied to the embedded portions of beams and joists, to joists of stables, wash-houses, etc., to wainscoting of ground-floors, to flooring beneath parquet work, to the joints of tongues and rebates, and to railway sleepers.

Mons. de Lapparent applied the method on a large scale by the use of a gas jet passed all over the surface of the timber, but Mr. Laslett, who experimented on timbers thus treated, says—

" I should not myself be inclined to use it on timber for works of construction, except as a possible means of preventing the generation of moisture or fungus where two unseasoned pieces of wood are placed in juxtaposition."

Creosoting, known also as Bethell's process, is effected by extracting the moisture and air from the tubes of the timber, and then forcing in *kreasote* (oil of tar), generally called *creosote*, at a high pressure.

The timber after being dried is placed in a closed wrought-iron cylinder. The air is then extracted from the cylinder and pores of the wood by a pump.

Creosote (see p. 452) at a temperature of about 120° is then forced into the cylinder, and penetrates the wood under a pressure of about 170 lbs. per square inch.

The creosote should be thick, rich in naphthaline, and free from ammonia.[1] The amount of creosote pumped in depends upon the nature of the timber and the purpose for which it is intended. The sapwood absorbs it more readily than the heart.

Fir timber or other soft wood will take from 10 to 12 lbs. per cubic foot.

Mr. Bethell recommends 7 lbs. per cubic foot for railway works and 10 lbs. for marine works.

Somewhat larger quantities than these are now generally used.

Into oak and other hard woods it is difficult to force more than 2 or 3 lbs. per cubic foot.[1]

To soft woods an imperfect form of this process may be applied by drying the timber over fires, and placing it while warm in hot creosote.

Of all the preservative processes at present known, creosoting seems to be the most successful ; it coagulates the albumen of the wood, fills its pores with an oily liquid, destroys insects and fungi, repels worms, excludes moisture, and prevents dry rot.

Experience seems to show that creosote will, to a certain extent, render timber proof against sea-worms, and even the white ant.

Several years ago a Commission was appointed by the Dutch Government to report upon the best method of protecting timber from the attacks of the sea-worm, known as the *teredo* (see p. 399).

This Commission tried every preservative means then known, including, among others—charring the surface, covering with paraffin, with sheet metals, nails (see p. 400), impregnation with all sorts of chemical substances, creosoting, and kyanising.

The conclusion they arrived at was that " the only process that could be relied upon for protecting wood from the attacks of the teredo was that of creosoting, and that this fails if not properly carried out." [2]

Kyan's Process consists in injecting corrosive sublimate (bichloride of mercury) in the proportion of 1 pound of sublimate to 15 gallons of water.

The Dutch experiments showed that this process did to a certain extent, though not altogether, repel the sea-worm, and it is said that it has some effect in retarding dry rot. It is now, however, seldom if ever used.

Boucherie's Process consists in impregnating the timber with sulphate of copper by a very simple process.

A reservoir filled with the solution (about 1 lb. of sulphate copper to 12½ gallons of water) is placed at a height of from 20 to 30 feet above the ground.

From this reservoir leads a pipe into a deep incision in the wood, so arranged that the liquid may reach the centre of the log. Thence it forces its way (under the pressure caused by the height of the tank) along the sap tubes, forces the sap out, and takes its place.

To see if the solution has passed right through the timber the far end is rubbed with prussiate of potash, which upon coming in contact with the sulphate of copper makes a brown stain.

---

[1] Britton.   [2] Dent.

Gardner's Process is one that has been lately introduced. It is said[1] to season timber more rapidly than any other process, to preserve it from decay and from the attacks of all kinds of worms and insects. It is also found to strengthen the timber, and render it uninflammable, and by it the timber may be permanently coloured to a variety of shades.

The process takes from 4 to 14 days according to the bulk and density of the timber. It consists in dissolving the sap (by chemicals in open tanks), driving out the remaining moisture, leaving the fibre only.

A further injection of chemical substances adds to the durability, or will make the timber uninflammable.

The process has been satisfactorily tested in mine props, railway sleepers, logs of mahogany for cabinet work, and in smaller scantlings of fir and pine.

The experiments showed that the sap was removed, that the resistance of the timber to crushing was augmented from 40 to 90 per cent, and its density was considerably increased.

Margary's Process was to soak the wood in acetate or sulphate of copper. It does not seem to have been successful.

Sir William Burnet's System consists in steeping the timber in a solution composed of 1 lb. of chloride of zinc to 4 gallons of water.

Payne's Process involved two injections into the pores of the timber, the first being sulphate of iron, the other sulphate of zinc. It is said to make the timber incombustible but brittle.

Combined Process.—In cases where the complete preservation of the timber is of vital importance, and expense no object, Mr. Britton recommends that the timber should first be injected with metallic salt (as in Burnett's system), dried, and then creosoted. By this means the whole is preserved; the salts protect the heart, and the creosote the sapwood.

Oak Casings may be preserved from injury done by weather by two coats of boiled oil applied cold.

Preservation from Fire.—Several methods for preserving timber from fire have been proposed from time to time.

It is said that timber that has been thoroughly Burnetised will only become charred and not burnt by fire.

Some years ago the following means of protection was recommended by Sir F. Abel.

The wood having a smooth and clean surface is first painted over with a dilute solution of the silicate, then with slaked fat lime of the consistency of cream, then with a stronger solution of silicate.

Cyanite is a fireproof solution, probably containing a soluble silicate, which has been frequently tried lately, and apparently with success. It is stated that it will cover twice as much as an equal quantity of priming.

Asbestos Paint (see p. 426) affords some slight protection against fire.

Tungstate of Soda imparts fireproof qualities to timber or fabrics covered with repeated coats of the solution.

CONVERSION OF TIMBER.

In reducing timber from the log or baulk to scantlings, the dimensions and form that the timber ought to possess when actually in use should be borne in mind, in order that proper allowance may be made for the alteration that will take place in consequence of the action of the atmosphere, which has an influence more or less even upon well-seasoned timber.

[1] Paper read before the Philosophical Society of Glasgow, by Jas. Deas, Esq., M.I.C.E.

Atmospheric Influence.—In straight-grained woods the changes in length caused by the effects of the atmosphere are very slight ; but the variations in width and depth are very great, especially in new timber.

Rondelet found that the usual changes of weather produced the following expansion and contraction in wood of average dryness :—

In fir from $\frac{1}{360}$ to $\frac{1}{75}$ of width ; mean $\frac{1}{124}$.

In oak from $\frac{1}{412}$ to $\frac{1}{80}$ of width ; mean $\frac{1}{140}$.

Mr. Hurst makes a practical allowance for shrinkage in 9-inch deals amounting to $\frac{1}{4}$ inch for " northern pine " deals, and $\frac{1}{8}$ for " white deals."

The first effect of atmospheric influence upon a log is that the external portions which are exposed to the air shrink ; but the interior, which is protected from the air, remains at its original bulk. The consequence is that the exterior splits, as shown in Fig. 156.

The following extract, taken by permission from Dr. Anderson's lecture on applied mechanics given before the Society of Arts, explains very clearly the manner in which timber shrinks when cut into scantling :—

" Notwithstanding the extent to which timber is used in the mechanical arts, it is singular that the natural law by which the contraction or shrinking of wood is governed is too much disregarded in practical operations. It is a subject which seems to have been entirely neglected by writers on the subject. . . .

" An examination of the end section of any exogenous tree, such as the beech or oak, will show the general arrangement of its structure. It consists of a mass of longitudinal fibrous tubes arranged in irregular circles that are bound together by means of radical strings or shoots which have been variously named. They are the " silver grains " of the carpenter, or the " medullary rays " of the botanist, and are in reality the same as end wood, and have to be considered as such, just as much so as the longitudinal woody fibre, in order to understand its action. From this it will be seen that the lateral contraction or collapsing of the longitudinal porous or tubular part of the structure cannot take place without first crushing the medullary rays ; hence the effect of the shrinking finds relief by splitting in another direction, namely, in radial lines from the centre, parallel with the medullary rays, thereby enabling the tree to maintain its full diameter, as shown in Fig. 156.

Fig. 156.

" If the entire mass of the tubular fibre composing the tree were to contract bodily, then the medullary rays would of necessity have to be crushed in the radial direction, to enable it to take place, and the timber would thus be as much injured in proportion as would be the case in crushing the wood in the longitudinal direction. If such an oak or beech tree is cut into four quarters by passing the saw twice through the centre at right angles, before contracting and splitting has commenced, the lines $a\,c$ and $c\,b$ in Fig. 157 would be of the same length, and at right angles to each other, or in the technical language of the workshop they would be square ; but after being stored in a dry place, say for a year, it would then be seen that a great change had taken place both in the form and in some of the dimensions :

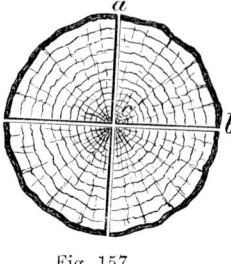

Fig. 157.

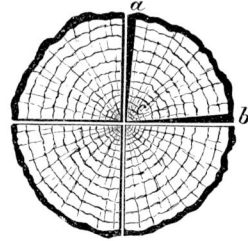

Fig. 158.

the lines *c a* and *c b* would be the same length as before, but it would have contracted from *a* to *b* very considerably, and the two lines *c a* and *c b* would not be at right angles to each other by the portion shown here in black in Fig. 158. The medullary rays are thus brought closer by the collapsing of the vertical figure.

"But, supposing that four parallel saw cuts are passed through the tree so as to form it into five planks, let us see what would be the behaviour of the several planks. Take the centre plank first. After due seasoning and contracting it would then be found that the middle of the board would still retain the original thickness from the resistance of the medullary rays, while it would be gradually reduced in thickness towards the edges for want of support, and the entire breadth of the plank would be the same as it was at first, for the foregoing reasons, and as shown in Fig. 159. Then taking

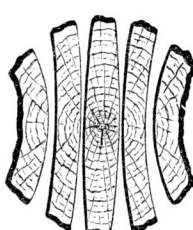

the planks at each side of the centre, by the same law their change and behaviour would be quite different. They would still retain their original thickness at the centre, but would be a little reduced on each edge throughout, but the side next to the heart of the tree would be pulled round, or partly cylindrical, while the outside would be the reverse, or hollow, and the plank would be considerably narrower throughout its entire length, more especially on the face of the hollow side, all due to the want of support. Selecting the next two planks, they would be found to have lost none of their thickness at the

Fig. 159.

centre, and very little of their thickness at the edges, but very much of

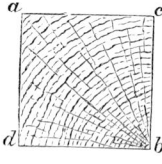

Fig. 160.

their breadth as planks, and would be curved round on the heart side, and made hollow on the outside. Supposing some of these planks to be cut up into squares when in the green state, the shape that these squares would assume after a period of seasoning would entirely depend on the part of the tree to which they belonged ;

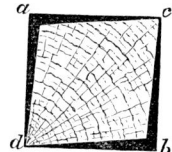

Fig. 161.

the greatest alteration would be parallel with the medullary rays. Thus, if the square were near the outside, as in Fig. 160, the effect would be that it would contract in the direction from *a* to *b*, and after a year or two it would be as in Fig. 161, the distance between *c* and *a* being nearly the same as it was before, but the other two angles *a* and *b* brought by the amount of their contraction closer together. By understanding this natural law, it is comparatively easy to know the future behaviour of a wood or plank by carefully examining the end wood in order to ascertain the part of the log from which it has been cut, as the angle of the ring growths and the medullary rays will show thus, as in Fig. 162. If a plank has this appearance it will evidently show to have been cut from the outside, and for many years it will gradually shrink all to the breadth, while the next plank, shown in Fig. 163, clearly points close to the centre or heart of the

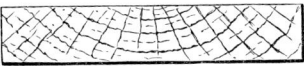

Fig. 162.

Fig. 163.

tree, where it will not shrink to the breadth, but to a varying thickness, with the full dimensions in the middle, but tapering to the edges, and the planks on the right and left will give a mean, but with the centre sides curved round, and the outside still more hollow.

"The foregoing remarks apply more especially to the stronger exogenous woods, such as beech, oak, and the stronger home firs. The softer woods, such as yellow pine, are governed by the same law, but in virtue of their softness another law comes into force, which to some degree affects their behaviour, as the contracting power of the tubular wood has sufficient strength to crush the softer medullary rays to some extent, and hence the primary law is so far modified. But even with the softer woods, such as are commonly used in the construction of houses, if the law is carefully obeyed, the greater part of the shrinking, which we are all too familiar with, would be obviated." . . . .

Experiments have shown that timber beams having the annual rings parallel to their depth are stronger than those which have the rings parallel to their width. Thus, in the log shown in Fig. 164 the piece cut from A will be stronger than that cut from B.

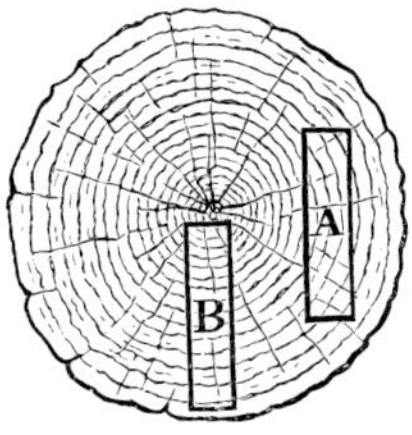

Again, the purpose for which the timber is intended should be borne in mind. Thus, in preparing floor boards, care should be taken that the hearts should not appear on the sur-

Fig. 164.

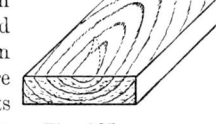

Fig. 165.

face of the finished board. If they are allowed to do so, as in Fig. 165, the central portions will soon become loose, will be kicked up, as shown in dotted lines, and will form a rough and unpleasant floor.

When planks which have shrunk to a curved form have to be used to form a flat board, they are sometimes sawn down the middle and glued together, the alternate pieces being reversed as in Fig. 166; thus the curvature in each

piece is so slight as to be almost inappreciable, and the reversal of the alternate pieces causes each to be a check upon the shrinkage of its neighbours.

Fig. 166.

Conversion of Oak.—There are several methods of converting oak described in Gwilt's *Encyclopædia of Architecture*, from which the following is taken; Fig. 167 being very slightly modified.

The log is first cut into four quarters.

Each quarter may then be converted in either of the following methods:—
The best method is shown at A in Fig. 167, "in which there is no waste, as the triangular portions form feather-edged laths for tiling and other purposes."

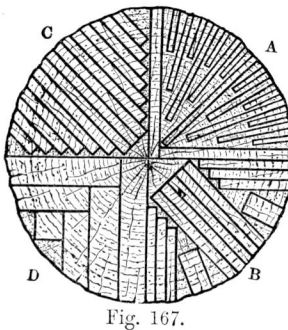   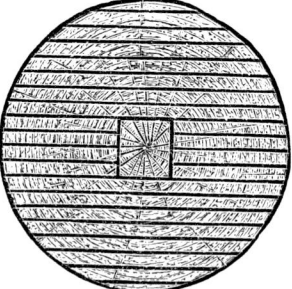

Fig. 167.   Fig. 167*a*

This method also cuts very obliquely across the medullary rays, and thus exhibits well the *silver grain* of the wood, which is so much admired for cabinet work and other ornamental purposes.

The next best method is at B. The method shown at C is inferior to the others; that at D is the most economical where thick stuff is required.

A good practical method adopted for cutting oak logs so as to get wide boards is to cut all the boards parallel to the same diameter, but leaving the heart to be used for quartering. **See Fig. 167*a*.**

Conversion of Fir.—At the great saw-mills in Sweden and Norway each log is carefully inspected before it is sawn, to find out how many of the most *marketable* sizes can be made out of it. Thus if 4-inch deals are in demand, or battens, they will arrange so as to cut more of these sizes, and fewer of the regular 3-inch deals, and *vice versâ*.

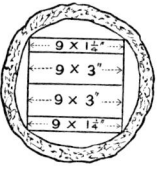   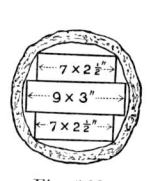

Fig. 168.   Fig. 169.

Two methods are shown in the accompanying figures, taken from Mr. Britton's work upon *Dry Rot*.

Fig. 168 shows an arrangement generally adopted at the present time. The 9 × 3 inch deals go into the English market; those 9 × 1¼ inches into the French market.

Fig. 169 shows the method that was adopted until the French market improved. It will be observed that the centre deal would include the pith, and it is in such a case subject to dry rot.

DESTRUCTION OF TIMBER BY WORMS AND INSECTS.

Timber both in its growing and converted states is subject to the attacks of worms and insects ; when these exist in large numbers they remove so much of the wood as seriously to impair the strength of any structure depending upon the timber, and in some cases they destroy the balks altogether.

It will not be necessary to describe these worms and insects in detail. But a brief notice of a few of the most important, gleaned chiefly from the works of Tredgold and Britton, may be useful.

Worms.—The TEREDO NAVALIS is the most common enemy to timber used in submarine work.

It is found in warm and cold climates, and in nearly every English port. It avoids fresh water and prefers clear water to that which is muddy. This is one reason why wood placed at the mouth of a river, or in turbid water, is not so liable to be attacked as when it is in clear salt water.

The *Teredo* is first deposited upon the timber in the shape of an egg, from which in time emerges a small worm ; this worm soon becomes larger, and commences its depredations.

Furnished with a shelly substance in its head, shaped like an auger, it bores into the wood, chiefly with the grain ; at the same time it lines the hole it makes with a thin coating of carbonate of lime, and closes the opening with two small lids.

As the work of the *Teredo* advances its size increases. Worms two feet long and ¾ inch in diameter have been found at Sheerness, and even larger ones are stated to exist.

The *Teredo* penetrates nearly all kinds of timber, but is most successful in fir.

The general opinion seems to be that the boring is mechanical, but some authorities think that it is done or assisted chemically by the aid of an acid secretion.

The *Xylophaga dorsalis* is of the same family as the *Teredo*, not so common, but more destructive ; it bores in all directions (some say only with the grain), and does not line its hole with shell.

The LIMNORIA TEREBRANS is a marine insect, resembling in appearance a very small woodlouse.

It is very abundant in British (salt) waters, and makes up for its diminutive size by the numbers in which it attacks timber : " as many as twenty thousand will appear on the surface of a piece of pile only 12 inches square." [2]

Mr. Stevenson found that Memel timber was destroyed by the *Limnoria* at the Bell Rock at the rate of about 1 inch inwards per annum. At Lowestoft, piles were eaten at the rate of 3 inches inwards per annum.

This insect prefers soft woods, avoids knots, but will attack all woods except teak and greenheart.

" The *Limnoria* almost always works just under neap tides. It cannot live in fresh water (or under the sand), and whilst it is destroying the surface of a pile, the *Teredo* is attacking the interior."

The *Tanais vittatus*, a species of the same family as the *Limnoria*, in appearance like a very small caterpillar with enormous foreclaws, was found by Mr. Hurst in beech piles.

The CHELURA TEREBRANS, or *wood-boring shrimp*, is also an inhabitant of British seas. It tunnels close below the surface of timber, the waves wash

---

[1] Dent.   [2] Britton.

away the thin covering of the tunnel, and then the shrimp drives another below, so that the timber is removed in successive flakes.

" The *Limnoria* will exist in comparatively foul water if salt, but the *Chelura* must have sea water comparatively pure, hence the former is most frequently found in harbours and the latter along the sea coast.

" The *Limnoria* and *Chelura terebrans* do not attack wood more than a few inches above high water or neap tides." [1]

The LYCORIS FUCATA is the enemy of the *Teredo*.

A little worm with legs, something like a centipede, it lives in the mud, crawls up the pile inhabited by the *Teredo*, enters the tunnel in which it is ensconced, eats the *Teredo*, enlarges the entrance to the tunnel, and then lives in it.

Protection against Worms.—A great many different plans have been tried in order to protect timber in marine works from the ravages of worms.

Copper sheathing is not effectual. The worm gets in between the copper and the timber, and moreover the sheathing decays.

Broad-head scupper nails driven in close together rust into a mass, and so form a good protection, but the process is expensive.

Creosoting by Bethell's process when properly carried out is more successful than other chemical processes.

Ants.—Of the ants proper, or those belonging to the order *Hymenoptera*, there are three species in particular which attack timber, viz.—[1]

1. *The Black Carpenter Ant (Formica fuliginosa)*, which prefers hard and tough wood, rather in standing trees than in seasoned timber. A tinge of black is seen round the holes it makes, caused by iron in its saliva acting upon gallic acid in the wood.

2. *The Dusky Ant (Formica fusca).*

3. *The Yellow Ant (Formica flava).*

The two last-mentioned species prefer soft woods.

The WHITE ANT (genus *Termes*) is a disagreeable-looking cream-coloured insect of fatty substance not quite a $\frac{1}{4}$ inch long, with a black head and lobster-like claws. It grows wings at the last stage of its existence in the nest and flies away to die.

It is found sometimes in Europe, but chiefly in tropical climates, more especially in Africa, the East Indies, the Mauritius, and St. Helena, generally in damp soils near the sea or rivers. Its nests are in the ground or in timber, but always where there is no vibration to destroy the cells.

White ants will eat the whole timber work of a house without noise. They bore close to the surface of the wood, but without destroying it, so that there is no visible indication of what they are doing.

They will even bore through the boards of a floor and up the legs of a table, leaving the latter a mere shell.

No timber has yet been found which is sure to resist them. Teak is riddled by them. Jarrah and greenheart are said to be more successful, but this is doubtful. Cedar while new keeps them at bay by its smell, but when this passes off they devour it eagerly. Oregon pine particularly attracts them. The natives of the countries infected by them use common unsawn yellow pine of long fibre with more success against them.

Protection against the white ant.—Creosoting with bone oil is the best preservative against white ants, but on account of its smell is only adapted for

---

[1] Hurst's Tredgold.

out-door work, and can hardly be applied to very dense tropical timbers. Kerosene is effective while its smell remains. The use of arsenic to guard against them has been abandoned as ineffectual.

Other Insects besides those above mentioned attack wood, among which may be mentioned the *Carpenter bee* of South Africa and the East Indies, and *wood beetles* in Ceylon.

There are also two or three kinds of small beetles in this country which destroy furniture, carvings, etc., and burrow into books in libraries. The best way of destroying them is by subjecting them to the vapour of chloroform or benzine.[1]

VARIETIES OF TIMBER USEFUL FOR DIFFERENT PURPOSES.

The undermentioned are the best of the ordinary descriptions of timber to use for the purposes named.

Piles.—Oak, beech, elm.

Posts.—Chestnut, acacia, larch.

Great Strength in Construction.—Teak, oak, greenheart, Dantzic fir, pitch pine.

Durable in Wet Positions.—Oak, beech, elm, teak, alder, plane, acacia, greenheart.

Large Timbers in Carpentry.—Memel, Dantzic, and Riga fir. Oak, chestnut, Bay mahogany, pitch pine, or teak, may be used if easily obtainable.

Floors.—Christiania, St. Petersburg, Onega, Archangel, make the best; Gefle and spruce inferior kinds; Dram battens wear well; pitch pine, oak, or teak, where readily procurable, for floors to withstand great wear.

Panelling.—American yellow pine for the best; Christiania white deals are also used.

Interior Joinery.—American red and yellow pine; oak, pitch pine, and mahogany for superior or ornamental work.

Window Sills, Sleepers.—Oak; mahogany where cheaply procurable.

Treads of Stairs.—Oak, teak.

Handles.—Ash, beech.

Patterns.—American yellow pine, alder, mahogany, Cowslie pine.

STRENGTH OF TIMBER.

The following Table, showing the strength and weight of timber, is gleaned from the records of many experiments, chiefly those given by Hodgkinson, Tredgold, Barlow, Rankine, and Laslett. Some of these, in their turn, have embodied the results of experiments made by Buffon, Muschenhoek, Rondelet, etc.

It will be seen that the figures given vary throughout a very wide range. This is quite in accordance with practice.

Experiments made upon selected pieces of good quality show results differing greatly from one another, the difference being caused by variety in the age or state of dryness of the specimen, the size and form of the piece tested, the method in which the test has been applied, and the skill of the experimenter.

---

[1] Britton.

TABLE showing the WEIGHT, STRENGTH, etc., of VARIOUS WOODS.

Wood seasoned.	Weight of a cubic foot (dry).	Tenacity per sq. inch, lengthways of the grain.	Modulus of Rupture.	Modulus of Elasticity.	Resistance to Crushing in direction of fibres.[1]		Comparative Stiffness and Strength, according to Tredgold. Oak being 100.[2]	
	Lbs.	Tons. *From To*	Lbs.	Lbs.	Tons per sq. inch. Moderately dry.	Thoroughly dry.	Stiffness.	Strength.
Acacia	48	5·0 8·1	...	1,152,000 to 1,687,500			98	95
Alder	50	4·5 6·3	...	1,086,750			63	80
Ash, English	43 to 53	1·8 7·6	12,000 to 14,000	1,525,500 to 2,290,000	3·8	4·2	89	119
„ Canadian	30	2·45	10,050	1,380,000		2·5	77	79
Beech	43 to 53	2·1 6·6	9,000 to 12,000	1,350,000	3·4	4·2	77	103
Birch	45 to 49	6·7	11,700	1,645,000	1·5	2·8		
Cedar	35 to 47	1·3 5·1	7,400 to 8,000	486,000	2·5	2·6	28	62
Chestnut	35 to 41	4·5 5·8	10,660	1,140,000	...		67	89
Elm, English	34 to 37	2·4 6·3	6,000 to 9,700	700,000 to 1,340,000	2·6	4·6	78	82
„ Canadian	47	4·1	14,490	2,470,000		4·1	139	114
Fir, Spruce	29 to 32	1·3 4·5	9,900 to 12,300	1,400,000 to 1,800,000	2·9	3·0	72	86
„ Dantzic	36	1·4 4·5	13,806	2,300,000		3·1	130	108
„ American red pine	34	1·2 6·0	7,100 to 10,290	1,460,000 to 2,350,000		2·1	132	81
„ American yellow pine	32	0·9	8,454	1,600,000 to 2,480,000		1·8	139	66
„ Memel	34	4·2 4·9	...	1,536,000 to 1,957,750		6	114	80
„ Kaurie	34	2·0	11,334	2,880,000		2·6	162	89
„ Pitch pine	41 to 58	2·1 4·4	14,088	1,252,000 to 3,000,000		3·0	73	82
„ Riga	34 to 47	1·8 5·5	6,600 to 9,450	1,328,000 to 3,000,000		2·1	62	83
Greenheart	58 to 72	3·9 4·1	16,500 to 27,500	1,700,000	5·8	6·8	98	165
Jarrah	63	1·3	10,800	1,187,000		3·2	67	85
Larch	32 to 38	1·9 5·3	5,000 to 10,000	1,360,000		2·6	79	103
Mahogany, Spanish	53	1·7 7·3	7,600	1,255,000 to 3,000,000		3·2	73	67
„ Honduras	35	1·3 8·4	11,500 to 12,600	1,596,000 to 1,970,000		2·7	93	96
Mora	57 to 68	4·1	21,000 to 22,000	1,860,000	...		105	164
Oak, English	49 to 58	3·4 8·8	10,000 to 13,600	1,200,000 to 1,750,000	2·9	4·5	100	100
„ American	61	3·0 4·6	12,600	2,100,000		3·1	114	86
Plane	40	5·4	...	1,343,250	...		78	92
Poplar	23 to 26	2·68	...	763,000	1·4	2·3	44	50
Sycamore	36 to 43	4·3 5·8	9,600	1,040,000		3·1	82	111
Teak	41 to 52	1·47 6·7	12,000 to 19,000	2,167,000 to 2,414,000	2·3	5·4	126	109
Willow	24 to 35	6·25	6,600	...	1·3	2·7		
Hornbeam	47·5	9·1	...	...		3·7	...	108

[1] From Hodgkinson's experiments on short pillars 1 inch diameter, 2 inches high, flat ends, and Laslett's on 2-inch cubes.

[2] This ratio is not always confirmed by the values of the moduli of elasticity as found by more recent experiments, and given in the fifth column of the above table.

Practical experiment upon material similar to that about to be used in any particular case is preferable to information extracted from tables ; but if it is necessary to use the latter, the engineer should be inclined to credit his material with the lowest of the figures recorded, and then to apply a good factor of safety to cover defects in the pieces used, which defects may not have existed in the specimens experimented upon.

Mr. Hodgkinson found that timber when wet had not half the strength of the same timber when dry. This is an important point to consider in sub-aqueous structures.

Resistance to Crushing across the Fibres.—When a vertical piece of timber stands upon a horizontal piece, the latter is compressed at right angles to the length of the fibres, and in this position it will not withstand so great a compressive force per square inch as does the vertical piece, whose fibres are compressed in the direction of their length.

Not many experiments have been made on this point.

Tredgold found that Memel fir was distinctly indented with a pressure of 1000 lbs. per square inch, and English oak with 1400 lbs. per square inch.

Mr. Hatfield's experiments chiefly on American woods, are quoted in Hurst's *Tredgold*, and form the basis of a table in Hurst's *Pocket-Book*, from which the following are taken :—

	Force per sq. inch required to crush the fibres transversely $\frac{1}{20}$ inch deep.
Fir, spruce	·22 tons.
Pine, Northern Memel	·60 ,,
,, White (*P. strobus*) American	·27 ,,
Mahogany, Honduras	·58 ,,
,, St. Domingo	1·92 ,,
Oak, English	·90 ,,
,, American	·84 ,,
Ash, American	1·03 ,,
Chestnut	·42 ,,

Resistance to Shearing.—On this point also but few experiments have been made.

The resistance to shearing in direction of the fibres of the wood is of course much less than that across the fibres.

Wood.	Resistance to Shearing per sq. inch in lbs.	
	Along Fibres.	Across Fibres.
Fir	556 to 634 [1]	
Oak	2300 [2]	4000 [2]
American oak	780 [3]	
Ash and elm	1400 [2]	
Spruce	600 [2]	
Red pine	500 to 800 [2]	

[1] Barlow *On Strength of Materials*, p. 23.　　[2] Rankine's *Civil Engineering*.
[3] Hatfield, quoted in Hurst's *Tredgold*.

Chapter VI

PAINTS AND VARNISHES.

PAINTS and Varnishes are used by the engineer and builder for covering the surfaces of wood, iron, and other materials, in order to protect them from the action of the atmosphere, or to improve their appearance.

The preparation of surfaces and the different processes involved in painting and varnishing materials of different kinds have already been briefly described in Part II.

It will now be necessary only to give a few particulars regarding the paints and varnishes in common use on engineering and building works.

The paints used by the engineer and builder as a rule consist of a *base* [1] (generally a metallic oxide) mixed with some liquid substance known as the *vehicle ;* upon this, permanency of the paint depends.

In most cases a *drier* is added to cause the vehicle to dry more quickly, and a *solvent* is sometimes required to make it work more freely.

When the final colour required differs from that of the base used, the desired tint is obtained by adding a *stainer* or *colouring pigment.*

It will be an advantage to glance at the properties of the substances used to effect the various objects above mentioned before describing the paints most commonly made from those substances.

The materials most commonly used for the purposes above mentioned are as follows :—

Bases.—White lead, red lead, zinc white, oxide of iron.

Vehicles.—Water, oils, spirits of turpentine.

Solvents.—Spirits of turpentine.

[1] Sometimes called a *pigment,* but here called the *base* in order to avoid confusion with the pigment added to give the colour ; see p. 411.

Driers.—Litharge, acetate of lead, sulphate of zinc, and binoxide of manganese, red lead, etc.

Colouring Pigments.—Ochres, lampblack, umber, sienna, and many metallic salts, the principal of which are mentioned at pages 411 to 415.

BASES.

White Lead is a carbonate of the metal. The best is produced by the Dutch process, which consists in placing gratings of pure lead in tan, and exposing them to the fumes of acetic acid ; by these they are corroded, and covered with a crust of carbonate, which is removed and ground to a fine powder.

There are other processes for manufacturing white lead, in which it is precipitated by passing carbonic acid through solutions of different salts of lead.

Clichy White is produced in this way by the action of carbonic acid gas upon acetate of lead.[1]

The white lead produced by precipitation is generally considered inferior to that prepared by corrosion. It is wanting in density or body, and absorbs more oil—it however does not require grinding.

Pure white lead is a heavy powder, white when first made ; if exposed to the air it soon becomes grey by the action of sulphuretted hydrogen.

It is insoluble in water, effervesces with dilute hydrochloric acid, dissolving when heated, and is easily soluble in dilute nitric acid.

When heated on a slip of glass it becomes yellow.

This substance may be used as the basis of paints of all colours.

Adulteration.—White lead may be purchased either *pure* or mixed with various substances—such as sulphate of baryta, sulphate of lead, sulphate of lime,[2] whiting (see p. 256,), chalk, zinc white, etc. These substances do not combine with oil so well as does white lead, nor do they so well protect any surface to which they are applied.

Sulphate of baryta, the most common adulterant, is a dense, heavy, white substance, very like white lead in appearance. It absorbs very little oil, and may frequently be detected by the gritty feeling it produces when the paint is rubbed between the finger and thumb.

MARKET FORMS.—White lead is sold either *dry* in powder or lump, or else *ground in oil* in a paste " containing from 7 to 9 per cent of linseed oil, and more or less adulterated, unless specially marked *genuine*."

When sulphate of baryta has been added, its presence is in most cases avowed ; the mixture is called by a particular name, which indicates to the initiated the proportion of sulphate of baryta that it contains. Thus—

Genuine Dry White Lead, Newcastle White, Nottingham White, Roman White, London White, are all names for *pure* white lead.

Kremnitz, or *Krems White,* known also as *Vienna White,* imported from Austria in small cubes ; *French White,* or *Silver White,* in drops, from Paris; and *Flake White,* made in England in small scales, should also all be *pure* white lead, but they differ considerably in density.

---

[1] Dent. [2] Barium sulphate, lead sulphate, calcium sulphate.

Venice White   contains 1 part white lead to 1 part sulphate of baryta.
Hamburg White   ,,   1   ,,   ,,   2   ,,   ,,
Dutch White, or ⎱
Holland White ⎰   ,,   1   ,,   ,,   3   ,,   ,,

" When the sulphate of baryta is very white, like that of the Tyrol, these mixtures are considered preferable for certain kinds of painting, as the barytes communicates opacity to the colour, and protects the lead from being speedily darkened by sulphurous smoke or vapours." [1]

Old White Lead.—White lead improves by keeping. It should not be exposed to the air, or it will turn grey (see p. 405). Old white lead of good quality goes further and lasts better than if it is used when fresh ; moreover, the paint made with fresh lead has a tendency to become yellow.

Fresh white lead often has a yellowish tinge, caused by the presence of iron.

Uses, Advantages, and Disadvantages.—Of all the bases for paints white lead is the most commonly used, and for surfaces of wood it affords in most cases the best protection, being dense, of good body, and permanent. It has the disadvantage, however, of blackening when exposed to sulphur acids, and of being injurious to those who handle it.

Test for Sulphate of Baryta.—" The testing of the quality of white lead is a very simple operation, as it is only necessary in the case of dry white lead to digest it with nitric acid, in which it dissolves readily on boiling. When ground with oil, the oil should be burnt off, and the residue treated with nitric acid ; or

" The ground white lead with the oil may be boiled for some little time with strong nitric acid, which destroys the oil, and dissolves the lead on the addition of water.

" The sulphate of baryta being insoluble in acid remains behind, and can be collected on a filter, washed with hot distilled water, and weighed."

Red Lead is produced by raising *massicot* (the commercial name for oxide of lead) to a high temperature, short of fusion, during which it absorbs oxygen from the air, and is converted into red lead or *minium*, an oxide of lead.

It is usually in the form of a bright red powder. Ground by itself in oil or varnish, it is durable and unaffected by light when the red lead is pure and used alone, but any preparation containing lead, or metallic salts mixed with it, deprive it of colour, and impure air makes it black.

Uses.—Red lead is used as a drier (see p. 410), also for painting iron (see p. 334) ; and in the priming coat for painting wood (see p. 417).

Adulteration and Tests.—Red lead is sometimes adulterated with brick dust, which may be detected by heating the red lead in a crucible, and treating it with dilute nitric acid ; the lead will be dissolved, but the brick dust will remain. [2]

Red lead may also be adulterated with *colcothar*, a sesquioxide of iron.

Antimony Vermilion, *Sulphide of Antimony*, produced from antimony ore, has been proposed as a substitute for red lead.

It is sold in a very fine powder, without taste or smell, and which is insoluble in water, alcohol, or essential oils.

It is but little acted upon by acids, and is stated to be unaffected by air or light. It is adapted for mixing with white lead, and affords an intensely bright colour when ground in oil. [3]

---

¹ **Ure.**   ² Davidson.   ³ *Proceedings Society of Engineers*, 1875.

Oxide of Zinc is the basis of ordinary zinc paint (see p. 419).

It is prepared by distilling metallic zinc in retorts, under a current of air ; the metal is volatilised, and white oxide is condensed. It is filled into canvas bags, and pressed to increase its density.

Zinc white is durable in water and oil ; it dissolves in hydrochloric acid ; it does not blacken in the presence of sulphuretted hydrogen (the sulphide of zinc being white) ; and it is not injurious to the men who make it, or to the painters who use it.

On the other hand, it does not combine so well with oil, and is wanting in body and covering power, and is difficult to work (see p. 419).

"The want of density is a great drawback to the use of zinc white, and the purest zinc oxide is not always the best for paint on account of its low specific gravity ; and in this respect the American zinc whites, which are frequently very pure, do not generally compete with the zinc white supplied by the Vieille Montagne Company, as made in Belgium." [1]

Uses, etc.—Oxide of zinc is the basis of zinc paint. It has considerable advantages in certain positions, as mentioned at p. 419.

Oxy-Sulphide of Zinc is used as the basis of *Griffith's patent white paint.* It is stated by Dr. Phipson to be prepared by precipitating chloride or sulphide of zinc by means of a soluble sulphate—of sodium, barium, or calcium. The precipitate is dried ; and levigated, while hot, in cold water.

The paint made with this substance for a base has several valuable characteristics, which are described at p. 422.

Oxide of Iron is produced from a brown hæmatite ore found at Torbay in Devonshire, and at other places. It forms the basis of a large class of paints of some importance (see p. 423).

The ore is roasted, separated from impurities, and then ground. Tints, varying from yellowish brown to black, may be obtained by altering the temperature and other conditions under which it is roasted

Oxide of iron is also produced as a bye product in the manufacture of aniline dyes.[1]

VEHICLES.

Oils are divided into two classes—Fixed oils and volatile oils.

FIXED OILS are extracted by pressure from vegetable substances ; they are of a fatty nature, do not evaporate on drying, and will bear a temperature short of 500° Fahr. without decomposing. They are subdivided into

Drying Oils, which become thick upon exposure to air. Of these, linseed oil is most commonly used as an ingredient for paint ; nut oil and poppy oil are also used (see p. 409).

Non-Drying Oils, which become rancid under similar atmospheric influences. These are not used in preparing paint.

VOLATILE or ESSENTIAL OILS are generally obtained by distillation, and have an odour resembling that of the plant from which they are obtained. They are, as a rule, colourless at first, but upon exposure to air and light they become darker, thicker, and eventually are converted into a kind of resin.

Oil of turpentine, commonly called spirits of turpentine, is the only variety of this class that is much used for ordinary paint.

---

[1] Dent.

Mineral Turpentine or *Petroleum Oil* is often used as a cheap vehicle instead of ordinary turpentine.

Coal Naphtha is one of the products of the distillation of coal tar. It is purified in a mill with sulphuric acid ; the sediment and water drawn off, the pure washed spirit remains.

Petroleum, a mineral oil, comes from America in casks. It is then distilled, and from it oils of various density are obtained, and used for burning in lamps, etc.

Benzoline is one of the products obtained from petroleum, and is much used as a solvent for bituminous paints. Paints mixed with benzoline or the heavier oils from petroleum do not set nearly so well, nor do they dry with so much cohesion as those in which naphtha is the solvent, but benzoline is much cheaper, and is therefore often sold as naphtha, and used instead of it.

Linseed Oil, produced by compressing flax seed, is the most commonly used, and by far the best of the oils used as an ingredient of paint, putty, and other similar substances.

It oxidises and becomes thick upon exposure to the air. This property is is very much increased by adding other substances to it and boiling them together (see Boiled Oil).

It is superior in drying powers, tenacity, and body to the other fixed oils.

The best oil comes from the Black Sea and the Baltic ; that from East Indian seed is inferior, as the seed is less carefully cleaned, and contains too much stearine.

Uses.—Raw linseed oil is clear and light in colour, works smoothly, and is used for internal work, for delicate tints, and for grinding up colours. Boiled oil is much thicker, darker, and more apt to clog. It is used for outside work, as its greater body and rapidity in drying make it a quicker and more efficient protection.

RAW LINSEED OIL is obtained by allowing the oil, as first expressed from the seed, to settle until it can be drawn off clear.

When of good quality it should be pale in colour, perfectly transparent, almost free from smell, and sweet in taste.

When it is to be used for delicate tints, it is sometimes clarified by adding an acid (such as oil of vitriol), which is afterwards carefully washed out.

This clarification is stated to be of no permanent advantage, for the oil in drying recovers its original colour.

Darkness in colour and slowness in drying are defects in inferior linseed oil.

These, however, are greatly diminished. and the substance of the oil is improved by keeping.

The oil should never be used within six months after being expressed from the seed, and it is better if kept for several years.

Raw oil is more suited for delicate work than boiled oil, as it is thinner and lighter in colour.

The drying of raw linseed oil " may be improved by adding about 1 lb. of white lead to every gallon of oil and allowing it to settle for at least a week ; this also improves the colour of the oil, whilst the lead can be used afterwards for common work." [1]

Raw oil spread in a film upon glass, or other smooth non-absorbent

---

[1] Seddon.

material, takes from two to three days to dry, according to the state of the weather.[2]

BOILED LINSEED OIL, frequently called *Drying Oil*, is prepared by heating raw oil with certain driers or by passing a current of air through raw oil.

The drying qualities of the raw oil can be greatly improved by boiling it alone, but other substances, such as those mentioned below, are generally added to it, which make it dry still more quickly.

When boiled it becomes much thicker, and not so suitable for indoor or delicate work, nor will it do for grinding colours, as it clogs and thickens too rapidly.

Boiled oil of a pale colour is necessary for use with light tints, but for deep colours a dark oil seems to be generally preferred, though apparently without much reason.

Dark Drying Oil may be made from the following ingredients :—

1 gallon linseed oil.
1 lb. red lead.
1 lb. umber.
1 lb. litharge.

The linseed oil is heated to about 200° Fahr. ; when it looks brown and the scum is all burnt off the other substances are added ; the whole is then raised to about 400° Fahr., and kept at that temperature for two or three hours. The oil is then drawn off, the albuminous matter being allowed to deposit, and is now clear and ready for use.

The umber is added simply to give the oil a dark colour.

Acetate of lead is sometimes used instead of the red lead and litharge, and tends to make the oil lighter in tint. A little resin is sometimes added.

Chevreuil states that oil heated to 160° with $\frac{1}{10}$ its weight of oxide of manganese has powerful drying properties.[2]

Good boiled oil spread in a film upon glass should be dry in from 12 to 24 hours,[3] if raw it would take from 2 to 17 days, according to the atmosphere.[4]

Pale Drying Oil may consist of 1 gallon of linseed oil mixed with about 7 lbs. litharge or acetate of lead, and raised to a moderate warmth.

Boiled Oil to be used with zinc paint must be free from oxides of lead. About 5 per cent by weight of powdered peroxide of manganese is boiled in the oil for five or six hours. The mixture is then allowed to cool, and filtered.

Drying Oil for common work may be made by boiling 1½ lb. red lead in a gallon of raw linseed oil, and allowing the mixture to settle.

Poppy Oil is extracted by pressure from the seeds of the common poppy. It should be colourless, or of a very light yellow tinge, sweet, and free from smell. Being very pale it is sometimes used for light tints, but though its colour stands longer than that of linseed oil, it eventually becomes of a brownish hue, and in drying and other qualities it is far inferior to linseed oil.

Nut Oil is expressed from walnuts. It should be nearly colourless, and therefore adapted for white and any light tints. It dries more rapidly than linseed oil, but is not durable, and is used only for common work, being cheap.

" Oil of Turpentine," " *Spirits of Turpentine,* or *Turps,*" is an essential or volatile oil, produced by distilling turpentine tapped from pines or larches. The residuum left after distillation is common *rosin.*

[1] Dent.   [2] Miller's *Organic Chemistry.*
[3] *Proceedings Society of Engineers,* 1875.   [4] Seddon.

" The best oil of turpentine comes from America."

" The gummy material known as *Canada Balsam* is produced by the *Pinus Canadiensis*, *Venice Turpentine* by the larch (*Pinus larix*), and *French Turpentine* by the *Pinus maritima*, which is extensively grown in the south of France." [1]

Characteristics and Qualities.—Ordinary oil of turpentine has a specific gravity of about ·86 to ·87, and boils at a temperature of 320° Fahr. If pure it should completely distil over at this temperature.

" On exposure to the air it oxidises, and is converted into a resinous substance."

" When spread upon any surface in a thin layer, as is the case when used for paint, it should dry in 24 hours, leaving a hard dry varnish." [1]

Good spirits of turpentine is lighter in weight and more inflammable than bad. It is colourless, and has a pleasant pungent smell, whereas the smell of inferior qualities is disagreeable.

Good spirits of turpentine should leave a very slight residue when evaporated.

Spirits of turpentine is often adulterated with mineral oil. The purer vegetable turpentine loses bulk by evaporation, and gains weight upon exposure to the air ; the spirit from the mineral oil flies off, leaving the oil without any assistance in hardening. [2]

Turpentine sometimes contains pyroligneous acid, and is the better for being kept and allowed to settle a long time before use.

Uses.—Spirits of turpentine is used as a solvent for resins and other substances in making varnishes ; also in paint to make it work more smoothly. It is useful also in flatting coats (see Part II.), but will not stand exposure to the weather.

DRIERS.

Driers are substances added to paint in order to cause the oil to thicken and solidify more rapidly.

The action of these substances is not thoroughly understood. Chevreuil has shown that the drying of linseed oil is caused by the absorption of oxygen, and there can be no doubt that for the most part driers act as carriers of oxygen to the oil, a very small quantity producing considerable effects.

The best driers are those which contain a large proportion of oxygen, such as litharge, acetate of lead, red lead, sulphate of zinc, verdigris, etc.

They are sometimes used to improve the drying qualities of the oil with which the paint is mixed, as explained at page 408, or they may themselves be ground up with a small quantity of oil and added to the paint just before it is used.

Litharge, or *oxide of lead*, is the drier most commonly used, and is produced in the oxidation of lead containing silver. It can be procured on a small scale by scraping off the dross which forms on molten lead exposed to a current of air. *Massicot* is a superior kind of litharge, being produced by heating lead to an extent insufficient to fuse the oxide.

Sugar of Lead (*acetate of lead*) ground in oil, and **Copperas** and **White vitriol** (*sulphate of zinc*), are also used as driers, especially for light tints.

---

[1] Dent.   [2] Cresy's *Encyclopædia.*

Oxide of Manganese is quicker in its effects, but is of a very dark colour, and seldom used except for deep tints.

Japanners' Gold Size and **Verdigris** (*acetate of copper*) are also used for dark colours. Care must be taken not to apply too much of the size, or it will make the paint brittle.

Red Lead (*oxide of lead*) is often used as a drier when its colour will not interfere with the tint required. It is not so rapid in its action as litharge or massicot.

Sulphate of Manganese is the best drier for zinc white, about 6 or 8 ounces only being used for 1 cwt. of ground zinc paint. The manganese should be mixed with a small quantity of the paint first, and then added to the bulk. If great care be not taken in mixing the drier the work will be spotted.[1]

Sulphate of Zinc is also a good drier for zinc paint.

Patent Driers contain oxidising agents, such as litharge or acetate of lead, ground and mixed in oil, and therefore in a convenient form for immediate use.

There is great danger, however, in using such driers, unless they are of the best quality from a reliable maker. Some of the inferior descriptions depend for their drying qualities upon lime.

Terebine consists of a powerful drier dissolved in spirits of turpentine ; it is used as a substitute for patent and other driers, and is used in the proportion of 1 oz. to 1 lb. of paint. Alone it will dry in about half an hour.

Xerotine Siccative is a species of terebine, but differs from it in that when mixed with oil the mixture does not become cloudy. The siccative becomes dangerously explosive when stored.

Precautions in using Driers.—"The following points should be observed in using driers :—

" 1*st*. Not to use them unnecessarily with pigments which dry well in oil colour."

" 2*d*. Not to employ them in excess, which would only retard the drying" and tend to destroy the paint.

" 3*d*. Not to add them to the colour until about to be used."

" 4*th*. Not to use more than one drier to the same colour." [2]

COLOURING PIGMENTS.

It is unnecessary to give anything like a complete list of the pigments used to produce the colours and tints used by the house painter and decorator. A few of the most useful may, however, be mentioned.

It is not proposed to give a detailed description of them, but merely sufficient to distinguish those that are injurious from the others.

Many of these colouring pigments, such as the ochres, umbers, etc., are from natural earths ; others are artificially made.

They may generally be purchased either in the form of dry powder or ground in oil.

Blacks.—LAMPBLACK is the soot produced by burning oil, rosin, small coal, resinous woods, coal tar, or tallow.

It is in the state of very fine powder ; works smoothly ; is of a dense

---

[1] Dent. [2] Seddon.

black colour when dry, and durable, but dries badly in oil. It gives a grey·ish black colour to paint, as compared with the deep hue produced by vegetable black of good quality.

VEGETABLE BLACK is a better kind of lampblack made from oil. It is very light, free from grit, and of a good colour. It should be used with boiled oil, driers, and a little varnish. Linseed oil or turps keeps it from drying.[1]

IVORY BLACK is obtained by calcining waste ivory in close vessels, and then grinding. It is intensely black when properly burnt.

BONE BLACK is inferior to ivory black, and prepared in a similar manner from bones.

BLUE BLACK and FRANKFORT BLACK of the best quality are made from vine twigs ; inferior qualities from other woods charred and reduced to powder.

GRANT'S BLACK, or *Bideford Black*, is a mineral substance found near Bideford. It contains a large proportion of siliceous matter. It is denser than lampblack, but has not so much staining power.[2]

Blues.—PRUSSIAN BLUE is made by mixing prussiate of potash (*Ferrocyanide of potassium*) with a salt of iron. The prussiate of potash is obtained by calcining and digesting old leather, blood, hoofs, or other animal matter with carbonate of potash and iron filings.

This pigment is much used, especially for dark blues, making purples, and intensifying black. It dries well with oil.

Slight differences in the manufacture cause considerable variation in tint and colour, which leads to the material being known by different names—such as *Antwerp Blue, Berlin Blue, Haerlem Blue, Chinese Blue*, etc.

INDIGO is produced by steeping certain plants, from Asia and America, in water, and allowing them to ferment.

It is a transparent colour ; works well in oil or water, but is not durable, especially when mixed with white lead.

ULTRAMARINE was originally made by grinding the valuable mineral *Lapis lazuli*. Genuine ultramarine so made is very expensive, but artificial *French and German Ultramarines* are made of better colour, and cheaply, by fusing and washing, and reheating, a mixture of soda, silica, alum, and sulphur.

It is used chiefly for colouring wall papers.

COBALT BLUE is an oxide of cobalt made by roasting cobalt ore. It is a beautiful pigment, and works well in water.

SMALT, SAXON BLUE, and ROYAL BLUE are coloured by oxides of cobalt.

CELESTIAL BLUE, or *Brunswick Blue*, and DAMP BLUE are chemical compounds (containing alum and other substances), which need not be described in detail.

BREMEN BLUE, or *Verditer*, is a compound of copper and lime of a greenish tint.

BLUE OCHRE is a natural coloured clay.

Yellows.—CHROME YELLOWS are chromates of lead, produced by mixing dilute solutions of acetate or nitrate of lead and bichromate of potash.

This makes a medium tint known as *Middle chrome*. The addition of sulphate of lead makes this paler, when it is known as *Lemon chrome*, whereas the addition of caustic lime makes it *Orange chrome* of a darker colour.

The chromes mix well with oil and with white lead either in oil or water. They stand the sun well, but, like other lead salts, become dark in bad air.

Chrome yellow is frequently adulterated with terra alba (*gypsum*).

NAPLES YELLOW is a salt of lead and antimony, supposed to have been originally made from a natural volcanic product at Naples. It is not so brilliant as chrome, but has the same characteristics, and is very difficult to grind.

---

[1] Davidson.   [2] Dent.

KING'S YELLOW is made from arsenic, and is therefore a dangerous pigment to use in internal work. It is not durable, and injures several other colours when mixed with them. *Chinese Yellow, Arsenic Yellow,* and *Yellow Orpiment* are other names for king's yellow.[1]

Turner's, Cassel's, Verona, Montpellier, and *Patent Yellow* are all oxychlorides of lead ; *Cadmium Yellow* a sulphide of cadmium. [1]

YELLOW OCHRE is a natural clay coloured by oxide of iron, and found abundantly in many parts of England.

It is not very brilliant, but is well suited for distemper work, as it is not affected by light or air. It does not lose its colour when mixed with lime as some other pigments do.

SPRUCE OCHRE is a variety of the above of a brownish-yellow colour.

OXFORD OCHRE is of a warm yellow colour and soft texture, absorbent of both oil and water.[1]

STONE OCHRE is found in the form of balls imbedded in the stone of the Cotswold hills. It varies in tint from yellow to brown.

TERRA DE SIENNA, or *Raw Sienna,* is also a clay, stained with oxides of iron and manganese, and of a dull yellow colour. It is durable both in oil and water, and is useful in all work, especially in graining.

YELLOW LAKE is a pigment made from turmeric, alum, etc. It is not durable, and does not mix well with oil or metallic colours.[1]

Browns generally owe their colour to oxide of iron.

RAW UMBER is a clay coloured by oxide of iron. The best comes from Turkey.

It is very durable both in water and in oil ; does not injure other pigments when mixed with them.

BURNT UMBER is the last-mentioned pigment burnt to give it a darker colour. It is useful as a drier, and in mixing with white lead to make stone colour.

VANDYKE BROWN is an earthy mineral pigment of dark-brown colour. It is durable both in oil and in water, and is useful for graining.

PURPLE BROWN is of a reddish-brown colour. It should be used with boiled oil ; and a little varnish and driers for outside work.

BURNT SIENNA is produced by burning raw sienna (see above). It is "the best colour for shading gold."[1]

BROWN OCHRE is another name for spruce ochre (see above).

SPANISH BROWN is also an ochre.

BROWN PINK is a vegetable pigment often of a greenish hue. It works well in water and oil, but dries badly, and will not keep its colour when mixed with white lead.

Bistre is from wood or peat soot. *Vandyke Brown, Cassel Earth, Egyptian Brown* are bituminous earths. *Asphaltum* is bitumen, and *Sepia* comes from the cuttle fish. *Light Cappagh Brown* or *Euchrome* and *Deep Cappagh Brown* or *Mineral Brown* are from bog earth and manganese.[2]

Reds.—CARMINE, made from the cochineal insect, is the most brilliant red pigment known. It is, however, too expensive for ordinary house painting, and is not durable. It is sometimes used for internal decoration.

---

[1] Davidson.   [2] Seddon.

RED LEAD has already been described (see p. 406). Ground by itself in oil or varnish it forms a durable pigment, or it may be mixed with ochres. White lead and metallic salts generally destroy its colour.

VERMILION is a sulphide of mercury found in a natural state as *Cinnabar* The best comes from China.

Artificial vermilion is also made both in China and on the Continent from a mixture of sulphur and mercury.

Genuine vermilion is very durable, but this pigment is sometimes adulterated with red lead, etc., and then will not weather.

Tests.—Vermilion can be tested by heating it in a test tube. If genuine it should entirely volatilise.

Pure powdered vermilion crushed between sheets of paper should not change colour.

Antimony Vermilion (see p. 406).

GERMAN VERMILION is the tersulphide of antimony and of an orange-red colour.

INDIAN RED is a ground hæmatite ore brought from Bengal. It is sometimes artificially made by calcining sulphate of iron. The tints vary, but a rosy hue is considered the best.

It may be used with turpentine and a little varnish to produce a dull surface, drying rapidly, or with boiled oil and a little driers, in which case a glossy surface will be produced, drying more slowly.

CHINESE RED and PERSIAN RED are chromates of lead, produced by boiling white lead with a solution of bichromate of potash. The tint of Persian red is obtained by the employment of sulphuric acid. These paints are much used for painting pillar post boxes.[1]

LIGHT RED is a burnt ochre. It shares the characteristics of raw ochres described at p. 413.

VENETIAN RED is obtained by heating sulphate of iron produced as a waste product at tin and copper works. It is often adulterated by mixing sulphate of lime with it during the manufacture. When pure it is known as *Bright Red*.[2]

"Special tints of purple and brown are frequently required which greatly enhance the value of the material. These tints should be obtained in the process of manufacture, and not produced by mixing together a variety of different shades of colour. When the tint desired is attempted to be obtained by this latter course it is never so good, and the pigments produced are known in the trade as 'faced colours,' and are of inferior value." [1]

ROSE PINK is a chalk or whiting stained with a tincture of Brazil wood. It fades very quickly, but is used for paperhangings, common distemper, and for staining cheap furniture.

DUTCH PINK is a similar substance made from quercitron bark.[1]

Lakes are made by precipitating coloured vegetable tinctures by means of alum and carbonate of potash. The alumina combines with the organic colouring matter and separates it from the solution.[3]

The tincture used varies in the different descriptions of lake. The best, made from cochineal or madder, is very expensive.

The colour is not a durable one, and dries slowly. It mixes well with white lead, and is used for internal work.

DROP LAKE is made by dropping a mixture of Brazil wood through a funnel on to a slab. The drops are dried and mixed into paste with gum water. It is sometimes called *Brazil Wood Lake.*

SCARLET LAKE is made from cochineal, and so are *Florentine Lake, Hamburg Lake, Chinese Lake, Roman Lake, Venetian Lake,* and *Carminated Lake.*

---

[1] Dent.   [2] Seddon.   [3] Ure.

Oranges.—CHROME ORANGE is a chromate of lead, brighter than ver- milion, but less durable.

ORANGE OCHRE is a bright yellow ochre burnt to give it warmth of tint. It dries and works well in water and oil, and is very durable.[1] It is known also as *Spanish Ochre.*

MARS ORANGE is also an ochre.

ORANGE RED is produced by a further oxidation than is required for red lead. It is a brighter and better pigment.[2]

Greens may of course be made by mixing blue and yellow pigments, but such mixtures are less durable than those produced direct from copper, arsenic, etc. The latter are, however, very objectionable for use in distemper, or on wall papers, etc., as they are injurious to health.

BRUNSWICK GREEN of the best kind is made by treating copper with sal-ammoniac. Chalk, lead, and alum are sometimes added. It has rather a bluish tinge ; dries well in oil, is durable, and not poisonous.

Ordinary Brunswick green is made by mixing chromate of lead and Prussian blue with sulphate of baryta.

MINERAL GREEN is made from bi-basic carbonate of copper. It weathers well.

VERDIGRIS is acetate of copper. It furnishes a bluish-green colour, durable in oil or varnish, but not in water. It dries rapidly, but is not a safe pigment to use.[1]

GREEN VERDITER is a carbonate of copper and lime.

PRUSSIAN GREEN is made by mixing different substances with Prussian blue.

There are several other greens made from copper, such as *Brighton Green, Malachite, Mountain Green, Marine Green, Saxon, African, French Greens, Patent Green,* etc. etc.

EMERALD GREEN is made of verdigris mixed with a solution of arsenious acid. It is of a very brilliant colour, but is very poisonous, is difficult to grind, and dries badly in oil. It should be purchased ready ground in oil, in which case the poisonous par- ticles do not fly about, and the difficulty of grinding is avoided.[3]

SCHEELE'S or MITIS-GREEN and VIENNA GREEN are also arsenites of copper, and highly poisonous.

CHROME GREEN should be made from the oxide of chromium, and is very durable.

An inferior chrome green is made, however, by mixing chromate of lead and Prussian blue as above mentioned, and is called Brunswick green.

The chrome should be free from acid, or the colour will fade. It may be tested by placing it for several days in strong sunlight.

TERRE VERTE is a natural coloured tint.

RINMAN'S GREEN is composed of cobalt and ferrous oxide of zinc.

Uses of Pigments.—The uses for which the pigments above mentioned are suitable may be classified as follows—[4]

(*a*). *More or less transparent, and fit for graining and finishing.*— Blacks (except mineral black and Indian ink), umbers, chrome greens, cadmium yellow, raw and burnt sienna, ochre, French ultramarine, mars orange, bistre and the bituminous browns, sepia.

(*b*). *Little if at all affected by heat or fire.*—Whites, ochres, or natural clays.

(*c*). *Fit for fresco or distemper work.*—Whites from sulphate of baryta or car- bonate or sulphate of lime, all the ochres, the reds, blues, browns, and blacks.

(*d*). *Injured by damp and impure air, especially sulphuretted hydrogen, unfit to use in distemper.*—White lead, all the yellows except the ochres, red lead, Chinese and Persian lead, Prussian and cobalt blues, orange salts of lead, and all greens.

(*e*). *Fade in strong lights.*—All vegetable colours more or less—including the yellows—Prussian blue, indigo, the peaty browns, and in less degrees the madders.[5]

---

[1] Davidson.   [2] Dent.

[3] *The Paperhanger's, Painter's, and Decorator's Assistant.*

[4] Modified from a table in Seddon's *Builder's Work.*   [5] Ure.

PROPORTIONS OF INGREDIENTS IN MIXED PAINTS.

The exact proportions of the ingredients to be used in mixing paints vary considerably according to circumstances.

The composition of paints should be governed by the *nature of the material to be painted.* Thus the paints respectively best adapted for protecting wood and iron differ considerably. *The kind of surface to be covered, i.e.* a porous surface, requires more oil than one that is impervious. *The nature and appearance of the work to be done.* Delicate tints require colourless oil ; a flatted surface must be painted without oil, which gives gloss to a shining surface. Again, paint used for surfaces to be varnished must contain a minimum of oil (see p. 431). *The climate, and the degree of exposure to which the work will be subjected ;* thus for outside work boiled oil is used, because it weathers better than raw oil. Turps is avoided as much as possible, because it evaporates and does not last ; if, however, the work is to be exposed to the sun, turps are necessary to prevent the paint from blistering. *The skill of the painter* also affects the composition ; a good workman can lay on even coats with a smaller quantity of oil and turps than a man who is unskilful ; extra turps, especially, are often added to save labour. *The quality of the materials* makes an important difference in the proportions used. Thus more oil and turps will combine with pure, than with impure white lead ; thick oil must be used in greater quantity than thin oil. When paint is purchased ready ground in oil, a soft paste will require less turps and oil for thinning than a thick paste. Lastly, the *different coats of paint* vary in their composition : the first coat laid on to new work requires a good deal of oil to soak into the material ; on old work the first coat requires turps to make it adhere; the intermediate coats contain a proportion of turps to make them work smoothly, and to the final coats the colouring pigment is added, the remainder of the ingredients being varied as already described, according as the surface is to be glossy or flatted.

Lead Paint.—Ordinary white paint is generally composed of white lead, linseed oil, driers, and spirits of turpentine.

A coloured lead paint is produced by adding a pigment to the above.

In the mixture each constituent plays a part.

The oil soaks into the pores of the material painted, and then dries into a resinous compound, keeping out the air, and preventing decay.

The drier causes the oil to oxidise and solidify more quickly.

The white lead gives body and opacity to the mixture. It does not merely mix with the oil, but combines with it to form a creamy compound which dries into a soapy substance.

The spirit of turpentine is merely a solvent added to make the paint work more freely ; it eventually evaporates and plays no permanent part.

PROPORTIONS OF INGREDIENTS.—The exact proportion of the ingredients best to be used in mixing paints varies according to their quality, the nature of the work required, the climate, and other considerations.

The composition of the paint for the different coats also varies considerably.

The proportions given below must therefore only be taken as an approximate guide when the materials are of good quality :—

Table showing the Composition of the different Coats of White Paint, and the Quantities required to cover 100 Square Yards of New Wrought Deal.

	Red lead.	White lead.	Raw linseed oil.	Boiled linseed oil.	Turpentine.	Driers.	Remarks.
INSIDE WORK.							
4 *coats not flatted.*	Lbs.	Lbs.	Pints.	Pints.	Pints.	Lbs.	
Priming . .	½	16	6	...	...	¼	Sometimes more red lead is used, and less drier.
2d coat . .	*	15	3½	...	1½	¼	* Sometimes just enough red lead to give a flesh-coloured tint.
3d coat . .	...	13	2½	...	1½	¼	
4th coat . .	...	13	2½	...	1½	¼	
INSIDE WORK.							
4 *coats and flatting.*							
Priming . .	1½	16	6	...	½	⅛	Some painters make these coats of the same composition as those for non-flatted work.
2d coat . .	...	12	4	...	1½	$\frac{1}{10}$	
3d coat . .	...	12	4	...	0	$\frac{1}{10}$	
4th coat . .	...	12	4	...	0	$\frac{1}{10}$	
Flatting . .	...	9	0	...	3½	$\frac{1}{10}$	
OUTSIDE WORK.							
4 *coats not flatted.*							
Priming . .	2	18½	2	2	...	⅛	When the finished colour is not to be pure white, it is better to have nearly all the oil boiled oil. *All* boiled oil does not work well. For pure white a larger proportion of raw oil is necessary, because boiled oil is too dark.
2d coat . .	...	15	2	2	½	$\frac{1}{10}$	
3d coat . .	...	15	2	2	½	$\frac{1}{10}$	
4th coat . .	...	15	3	2½	0	$\frac{1}{10}$	

For every 100 square yards, besides the materials enumerated above, 2½ lbs. white lead and 5 lbs. putty will be required for *stopping* (see Part II. Chap. x.)

The area which a given quantity of paint will cover depends upon the nature of the surface to which it is applied, the proportion of the ingredients, and the state of the weather.

When the work is required to dry quickly, more turpentine is added to all the coats.

In *repainting old work*, the surface (after the necessary preparation, see Part II.) is considered as if it were primed. Only two more coats are generally applied, of which the first is called the second colouring ; a fourth coat is seldom required. The second and third coats contain equal parts of oil and turps ; all the remaining ingredients are as shown in the Table above.

For *outside work exposed to the sun*, the second and third coats each contain 1 pint turpentine and 4 pints of boiled oil, the remaining ingredients being as stated in the Table. The extra turpentine is introduced to prevent blistering.

In *cold weather* more turps is used to make the paint work freely.

WHITE LEAD PAINT.—Good paint of this description should be made of pure white lead. If it is to be untinted, care must be taken to exclude any substance which will detract from the brightness of the white, and it must be kept in closed vessels, or the action of the air will give it a brown shade.

Uses, Advantages, and Disadvantages.—White lead paint itself, and also as a basis for coloured paints, is one of the commonest and best protecting coverings that can be applied to surfaces of wood. Where it is exposed, however, to the fumes of sulphur acids, such as are evolved from decaying animal matter, in laboratories, and in some manufacturing towns, it soon becomes darkened by the formation of black sulphide of lead. It has also the disadvantage of producing numbness and painters' colic in those who use it.

COLOURED LEAD PAINTS are made by adding to a basis of white lead paint certain stainers or colouring pigments described at p. 411.

These pigments should be separately ground in oil, and small portions carefully added to the last two coats that are applied until the required colour is obtained.

A list of some of the pigments used to produce different tints is given at page 420.

It is better to ascertain the proportion required by experimenting at first upon a small sample.

Where the colour is very deep, the amount of pigment becomes very large in proportion to the white lead ; and in some cases, as in very common black paint, the white lead is omitted altogether, to the great detriment of the protecting qualities of the paint.

MIXING LEAD PAINT.—Dry white lead is ground by machinery in oil for general paints. But for hard colours and filling up compositions it is ground in turps with a portion of Japan gold size or varnish added to bind it.

The paste is softened and made smooth by adding a small quantity of oil and turps, and working it well with a palette knife.

The colouring pigments, if any, are then added, and the paint is brought to the consistency of cream by adding more oil and turps.

It is then cleared by passing it through a canvas or tin strainer.

When about to be used, the paint is thinned to the consistency necessary to enable it to work freely, by adding more oil and turps, called *thinnings*, and the driers are also added.

If the paint is too thick, it will be difficult to work, and will make an uneven surface. If too thin, it will not have body enough, and more coats will be required.

As the paint becomes thicker during use, or when put upon one side for a time, it will require further thinning, and perhaps repeated straining to clear it from skin and dirt.

To prevent mixed paints from "skinning over," or drying up, they should be kept constantly covered with water or with a thin film of linseed oil.

INJURIOUS EFFECT OF LEAD PAINT.—Lead paint produces most injurious effects upon those who use it.

Entering the pores of the skin, it is absorbed by the system, and leads to numbness and a kind of paralysis. It also produces a complaint known as "painters' colic."

Zinc Paint, ordinarily so called, is made with oxide of zinc (see p. 407), instead of white lead, as a basis.

Characteristics and Uses.—Zinc white does not combine with oil so readily as white lead. Its covering properties are therefore inferior, and it takes a long time to harden.

It is acted upon by the carbonic acid in rain water, which dissolves the oxide, and it therefore weathers badly.

The acids contained in unseasoned wood have a great effect upon it.[1]

[1] Dent.

Zinc paint may be used without fear of painters' paralysis, and as it has no smell, places in which it has been used may be occupied directly it is dry.

" Zinc white paint when pure retains its colour well, and will stand washing for several years without losing any of its freshness. When dry it becomes very hard, and will take a fine polish." [1]

This paint is suitable in large manufacturing towns where it is subjected to vapours containing sulphur, or in places where foul air is emanated from decaying animal matter. The zinc is not (like white lead) blackened by exposure to sulphuretted hydrogen.

In such positions of course zinc paint should not be mixed with "patent" or other driers which contain lead. The best driers to be used with it are sulphate of manganese and sulphate of zinc (see p. 411).

Zinc white is recommended as being preferable to white lead for painting on a dark ground. The reason for this is that the soap formed by the combination of the lead and oil in lead paints is semi-transparent, and the dark ground shows through it Another form of zinc paint is described at page 422.

Coloured Paints.—It has already been mentioned that coloured lead paints are produced by adding a suitable pigment to a white lead paint until the required tint is obtained.

It would of course be impossible to give instructions for the composition of the great variety of colours and tints in which paint may be required.

A few, however, of the most common tints produced by mixing two or more colours may be mentioned.

The colours used are generally divided into classes as follows :—

Common Colours, including greys, buffs, and stone colours.

Superior, or Fine Colours, such as bright yellows, warm tints, cloud colours, and common greens.

Delicate tints, such as blue verditer, pea-greens, pinks, etc.

The following list shows the pigments that may be added to white lead paint [2] to produce a few of the most frequently used compound colours.

The same pigments, except those containing lead, may be used with a zinc-white basis for coloured zinc paints.

PIGMENTS FOR COLOURED PAINTS.

COMMON COLOURS :—*Stone Colour.*—Burnt umber.
Raw umber.
Yellow ochre.
Drabs.—Burnt umber.
Burnt umber and yellow ochre for a warm tint.

---

[1] Seddon. [2] Or to white distemper ; see p. 256.

Buffs.—Yellow ochre.

Yellow   „   and Venetian red.

Greys.—Lampblack.

Indian red—indigo—for a warm shade.

Egyptian blue—or French ultramarine—and
vermilion—for a warm shade.

Brown.—Burnt sienna, indigo.

Lake, Prussian blue (or indigo) and yellow ochre

Superior Colours :—*Yellows.*—Chrome yellow.

Green.—Prussian blue, chrome yellow.

Indigo, burnt sienna (or raw umber).

Prussian blue, raw umber.

Avoid arsenical greens.

Salmon.—Venetian red.

Vermilion.

Fawn.—Stone ochre and vermilion.

Delicate Tints :—*Sky-blue.*—Prussian blue.

Pea-green.—Brunswick green.

French   „

Prussian blue, chrome yellow.

SPECIAL PAINTS.

During the last few years a great many substances have been
proposed as bases for paint instead of white lead.

The paints made with these substances are called by special
names, and often have peculiar qualities which adapt them for
use under particular circumstances.

It would be almost impossible to give a complete list of all
these special paints, but it will be useful to mention a few of the
most prominent with their characteristics.

Inodorous Paint[1] is mixed without any turpentine, the evaporation of
which in ordinary paints causes a strong unpleasant smell, which in some
people produces headache, and even more injurious effects.

In this paint the ordinary white lead, or zinc white ground in oil, "instead of being thinned with oil and turpentine, is mixed with methylated
spirit in which shellac has been dissolved, together with a small quantity of
linseed and castor oil."

"This methylated spirit evaporates very rapidly, leaving behind the shellac,
which acts the part of the film of varnish left by the oil and turpentine in
the ordinary method of painting, protecting the wood or stone, and at the
same time attaching the pigment to the painted surface."

This paint dries very rapidly. The second coat can be applied an hour
after the first, and three-coat work can be finished in one day. The rapid

---

[1] Dent.

drying makes it difficult to paint a large, uninterrupted surface, without showing marks where one portion dried before the next was commenced.

For interior work in occupied buildings this paint has very great advantages ; also where rapidity in execution is required, but it is not so durable as paint mixed in oil and turps.

" In oak graining it is desirable, perhaps better, that the varnishing coat should be put on as usual ; but in this case the odour arising from two coats of paint work is at all events avoided, and the whole is finished in a day, instead of lasting over two or three days." [1]

Freeman's " Non-poisonous " White Lead " is prepared by grinding under considerable pressure a precipitated sulphate of lead with 25 per cent of zinc oxide, whereby the density of the mixture is greatly increased. This preparation possesses the advantage of a very simple and unobjectionable method of manufacture and of keeping its colour better than ordinary white lead when employed in situations exposed to air containing sulphur compounds, such as in railway tunnels. It is equal to the ordinary white lead in point of colour, and is reported to be so as regards ' body ' and ' durability,' but this last point can only be decided after the lapse of sufficient time." [2]

Charlton White is a mixture of sulphate of zinc with sulphate of baryta or strontia. It is more pulverulent than zinc white, and more opaque. Requires more oil than white lead, less than zinc white. Tested for body by saturation with blue, it shows itself to be 60 per cent stronger than zinc oxide and 30 per cent stronger than genuine white lead. It must be used with lead, less direct, and is not affected by sulphurous vapours. In outdoor work it must be " bound " by varnish, and in all cases it is perfectly harmless to makers or users.

Charlton Enamels are preparations of Charlton white, and gums, which dry with a smooth hard surface and do not crack or blister.

Duresco is a preparation of Charlton white worked up by a process which is a secret. It dries out perfectly " flatt," is quite solid, washable, and non-poisonous, is much less expensive than oil paints, and more easily applied ; all this makes it peculiarly valuable to internal wall decoration.

A patent White Sulphide of Zinc Paint is manufactured at Liverpool by the Sanitary Paint Co., which consists of a mixture of sulphate of zinc and sulphate of baryta. . . . This paint when not properly manufactured has sometimes been found to become discoloured under the influence of strong sunlight, the dark tinge which it assumes passing off again after a few hours." [2]

Griffith's Patent White Paint is a form of zinc paint which was introduced about 1876. Its basis, oxy-sulphide of zinc (see p. 407), is said to be cheaper than white lead. It has 25 per cent more covering power for the same weight, is not poisonous, is more stable, is of a brilliant white colour, dense, and opaque ; does not blister or yield to heat or gas, is not discoloured by sulphuretted hydrogen, is neutral towards iron, and will mix with colours which white lead destroys.

Albarine is a white enamel which is found to be very superior as regards hardness, enamel-like appearance, whiteness, and easy application ; one gallon will cover on an average 50 yards.

---

[1] Dent.   [2] Dent's *Cantor Lectures.*

Oxide of Iron Paints.—In these oxide of iron (see p. 407) forms the basis. They are free from injurious ingredients such as those of lead paints. For painting iron work they are said to be particularly suitable, on the ground that they do not set up any galvanic action such as is supposed to take place between lead paints and iron surfaces. It is very doubtful, however, whether any such galvanic action exists. When the surface of the iron is rusty, the rust becomes incorporated with the paint.

The paint must, however, be made from the sesquioxide or red oxide of iron. If made from the protoxide it is liable to rust in itself.[1]

The cost of good oxide of iron paints is about the same as that of lead paints, but in application they are cheaper, as weight for weight they cover a greater surface.

1 lb. oxide of iron paint mixed in the proportion of $\frac{2}{3}$ oxide to $\frac{1}{3}$ linseed oil should cover 21 square yards of sheet iron.[1]

To ensure this power of covering a large area with a small quantity of paint, the ingredients should be reduced to an impalpable powder before they are mixed with the oil. They are ground for seven or eight hours.

"When mixed with about one-third of white lead they form a very hard mastic similar to that made from red lead." [2]

TORBAY PAINT is produced from a brown hæmatite iron ore found in Devonshire. It contains from 50 to 65 per cent oxide of iron, the remainder being siliceous matter.

The colour of the oxide varies from yellowish brown to red and black.

Blue, green, and other tints are produced by adding pigments which are not oxides of iron, and which therefore alter the composition of the paint.

This paint has been in use for many years, it is especially suitable for painting iron work, and has borne a high character for durability under exposure to weather and fumes of manufactories.

An official report, quoted in the Manufacturers' Circular, says that " 62 lbs. of the Torbay iron paint effectually cover as much surface as 112 lbs. of either white or red lead paint."

There are several inferior imitations of this paint. A great deal of the so-called Torbay paint is, however, nothing more than sulphate of baryta coloured with oxide of iron, whereas sulphate of baryta is never found in the genuine paint in any appreciable quantity.

BLACK OXIDE OF IRON PAINT is made from the oxide obtained as a bye product in making dyes, ground in oil with about 15 per cent of terra alba, Paris white, or sulphate of baryta. It is said that without the addition of these substances the oxide of iron would set with the oil into a solid mass.

This paint is used for painting shot and shell.

PULFORD'S MAGNETIC PAINT is made from the magnetic or black oxide of iron.

PURPLE BROWN OXIDE is a hydrated peroxide of iron used as a basis for paint.

Silicate Oxide Paint is prepared from an iron ore in Devonshire by the Silicate Oxide Paint Company in three colours only—yellow, red, and black. It contains more oxide of iron and less siliceous matter than the Torbay paint.

---

[1] *Proc. Society of Engineers*, 1875.   [2] Seddon.

Titanic Paint is made by powdering a black iron ore, which contains oxide of iron and oxide of titanium in nearly equal proportions, mixed with other ores. It is said to harden without the aid of a drier, and to be particularly well adapted for withstanding heat.[1]

Anti-Corrosion Paint is a name given to different compositions, which consist chiefly of oil, some strong driers, and a pigment mixed with very fine sand.

They are sold dry, and require only to be mixed, not ground with oil.

They are used chiefly for external work, "lasting longer than white lead and costing less."

"The original makers of this paint are Messrs. Walter Carson and Sons, and if genuine, as supplied by this firm, it should consist of ground glass and white lead in about equal proportions.

"The rubbish which is frequently sold as anti-corrosion has greatly injured the reputation which this paint at one time possessed. It can be obtained as low as 6s. per cwt., whilst the price of the genuine is from 22s. to 24s. It is not at all uncommon to find anti-corrosion containing from 35 to 45 per cent of sulphate of baryta, a substance which I am assured is never employed by the original makers."[1]

"An anti-corrosive paint is also made of equal proportions of whiting and white lead, with half the quantity of sand, dust, and any required colouring matter. Being mixed with water, it can be used as a water colour, but is generally applied as an oil paint, the best oil for the purpose being 1 boiled to 12 of raw linseed and 3 of sulphate of lime, all by weight. One gallon of the oil will take 7 lbs. of the paint."[2]

Enamel Paint consists of a metallic oxide, such as oxide of zinc or oxide of lead, ground with a small quantity of oil, and mixed with petroleum spirit holding resinous matter in solution.[1]

This paint can be prepared to dry either with a firm glossy surface, like porcelain, or with the appearance of an ordinary flatted coat.

It can be made in any colour or tint, however delicate ; requires no oil, turpentine, driers, grinding, or mixing, as it is sent out ready for use.

It is about the same price as ordinary paint, but two coats of it are said to be sufficient.

This paint has been extensively employed in the metropolis, and is said to be particularly suitable for surfaces required to be hard and washable ; also for those exposed to the action of steam, acids, or alkalies, or to the fumes of gas (see Silicate Enamel Paint, p. 425).

Indestructible Paint is similar to enamel paint in composition and characteristics, except that it contains bitumen and is made in three colours only—viz. bronze-green, chocolate, and black.

Gay's Impenetrable Paint dries quickly with hard enamel face, is very durable, smells less than ordinary paint, and is said to resist heat and damp better. It is supplied ready for use, and is familiar to all as the covering used for the post pillar boxes.

Silicate Paints, made by the Silicate Paint Company, have for their basis a very pure silica obtained from a natural deposit in the west of England. This is levigated, calcined, and mixed with resinous substances.[3]

---

[1] Dent.   [2] Seddon.   [3] Phipson, International Congress.

These paints are stated to have no chemical action on metals, to stand 200° heat without blistering, to set quickly and dry with a hard surface, to be indestructible, and, weight for weight, to cover double the surface as compared with lead paint.

This paint is sold in the same form as lead paints, and must be used with special " silicate driers."

" The silicate paints supplied by the Silicate Paint Company are highly recommended by the architect of the London School Board for all internal work where health and cleanliness are aimed at." [1]

Griffith's Silicate Enamel Paint is stated in the patentee's circular to possess the following characteristics among others :—

It is supplied ready for use ; forms hard enamelled surfaces ; prevents the corrosion or oxidation of metal ; is proof against the penetration of damp ; dries rapidly ; is not injured by gases, fumes, hot or cold water, soap, or dilute acid ; requires no varnish.

One coat is sufficient for waterproofing, but two or more are required to produce a highly-glazed surface. The bulk is about three times that of ordinary paint for the same weight. On metal one gallon will cover 500 square feet ; the quantity required to cover other substances depends upon the porosity of the material to be covered.

Silicate Oxide Paint is prepared from an ore in Devonshire in three colours only—yellow, red, and black. It contains more oxide of iron and less siliceous matter than the Torbay paint.

Silicate Alumina Paint is of the same description.

Wood's Compo Paints " are coloured varnishes rather than paints, and very good for outdoor work, containing neither oil, turps, nor driers, and drying rapidly with a bright gloss. They neither crack nor blister in the sun, and one coat on bare iron, stone, or wood is equal to two of ordinary paints." [1]

Szerelmey's Compositions are as follows :—1. Stone liquid (see p. 80). 2. Iron paints. 3. Porcelain paints, varnishes, and encaustics.

The IRON PAINTS of several colours are sold in paste, ground in oil, or in liquid. They are tough and elastic, and prevent or stop rust and corrosion.

They dry in from 24 to 48 hours.

One pound of the paste will cover 4 square yards, and one pint of the liquid 10 to 12 square yards. Two coats are generally sufficient.

This paint was used for the iron roofs of the Houses of Parliament, and is applicable to dry surfaces of iron or wood.

SZERELMEY'S LIQUID ENAMELS are sold in a liquid state, and are applied with a brush.

They dry in from two to four hours. 1 lb. covers about 4 square yards. One coat is sufficient for iron, two are required for wood.

They can be applied to dry surfaces of wood, iron, tin, whitewash, or plaster.

Granitic Paint is said by the manufacturers to be proof against heat, wet, or frost ; to be more durable and cheaper than lead paints ; and to be specially adapted for painting or making joints in iron work.

One cwt. of the light colour will cover from 600 to 650 yards, and one

---

[1] Seddon.

cwt. of the dark colour will cover from 1000 to 1200 yards—one cwt. on wood, on stone or iron much more.[1]

The paint is sold in powder or ground in oil ; the latter only should be used for flatting.[2]

Bituminous Paints are made from vegetable bitumen, asphalte, and mineral pitches dissolved in paraffin, petroleum or naphtha, various oils, and other substances.

They are also " largely made from the products of coal and other mineral oils.

" They have various degrees of fineness, the cheapest kinds having a great resemblance to tar, and they are admirably suited for painting the inside of pipes, or for iron work fixed under water, such as bridge cylinders and screw piles.

" The fine sorts, while possessing the same properties, give a smoother surface, and can be used in ordinary situations, especially where water or foul vapours have to be resisted.

" The price varies from 18s. to 30s. per cwt., and the paint is mixed for use with specially prepared mineral oil.

" A paint made from bitumen dissolved in paraffin and linseed oils while in a state of great heat, is said to possess special qualities of durability, in that it can resist the action of ordinary detergents, and of all alkalies and acids.

" When mixed ready for use this paint costs from 40s. to 60s. per cwt., according to colour and fineness." [3]

A paint of this kind is also made by dissolving equal parts of asphalte and resin in common turpentine.[4]

Champion's Black Paint is a compound of lampblack, mineral matter, and oil.

Tar Paint.—The paint successfully used for the canvas roof over the tubes of the Britannia Bridge was composed as follows :—Coal tar, 9 gallons ; slaked lime, 13 lbs. ; turpentine or naphtha, 2 or 3 quarts—the whole being dredged over with sand. The addition of the quicklime is indispensable to neutralise the free acid that exists in the tar.

A tar paint recommended by Mr. Hurst as the best protection for iron consists of 1 gallon coal tar distilled to expel the watery vapour and naphtha, and afterwards mixed with $\frac{1}{2}$ pint naphtha and $\frac{1}{2}$ pint boiled oil.

ORDINARY TARRING.—Boil 6 gallons *Coal Tar* with 1 lb. resin, 1 lb. pitch, and apply hot ; or use *Stockholm Tar*, with the same proportion of pitch only. Yellow ochre may be added to give a brown tinge.

Silicate Zoppisa, sold by the Granitic Paint Company, is a washable distemper made in all colours ; it is said to dry hard and flat, and to render the surface to which it is applied damp-proof and durable.[1]

Asbestos Paints are much used for internal rough woodwork, which they will protect against sparks or light flames, but they cannot stand the weather. They are, as a rule, mixed with oil in two coats, and are thinned when necessary with warm water.

One gallon will cover 150 feet two coats.

---

[1] Circular.   [2] Seddon.
[3] Matheson.   [4] Davidson.

Asbestos Oil Paints are also made, and it is claimed that their covering power is greater than that of ordinary lead or zinc paints.

One cwt. in fact when thinned will cover 600 square yards. One cwt. sent out ready for use will cover 480 square yards.

Crease's Antiwater Enamel Paint for iron is a sort of silico-calcic cement. It adheres to iron fairly well, especially when the iron is nearly always submerged in water.

Crease's Anticorrosion is a black bitumen paint useful for coating submerged iron surfaces.

Granulated Cork Paint is applied over paint to protect and roughen surfaces upon which the condensed moisture of the atmosphere is likely to deposit, such, for example, as the asphalte floors of cellars, paved ceilings and the girders supporting them.

Coating floors.—For asphalte floors the composition is made and applied as follows :—4 parts (by weight) of Venetian red and 1 part of red lead are mixed into a stiff paste with Stockholm tar and well worked together. It is then laid on the surface of the asphalte about $\frac{1}{8}$ inch thick, and $\frac{3}{8}$ of the granulated cork is sprinkled over the paint and pressed in with a float. In about 48 hours the composition is hard, and the loose cork can be brushed off. As the granulated surface will not withstand wear, it must be protected by open boardings or by gratings.

For surfaces overhead, such as slabs or iron girders, ordinary red lead paint is used as a matrix for the granulated cork, which is forced into it and then painted over.

Coating Iron.—(1.) The surface of the iron is prepared and coated with two coats of red lead or oxide of iron paint. (2.) An adhesive composition composed of the ingredients mentioned below is then applied rather more thickly than ordinary paint, and well sprinkled with granulated cork. (3.) After 4 days a coat of white zinc paint much thinner than ordinary paint is applied, then one coat of distemper.

Proportions of adhesive composition to make 112 lbs :—

White lead, 22 lbs. ; driers, $10\frac{1}{2}$ lbs. ; boiled oil, $3\frac{3}{4}$ gallons ; sperm yellow, 44 lbs. ; resin, $2\frac{1}{2}$ lbs.

Luminous Paint is a preparation of sulphide of calcium made up with varnish. Oil destroys its properties, and care should be taken to apply it only to perfectly clean surfaces free from lead paint or corrosion.

The characteristic of this paint in presenting a luminous surface for many hours or even days after the source of light has been cut off is well known, and it is capable of various obviously useful applications.

The action is supposed to be due to "molecular vibration set up in the body by waves of light rich in actinic rays, which vibration is maintained in the dark, and is the cause of luminosity so long as the energy remains active and not absorbed."

VARNISH.

Varnish is a solution of resin in either oil, turpentine, or alcohol.

The oil dries and the other two solvents evaporate, in either case leaving a solid transparent film of resin over the surface varnished.

In estimating the quality of a varnish the following points must be considered :—1. Quickness in drying; 2. Hardness of film or coating; 3. Toughness of film; 4. Amount of gloss; 5. Permanence of gloss of film; and 6. Durability on exposure to weather.[1]

The quality of a varnish depends almost entirely upon that of the ingredients it contains.

Much skill is, however, required in mixing and boiling the ingredients together.

Uses.—Varnish is used to give brilliancy to painted surfaces and to protect them from the action of the atmosphere, or from slight friction.

Varnish is often applied to plain unpainted wood surfaces in the roofs, joinery, and fittings of houses, and to intensify and brighten the ornamental appearance of the grain.   It is also applied to painted and to papered walls.

In the former case it is sometimes flatted so as to give a dead appearance, similar to that of a flatted coat of paint.

INGREDIENTS OF VARNISH.

The *Gums* are exudations from trees.   At first they are generally mixed with some essential oil.   They are then soft and viscous, and are known as *Balsams;* the oil evaporates and leaves the *Resin,* which is solid and brittle.

Resins are often called "*gums*" in practice, but a *gum,* properly speaking, is soluble in water, and therefore unfit for varnishes, while *resins* dissolve only in spirits or oil.[2]

Gum Resins are a natural mixture of gum with resin, and sometimes with essential oil found in the milky juices of plants.   When rubbed up with water the gum is dissolved, and the oil and resin remain suspended.

Resins.—The quality of the resin greatly influences that of the varnish. The softer varieties dissolve more readily than the others, but are not so hard, tough, or durable.

COMMON ROSIN or *Colophony* is either brown or white.   The brown variety is obtained by distilling the turpentine of spruce fir in water ; the white is distilled from Bordeaux turpentine.

The principal resins used in good work are as follows :—

AMBER, obtained chiefly from Prussia, is a light yellow transparent substance found between beds of wood coal, or, after storms, on the coasts of the Baltic.   It is the hardest and most durable of the gums, keeps its colour well. and is tough, but difficult to dissolve, costly, and slow in drying.

---

[1] Dent.   [2] Seddon.

GUM ANIMÉ is imported from the East Indies ; is nearly as insoluble, hard, and durable as amber, but not so tough. It makes a varnish quick in drying, but apt to crack, and the colour deepens by exposure.

COPAL is imported from the East and West Indies and America, etc., in three qualities, according to colour, the palest being kept for the highest class of varnish. These become lighter by exposure.

MASTIC is a resinous gum from the Mediterranean ; it is soft and works easily.

GUM DAMMAR is extracted from the Kawrie pine of New Zealand, and also from India. It makes a softer varnish than mastic, and the tint is nearly colourless. Much of the Kawrie gum is dug out of the ground.

GUM ELEMI comes from the West Indies, and somewhat resembles copal.

LAC is a resinous substance which exudes from several trees found in the East Indies. It is more soluble than the gums above mentioned.

Stick Lac consists of the twigs covered with the gum. *Seed Lac* is the insoluble portion left after pounding and digesting stick lac. When seed lac is melted, strained, and compressed into sheets, it becomes *Shell Lac.* Of these three varieties shell lac is the softest, palest, and purest, and it is therefore used for making lacquers.

SANDARACH is a substance said to exude from the juniper tree. It resembles lac, but is softer, less brilliant, and lighter in colour, and is used for pale varnish.

DRAGON'S BLOOD is a resinous substance imported from various places in dark brown-red lumps, in bright red powder, and in other forms. It is used chiefly for colouring varnishes and lacquers.

Solvents.—These must be suited to the description of gum they are to dissolve.

BOILING LINSEED OIL (and sometimes other oils, such as rosemary) is used to dissolve amber, gum animé, or copal.

TURPENTINE for mastic, dammar, and common resin.

METHYLATED SPIRITS OF WINE [1] for lac and sandarach.

Wood Naphtha is frequently used for cheap varnishes. " It dissolves the resins more readily than ordinary spirits of wine, but the varnish is less brilliant, and the smell of the naphtha is very offensive. It is therefore never employed for the best work.[2]

Driers are generally added to varnish in the form of *Litharge, Sugar of Lead,* or *White Copperas.*

The sugar of lead not only hardens but combines with the varnish.

A large proportion of driers injures the durability of the varnish, though it causes it to dry more quickly.

DIFFERENT KINDS OF VARNISH.

Varnishes are classified as oil varnish, turpentine varnish, spirit varnish, or water varnish, according to the solvent used. They are generally called by the name of the gum dissolved in them.

Oil Varnishes, made from the hardest gums (amber, gum

[1] Spirits of wine to which a little wood naphtha has been added to make it undrink-able, and therefore not liable to duty.   [2] Holtzapffel.

animé, and copal) dissolved in oil, require some time to dry, but are the hardest and most durable of all varnishes. They are specially adapted for work exposed to the weather, and for such as requires polishing or frequent cleaning. They are used for coaches, japan work, for the best joinery and fittings of houses, and for all outside work.

Turpentine Varnishes are also made from soft gums (mastic, dammar, common resin) dissolved in the best turpentine. They are cheaper, more flexible, dry more quickly, and are lighter in colour than oil varnishes, but are not so tough or durable.

Spirit Varnishes or **Lacquers** are made with softer gums (lac and sandarach) dissolved in spirits of wine or pyroligneous spirit. They dry more quickly, and become harder and more brilliant than turpentine varnishes, but are apt to crack and scale off, and are used for cabinet and other work not exposed to the weather.

Water Varnishes consist of lac dissolved in hot water, mixed with just so much ammonia, borax, potash, or soda, as will dissolve the lac. The solution makes a varnish which will just bear washing. The alkalis darken the colour of the lac.

Mixing Varnishes requires great skill and care. Full details of the process are given in Holtzapffel's *Manipulation* and other works.

Space does not permit here of more than the mention of one or two points that may be useful in mixing varnishes on a small scale. As a rule, it is better to buy varnish ready mixed when possible.

MIXING OIL VARNISHES.—The gum must first be melted alone till it is quite fluid, and then the clarified oil is poured in very slowly. The mixture must be kept over a strong fire until a drop pinched between the finger and thumb will, on separating them, draw out into filaments. The pot is then put upon a bed of hot ashes and left for 15 or 20 minutes, after which the turpentine is poured in, being carefully stirred near the surface. The mixture is finally strained into jars and left to settle.

Copal varnishes should be made at least three months before use ; the longer they are kept the better they become. When it is necessary to use the varnishes before they are of sufficient age, they should be left thicker than usual.[1]

The more thoroughly the gum is fused, the stronger the varnish and the greater the quantity.

The longer and more regular the boiling, the more fluid the varnish.

If brought to the stringy state too quickly more turpentine will be required, which makes the varnish less durable.

MIXING SPIRIT AND TURPENTINE VARNISHES simply consists in stirring or otherwise agitating the resins and solvent together. The agitation must be continued till the resin is all dissolved, or it will agglutinate into lumps. Heat is not necessary, but is sometimes used to hasten the solution of the resin. The varnish is allowed to settle, and is then strained through muslin.

---

[1] Holtzapffel.

Mixing Turpentine Varnishes.—In many cases the resin, such as mastic, dammar, or common resin, is simply mixed with turpentine alone, cold or with slight heat. Care must in such cases be taken to exclude all oil.

Application of Varnish —In using varnish great care should be taken to have everything quite clean, the cans should be kept corked, the brushes free from oil or dirt, and the work protected from dust or smoke.

Varnish should be uniformly applied, in very thin coats, sparingly at the angles.

Good varnish should dry so quickly as to be free from stickiness in one or two days. Its drying will be greatly facilitated by the influence of light ; but all draughts of cold air and damp must be avoided.

No second or subsequent coat of varnish should be applied till the last is permanently hard, otherwise the drying of the under coats will be stopped.

The time required for this depends not only upon the kind of varnish but also upon the state of the atmosphere.

Under ordinary circumstances spirit varnishes require from two to three hours between every coat ; turpentine varnishes require six or eight hours ; and oil varnishes still longer, sometimes as much as twenty-four hours.

Oil varnishes are easier to apply than spirit varnishes, in consequence of their not drying so quickly.

Porous surfaces should be sized before the varnish is applied, to prevent it from being wasted by sinking into the pores of the material.

Varnish applied to painted work is likely to crack if the oil in the paint is not good ; also, if there is much oil of any kind, the varnish hardens more quickly than the paint, and forms a rigid skin over it, which cracks when the paint contracts.

The more oil a varnish contains the less likely it is to crack.

All varnishes improve by being kept in a dry place.

One pint of varnish will cover about 16 square yards with a single coat.[1]

RECIPES FOR VARNISHES.

The following recipes give the proportions of ingredients for a few varnishes in connection with house-painting :—

Oil Varnishes.—Copal Varnishes.—*Best Body Copal Varnish.*[2]—Fuse 8 lbs. of fine African gum copal, add 2 gallons clarified oil. Boil very slowly for four or five hours till quite stringy, and mix with $3\frac{1}{2}$ gallons turpentine.

This is used for the body part of coaches, and for other objects intended to be polished. The above makes the palest and best copal varnish, possessing great fluidity and pliability, but it is very slow in drying, and, for months, is too soft to polish.

Driers are therefore added, but they are injurious (see p. 429).

To avoid the use of driers, gum animé is used instead of copal, but it is less durable and becomes darker by age.

The copal and animé varnishes are sometimes mixed ; one pot of the latter to two of the former for a moderately quick drying varnish of good quality, and two pots of the animé to one of the copal for quicker drying varnish of common quality.

Best Pale Carriage Copal Varnish.[3]—Fuse 8 lbs. of second sorted African copal, add $2\frac{1}{2}$ gallons of clarified oil. Boil slowly together for 4 or 5 hours until quite stringy ; add $5\frac{1}{2}$ gallons of turpentine mixed with $\frac{1}{4}$ lb. dried copperas, $\frac{1}{4}$ lb. litharge ; strain, and pour off.

---

[1] Seddon. [2] Holtzapffel. [3] Ure, Spon.

In order to hasten drying, mix with the above while hot 8 lbs. of second sorted gum animé, 2½ gallons clarified oil, ¼ lb. dried sugar lead, ¼ lb. litharge, 5½ gallons of turpentine.

This varnish will, if well boiled, dry hard in 4 hours in summer or 6 in winter. Some copal varnish takes, however, 12 hours to dry.

This varnish is used for carriages, and also in house painting for the best grained work, as it dries well and has a good gloss.

A stronger varnish is made for carriages, known as Best Body Copal Varnish.

Second Carriage Varnish.—8 lbs. of second sorted gum animé, 2¾ gallons fine clarified oil, 5¼ gallons turpentine, ¼ lb. litharge, ¼ lb. dried sugar of lead, ¼ lb. dried copperas, boiled and mixed as before. Used for varnishing black japan or dark house painting.[1]

Pale Amber Varnish.—Pour 2 gallons of hot clarified oil on 6 lbs. of very pale transparent amber. Boil till strongly stringy, and mix with 4 gallons turpentine. This will work very well, be very hard, and the most durable of all varnishes, and improves other copal varnishes when mixed with them ; but it dries very slowly, and is but little used on account of its expense.[2]

White Coburg Varnish is of a very pale colour, dries in about 10 hours, and in a few days is hard enough to polish.

WAINSCOT VARNISH is made of 8 lbs. gum animé (second quality), 3 gallons clarified oil, ¼ lb. litharge, ½ lb. sugar of lead, ¼ lb. copperas, boiled together till strongly stringy, and then mixed with 5½ gallons turpentine.

It may be darkened by adding a little gold size.

This varnish dries in two hours in summer, and is used chiefly for house painting and japanning.[2]

Spirit Varnishes.—CHEAP OAK VARNISH.—Dissolve 3½ lbs. of clear good resin in 1 gallon of oil of turpentine. Darken, if required, by adding well-ground umber or fine lampblack.[3]

Oak varnish is used for common work. It dries generally in about 10 hours, though some is made to dry in half the time, and known as *Quick Oak Varnish ;* another variety is called *Hard Oak Varnish*, and is used for seats.[4]

"*Copal Varnish* (spirit).—By slow heat in an iron pot melt ½ lb. of powdered copal gum, 2 oz. of balsam of copivi, previously heated and added. When melted, remove from the fire and pour in 10 oz. of spirits of turpentine, also previously warmed. Copal will more easily melt by powdering the crude gum, and let it stand for a time covered loosely."[3]

WHITE HARD SPIRIT VARNISH may be made by dissolving 3½ lbs. gum sandarach in 1 gallon spirits of wine ; when solution is complete adding 1 pint of pale turpentine and shaking well together.

BROWN HARD SPIRIT VARNISH is made like the white, but shellac is substituted for the sandarach. It will bear polishing.

FRENCH POLISH.—The simplest and probably the best is made by dissolving 1½ lb. of shellac in 1 gallon spirits of wine without heat.

Other gums are sometimes used, and the polish may be darkened by adding benzine, or it may be coloured with dragon's blood.

It is used chiefly for mahogany work, in joinery, hand-rails, etc., and is applied by rubbing it well into the surface of the wood, which has been previously made smooth with sandpaper, etc.

HARDWOOD LACQUER is made by dissolving 2 lbs. shellac in 1 gallon spirits of wine. It is generally used for turned articles, being applied to them with a rag while they are on the lathe.

LACQUER FOR BRASS.—The simplest and best lacquer for work not requiring to be coloured is made by dissolving with agitation ½ lb. of the best pale shellac in 1 gallon cold spirits of wine. The mixture is allowed to stand, filtered, and kept out of the influence of light, which would make it darker.

Turpentine Varnishes.—TURPENTINE VARNISH consists of 4 lbs. of common (or bleached) resin dissolved in 1 gallon of oil of turpentine, under slight warmth.

---

[1] Spon.   [2] Holtzapffel.   [3] *Painter, Paperhanger, and Decorator's Assistant*
[4] Mr. Manders' Circular.

It is used for indoor painted work, and also to add to other varnishes to give them greater body, hardness, brilliancy.[1]

BLACK VARNISH for Metal Work.—Fuse 3 lbs. of Egyptian asphaltum ; when it is liquid add ½ lb. shellac and 1 gallon turpentine.[1]

Brunswick Black.—Boil 45 lbs. asphaltum for 6 hours over a slow fire. During the same time boil 6 gallons oil which has been previously boiled, introducing litharge gradually until stringy, then pour the oil into the boiling asphaltum. Boil the mixture until it can be rolled into hard pills, let it cool, and then mix with 25 gallons turpentine, or as much as will give it proper consistency.[2]

Varnish for Iron Work.—The following is recommended by Mr. Matheson as very effective :—30 gallons of coal tar, fresh, with all its naphtha retained ; 6 lbs. tallow ; 1½ lb. resin ; 3 lbs. lampblack ; 30 lbs. fresh slaked lime, finely sifted—mixed intimately and applied hot. "When hard, this varnish can be painted on by ordinary oil paint if desired."

Crystal Varnish consists of melted Canada balsam thoroughly mixed with an equal quantity of oil of turpentine. A coating of it will convert good thin paper into tracing paper.

Water Varnish.—*Light Coloured.*—Mix 16 oz. ordinary water of ammonia with 7 pints water, 2 oz. pale (or white) shellac, and 4 oz. gum arabic.

Ordinary.—Mix 6 oz. borax, 2 lb. shellac, and 4 oz. gum arabic with 1 gallon water.

Varnish for Paper consists of 4 lbs. of dammar dissolved in 1 gallon of turpentine, with moderate agitation or gentle heat. It is suitable for paperhangings and similar purposes.[1]

Japanning consists in applying successive coats of *japan*, *i.e* ordinary lead paint, ground in oil and mixed with copal or animé varnish. Each coat is dried in turn at the highest temperature it will bear without melting. The surface is then treated with from two to six coats of the best copal or animé varnish without driers.

Common so-called japanned work is not dried by heat, but merely painted and varnished.

Proper japanning will stand a very high temperature, and may therefore be used for baths and other metal work subject to considerable heat.

Stains are liquid preparations of different tints applied to the carefully-prepared smooth unpainted surface of common light-coloured wood, such as fir, in order to give it the appearance of more rare and highly coloured woods, such as rosewood, mahogany, walnut, etc.

Liquid Stains are prepared in all colours to imitate different woods, such as rosewood, walnut, ebony, oak, maple, etc., and sold in powder, or in the liquid state ready for immediate application.

The powder is dissolved in hot water before use.

The liquid stain or the solution from the powder is laid on plentifully with a brush or sponge in one or two coats according to depth of tint required.

When the wood is thoroughly dry it must be twice sized with a very strong solution of size, and then varnished.

---

[1] Holtzapffel.   [2] Ure.

When stains ready prepared are not procurable, they may be made without much difficulty.

The following are a few receipts :—

Mahogany Stain.—A thin mixture of burnt sienna ground in vinegar may be used, grained and shaded while wet with the same, thickened with more sienna.[1]

Black Walnut.—Same as above, but using burnt umber.[1]

Walnut Stain.—Boil together for ten minutes 1 quart water, 1½ oz. washing soda, 2½ oz. Vandyke brown, ¼ oz. bichromate potash.[2]

Oak Stain.—Dissolve 2 oz. of American potash, 2 oz. pearlash, in about a quart of water ; keep corked, and dilute with water for lighter tints.

Black Stain.—Boil ½ lb. logwood in 2 quarts water, add 1 oz. pearlash, and apply hot ; then boil ½ lb. logwood in 2 quarts water, add ½ oz. verdigris and ½ oz. copperas ; strain, put in ½ lb. rusty filings ; with this apply a second coat.

Red Stain.—Use a solution of dragon's blood in spirits of wine.

Wash for Removing Paint.—Dissolve 2 oz. soft soap, 4 oz. potash, in boiling water, add ½ lb. quicklime. Apply hot, and leave for twelve to twenty-four hours. This will enable the old paint to be washed off with hot water.

This is a quicker and neater process than either burning or scraping off (see Chap. x., Part II.).

Cleaning Old Paint is effected by washing with a solution of pearlash in water. If the surface is greasy it should be treated with fresh quicklime mixed in water, washed off, and reapplied repeatedly.

Extract of Lethirium is a ready-made preparation which removes old paint very quickly.

For this purpose the pure extract must be thinly brushed over the surface twice or thrice.

To remove a single coat of paint the extract is diluted with thirty times its bulk of water

To clean painted surfaces it is diluted with 200 or 300 parts of water.

The extract must be carefully washed off with vinegar and water before laying on another coat of paint.

Marvel Fluid is another patent preparation for washing off old paint.

Mordant to make paint adhere to zinc surfaces is composed as follows :—

Soft water,	64 parts.	
Chloride of copper,	1 part.	
Nitrate of	,,	1 ,,
Sal-ammoniac,	1	,,
Hydrochloric acid,	1	,,

---

[1] *The Paperhanger, Painter, Grainer, and Decorators' Assistant.*   [2] Spon.

Chapter VII.

GLASS.

General Remarks.—Glass of the kind used in buildings is a mixture of pure sand, soda, and chalk, with a proportion of broken glass,[1] etc. These are melted together at a very high temperature, and brought by different processes into convenient forms for use.

It is not of importance to the engineer or builder to know the exact nature or proportion of the constituents in different kinds of glass, as he can never be called upon to make these for himself. A knowledge of the processes involved is useful only so far as it enables him to distinguish one kind of glass from the other.

The different varieties of glass in ordinary use will now be mentioned in turn, with brief notes as to the qualities sold and the purposes for which they are used.

Before considering the various descriptions of glass used by the builder, a few points may be noticed which are common to all kinds of glass.

Large panes are more expensive than small ones, as it is more difficult to preserve the entire sheet of glass in making, whereas the smaller panes can be cut from what is technically called " breakage."

An extra price is charged for moderate curves in one direction, and further

---

[1] The following are the proportions (roughly) for a few different kinds of glass :—

	PERCENTAGE IN		
	Common Glass.	Crown Glass.	Plate Glass.
Fine white sand .	60	38	40
Sulphate of soda .	20	19	13
Chalk . . .	20	5	7
Broken glass . .	..	38	40
Manganese . .	...	0	A trace ·15

extras on double curves ; also for obscuring, polishing, and grinding sides or edges.

All glass differing from that in ordinary consumption, however trifling the difference, is also charged extra. The extra labour and risk in carrying out exceptional work is charged for. Triangular and other irregular shapes are charged as square—*i.e.* the area measured is that of the circumscribed rectangle.

The various descriptions of sheet glass are identified by their weight per foot superficial in ounces.

The different descriptions of rolled glass have their thickness stated in fractions of an inch.

In bending rolled plate the smooth side is outside unless ordered to the contrary.

In fixing glass those varieties, such as crown glass, that are slightly convex, should have the convexity outwards.

In the case of glass having only one smooth side, it is generally recommended that the smooth side should be placed outwards. It is better, however, to place the rough side outwards, for the rays of light are then retained and the surface appears flat ; if the smooth side is outwards, the rays are reflected, and the slightest undulation in the glass is easily perceived.

Crown Glass is made as follows :—A blowpipe is dipped into melted glass, which is then blown into the form of a large globular bottle. A rod tipped with a blob of hot glass is so placed that the blob or "punty" sticks to the centre of the bottom of the blown globe. The globe is then detached from the blowpipe, heated, and rotated vigorously until it whirls out by centrifugal force into a flat disc or "*table*," having a blob or "bullion" of glass in the centre.

It will be seen that this process of manufacture tends to make the disc gradually thicker from the circumference to centre. In cutting the disc into panes the boss in the centre must be avoided, so that there is a good deal of waste.

The area of panes that can be produced from a table varies according to circumstances.

Of course the centre bullion must be cut out in a small pane. This pane varies in size from 5 to 10 inches square, and is often used for stables or very common cottages. Lately, however, such panes have been in demand for use in better houses built in the Queen Anne style of architecture.

If the remainder of the table be cut into panes of the most advantageous size to produce a maximum quantity, it may yield about 13 feet superficial. But if the panes are cut as required, they will amount to only 10 or 11 square feet. The largest "squares" produced are about 33 × 25 inches.

The portion containing the bullion cannot be flattened ; the smaller halves of the disc (which do not include the boss or "bullion") may be flattened, if desired, at an extra cost, so as to correct the slight convexity that exists in the tables.

Market Forms.—Crown glass is sold in crates of tables, *i.e.* half discs ; crates of slabs, flattened or unflattened ; and in squares, *i.e.* rectangular pieces cut to various dimensions.

Thicknesses.—There are two thicknesses—

The *usual*, about $\frac{1}{20}$th inch thick, and weighing some 10 oz. per square foot ; and the *extra*, about $\frac{1}{15}$th inch thick, and weighing some 16 oz. per square foot.

The Quantity in Crates varies according to the thickness and kind of glass, and is shown in the following Table :—

	Usual Thickness.	Extra Thickness
Crate of Tables	18 Tables, averaging 53 inches diameter.	12 Tables, averaging 52 inches diameter.
Crate of Slabs	36 Slabs, averaging { 24 inches. in extreme { 22¾ ,, width { 21½ ,,	24 Slabs, averaging { 24 inches. in extreme { 22¾ ,, width { 21½ ,,

The extreme widths of slabs given as 21½, 22¾, 24, etc., refer to the distance from the line where the disc is cut in two to the farthest point of the circumference. Extra sized slabs, flattened and unflattened, are made in 12 sizes, each increasing ½ inch in width from 24½ to 30 inches.

Sizes.—The maximum area of the squares kept in stock is 5 superficial feet.

Qualities.—There are four qualities classed as mentioned below, which may be used where comparatively small panes or squares are required.

Picture qualities. { A } These may be used for pictures, or for the very best window
                     { B } glazing.

Glazing qualities. Best.   For the best class of dwelling-houses.
                    Seconds.   ,,   second ,,   ,,
                    Thirds.   ,,   third   ,,   ,,
                    Fourths, or coarse, for agricultural cottages.

Characteristics.—Crown glass is said to be more free from colour than sheet glass, and it has a finer surface, as it does not come into contact with any other substance during the process of manufacture ; but it is being rapidly superseded by the latter, in consequence of the demand for large sizes, and some of the principal manufactories have ceased to make crown glass altogether.

Unflattened glass, " unless specially selected, is so much curved as to necessitate cutting the sash bars, or using a large amount of putty."

Sheet Glass is first blown in the form of a large hollow cylinder. The ends of the cylinder are then cut off, and it is split down one side with a diamond, after which it is placed in a flattening kiln, where, under the influence of heat, it opens out into a flat sheet, which is carefully annealed and then cooled very gradually.

Qualities.—The qualities of sheet glass are as follows, and may be used for the purposes mentioned :—

A.   For pictures (the best).
B.   Do.   (ordinary).
Best.   For the best glazing in first-class dwelling-houses.
Seconds. Good glazing.
Thirds.   Ordinary glazing.
Fourths. Coarse. Unfit for most purposes. The supply is limited.

The different classes may be used for the same purposes as the corresponding qualities of crown glass, as given above, but are available for large panes.

Thickness and Weight.—The following are the weights of glass made, and the thicknesses which correspond to the respective weights :—

Weight.	Thickness in inches.
15 ounce [1]	$\frac{1}{16}$
21 ,,	$\frac{1}{10}$
26 ,,	$\frac{1}{9}$
32 ,,	$\frac{1}{7}$
36 ,,	$\frac{1}{6}$
42 ,,	$\frac{1}{5}$

Every $\frac{1}{16}$ inch adds 13 oz. to the weight per foot superficial.

Sizes.—The largest sizes which, for all practical purposes, are made in the various substances of sheet glass are as follows :—

Weight per foot superficial in ounces.	Maximum length. Inches.	Maximum width. Inches.	Maximum area in feet superficial.
15	55	38	13
21	85	49	22
26	85	49	22
32	85	49	22
36	70	44	19
42	70	44	19

It will be understood that the size is governed by the maximum area. A sheet may be of the maximum length *or* of the maximum breadth, but no combination of length and breadth must exceed the area given in the last column.

The usual stock sizes of sheet glass are from 48 inches × 34 inches up to 50 inches × 36 inches. Any size above these comes under a special tariff of prices.

The variation of price, according to weight per foot superficial and size, is given in the ordinary builder's Price Books.

Market Forms.—Sheet glass is generally sold in crates. The number of sheets in a crate varies according to the thickness of the glass, and is as follows :—

    15 oz. glass   .   40 sheets ⎫
    21 ,,   .   .   .   34 ,,   ⎬ of stock sizes.
    26 ,,   .   .   .   28   ,   ⎭

Characteristics.—Sheet glass has a somewhat duller surface than crown glass, but can be made thicker and to yield larger panes.

CYLINDER GLASS, GERMAN PLATE GLASS, and BRITISH SHEET GLASS, are various names given to sheet glass.

FLUTED SHEET GLASS is of a wavy section, having flutes or corrugations on both sides.

The sizes kept in stock do not exceed 13 feet in area, or 55 inches in length, or 38 inches in width. It is not advisable to make fluted sheets larger than this, but, if ordered, they can be made of the same size as ordinary sheets.

---

[1] This glass, though sold as 15 oz., generally weighs 16 oz. per foot superficial, and is $\frac{1}{12}$ inch in thickness.

This glass is used in situations where it is necessary to secure privacy, without so much obstruction to light as is offered by obscured glass.

PATENT PLATE GLASS, or *Blown Plate*, is made by polishing sheet glass on both sides.

It must not be confounded with British plate glass, which is a better and more expensive material.

Patent plate may be distinguished from British plate by the bubbles in the glass. In the former case these are elongated and irregular, in consequence of the glass having been blown after the bubbles were formed. In British plate the bubbles are circular. The surface of the patent plate is also more wavy than that of British plate.[1]

Qualities.—Patent plate is made in the three qualities which are respectively used for the purposes mentioned below.

Best. B, For engravings or very good glazing.
Second. C, For good glazing.
Third. C C, For ordinary glazing.

Colour.—Messrs. Chance of Birmingham make each of the qualities above mentioned in two colours—the *Usual* (or *Crystal*), and the *Extra white*. The usual is the better for glazing, as it is harder, more lustrous, and less liable to be scratched in cleaning. The extra white is better for engravings and water-colour drawings, etc.

Thickness and Weight.—Each quality (with the exception stated in Table) is made in the following gradations of thickness and weight, identified as Nos. 1 to 4 :—

	Average thickness.	Average weight per foot superficial.	Remarks.
No. 1	$\frac{1}{16}$ inch	13 ounces	Extra white is made in Nos. 1 and 2 thicknesses only.
No. 2	$\frac{1}{12}$,,	17 ,,	
No. 3	$\frac{1}{10}$,,	21 ,,	
No. 4	$\frac{1}{5}$ to $\frac{1}{8}$ in.	24 ,,	

Sizes.—The squares kept in stock do not exceed 10 or 12 feet in area, the length being not greater than 50 inches, or the width than 36 inches.

" Flattened sheet glass and patent plate should be cut with the convex side of the air bubbles downwards, or it will be liable to crack starwise, and it should be glazed with the convex face outwards, or it will present the appearance of being hammered on the face." [1]

British Plate Glass, ordinarily known as *Plate Glass,* is made by pouring white-hot glass on to an iron table, and rolling it out under a heavy metal roller.

The surface is either left rough, or polished, or indented by a pattern cut upon the surface of the table. The several varieties of plate glass differ from one another according to the nature of the surface thus formed, and are named *Rough-cast Plate, Rolled Plate,* or *Polished Plate,* accordingly.

Advantages.—All plate glass has the advantage of being strong. If of sufficient thickness it keeps out the cold, and, moreover, is a " preventive to robbery, as it will not yield to the diamond and allow of being noiselessly

---

[1] Seddon.

removed." Other advantages are possessed by the different descriptions according to the nature of their surface.

ROUGH-CAST PLATE, or *Rough Plate,* is the glass cast as above described and rolled upon a smooth iron table.

One side has a wavy but polished appearance ; the other side is also wavy but dull.

Quality.—This is the cheapest plate glass made, and there is only one quality.

Size and Thickness.—The plates kept in stock range as high as those containing 60 superficial feet.

The thicknesses made are $\frac{1}{4}$, $\frac{3}{8}$, $\frac{1}{2}$, $\frac{3}{4}$, and 1 inch.

Uses.—Rough plate may be used in all situations where a certain amount of light is required, combined with strength—such as lights in pavements, in risers to steps, in windows close to the ground, etc. etc.

ROUGH ROLLED PLATE, or *Rolled Plate,* is made after the patent of Messrs. Hartley and Co., Sunderland, and is often called *Hartley's Rolled Plate,* or *Hartley's Patent Rough Plate.*

The melted glass is rolled as before, but upon a table having lines, or, in some varieties, flutes, cut upon its surface.

Glass of this description is wavy, but smooth on one side ; the other side being marked with parallel ridge lines, or with flutes.

Rough rolled plate is divided into

Plain, which has very narrow parallel ridge lines close together.

Fluted.—Small, with about 11 flutes to the inch.

Large, with about 4 flutes to the inch.

Sizes.—Those kept in stock range as high as 30 feet in area, the length not exceeding 110 inches, or the width 36 inches.

Thickness.—Both plain and fluted (large and small) are made in the following thicknesses :—$\frac{1}{8}$, $\frac{3}{16}$, $\frac{1}{4}$, $\frac{3}{8}$ inch. The $\frac{1}{8}$ thickness weighs about 2 lbs. per square foot, and the other thicknesses in proportion.

Uses.—This glass is suitable to any position where coarse, strong, translucent material is required. The light is admitted without scorching or glare.

It is used for skylights, conservatories, cupolas, roofs of all kinds (the large-fluted form in especially large roofs). It is also used for the windows of railway stations, factories, etc.

BRITISH POLISHED PLATE GLASS is made from material of a superior description, cast and rolled in the same way as rough plate, and then carefully ground down to a plane surface, and polished on both sides.

Qualities.—There are three qualities :—
Ordinary glazing } for windows, etc.
Best glazing
Silvering quality for looking-glasses.

Thickness.—The usual thickness of polished plate glass is about $\frac{1}{4}$ inch, but special thicknesses are made as follows :—$\frac{3}{16}$, $\frac{1}{2}$, $\frac{5}{8}$, $\frac{3}{4}$, $\frac{7}{8}$, 1 inch.

Size.—The plates kept in stock ($\frac{1}{4}$ inch thick) range as high as 100 feet superficial ; larger plates, or plates exceeding 160 inches long, or 96 inches wide, are charged an extra price.

The limit of area for special thicknesses is as follows :—

Thickness (inches) .	.	.	.	.	$\frac{3}{16}$	$\frac{1}{2}$	$\frac{5}{8}$	$\frac{3}{4}$	$\frac{7}{8}$	1
Maximum area in square feet	.	.	.		25	50	50	40	20	20

Uses.—Polished plate is used for large windows and glass doors in the best houses. It has all the advantages of other thick plate glass, and in addition is very clear and colourless, and transmits a large proportion of light. When scratched on the face it can be repolished.

PATENT DIAMOND RCUGH PLATE GLASS has one side smooth but slightly wavy ; the other side with a raised oblique lozenge-shape pattern filled in with narrow ridge lines.

PATENT QUARRY ROUGH PLATE GLASS is similar to the above, but the pattern is larger to imitate the quarries or small panes used in leaded quarry work.

There are two sizes, the measurement of the quarries, from point to point, both ways being as follows :—

Large size, 6 × 4⅛ inches.
Small ,, 3 × 2$\frac{1}{16}$,,

The large size is used for churches, chapels, etc. ; the other for schools, staircase windows, waiting rooms, etc. etc.

In glazing, the smooth side of the glass should be inside.

Perforated Glass.—Patent rough plate ⅛ and ¼ inch thick, and 26 oz. sheet glass, are both made in panes containing up to 8 feet superficial. The perforations run across the width of the pane, and are useful for purposes of ventilation.

" There are two kinds of perforated glass : one having the perforations manufactured in the glass, the other having them afterwards cut. The latter is the best, as the former break very readily." [1]

Cathedral Glass is generally rolled or sheet glass of a neutral tint. It is much used for ecclesiastical work.

Patent Rolled Cathedral is a species of thin rolled plate ⅛ inch thick, wavy on both sides, and tinted ; and rolled white cathedral is of the same colour as ordinary glass without the lines.

Sheet Cathedral is also tinted and used for the same purposes. One variety has sand thrown upon its surface when hot, so that it fuses in, giving an appearance which is useful for artistic purposes. This is known as *Sanded Sheet Cathedral.*

Ground Glass, or Obscured Glass, has one side covered with an opaque film, formed either by grinding the surface or by melting powdered glass upon it.

The names for this glass seem to be used indiscriminately, without reference to the process by which it is made. Such glass is useful wherever light is required without transparency.

Enamelled Glass is obscured in parts to a design which is stencilled upon it. Powdered glass, or enamel, is placed so as to form the pattern, and is then fluxed in by heat as before.

STAINED ENAMELLED GLASS is made as follows :—The whole is first covered with enamel ; the parts to be coloured are rubbed off with the aid of stencil plates, and then treated with chemical substances ; these, when subjected to the heat of the kiln, produce the colour required.

Embossed Glass is also obscured in parts so as to form a pattern, as follows :—The design is drawn or stencilled on the glass to be enamelled,

---

[1] Seddon.

and the remainder of the surface covered with Brunswick black. The whole is then covered with fluoric acid, which eats into the unprotected portions, obscuring them in the form of the pattern drawn.

Coloured Glass can be made in every variety of tint by adding metallic oxides and other substances to the materials before fusion.

FLASHED COLOURS are those in which plain sheet glass is covered on one side only with a thin layer of coloured glass.

Designs may be formed in this glass by eating off the coloured layer, where it is not required, with fluoric acid.

POT METALS are those in which the glass is coloured throughout its thickness.

Special kinds of Glass are made for painted windows and other work of an artistic kind, but the description of such glass falls outside the scope of these Notes.

Glass Tiles are made both in rough plate and sheet glass, either plain, fluted, or to correspond with the various shapes of earthenware tiles, so as to be worked in with them in roofs, and admit light without the expense of skylights, etc.

Glass Slates are also made both in rough plate and in sheet glass : the former in thicknesses from $\frac{1}{8}$ to $\frac{1}{2}$ inch ; the latter of glass varying from 16 to 32 oz. The areas of the glass slates correspond with those of ordinary building slates as given at p. 27.

Wire-rolled Glass is manufactured about $\frac{1}{4}$ inch thick, with a wire mesh embedded within the thickness of the glass, which can be obtained plain on one side and fine-lined on the other, with varying dimensions of mesh. The additional strength imparted by the wire is of value in the event of fracture of the glass by hail or otherwise.

Interception of Light by Glass.—The effect of different descriptions of glass on the diminution of light has been shown by experiment[1] to be as follows :—

British polished plate $\frac{1}{4}$ inch thick intercepts 13 per cent of the light.

Rough-cast plate	„ „ 30	„
Do., rolled, four flutes to an inch	„ 53	„
Sheet glass, 32 oz.	„ 22	„

---

[1] Galton.

CHAPTER VIII.

PAPERHANGING.

WALL papers may be divided into three classes :—
COMMON OR PULP PAPERS, in which the ground is the natural colour of the paper as first made, the pattern being printed upon it.

SATIN PAPERS, of which either the whole ground, or the pattern, or both, are of a polished lustre, having somewhat the appearance of satin. They are made by painting the paper over with the required colour, mixed with Spanish white, etc., after which it is polished with a burnisher. Or the colour is mixed with plaster of Paris, laid on, sprinkled with powdered French chalk, and then rubbed over with a hard brush to give the appearance of satin.

Satin papers are very susceptible to damp, even from the paste used in hanging them ; they require to be hung with care, on dry walls, and should be protected by a lining paper. When once hung, if thoroughly dry, they can be kept clean for a long time, as the smooth surface of the paper prevents dust and dirt from adhering to it.

FLOCK PAPERS, the design on which is formed by the adhesion of flock sheared off from the surface of woollen cloth. The pattern is first printed on the paper in size, next in varnish, the flock is then thickly sprinkled on, and adheres to the varnish, thus forming the pattern.

Printing.—The pattern on the best papers is printed from wood blocks. The position of each block is guided by four pins in its corners, and a separate block is required for each colour.

Wall papers are printed also in large quantities, and very cheaply, by machinery, the patterns being engraved on metal rollers, one for each colour required, and printed on continuous bands of paper several hundred yards long.

Machine-printed papers are inferior to those printed by hand ; the colours of the former often wear off from not being properly set.

Some of the common grained, marbled, and granite papers are roughly coloured by hand, and elaborate papers of the highest class have to be painted by artists.

Distinction in Appearance between Different Classes of Paper.— Pulp papers can easily be distinguished from others, as the back is of the same colour as the ground of the front.

Hand-printed papers can be distinguished from those that are machine-printed, as the former retain the marks of the pins used as guides for the position of the wood blocks.

---

[1] Galton.

Market Forms.—Wall papers are sold by the *piece*, except in the case of borders, which are sold by the yard, or dozen-yards run.

The price varies according to the description and quality of the paper, and the nature of the pattern, an extra being charged for every additional colour included. The introduction of gold or silver in the pattern also enhances the price considerably, in proportion to the amount used.

Down each side of the paper is a blank margin about ½ inch wide. In hanging good papers both these margins are cut off, and the adjacent pieces are placed edge to edge. In common papers, however, only one margin is cut off, and the cut edge of one piece of paper overlaps the margin of the adjacent piece.

ENGLISH PAPERS.—In these each *piece* is generally 12 yards long and 21 inches wide. It therefore contains 7 square yards.

After the margins are removed the paper is 20 inches wide.

Each yard in length of the paper then contains 36 × 20 inches = 5 feet superficial, and each piece 12 × 5 = 60 feet superficial.

The number of pieces of paper required for a room is therefore equal to the number of superficial feet to be covered divided by 60.

An allowance of from ⅛ to 1⁄10 must, however, be made for waste. This allowance is greater for good papers and large patterns than for common papers and small patterns.

Some manufacturers make papers of special widths differing from those mentioned above.

FRENCH PAPERS are made in *pieces* containing 4½ square yards. The length and breadth of a piece vary considerably, according to quality, but they often run about 9 yards long and 18 inches wide.

BORDERS are sold in pieces containing 12 yards, technically known as *dozens*.

LINING PAPER is common uncoloured paper placed under the better classes of paper, in order to protect them against damp and stains from the wall below, and to obtain a smoother surface to work upon.

Colours.—The colouring pigments used for wall papers are as a rule harmless, being pretty much the same as those given at page 420.

Some of the white grounds contain, however, a proportion of white lead, and in some red papers arsenic is used to fix the dye.[1] Papers containing green are as a rule very objectionable, because they are often coloured with pigments containing arsenic, mercury, copper, arsenite of copper (Scheele's green), and other deleterious substances. These fly off in the form of dust, and may poison the occupants of the room in which the paper is hung.

"Green is by no means the only dangerous colour, others are fully as harmful. Blue, mauve, red, and brown have been found to contain great quantities of arsenic. Even the delicate French greys yield it very considerably."[2]

"Arsenic is often found to the extent of from 6 to 14 grains to the superficial foot of wall ; and Dr. A. S. Taylor has stated that he found some deep green papers which contained from 20 to 70 grains per superficial foot."[3]

Test for Arsenite of Copper.—"The presence of arsenite of copper in a sample of such paper is readily proved by soaking it in a little ammonia, which will dissolve the arsenite of copper to a blue liquid, the presence of arsenic in which may be shown by acidifying it with a little pure hydrochloric acid and boiling with one or two strips of pure copper, which will become covered with a steel-grey coating of arsenite of copper.

"On washing the copper, drying it on filter paper, and heating it in a small tube, the arsenic will be converted into arsenious acid, which will deposit in brilliant octahedral crystals on the cool part of the tube. It is obvious that, to avoid mistakes, the ammonia, hydrochloric acid, and copper should be examined in precisely the same way, so as to render it certain that the arsenic is not derived from them."[4]

---

[1] Ure.   [2] Morris, *Healthy Homes.*   [3] Hurst.   [4] Bloxam.

Lincrusta Walton is a mixture of boiled linseed oil with dryers and fibre rolled on to a textile material and subjected by machinery to pressure, under which designs are formed upon it in relief.

It is made in lengths like wall paper, and in five colours—green, drab, red, brown, and buff.

The surface is hard, and can be washed or scrubbed without injury. It is a non-conductor of heat, and very durable.

It is fixed to walls by a thick mixture of $\frac{1}{3}$ glue to $\frac{2}{3}$ paste. Where the wall is very damp it should receive two coats of lincrusta varnish before the material is hung—and if the weather is cold the lincrusta should be put in a warm place before it is used, as it will then hang better.[1]

Damp Walls should be covered with a thin sheet of some waterproof material before the wall paper is hung.

Thin sheet lead, tinfoil, indiarubber, gutta percha, and thick brown paper have all been used for this purpose, the metals being the best but most expensive. The foil is made so thin that it may be fastened to the wall with paste.

Varnishing and Painting Wall Papers.—Wall papers (except the most delicate) may be finished with good copal varnish over two coats of size, or they may be bought ready varnished.

Flock papers may be painted (after well sizing) when they become shabby. In some cases they have a roller covered with wet paint passed over them, so that the raised pattern only receives the paint.

Washable Paperhangings, made by Messrs. Wilkinson and Son, of London, are said to become as hard as stone when hung, to withstand washing, and to be non-absorbent of the contagion of infectious disorders.

Such papers would of course be better than those of the ordinary description for a sick room. The walls of hospital wards, however, are generally rendered in cement, and brought to a highly polished non-absorbent surface, thus avoiding the use of paper altogether.

Paperhanging.—The points to be attended to in hanging wall papers have been mentioned in Part II.

Expensive papers require to be hung with the most skill and care. At the same time, common papers are more difficult to hang well, as they are very apt to tear with their own weight when saturated with paste.

In hanging flock or other thick papers, the paste should be applied some time before they are hung, in order that it may soak well into them.

The ceilings should be finished before the paperhanging begins.

Uses.—Wall papers are intended chiefly for ornament ; they relieve the bareness of the walls, and give the room a bright cheerful appearance.

A plain white paper may sometimes be applied with advantage to ceilings, especially where, from want of stiffness in the floor above, or from some defect in the plastering, the ceiling is inclined to crack.

[1] *Journal of Decorative Art,* March 1884.

CHAPTER IX.

MISCELLANEOUS.

THIS Chapter will include the description of a few materials which could not be conveniently brought under any of the heads comprised in the former chapters.

GLUE.

Glue is prepared from waste pieces of skins, horns, hoofs, and other animal offal.

These are steeped, washed, boiled, strained, melted, reboiled, and cast into square cakes, which are then dried.

The strongest kind of glue is made from the hides of oxen; that from the bones and sinews is weaker. The older the animal the stronger the glue.

Characteristics of Good Glue.—Good glue should be hard in the cake, of a strong dark colour, almost transparent, free from black or cloudy spots, and with little or no smell.

The best sorts are transparent, and of a clear amber colour.

Inferior kinds are sometimes contaminated with the lime used for removing the hair from the skins of which they are made.

The best glue swells considerably (the more the better) when immersed in cold water, but does not dissolve, and returns to its former size when dry.

Inferior glue, made from bones, will, however, dissolve almost entirely in cold water.

Preparation of Glue.—To prepare glue for use it should be broken up into small pieces, and soaked in as much cold water as will cover it, for about twelve hours.

It should then be melted in a double glue pot, covered, to keep the glue from dirt. Care must be taken that the outer vessel is full of water, so that the glue shall not burn, or be brought to a temperature higher than that of boiling water.

The glue is allowed to simmer for two or three hours, then gradually melted, so much hot water being added as will make it liquid enough just to run off a brush, in a continuous stream, without breaking into drops.

When the glue is done with, some boiling water should be added to make it very thin before it is put away.

Freshly made glue is stronger than that which has been repeatedly remelted. Too large a quantity should not therefore be made at a time.

"Glue may be freed from the foreign animal matters generally in it by softening it in cold water, washing it with the same several times till it no longer gives out any colour, then bruising it with the hand and suspending it in a linen bag beneath the surface of a large quantity of water at 60° Fahr."

By doing this the pure glue is retained in the bag, and the soluble impurities pass through. If the softened glue be heated to 122° without water, and filtered, some other impurities will be retained by the filter, and a colourless solution of glue obtained.[1]

Uses.—Glue is used chiefly by the joiner for joints, veneering, etc.

The precautions to be attended to in using glue have already been mentioned in Chap. vi., Part II.

A minimum amount of glue should be used in good work, and it should be applied as hot as possible. The surfaces of wood to be united should be clean, dry, and true ; they should be brought together as tightly as possible, so that the superfluous glue is squeezed out.

Strength of Glue.—"The cohesion of a piece of solid glue, or the force required to separate one square inch, Mr. Bevan found to be 4000 lbs."

From other experiments Mr. Bevan found that the adhesion of two pieces of ash glued end to end amounted to at least 715 lbs. per square inch.

"The lateral adhesion of a piece of board cut out of Scotch fir, which had been quite dry and seasoned, was 562 lbs. to the square inch. Therefore, if two pieces of this board had been well glued together the wood would have yielded in its substance before the glue."

"The strength of common glue for coarse work is increased by the addition of a little powdered chalk." [2]

GLUES TO RESIST MOISTURE.—"A good glue for outside work is sometimes made by grinding as much white lead with linseed oil as will just make the liquid of a whitish colour and strong, but not too thick." [2]

"Mix a handful of quicklime in 4 oz. of linseed oil ; boil them to a good thickness, then spread it on tin plates in the shade, and it will become very hard, but may be easily dissolved over the fire as glue." [3]

"Skimmed milk, in the proportion 1 lb. glue to 2 quarts of milk, is sometimes used to dissolve glue, with the view of increasing its capability of resisting moisture." [3]

"Ordinary glue can be rendered insoluble in water by adding to the water with which it is mixed a small quantity of bichromate of potash ; the exact proportion must be ascertained by experiment, but for most purposes $\frac{1}{50}$th the amount of glue will be sufficient." [4]

MARINE GLUE.—One part of indiarubber is dissolved under gentle heat in 12 parts of mineral naphtha or coal tar. When melted, 20 parts of powdered shellac are added, and the mixture is poured out on metal plates to cool. It is applied by a brush in a melted state, and is specially suitable for all work exposed to wet or moisture.

SIZE.

Size is, or should be, made from the best glue. The glue is prepared by boiling down the skin and horny parts of animals, parchment clippings, etc.

Inferior glue is said to remain damp and to become mildewed.

To make size, a piece of glue is placed in the pot and covered over with water. When melted, it is thinned by adding more water.

A pound of glue makes about a gallon of size.

Very good size may easily be made by boiling parchment clippings for several hours and straining them through a cloth.

Size is used with earthy colouring matters to make them adhere to surfaces, also for clear cole, as described below.

Double Size is merely size of double the strength of ordinary size.

Patent Size "is a gelatine, and can be used without any soaking as required for glue size." [4]

Kilvin Dry Size is said to be colourless and odourless. It is sold in powder, and becomes gelatinous on cooling after a minute's boiling. It will keep several days in the hottest weather, and will not affect the most delicate tints. [4]

---

[1] Ure. [2] Tredgold. [3] Spon. [4] Seddon.

Clear Cole is the name given to a coating of size which is often used to fill up the pores of wood or plaster in order to prepare them to receive varnish, colour, etc., without absorbing too much.

Parchment Size is used by gilders. It is made by dissolving shreds of fine parchment in warm water.

Gold Size, of different kinds, is applied to surfaces to be gilded, as a basis to receive and secure the gold leaf.

OIL GOLD SIZE is made by grinding ochre in boiled linseed oil. The mixture is made as stiff as possible, kept for several years, if possible, and thinned down with boiled linseed and fat oil for use.

This is the best size to use for outside work, and for any work that is not burnished. It is, however, rather slow setting, and must be applied some 12 to 18 hours before the leaf is laid on.

BURNISH GOLD SIZE is laid over a basis of size and whiting to secure gold leaf that is to be rubbed bright with a burnisher.

It may be made with a mixture of "black lead, deer suet, and red chalk, 1 oz. each, with 1 lb. of pipeclay, ground together to a stiff paste," but it is generally purchased ready made.

JAPANNERS' GOLD SIZE is made by boiling gum animé in linseed oil with driers. The process is an elaborate one, and is fully described in Spon's *Workshop Receipts.*

This size sets very quickly (in from 20 to 30 minutes when pure), and is used for gilding when but short time is available, also for repairs.

It is not so durable, nor does it make such good work, as oil size.

In gilding or japan work it is used as a basis to secure gold leaf or gold powder.

KNOTTING.

Knotting is the material used by painters to cover over the surfaces of knots in wood before painting.

The object is to prevent the exudation of turpentine, etc., from the knots, or, on the other hand, to prevent the knots from absorbing the paint, and thus leaving marks on the painted surface.

Ordinary Knotting is often applied in two coats.

First Size Knotting is made by grinding red lead in water and mixing it with strong glue size. It is used hot, dries in about ten minutes, and prevents exudation.

Second Knotting consists of red lead ground in oil, and thinned with boiled oil and turpentine.

Patent Knotting is chiefly shellac dissolved in naphtha.

The following is a receipt for a similar knotting :—

"Add together ¼ pint japanners' gold size, 1 teaspoonful red lead, 1 pint vegetable naphtha, 7 oz. orange shellac. This mixture is to be kept in a warm place whilst the shellac dissolves, and must be frequently shaken." [1]

Hot Lime is sometimes used for killing knots. It is left on them for about 24 hours, then scraped off, and the surface coated with size knotting ; or if this does not kill the knots, they are then painted with red and white lead ground in oil, and when dry rubbed smooth with pumice stone.

Sometimes after application of the lime the knots are passed over with a hot iron, and then rubbed smooth (see Part II.)

When the knots are very bad they may be cut out, or covered with silver leaf, as described in Part II.

---

[1] Davidson.

PASTE.

Paste is required by the paperhanger, in different degrees of strength, according to the thickness and weight of the paper to be hung with it.

Paste should be made with the best white wheat flour.

The following receipts[1] are for paste of different strengths, the strongest being the last :—

No. 1.—Beat up 4 lbs. of good white sifted wheat flour in cold water to form a stiff batter, taking care to get rid of all lumps ; then add enough cold water to bring it to the consistence of pudding batter.

Pour boiling water over the batter, stirring rapidly. When the mixture swells and loses the white colour of the flour it is ready.

This makes about ¾ pailful of paste, enough for a day's work. It should be used cold, and is adapted for ordinary work.

No. 2 is made in the same way as No. 1, except that just before the boiling water is poured on 2 oz. of alum are mixed with the batter.

The alum imparts strength to the paste, but is said to make it more difficult to lay on. This paste is used for hanging flock papers.

No. 3.—Make a batter as in No. 1, but of less consistency, and to 2 quarts of batter add ½ oz. of pounded rosin ; set the mixture over a moderate fire, stirring till it boils and thickens, then allow it to cool, and thin with thin gum arabic water.

This paste is used only where strong adhesiveness is required, and is indispensable in papering over varnished or painted surfaces.

No. 4 is the same as No. 3, but without gum, and is used for securing the edges of flock papers.

GOLD LEAF.

Gold leaf is required for gilding, in order to ornament different parts of buildings, more especially the internal fittings, such as the mouldings of the joinery or the decorations of the ceilings or walls.

It is classed as *singles, doubles,* or *trebles,* according to thickness, and sold in books, each containing 25 pieces, whose dimensions are $3\frac{1}{4}$ by $3\frac{1}{4}$ inches. They are placed between the paper leaves of the book, which are rubbed with red chalk to prevent the gold from adhering.

The book should be warmed before use, so as to make the leaves quite dry and easy to detach from one another.

There are several different tints of gold leaf, varying from deep orange red down to a pale silvery hue.

Foreign Gold Leaf is thinner than that made in England, and the area of the leaves is smaller.

Pale Leaf Gold is an alloy of silver and gold beaten into leaf.

Dutch Gold is copper leaf coloured yellow by the fumes of molten zinc. It is much cheaper than genuine gold leaf, and useful for large surfaces, where it can be protected by varnish. Without such protection it becomes discoloured.

Bessemer's Gold Paint is in the form of powder. It is mixed with a little transparent varnish, and laid on with a brush.

---

[1] Slightly modified from those given in the *Paperhanger, Painter, Grainer, and Decorator's Assistant.*

PUTTY.

Painters' and Glaziers' Putty is made with whiting (see p. 256) and oil. The whiting is reduced to very fine powder, carefully dried, passed through a fine sieve (about 45 meshes to the inch), mixed with raw linseed oil into a stiff paste, well kneaded, left for 12 hours, and worked up in small pieces till quite smooth.

For particular purposes, such as in fanlights, where the lap or hold is very narrow, a little white lead may be added with advantage.[1]

HARD PUTTY may be made by substituting turps for part of the oil.
VERY HARD PUTTY, from oil, red lead, white lead, and sand.
SOFT PUTTY, from 10 lbs. whiting, 1 lb. white lead, mixed with $\frac{1}{2}$ gill best salad oil and enough boiled linseed oil to bring it to the proper consistence.[1]
The harder kinds crack unless they are soon painted.

Plasterers' Putty (see p. 244).

Thermo-Plastic Putty contains tallow, which keeps the putty pliable, so that it is not loosened by the expansion and contraction of large panes of glass under changes of temperature.[2]

RUST CEMENT.

Rust Cement, known also as *Cast Iron Cement,* and by other names, is used for caulking the joints of cast iron tanks, pipes, etc.

It is composed of cast iron turnings, pounded so that they will pass through a sieve of eight meshes to the inch; to these are added powdered sal-ammoniac, and sometimes flour of sulphur.

The ingredients having been mixed are damped, and soon begin to heat. They are then again well mixed and covered with water.

The exact proportions of the ingredients vary. A simple form is 1 oz. of sal-ammoniac to 1 cwt. iron turnings.

The following are recommended by Mr Molesworth :—

Quick-setting Cement—1 sal-ammoniac by weight.
      2 flour of sulphur.
      80 iron borings.
Slow-setting Cement—2 sal-ammoniac.
      1 flour of sulphur.
      200 iron borings.

The latter cement being the best if the joint is not required for immediate use. In the absence of sal-ammoniac the urine of an animal may be substituted.

The cement will keep for a long time under water. Its efficacy depends upon the expansion of the iron in combining with the sal-ammoniac.

---

[1] Spon.   [2] Seddon.

LATHS.

The laths principally required by the builder are of two kinds—those used for plastering, and those used for roofs to support the covering of slates or tiles.

Plasterers' Laths are thin strips of wood, about an inch wide generally, 3 or 4 feet long, and of a thickness varying according to the work for which they are to be used (see Part II.)

They should be straight ; free from large dead knots, which fall out and weaken them ; from splits ; and from sap, which leads to decay.

They are sometimes made by hand, sometimes by machinery. In either case they should be split or rent from the log, so that each lath has its longitudinal fibres intact. In sawn laths the fibres are generally cut across in places, which makes the laths weak and apt to break across.

Oak laths are sometimes used, but for ordinary work laths should be of the best Baltic fir.

Thickness.—Plasterers' laths are made in three thicknesses classified as follows :—

Single laths	.	.	.	$\frac{1}{8}$ to $\frac{3}{16}$ inch thick.
Lath and half laths .	.	.	$\frac{1}{4}$	,,
Double laths	.	.	.	$\frac{1}{2}$	,,

They are made also in various lengths, varying from 2 to 5 feet, but the lengths most commonly used are 3 feet and 4 feet.

Market forms.—Laths are split in this country, and are also imported from the Baltic and from America, and sold in bundles, round or half round, being either the whole or half of a young tree split up.

The bundles generally contain 360 lineal feet, but sometimes as much as 500 feet run of laths.

Metal Laths are manufactured from 28 BWG iron in any lengths up to 36 inches. They are fixed in the same way as ordinary laths, and the key for the plaster is afforded by the dovetail form into which the metal is bent. They are of common fireproof, and are very useful in special circumstances.[1]

Slate or Tiling Laths, or *Battens* as they are often called, are generally sawn out of boards and sold in 10-feet lengths, the width and thickness varying from $1\frac{1}{4}$ inch × $\frac{3}{4}$ inch to 2 inches × 1 inch, or even 3 inches × 1 inch.

VULCANISED INDIARUBBER.

Vulcanised Indiarubber consists of indiarubber mixed with 44 per cent oxide of zinc and 4 per cent of sulphur. An excess of sulphur injures the material, causing it to become brittle with age.

This material is used chiefly for jointing pipes, for valves, etc.

A rough way of testing its quality is to throw a piece into water ; if it sinks, it probably contains an injurious excess of sulphur.

A good sample should stand a dry heat of 270° Fahr. for 1 hour and a moist heat of 320° Fahr. for 3 hours.

---

[1] Patentees' Circular.

TAR.

Coal Tar is produced by heating coal in close iron vessels, and is a bye product in the manufacture of gas. When itself distilled it produces, in various stages—first, *coal naphtha*, which is useful for dissolving indiarubber, etc. ; then *dead oil* or *creosote*, used (see p. 392) for preserving timber ; and, lastly, *pitch*, which is used for asphalte work (see p. 255), also as an ingredient of varnishes, etc.

Wood Tar is produced by the distillation of pine and other resinous trees. It has strong preservative qualities, owing to the creosote it contains. It is imported in barrels containing about 30 gallons, from the north of Europe as *Stockholm* and *Archangel* tar, and from the United States as *American* tar. Of these varieties Stockholm is considered the best ; the residue left after distillation is *pitch* (see p. 255).

Mineral Tar is a natural substance found in Burmah, and also obtained by distilling bituminous shales, such as those found in Dorsetshire and elsewhere. It contains less volatile matter than the other kinds of tar, but is otherwise of similar composition.

CREOSOTE.

Creosote is a product obtained in distilling tar. It is an oily, dark liquid, varying in composition according to the quality of the coal from which it is obtained, and containing hydrocarbons of different degrees of volatility and varying antiseptic qualities. Until lately the portions of low specific gravity were considered the best, but experience shows that the lighter portions are volatile and soluble in water, so that the valuable acids may be washed out ; a heavy oil, well heated, and with high pressure, gives a better result. The naphthaline is dissolved by the heat, and afterwards fills the pores of the wood and then solidifies.[1]

" The minute glistening scales generally observable on newly creosoted wood consist of naphthaline, a substance that possesses considerable antiseptic properties ; when this substance exists in the liquor in moderate quantities it thickens and confirms its consistency, but when there is a very large proportion . . . it makes the liquor too solid." [2]

Dr. Tidy's specification for creosote is here summarised.[2]

1. To be quite liquid at 100° without deposit until the temperature falls to 95°.
2. One-fourth not to distil over in a retort at less temperature than 600°, and this fourth to be heavier than water.
3. To contain 8 per cent of tar acids by analysis with caustic soda and sulphuric acid.
4. No bone oil or shale oil or any oil not distilled from coal tar.

There are two classes of creosoting oils, known in the trade as London oils and country oils.

"The London oils, which consist of those obtained from the gas tar derived from Newcastle coal, contain a large proportion of naphthaline, and are heavier and thicker than the country oils of the Midland districts, which yield a large proportion of tar acids, as they are called."

Previous to 1863 but little of this thin country oil was used, but since that they became more in demand, under the impression that the tar acids were the most valuable part of the oil. Subsequent experiments have shown, however, that the "so called green oils distilling over at a high temperature formed the best portion of the creosoting liquor, and that the importance of the tar acids had been much overrated."

The specific gravity of creosote depends upon the locality in which it is

---

[1] Dent's *Cantor Lectures.*   [2] *R.E. Journal.*

distilled. The material is sold in casks containing from 36 to 38 gallons each.

Hygeian Rock Building Composition is a bituminous substance used for keeping damp out of houses.

The walls are built in two thicknesses, with a space of about ½ inch or more between them, into which as the wall is carried up the composition is run in a liquid state. Existing walls are made damp-proof by adding a lining of tiles or bricks with the composition between. The material is said not only to keep out damp and vermin, but to add to the strength of the wall. It is sold in bags of 1 cwt., which will cover about 2½ square yards ½ inch thick." [1]

FELT.

Felt, generally saturated with bitumen and other substances, is sold in various forms useful to the engineer and builder. The following information regarding the different descriptions is from the circular of Messrs. Engert and Rolfe :—

Asphalted Roofing Felt is nearly black in colour, has a strong odour of asphalte, and is about ⅛ inch thick.

It is made 32 inches wide, and in any lengths up to 35 yards ; and is used as a roof covering for temporary buildings, the lining under slates, etc., on roofs, etc.

A coat of lime whiting or|whiting and size is recommended where the smell of the asphalte would be objectionable.

Sarking Felt is like the above, but only about $\frac{1}{12}$ inch thick. It is made in rolls 32 inches wide and 30 yards long, and is used as a roof covering for temporary sheds, and under slates.

Inodorous Bitumen Felt is of a brown colour, about ⅛ inch thick. It is made 32 inches wide, and in lengths up to 35 yards. It is used for damp walls, for lining iron houses, under slates or roofs, for laying under floors to deaden sound ; for bedding girders, columns, and heavy iron work.

Fibrous Asphalte is a sort of felt well impregnated with asphalte mixed with grit.

It is made in slabs 32 inches long, and either 4½, 9, 13½, 18, 23, 27, 30, or 36 inches wide.

These slabs are very tough and waterproof. They are used for damp-proof courses (see Part I.), being bedded in cement or mortar ; the joints overlap 2 inches, and are kept clear of mortar.

Hair Felt, for preventing the escape of heat from boilers, pipes, etc., is dry, and not impregnated with asphalte, etc.

It is made 3 feet wide, and in lengths up to 20 yards ; also in sheets 34 inches by 20 inches.

The felt is classed by numbers, according to weight per sheet, as follows :—

Nos.		0	1	2	3	4	5
Weight of sheet	.	Thin	16	24	32	40	48 ounces.
Thickness of sheet	.	...	¼	⅜	½	⅝	¾ inch.

This felt is attached to the boiler by a cement composed as described below, then covered with canvas and painted.

Cement *for attaching Hair Felt to Boilers.*—1 lb. red lead, 3 lbs. white lead, and 8 lbs. whiting, are thoroughly mixed with boiled linseed oil to the consistency of treacle, and spread over the edges of the felt and on the side next to the boiler.

---

[1] Patentees' Circular.

Tarring *Felt.*—Three parts coal tar are boiled with one part slaked lime, powdered chalk, or whiting. The mixture is applied warm, and dusted with as much sand or grit as it will absorb. Stockholm, Archangel, or thick purified coal tar may be used after merely warming, not boiling.

Painting *Felt for Exterior Work.*—First prepare with a coat of lime whiting, then paint with red lead, boiled oil, and driers (no turps), on which sprinkle fine white silver sand ; over this any paint may be used.

ASBESTOS.

Asbestos, the well-known fireproof and acidproof fibrous mineral, is the basis of several substances useful to the builder.[1]

The raw material comes from Italy, Canada, California, Australia, etc. The two first are the best in the order given. Italian asbestos is grey or brown in colour, Canadian white, *Asbestos Paints* (see p. 426).

Asbestos Concrete Coating is of a drab colour, and is used to cover beams to retard the action of fire upon them ; 100 lbs. will cover 200 square feet $\frac{1}{8}$ inch thick.

Asbestos Roofing is made from canvas cemented to a surface layer of felt and a Manilla lining in compact flexible sheets resembling leather.

It is supplied in rolls about $38\frac{1}{2}$ inches wide, containing 200 square feet, and weighs about 85 lbs. per square.

Asbestos Sheathing is fireproof, and used for lining wooden partitions, ceilings, etc. It is made in rolls of from 60 to 100 lbs. 42 inches wide, weighing about 6 lbs. per square ; also " double thick," weighing about 10 lbs. per square.

Asbestos Building Felt is fireproof ; it is made in rolls of about 70 lbs. weight, 36 inches wide, weighing about 60 lbs. per square ; also " extra heavy," weighing about 16 lbs. per square.

WILLESDEN FABRICS.

Willesden Fabrics [1] are vegetable substances which have been treated with certain compounds of copper and ammonia, the effect of which is to coat and impregnate them with cupro-cellulose, a varnish-like substance which not only protects the surfaces but adds strength to the fibres by cementing them together. This enables the fibre to resist the weather, and renders it less liable to catch fire. Ropes, cordage, and netting are treated in this way, but the fabrics most useful to the builder are the Willesden paper and canvas.

Willesden Paper is of three classes.

UNWELDED (marked WPG 1), or "one ply" paper, is made 54 inches wide, of indefinite length, and is chiefly used for packing.

WELDED, which consists of several "plys" or thicknesses of paper formed (while they are still gelatinised by the action of the cupro-ammonia solution) into a compact sheet or thickness.

The different classes of this paper are known as follows. They are all made in brown and neutral green colours.

Willesden 2 ply (WPG 2) is in continuous lengths, 54 inches wide. It is useful for underlining slates, tiles, internal decorations, floors, damp walls, leaky roofs, etc.

Willesden 4 ply (WPG 4) is weatherproof and strong, a bad conductor of heat, free from condensation, does not easily catch fire, does not require painting, and is said to be proof against the white ant ; it is useful for roofing, sides of huts, etc.

The relative covering powers of this and good galvanised iron are stated by the manufacturers as follows :—

	Willesden paper. WPG 4	Galvanised iron.
Weight of one square in lbs. .	15 to 18	103 to 280
Area covered by one ton in squares .	125 to 150	8 to 22

Willesden 8 ply may be used as panel board.

Willesden Canvas is prepared in a similar way to the paper, and can be used with advantage for most purposes to which canvas is applied, including making hose.

---

[1] Patentees' Circulars.

WIRE WOVE ROOFING.

Wire Wove Roofing consists of a semi-transparent substance, apparently some preparation of linseed oil upon a basis of very fine wire mesh. It is said to be waterproof, tough, elastic, strong, and not affected by atmospheric changes, and is made in sheets of 10 feet by 4 feet and of 10 feet by 2 feet.

EMERY.

Emery Cloth,[1] consisting of ground emery of different degrees of fineness attached to calico by glue, is used for finishing and polishing metal surfaces. *Emery Paper* is not much used for builder's work.

Glass Cloth and Glass Paper are made respectively from calico and paper coated with ground glass, and are used for producing a smooth surface on wood or for rubbing down painted surfaces.

SILICATE COTTON.

Silicate Cotton[1] or *Slag Wood* is a glasslike fibre blown from blast furnace slag. It is incombustible, vermin-proof, and very light, one ton covering 1800 square feet 1 inch thick, and is therefore useful for making floors and ceilings sound and fireproof.

NAILS.

There are some 300 varieties of nails, named chiefly from the shape of their heads and points, or according to the particular use for which they are intended.

No attempt will be made to describe them all, but it may be useful to the student to know the names and characteristics of some of those in most common use for building and engineering work.

The thickness of different classes is expressed by the terms "*fine,*" "*bastard,*" "*strong;*" and their weight is generally given in lbs. per 1000, and their length in inches.

In former times nails were described according to their price per 100— thus, "tenpenny nails" and "fourpenny nails" were those costing 10d. and 4d. per 100 respectively. These terms are still sometimes used, but their meaning is now indefinite. It varies in different localities, and no longer refers to the price of the nails. The term "Tenpenny nails" now generally means nails about $2\frac{3}{4}$ inches long, not nails at 10d. per 100. In the same way "Sixpenny nails" are generally $1\frac{1}{2}$ inches long, "Eightpenny" $2\frac{1}{4}$ inches, and "Twentypenny" $3\frac{1}{2}$ inches. Makers differ, however, as to the lengths corresponding to the different names.

Cast Nails, made by running molten iron into moulds, are brittle and inferior in strength, but cheap. They are used for horticultural purposes, for lathing, and for many other purposes in common work.

Malleable Nails are made in the same way as cast nails, but are afterwards rendered malleable by the process described at page 268. They can be made thinner than the common cast nails.

Hand-Wrought Nails are forged by manual labour. They are tougher and stronger than other varieties, and will bear clenching, but are more expensive. Their angles are sharp and clear, and the shanks are slightly compressed just under the heads.

Cut Nails are of a cheaper description, cut by machinery out of sheets of iron.

---

[1] Manufacturers' Circulars.

Patent Machine-Wrought Nails are made out of wrought iron pressed while red-hot into shape by grooved rollers, then cut up, and the heads formed by a die. They have not such sharp clean angles as the hand-wrought nails, and are not so strong or elastic. The shank under the head is rather flattened out, and their grip is maintained by the shank being slightly thicker near the point than in 'the centre. They are slightly cheaper than hand-wrought nails, and at present Rose and Clasp nails are the chief varieties made.

Varieties of Nails in Common Use.—The following descriptions are of nails in common use :—

Rose Nails are either wrought, cut, or pressed. They take their distinctive name from the shape of their heads, and are divided into classes according to the nature of their points.

Rose Sharp Points are used for coopering, fencing, and for coarse purposes with hard woods. There are both wrought and stamped varieties. They are classed according to stoutness, as " *Fine* " (or " *Canada*") and " *Strong.*"

Rose Flat Points (Fig. 170) have chisel points, and are used when the ordinary points would act as wedges and split the wood. They are driven with the flat point along the grain, so as to prevent splitting and hold faster. These also are classed as "Fine" or "Strong."

Rose Clench are square ended, and easily punch through thin metal coverings without first boring a hole. They are used by boat-builders, and also for packing cases.

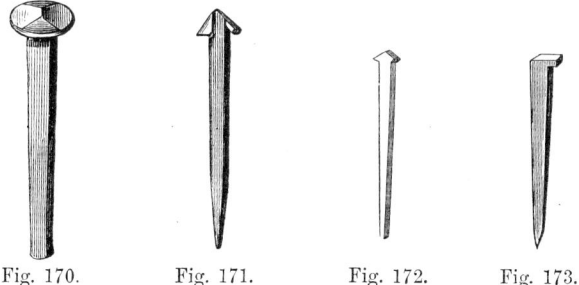

Fig. 170.   Fig. 171.   Fig. 172.   Fig. 173.

Clasp Nails are much used by carpenters in soft woods, such as fir. They have heads which project downwards and stick into the wood, holding it together. They are also easily driven below the surface, so as to allow the plane to pass over them.

Fig. 171 shows the shape of the wrought description. The cut clasp have heads nearly flat on both sides, as in Fig. 172.

Wrought Clasp are divided into two classes—*Fine* and *Strong*, and are used for ledges to doors and other work where the nail requires to be clenched.

To effect this a nail is selected of a length greater than the thickness of the wood through which it passes, and the projecting point is hammered, so as to be turned back into the wood.

Cut Clasp are used for fixing rafters, ceiling joists, also architraves, skirtings, linings, and other joinery.

Brads (Fig. 173) are flat-sided nails, either wrought or cut, with heads of the same thickness as the shank, of a shape known as *billed*, and being driven with the flat sides parallel to the grain, are not liable to split the wood.

The larger sizes are used for flooring ; the smaller for light work, such as fixing small mouldings, beads, etc.

The ends of cut brads are not pointed as in wrought brads.

The lighter varieties are called *Joiners' Brads* and *Cabinet Brads*.

Glaziers' Brads or *Sprigs*, used for securing large panes of glass, are of the shape shown in Fig. 174, and have no heads.

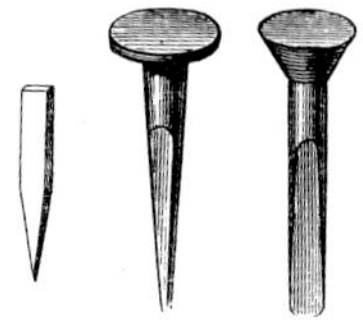

CLOUT NAILS (Fig. 175) have flat, circular heads ; shanks round under the head, and with points either tapered or flat. The smaller sizes are mostly sharp, and the larger have flat chisel points. They are used for fastening sheet metal, felt, nailing hoop iron to wood, etc., and are made in two varieties, *fine* and *strong*.

Countersunk Clouts (Fig. 176) have heads shaped so as to fit a counter-sinking, and are generally made with flat points.

Figs. 174. 175. 176.

They are much used by wheelwrights and smiths, and for securing iron plates, etc., to woodwork.

WIRE NAILS, known also as *French Nails* (or *Pointes de Paris*), are round or square in section, very tough and strong. They are said not to split the wood, and to require no hole bored for them. They are sold in lengths from $\frac{5}{8}$ to 4 inches, and of different thicknesses, varying from Nos. 5 to 18 B.W.G., and are used for packing-cases and other purposes.

DOG NAILS are made with solid and slightly countersunk heads. These are sometimes hemispherical ("die-heads") ; the shanks are generally round, at least under the head, and their points flat.

They are used for nailing down heavy ironwork, and for various other purposes when the heads are not required to be flush with the surface of the work.

SPIKES are very large wrought nails used for heavy work, when great strength is required, as in wood bridges, scuppering, etc. They range from 4 to 14 inches in length ; the smallest sizes have rose heads, but the larger ones have square heads with flat tops, as shown in the figure, which, it must be observed, is on half the scale of the sketches of the smaller nails.

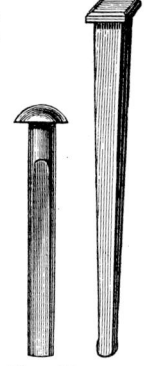

TACKS are small, short, and light nails, and are divided into three classes—Rose, Clout, and Flemish ; the two former are named according to the shape of their heads. Clout tacks resemble the nail shown in Fig. 175. Flemish are similar,

Figs. 177. 178.

but that the shank tapers throughout the upper portion, and is not finished in a cylindrical form as shown in Fig. 175. Tacks are used for close nailing very light sheet metal, but chiefly for upholsterers' work.

Tacks are generally wrought, but some of the smaller kinds are cut. They are either *blacked, blued*, or *tinned*.

Copper Nails are made of the same shapes as iron nails, and are used in positions where the latter would be subject to corrosion.

Composition Nails are those made of different alloys to avoid corrosion, or to prevent the galvanic action set up by iron when in contact with zinc or other metals. They are varied in shape according to the purpose for which they are to be used.

SLATING NAILS have circular flat heads and sharp-pointed shanks ; some are slightly countersunk, as in Fig. 179.

They are made of various metals. For temporary work *cast* nails may be used, for better work *malleable* nails ; these, however, soon corrode away unless galvanised or dipped hot in boiled oil. *Zinc* nails are cheap, and sometimes used, but are too soft. *Copper* nails are often used in superior work, but are also soft. *Composition* nails are cast from an alloy (about 7 copper to 4 zinc) which is hard and does not corrode. When made of a really good strong alloy they are the best for superior work.

179.

TILE PEGS is the name given to short cast-iron nails too thick for slating, and used for securing tiles to roofs.

Steel Nails, made from molten metal pressed in moulds, have lately been introduced, and used largely for the best class of work. They are finer and cleaner than ordinary nails, but much dearer.

LATH NAILS may be obtained either wrought, cut, or cast. The cast and the cut are the cheapest. The cut nails are generally used.

Wrought nails should be used for oak laths.

The length of the nails varies acording to the thickness of the lath, being

$\frac{3}{4}$ inch for single laths.
$\frac{7}{8}$,,   lath and half laths.
1   ,,   double laths.

MISCELLANEOUS.—Besides the above-mentioned there is an innumerable variety of patent nails of different descriptions and in different metals, also brass-headed and fancy-headed nails, and nails used for special purposes, unconnected with buildings. These need not be further referred to.

Weight of Nails.—The Table on the next page, which is taken chiefly from Government schedules, shows the weight per 1000 of some of the most useful sizes of different kinds of nails.

Spikes are generally sold by the cwt. Their weight may be taken as follows :—

Length	.	.	.	5	6	7	8	9 inches.
Weight per 1000	.	.		100	200	300	450	600 lbs.

Pound Nails are of a particularly heavy description, and are also sold by the cwt.

TABLE, SHOWING COMMON SIZES AND WEIGHTS, IN LBS. PER 1000, OF DIFFERENT KINDS OF NAILS, ETC.

Length from point to under side of head in inches.	3/16	1/4	5/16	3/8	7/16	1/2	9/16	5/8	11/16	3/4	7/8	1	1 1/8	1 1/4	1 3/8	1 3/16	1 1/2	1 3/4	2	2 1/4	2 1/2	2 3/4	3	3 1/8	3 1/4	3 3/8	3 1/2	3 3/4	4	4 1/2	5	6
Iron Nails.																																
Brads, Cut										2/6					1 1/4				6	8	12	12	14		32							
Wrought, Fine	5/16	1/4								3		1 4/16		1 1/4	1 3/4				4	12	16		24		20				40			
,, Flooring	1/2									4					5				9	10	15			20	18							
Clasp, Fine			6/16					1 4/16	1 1/6			1 3/4	1 5	2 1/2			4		8	10	10	14			25	32	32					
Strong			1 5/16				1 4/16	1 8/16			2	2	5	3			3 3/5		6		12											
Cut								1 8/16		1 1/4			4	3			4			10	15											
Clout, Fine	1 3/16							1 9/16					9	3			9		8		12	14	60	20			80					
Strong										1 3/4	1 8/16	2	10	7			10		23	32	40				76							
Countersunk											1 4/16		4				4		16	20	24		56		28				36			
Tinned	1/8							2		1 3/4			16																50 7/16	40		
Dog, Die-head												1									20	20								70	90	
Rose, Fine													7	4				2	10		16											
Strong																																
Sprigs, Glaziers'																																
Tacks, Flemish																																
Strong Cut																																
Fine Cut																																
CORDES' PATENT WROUGHT NAILS.[1]																																
Rose, Fine, Flat point												2 1/2		3 3/5				7	8–10	10	10	12 1/2–14	14–17			32		42			
Strong, do.												2 1/2		6			7	10	12	17	20	23–25					Sold by the cwt.				
Best or Cooper's sharp										1 4/16						7	10	12	17	20	23				32						
Best Canada, sharp	1 10/16							1 4/16		1 3/4			5 5/16			5	7	8–10	8–10	10	12 1/2	14–17									
Clasp, Fine																5	7	8	8	10		14									
Copper Nails.																																
Dog, Die-head													14 1/16			14 1/16	29 1/16		40													
Jagged												2 5/16	6 1/16		51 1/2					18 1/16	18 1/2							80 1/16		80 3/8	
Rose, Fine												3 1/4	6 1/16			10 3/16	14 3/4	11 1/4		10 3/16		30 1/16									
Strong												4 7/16	8 1/16			11 1/4											50 6/16				
Slating Nails.																																
Tacks												2 15/16	7 2/16																		
Composition							1 15/16			1 3/16		2 15/16		3 1/16				11													
Zinc		1 10/16																													

1 From manufacturer's list.

Adhesive Force of Nails.—The following abstract of records of experiments on the holding power of nails may be useful :—

HOLDING POWER OF WROUGHT IRON TENPENNY NAILS, 77 to the lb., about 3 inches long, nailed through a 1-inch board into a block, from which it was dragged in a direction perpendicular to length of nails.[1]

No. of Nails in Square Foot.	Kind of Plank.	Kind of Block.	Average Breaking Weight per Nail.	No. of Nails in Square Foot.	Kind of Plank.	Kind of Block.	Average Breaking Weight per Nail.
			lbs.				lbs.
8	Pine	Pine	380	12	Oak	Oak	542·5
8	Oak	,,	415	6	Pine	Pine	463·5
8	,,	Oak	465	6	Oak	,,	332·5
4	Pine	Pine	341	6	,,	Oak	437·5
4	Oak	,,	446	16	Pine	Pine	289
4	,,	Oak	551	16	Oak	,,	420
12	Pine	Pine	612	16	,,	Oak	433
12	Oak	,,	555·5				

The surfaces in contact were from 1 to 2 square feet. The average strength decreased generally with the increase of surface.

ADHESIVE FORCE of NAILS forced into dry CHRISTIANIA DEAL at right angles to grain of wood.[2]

	Number to the lb.	Length in inches.	Length forced into the Wood.	Force in lbs. required to extract.
Fine sprigs	4,560	0·44	0·40	22
,, ...	3,200	0·53	0·44	37
Threepenny brads . . .	618	1·25	0·50	58
Cast-iron nails . . .	380	1·00	0·50	72
Wrought iron 6d. nails . .	73	2·50	1·00	187
,, ,, ,, . .	...	...	1·50	327
,, ,, ,, . .	...	...	2·00	530
,, 5d. ,, . .	139	2·00	1·50	320

The relative adhesion when driven transversely and longitudinally is in deal about 2 to 1, in elm about 4 to 3.

To extract a common sixpenny nail from a depth of 1 inch required—

	lbs.		lbs.
Beech, dry, across grain . .	167	Elm, dry, across grain . .	327
Deal, Christiania, dry, do. .	187	Do. do., with grain . .	257
Do., do., with grain .	87	Oak, do., across grain . .	507
		Sycamore, green, do. . .	312

From experiments by Lieutenant Fraser, R.E., it appears that the holding power of spike nails in fir is 460 to 730 lbs. per inch in length.

---

[1] Haupt's *Military Bridges*.   [2] Tredgold, Bevan's Experiments.

SCREWS.[1]

Wood-Screws (for screwing into wood) are made of metal, with sharp or bevelled threads of different forms. The most usual is shown by the section Fig. 180.

The points are generally made sharp, so that they may penetrate the wood; the body of the screw is tapered, so that the deeper it is driven the more tightly it will fill the hole; the thread does not extend throughout the length of the screw, but a considerable portion below the head is left smooth; the thread is formed to an acute angle, and there is a considerable pitch, or distance between the threads.

Wood-screws are made in various sizes, and are divided as to strength into three classes—*Strong, Middling,* and *Fine.*

Each length is made in from 15 to 30 different thicknesses, identified by numbers.

The following are the thicknesses or diameters corresponding to some of the numbers. The thicknesses of the other numbers are interpolated between those given, varying in succession about $\frac{1}{64}$ inch :—

Number	00	0	1	5	10	14	18	22	27	32	40
Thickness or diameter in parts of an inch .	$\frac{1}{32}$	$\frac{3}{64}$	$\frac{1}{16}$	$\frac{1}{8}$	$\frac{3}{16}$	$\frac{1}{4}$	$\frac{5}{16}$	$\frac{3}{8}$	$\frac{7}{16}$	$\frac{1}{2}$	$\frac{5}{8}$

The following Table shows the numbers or thicknesses in which iron wood-screws of different lengths are made :—

Length from top of head to point in parts of an inch.

$\frac{1}{4}$	$\frac{3}{8}$	$\frac{1}{2}$	$\frac{5}{8}$	$\frac{3}{4}$	$\frac{7}{8}$	1	$1\frac{1}{4}$	$1\frac{1}{2}$	$1\frac{3}{4}$
					Numbers made.				
0 to 16	1 to 16	1 to 16	1 to 18	2 to 20	3 to 24	4 to 26	5 to 28	6 to 30	7 to 32

Length from top of head to point in parts of an inch.

2	$2\frac{1}{4}$	$2\frac{1}{2}$	$2\frac{3}{4}$	3	$3\frac{1}{2}$	4	$4\frac{1}{2}$	5	6
					Numbers made.				
8 to 36	9 to 38	10 to 40	11 to 40	12 to 40	14 to 40	16 to 40	16 to 40	18 to 40	20 to 40

They are also classified according to the shape of their heads, as round-headed, flat-headed (or countersunk), square-headed, cone-headed, ball-headed, hexagon-headed, Gothic-headed, etc. etc.

Wood-screws are sold by the dozen or by the gross of 12 dozen. Those varieties that are used for securing furniture to doors, etc., should be, and by some houses are, supplied with the furniture.

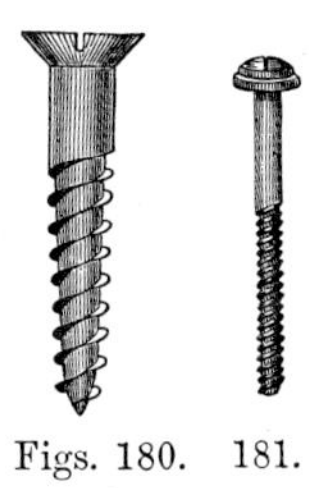

FLATHEADED SCREWS (Fig. 180) are used in wood for fixing all metal work or furniture whose thickness is sufficient to admit of the head of the screw being countersunk into them, so that the top of the head is flush with the face of the metal to be screwed on.

ROUND-HEADED SCREWS (Fig. 181) are used where the metal is too thin to be countersunk, as in some forms of locks, latches, etc.

Figs. 180.   181.

---

[1] Sc. *Screw-nails.*

Patent Pointed Screws are made with sharp points like that of a gimlet, as shown in Fig. 180. They resemble the general description given above, and are commonly used.

Fig. 182 shows an old-fashioned form of screw, with an angular thread and blunt point, formerly known as *Nettlefold's Patent Screw*. The advantage claimed for it was that the top side of the thread being horizontal or inclined upward, offers great resistance to the screw being dragged forcibly out.

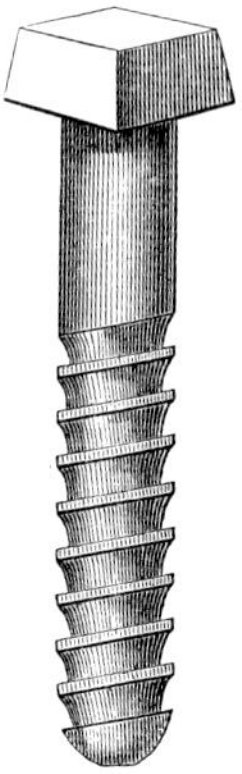

COACH SCREWS are large heavy screws used where great strength is required in heavy woodwork, and for fixing iron work to timber. They have square heads, so that they can be screwed home with a spanner or wrench, and a thread like that shown in Fig. 183.

Fig. 182.

HANDRAIL SCREWS are of a peculiar construction, and are intended for joining together two lengths of a staircase handrail, as shown in Figs. 184, 185.

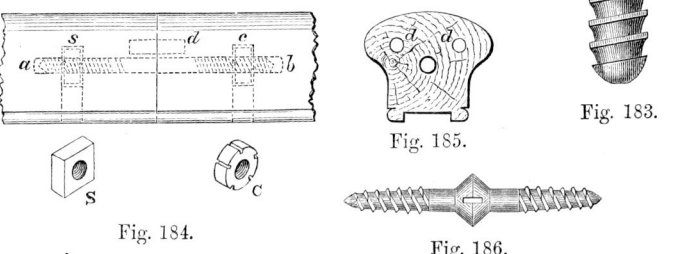

Fig. 183.

Fig. 185.

Fig. 184.

Fig. 186.

The screws are from about 3 to 6 inches long, and are threaded at each end.

A square nut *s* is made for one end, and for the other end a circular nut *c*, the latter having at intervals deep nicks in its circumference to receive the end of a screwdriver.

The sketches at *s* and *c* show the form of these nuts. Deep slots are cut from the under side to the centre of the handrail, through which they are dropped into the positions *s c* in Fig. 184. A longitudinal hole, *ab*, is bored in the handrail, in which the screw is placed so as to pass through the nut at each end. The circular nut is turned on the screw by means of a screwdriver, so that the portion of the handrail in which it is fixed is drawn toward the other until the joint between them is quite tight. *dd* are dowels inserted to strengthen the joint.

Fig. 186 [1] shows another form of handrail screw, known as a dowel-screw.

BRASS SCREWS may be obtained in nearly every form, at about three times the cost of iron screws. They are very useful for securing work which requires to be easily removable—such, for example, as the beads of sash frames (see Part I.)

Screws are made in several other forms besides those mentioned, for special purposes, which need not be further referred to.

Screws for Metal are made in different forms from wood screws ; the diameter of the screw is the same throughout ; the threads are close together, V-shaped, but with the points of the V's rounded off.

The great difference between screws for metal and those for wood is that the latter, by the pressure of their threads against the fibres make a hole into which they will fit exactly, whereas in metal such a hole has to be tapped of the exact size to receive the screw.

Unless the internal thread of the nut, or of other metal into which the screw is to be driven, exactly fits the thread of the screw, one or the other will become distorted in screwing ; they will bear unequally upon one another, and great loss of strength would ensue, together with difficulties in working.

---

[1] Knight's *Dictionary of Mechanics.*

Whitworth's Standard Thread.—Screws for bolts and nuts, and for metal work, are now generally made according to Sir J. Whitworth's standard, the same form of thread being used throughout, and the same pitch and depth of thread being always used for screws of the same diameter, so that both screws and nuts are always interchangeable, which is an immense advantage in case of loss or fracture.

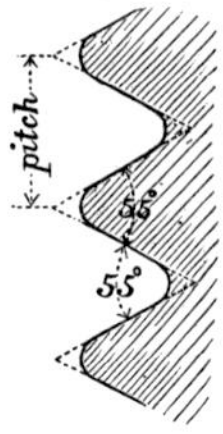

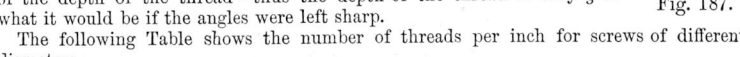

Whitworth's Standard Screw Thread is shown in section in Fig. 187. The sides of the thread are inclined at 55° to one another, and the sharp angles at the top and bottom are rounded off, each to a depth of about $\frac{1}{6}$ of the depth of the thread—thus the depth of the thread is only $\frac{2}{3}$ of what it would be if the angles were left sharp.

Fig. 187.

The following Table shows the number of threads per inch for screws of different diameters :—

Number of threads per inch.	Diameter of screw. Dec. of inch.	Number of threads per inch.	Diameter of screw. Dec. of inch.	Number of threads per inch.	Diameter of screw. Dec. of inch.
48	·100	12	·600	4	2·375
40	·125	11	·625	4	2·500
32	·150	11	·650	4	2·625
24	·175	11	·675	$3\frac{1}{2}$	2·750
24	·200	11	·700	$3\frac{1}{2}$	2·875
24	·225	10	·750	$3\frac{1}{2}$	3·000
20	·250	10	·800	$3\frac{1}{4}$	3·250
20	·275	9	·875	$3\frac{1}{4}$	3·500
18	·300	9	·900	3	3·750
18	·325	8	1·000	3	4·000
18	·350	7	1·125	$2\frac{7}{8}$	4·250
16	·375	7	1·250	$2\frac{7}{8}$	4·500
16	·400	6	1·375	$2\frac{3}{4}$	4·750
14	·425	6	1·500	$2\frac{3}{4}$	5·000
14	·450	5	1·625	$2\frac{5}{8}$	5·250
14	·475	5	1·750	$2\frac{5}{8}$	5·500
12	·500	$4\frac{1}{2}$	1·875	$2\frac{1}{2}$	5·750
12	·525	$4\frac{1}{2}$	2·000	$2\frac{1}{2}$	6·000
12	·550	$4\frac{1}{2}$	2·125		
12	·575	4	2·250		

Whitworth's Gas Threads.—For the screwed ends of wrought-iron gas-tubing and for common metal work a shallower thread is used, as shown in the following Table :—

Diameter in inches	$\frac{1}{8}$	$\frac{1}{4}$	$\frac{3}{8}$	$\frac{1}{2}$	$\frac{3}{4}$	1	$1\frac{1}{4}$	$1\frac{1}{2}$	$1\frac{3}{4}$	2
Number of threads per inch	28	19	19	14	14	11	11	11	11	11

STOVE SCREWS are small screws of the form shown in Fig. 188, used for uniting the different parts of stoves, grate fronts, etc. The heads are sometimes square, or cup-shaped, instead of being circular and flat as shown in Fig. 188.

Adhesive Power of Screws.—Mr. Bevan experimented on iron woodscrews 2 inches long, $\frac{22}{100}$ diameter at exterior of threads, threads $\frac{35}{1000}$ deep, 12 to the inch. These were driven into boards $\frac{1}{2}$ inch thick. The force required

Fig. 188.

to extract them was—from hard woods about 790 lbs., from soft woods about half that amount.[1]

MAKING SCREWS is a subject which is beyond the scope of these Notes.

Wood screws ordinarily used by the carpenter and joiner are made by machinery, the thread being turned in a sort of lathe. Very large screws are also turned in lathes in ordinary workshops.

Small metal screws are cut in *screw plates*, larger ones with *stocks* and *dies ;* and the threads to receive screws may be tapped by the aid of hard *master taps*, turned by means of a long double handle.

Bolts and Nuts are a good deal used by the carpenter for heavy work, and are also required in connection with iron roofs, etc.

They hardly come within the range of notes on materials, and it is impossible, for want of space, to describe them at all.

---

1 Tredgold.

APPENDIX.

SHORT NOTE ON THE

PHYSICAL PROPERTIES OF MATERIALS, AND ON THE LOADS AND STRESSES TO WHICH THEY ARE SUBJECTED.

A DETAILED description of the physical properties of materials, and of the loads and stresses to which they are subjected, would be beyond the province of this volume, especially as the subject will be entered upon in Part IV. The following short explanations of some of the terms employed in describing those properties and stresses may, however, be useful.

Load.—The combination of external forces acting upon any structure is called the load.

Dead load is that which is very gradually applied, and which remains steady.

Thus the weight of any structure is itself a dead load. Grain gradually poured on to a floor, or water run slowly into a tank, would also be dead loads.

Live load is that which is applied suddenly, or is accompanied by shocks or vibration.

Thus a fast train coming on to a bridge, or a sudden gust of wind upon a wall or roof, causes live loads.

Without going into the theory of the subject, it is sufficient to state that practically a live load produces in most cases very nearly twice the stress and strain which a dead load of the same weight would produce.

Therefore to find the dead load which would produce the same effect as a given live load, the latter must be multiplied by 2.

This is called converting the live load into an equivalent dead load.

Illustration.—A bridge may weigh one ton per foot of area (*i.e.* dead load), and carry a live load of two tons per foot of area ; the equivalent dead load would be $(1 + 2 \times 2) =$ 5 tons per foot of area.

The breaking load for any structure or piece of material is that dead load which will just produce fracture in the structure or material.

The FACTOR OF SAFETY is the ratio in which the breaking load exceeds the working load (*i.e.* the load which can be safely applied in practice). This ratio varies with the nature of the load and the nature of the material, and is found by experience.

For the reasons stated above, the factor of safety for a live load is generally taken at double that for a dead load.

The factors of safety for several different kinds of iron structures are given at p. 324. The following Table shows those recommended by Professor Rankine[1] for general practice :—

	FACTORS OF SAFETY.	
	Dead Load.	Live Load.
For perfect materials and workmanship	2	4
For good ordinary ma- { Metals .	3	6
terials and work- { Timber	4 to 5	8 to 10
manship { Masonry	4	8

When a load is mixed, *i.e.* partly live and partly dead, the live portion may be converted into an equivalent amount of dead load, and the factors of safety for dead load then applied to the whole ; or

A compound factor of safety may be deduced by applying the following rule :—

---

[1] Rankine's *Useful Rules and Tables.*

Multiply the factor of safety for dead load by the fraction that the dead load is of the whole load, and multiply the factor of safety for live load by the fraction that the live load is of the whole load. The sum of the results thus obtained will give the compound factor of safety.

For example :—In a certain iron bridge the dead load is 5 tons per bay, the live load 9 tons per bay ; the total load is therefore 14 tons per bay.

The dead load is $\frac{5}{14}$ of the whole.

,, live ,, $\frac{9}{14}$,,

The factor of safety for dead load is 3, and for live load is 6.

The compound factor of safety will be equal to

$$(\tfrac{5}{14} \times 3) + (\tfrac{9}{14} \times 6) = \tfrac{69}{14} = 4\tfrac{13}{14} = \text{say } 5.$$

The working load is the greatest dead load the material can with safety bear in practice. It is found by dividing the breaking load by that factor of safety which is found to be suitable to the particular case.

The proof load is the greatest load that can be applied to a piece of material to prove or test it by straining it to the utmost extent without producing permanent deformation or injury, *i.e.* not beyond the elastic limit (see p. 327).

The breaking load or working load may be either live or dead, or a combination of both, but for convenience it is usual to reduce it all to an equivalent dead load, by doubling the live load and adding it to the dead load.

Stresses.—STRESS and STRAIN are words often used indifferently, either to mean the alterations of figure produced in a body by any forces, or to mean the forces producing those alterations.

Of late years, however, the word *strain* has been taken to mean only the alterations of form caused by the forces, and *stress* to mean the forces producing these alterations.

Materials are subject to the under-mentioned stresses, which produce strains, and (when carried far enough) fracture as stated.

Stresses.	Strain.	Mode of Fracture.
Tensile or } Pulling	{ Stretching } Elongation	} Tearing.
Compressive or } Thrusting	{ Shortening } Squeezing	} Crushing.
Transverse or Bending	Bending	Breaking across.
Shearing	Distortion	Cutting asunder.
Torsional or } Twisting	Twisting.	{ Twisting or wrenching asunder.

Intensity of stress is the amount of stress on a given unit of surface, and is expressed in lbs., or sometimes in tons, per square inch.

The ultimate stress, or *breaking stress*, on any piece of material is the stress produced by the breaking load.

The proof stress is the stress produced by the proof load.

The working stress is that produced by the working load. It is always much smaller than the proof stress, in order to leave a margin of safety to cover defects in material, etc.

A bar of 1 square inch sectional area might have a breaking strength of twenty tons, but the working stress to which it was subjected might be only five tons. The factor of safety in that case would be four. Its proof strength might be ten tons, this being the weight the bar could bear without exceeding the elastic limit.

Strength.— *Tenacity* or *tensile strength* is the resistance offered by material to tension, that is to a stress tending to tear it asunder, as, for example, in the case of a vertical rod having a weight suspended from it, or in the tie rod of a roof, or the tension flange of a girder.

Strength to resist crushing is the resistance offered by material to a compressive stress, thrust, or pressure. Such a stress tends to make it shorten, and eventually to crush it. Examples of this stress occur in the case of a short column supporting a weight, or in a strut which keeps two tottering walls from falling toward each other, or in the compression flange of a girder.

It should be observed, however, that long columns and struts tend to fail by bending outwards in the centre and then breaking across. This form of failure is called *buckling*.

Transverse strength is the resistance offered by a body to forces acting across it, tending to bend it, and eventually to make it break across. Thus a beam supported at both ends and loaded over any part of its length, bends downward and tends to break across.

When a body is subjected to transverse stress, some parts of it are in compression, some in tension, and others are exposed to a shearing stress, therefore transverse stress is a combination of these three stresses. A beam secured at the ends, and subject to pressure from below, bends upwards and also tends to break across.

Shearing strength is the resistance offered by a body to being shorn, that is, to being distorted by one part of it sliding on another part. Thus, if two lapped plates united by a rivet be drawn longitudinally in opposite directions, the rivet would tend to shear by the upper plate sliding upon the lower.

Torsional strength is the resistance offered by a body to being broken by torsion, *i.e.* twisting. This stress frequently occurs in machinery, but not in structures connected with buildings.

Strength to resist bearing is the resistance offered by a material to being indented, or partially crushed by another body pressing upon it. Thus, the shank of a rivet may be indented by the plate bearing upon it, or the edge of the hole in the plate may be indented by the rivet ; again, a beam may be indented by the end of a post bearing upon it. Indentation by bearing is merely one form of crushing.

The *ultimate strength* of any material is the intensity of stress required to produce fracture in any specified way.

The *proof strength* is the intensity of stress required to produce the greatest strain of a specific kind without injuring the strength of the material.

Pliability is the tendency of a body to change its form temporarily under different stresses.

Stiffness or Rigidity is the reverse of pliability, and expresses the disinclination of some bodies to change their form under stresses.

Thus stones and bricks are rigid up to a certain point.

Elasticity[1] is the property which all bodies have (in greater or less degree of perfection) of returning to their original figure after being distorted (*i.e.* strained) by any kind of stress.

When the original figure is completely and quickly recovered, the elasticity is said to be *perfect.*[2]

When the original figure is not completely recovered, but remains permanently distorted to a certain extent. the elasticity is said to be *imperfect,*[3] and the distortion produced is called a *permanent set,* or *set.*

It has been found by experiment that the elasticity of most building materials is practically perfect up to a certain point. When stresses below this point are applied and removed, the strain, distortion, or change of figure is only temporary. There is no appreciable *set.* Stresses above this point, however, cause sets (see p. 328).

The *Elastic limit* of a material is the maximum intensity of stress that can be applied to it without causing an appreciable set.

A Modulus of Elasticity is a number representing the ratio of the intensity of stress (of any kind) to the intensity of strain (of any kind) produced by that stress, so long as the elastic limit is not passed.

The modulus of tensile elasticity is found by dividing the tensile stress in lbs. per square inch of sectional area by the elongation (produced by that stress) expressed as a fraction of the length of the body.

Thus, if a weight of one ton hung from an iron bar produce an elongation of $\frac{1}{12000}$ of the length of the bar, the modulus of elasticity of that bar will be 2240 lbs. $\div \frac{1}{12000} =$ 26,880,000 lbs. This is rather lower than the modulus of average wrought iron.

Similarly the modulus of compressive elasticity is found by dividing the compressive stress in lbs. per square inch of section by the contraction (or rather shortening) produced by that stress, expressed as a fraction of the length.

In most building materials the modulus of tensile and that of compressive elasticity are practically equal to one another so long as the stresses do not exceed the elastic limit.

---

[1] The elasticity here referred to is sometimes called elasticity of figure ; there is also an elasticity of volume, which need not be considered in connection with building materials.

[2] Mr. Eaton Hodgkinson's investigations seem to show that the elasticity of every solid is really imperfect, that the slightest strain produces a set. Up to a certain limit of stress, however, the sets produced are so small that they cannot be measured with ordinary instruments, and therefore within that limit the elasticity may be said to be *sensibly perfect* for all practical purposes (see p. 315).

[3] Because the elongations and shortenings under equal stresses are practically equal up to the elastic limit ; beyond that they are irregular.

The modulus is generally denoted by the letter E, and its value is given in the tables, because it is useful in calculating the stiffness of beams and girders.

In advanced works on applied mechanics several other moduli are used, which, however, are not required in ordinary calculations, and need not be referred to in these Notes.

Deflection is the bending caused by a transverse stress. If the intensity of stress be below the elastic limit the deflection will disappear when the stress is removed, but if the intensity of stress be in excess of the elastic limit a permanent *set* will remain.

Resilience is a term used to express the quantity of "work done" in deforming a piece of material (up to the elastic limit) by the application of any kind of stress. It is equal to the product of the alteration of figure into the mean load which acts to produce such alteration. Thus the resilience of a bar in tension is found by multiplying the proof load by half the corresponding elongation.[1]

Resilience may be tensile, compressive, transverse, shearing, etc., according to the nature of the stresses imposed.

Malleability is the property of being permanently extensible in all directions by hammering or rolling.

Ductility is the property of being permanently elongated or drawn out under a tensile stress higher than the elastic limit. The change of form remains after the force is removed. It is therefore the converse of elasticity.

Brittleness is the inclination to break suddenly under any stress.

Hardness is the property of resisting indentation, or wear by friction.

Softness is the converse of hardness.

Toughness is a term defined in several different ways.

Mr. Stoney defines it as the union of tenacity with ductility.

Ultimate toughness is defined by Professor Rankine as being the greatest strain which a body will bear without fracture ; proof toughness the greatest strain it will bear without injury. He points out that malleable and ductile solids have ultimate toughness greatly exceeding their proof toughness, but that brittle solids have their ultimate and proof toughness equal, or nearly equal.[2]

Fusibility is the property of becoming fluid when subject to heat. The temperature at which this is effected differs in each metal, and is called its *melting point.*

Weldability is the power possessed by some metals of adhering firmly to portions of the same—or to other metals—when the two pieces are raised to a high temperature and hammered together.

Hardening is the property of becoming very hard when heated and quenched.

Tempering is lowering the degree of hardness after the process just mentioned, by reheating and cooling at different temperatures (see p. 305).

---

1 Rankine's *Applied Mechanics.*   2 Rankine's *Useful Rules and Tables.*

INDEX.

A

Abercarne sandstone quarries, 41.
Aberdeen granite, resistance to wear, 85 ; strength of, 81.
Aberdeenshire granite quarries, 16, 20, 21.
      „   serpentine, 34.
Aberfoyle slate quarry, 32.
Aberllefenny slate quarries, 31.
Aboyne marble quarries, 53.
Absorption of bricks, 111, 114 ; of firebricks, 129 ; of granular and shelly limestones, 57 ; of lime-stones, 83 ; of sandstones, 36, 82 ; of slates, 25 ; of stone, 11, 81.
Acacia, common, 377 ; appearance, characteristics, and uses of, 377, 402 ; weight and strength of, 402.
Acetate of copper and lead as driers, 411.
Achscrabster sandstone quarry, 46.
Acid, ferro-silicic, as a preservative for stone, 80.
      „   test for stone, 11 ; steel, 308.
Ackworth sandstone quarries, 41.
Action of foreign constituents on limes and cements, 235.
      „   of water on lead, 340.
Adamantine clinkers, 110 ; size and weight of, 113.
Addistone slate quarry, 31.
Adhesive power of screws, 463.
Adie's cement-testing machines, No. 1, 187 ; No. 2, 189.
Admiralty tests for wrought iron, 282 ; for steel, 308 ; for steel plates, 308.
Adulteration in Portland cement, 192 ; of red lead, 406 ; of white lead, 405.

African green, 415.
      „   oak or teak, 374.
Agents which destroy stone, 10.
Aggloméré, Coignet's Béton, 225.
Aggregate for concrete, 215 ; materials used for, size and shape, 216 ; packing, 217.
Agricultural drain pipes, 135.
Air bricks, perforated, 139.
      „   flues, combined smoke and, 141.
Air-slaking of quicklime, 151.
Aish stone, position in quarry, 7, 61.
Aislaby sandstone quarries, 39.
Alburnum or sap wood, 357.
Alder, 375 ; appearance, characteris-tics, and uses of, 375.
Alexandra slate quarry, 31.
Alkalies in clay for brickmaking, 88 ; colouring action of, 90 ; hydraulic properties of, 153.
Alkaline silicates as a preservative for stone, 78.
      „   „   for artificially pro-ducing hydrauli-city, 185.
Alloys, 348-351; Babbit's metal, 349 ; bell metal, brass, bronze (aluminium and phosphor), gun metal, and Muntz metal, 348 ; sterro-metal and white brass, 348 ; table of com-position of various, 348.
      „   of lead and tin, 353.
Altmover sandstone quarries, 47.
Altnaveigh granite quarry, 21.
Alumina in clay for brickmaking, 86, 88 ; in fireclay, 126.
      „   soluble oxalate of, action of, on limestones, 80.
Aluminium bronze, 348.
Amber, 428 ; varnish pale, 432.

END OF PART III.

Printed by R. & R. CLARK, LIMITED, *Edinburgh.*